Solar-to-Chemical Conversion

Solar-to-Chemical Conversion

Photocatalytic and Photoelectrochemcial Processes

Edited by

Hongqi Sun

WILEY-VCH

Editor

Prof. Hongqi Sun
Edith Cowan University
School of Engineering
270 Joondalup Drive
6027 Joondalup
Australia

Library of Congress Card No.:
applied for

British Library Cataloguing-in-Publication Data
A catalogue record for this book is available from the British Library.

Bibliographic information published by the Deutsche Nationalbibliothek
The Deutsche Nationalbibliothek lists this publication in the Deutsche Nationalbibliografie; detailed bibliographic data are available on the Internet at <http://dnb.d-nb.de>.

Print ISBN: 978-3-527-34718-6
ePDF ISBN: 978-3-527-82509-7
ePub ISBN: 978-3-527-82508-0
oBook ISBN: 978-3-527-82507-3

Typesetting SPi Global, Chennai, India
Printed and bound CPI Group (UK) Ltd, Croydon, CR0 4YY

Printed on acid-free paper

10 9 8 7 6 5 4 3 2 1

Contents

1

Introduction: A Delicate Collection of Advances in Solar-to-Chemical Conversions

Hongqi Sun

Edith Cowan University, School of Engineering, 270 Joondalup Drive, Joondalup, WA 6027, Australia

Tremendous efforts have been made by worldwide researchers toward effective solar energy conversion and utilization. Compared with solar panels or solar cells that convert solar energy to electricity, photocatalytic process can store solar energy to chemical energies and then has attracted extensive attention. This book seizes this great timing to delicately collect the fundaments of solar-to-chemical technologies. Researchers, students, and broad readership would use this book to become experts from beginners. The researchers in the fields and the community may also find it useful for further advances to this exciting area.

This book will cover the fundamentals in solar energy conversion to chemicals, either fuels or chemical products. Natural photosynthesis will be firstly presented to give a tutorial introduction while main attention will be focused on artificial processes for solar energy conversion and utilization. The chemical processes of solar energy conversion via homogeneous and/or heterogeneous photocatalysis will be described with the mechanistic insights. Reaction systems afford a variety of applications, for example, water splitting for hydrogen or oxygen evolution, photocatalytic CO_2 reduction to fuels, and light-driven N_2 fixation, etc. Emerging photocatalysis in upgrading or reforming fossil fuels will also be covered. Design and theoretical fundamentals of solar energy conversion to chemicals will be explained in detail based on semiconductor photocatalysis. Enormous research outcomes in the individual field are related to solar-to-chemical conversion, while this might be the first delicate collection detailing the fundamentals of each catalytic process, along with most challenging issues that hinder the processes move to an industrial scale. Therefore, it is believed that this book will be unique and can offer the readers a broad view of solar energy utilization based on chemical processes and their perspectives for future sustainability.

This book includes 16 chapters and the brief information and highlights of each chapter are as follows:

Chapter 1: Introduction: A Delicate Collection of Advances in Solar-to-Chemical Conversions. This chapter briefs the background of this book, introduces the objectives, and provides the main information of each chapter of the book. It is

Solar-to-Chemical Conversion: Photocatalytic and Photoelectrochemcial Processes, First Edition.
Edited by Hongqi Sun.

expected that the readers would have a general idea of this book and can then directly move to the specific contents for perusing. It is anticipated that this chapter alone can work as the abstract of the book.

Chapter 2: Artificial Photosynthesis and Solar Fuels. This chapter contributes to the conceptual processes in the conversion of solar energy into chemical energies, i.e. solar fuels. The engineered processes mimicking natural photosynthesis mark the term of artificial photosynthesis. The basic principles for converting carbon dioxide and water into value-added solar fuels, which can be hydrogen, oxygen, and hydrocarbons, are outlined. The core of the technology is photocatalysis, typically being facilitated by semiconductor materials. Via the process, a variety of solar fuel products can be produced. They may include (i) hydrocarbons (methane, methanol, formaldehyde, formic acid, and C2), (ii) carbon monoxide, (iii) oxygen, and (iv) hydrogen. This is because of the multi-electron, multi-hole, and multi-proton reactions. The mechanism and selectivity were examined. The strategies for facing the challenges in this exciting area were proposed at the end of the chapter.

Chapter 3: Natural and Artificial Photosynthesis. This chapter provides an overview of the most important process, i.e. natural photosynthesis. Such a process has been literally powered the whole planet with various species and the ecosystem. It introduces the detailed processes, including light harvesting, charge separation and accumulation, water oxidation, and nitrogen fixation. With the insights into them, inspirations are achieved for artificial photosynthesis, which leads to the fantastic explorations on solar-to-chemical conversions.

Chapter 4: Photocatalytic Hydrogen Evolution. This chapter mainly introduces heterogeneous photocatalytic reactions for hydrogen evolution. The basic principles involved in photocatalytic hydrogen evolutions are introduced. Following that, photocatalytic hydrogen evolution reactions under ultraviolet (UV) light, visible light, and near-infrared light irradiations were investigated in detail. Because of the light absorption ability, which is determined by the band gap of a semiconductor, specific semiconductors only work better in a specific condition. As a result, titanium dioxide and its modified counterparts are mainly introduced in UV reactions, carbon nitride and various modified ones are presented in visible-light photocatalysis, and upconversion materials are discussed for near-infrared light reactions. Comprehensive survey on materials for the specific light regions was also available. For providing the insights into the reactions, the roles of sacrificial reagents and reaction pathways were also discussed in this chapter.

Chapter 5: Photoelectrochemical Hydrogen Evolution. The status, opportunities, and challenges in the hydrogen energy are discussed. Other than powdered photocatalysis, this chapter concerns photoelectrochemical (PEC) and photovoltaic-driven water electrocatalysis (PV + EC). The key of these processes is the efficient photoelectrode fabrication, which determines the light harvest, charge separation and transfer, and surface reactions. The configurations of PEC and the strategies for promoting charge transfer through the semiconductor film and providing strong driving force for carriers' separation are reviewed. The catalyst design for achieving such strategies was paid with extra attention.

Chapter 6: Photocatalytic Oxygen Evolution. Water oxidation half-reaction has been regarded to be the primary barrier in solar fuel production, because such a reaction requires multiple protons and electrons to be involved. This chapter identifies the critical challenge of water oxidation to be the design of efficient, low-cost catalysts with excellent stability. The morphology, structure, and photocatalytic water oxidation performances of various earth-abundant materials are reviewed. Both homogeneous and heterogeneous reactions are discussed. Moreover, quantum size effect, localized surface plasmon resonance, active facet exposure defect engineering, and heterojunction construction in low-dimensional materials are discussed in terms of enhanced photocatalytic water oxidation.

Chapter 7: Photoelectrochemical Oxygen Evolution. This chapter reviews the fundamentals of PEC oxygen evolution toward higher efficiencies as compared with powdered or homogeneous photocatalytic oxygen evolution. It discusses the factors affecting the photoanodic current, the relationship between electrode potentials and pH in the electrolyte, the evaluation method of PEC performance of photoanode materials, the relationship between flat band potential and photocurrent onset potential, the selection strategy of photoanode materials, and PEC device for water splitting. The determining parameters, such as nanostructuring, morphology control, donor doping, cocatalyst loading, heterojunction formation, and electron-conductive materials, are discussed for better design of photoanode materials. It concludes that the rational material design and the knowledge of thermodynamics and kinetics are required and the low-cost photoanode materials composed of inexpensive earth-abundant elements control the feasibility of this technology.

Chapter 8: Photocatalytic and Photoelectrochemical Overall Water Splitting. After the discussions of the individual hydrogen or oxygen evolution processes, this chapter focuses on the overall water splitting processes. The fundamental scientific requisites, mechanism aspects, and development horizons for overall water splitting are comprehensively covered. Then the development of photocatalytic technologies and hybrid systems integrating photovoltaic devices and photoelectrodes in photoelectrochemical platforms for overall water splitting are introduced. An extensive library of light-responsive semiconductor-based materials, attractive cocatalysts, and plasmonic nanostructures and assessed synthesis approaches, e.g. the construction of heterojunctions between state-of-the-art semiconductors, is presented. It is expected that the development of practical and adequate materials for solar-driven overall water splitting systems can further advance this exciting system.

Chapter 9: Photocatalytic CO_2 Reduction. After hydrogen/oxygen evolutions, another main topic in solar fuels, i.e. CO_2 reduction, starts in this chapter. Photocatalytic CO_2 reduction has been regarded to be economical, recyclable, and safe, therefore holding a great promise for addressing worldwide energy and environmental problems. In this chapter, the whole process of photocatalytic CO_2 reduction is analyzed in terms of energy and mass flow. Discussions are made to illustrate the effects of the flow of solar energy through concentrator, reactor, reaction solution, and photocatalyst and the mass flow from the adsorption and activation of reactants (such as CO_2) on the surface of the photocatalysts to the formation of CO, CH_4, and other products through photocatalytic reaction. Contributions are also made to present the energy loss and obstacles contained in the transformation processes and the possible strategies to improve the overall reaction efficiency.

Chapter 10: Photoelectrochemical CO_2 Reduction. As inspired by PEC water splitting, PEC reduction of CO_2 can also integrate and optimize both photocatalysis and electrocatalysis toward an improved CO_2 reduction system with a higher efficiency and stability. This chapter first introduces the fundamentals and reaction parameters in the concerned systems; then research advances on the development catalyst materials are surveyed. This includes the semiconductors, cocatalysts, and hybrid semiconductors. Novel reactor systems for PEC CO_2 reduction are introduced. At last, research opportunities and challenges are identified.

Chapter 11: Photocatalytic and Photoelectrochemical Nitrogen Fixation. Using solar energy to convert nitrogen in air to valuable chemicals is a direct mimicking system to natural photosynthesis. Artificial photocatalytic and PEC reduction of N_2 to ammonia emerged as a fantastic venue to this end. This chapter introduces the recent advances in photocatalytic and photoelectrochemical N_2 reduction to NH_3. The fundamental principles on N_2 reduction are presented. Then, the current design strategies, mainly including the defect engineering, structural regulation, interface control, heterojunction construction, cocatalyst engineering, and biomimetic engineering, for the preparation of heterogeneous catalysts are summarized. At last, the remaining challenges as well as future perspectives in this very exciting research field are outlined.

Chapter 12: Photocatalytic Production of Hydrogen Peroxide Using MOF Materials. In theory, oxygen and water can be employed to produce hydrogen peroxide using solar energy, which fortunately was demonstrated to be feasible through photocatalysis. It was reported that synthesis of H_2O_2 from O_2 reduction reaction using metal organic frameworks (MOFs) is attractive and promising. In this chapter, the H_2O_2 production through visible-light-induced O_2 reduction by an MOF, MIL-125-NH_2, coupled with oxidative reaction of benzyl alcohol was achieved in a single-phase system composed of acetonitrile and benzyl alcohol. These works creatively developed the application of MOFs in the field of new energy production.

Chapter 13: Photocatalytic and Photoelectrochemical Reforming of Methane. Solar energy can serve as the input energy for methane activation because of the wide distribution and large reserve. With that, photocatalysis and photoelectrocatalysis are recognized as effective approaches for methane reforming on which more attention has been put in the field of energy preparation. In this chapter, photocatalytic and photoelectrochemical methane reforming are introduced. The differences between these two catalytic processes are investigated in detail. After that, recent research progresses on the methane activation reactions via these two techniques are provided. The related reaction mechanisms are discussed insightfully. At last, promising perspectives on the methane upgrading via solar energy excitation are proposed.

Chapter 14: Photocatalytic and Photoelectrochemical Reforming of Biomass. It is very interesting to combine two renewable resources, e.g. solar energy and biomass, together in one process. Recently photocatalytic and photoelectrochemical reforming of biomass was demonstrated. This chapter discusses the photocatalytic conversion of processed and native lignin, carbohydrates, native lignocellulose, glycerides, and glycerol to hydrogen and value-added chemicals. Also, the photoelectrochemical (PEC) reforming of biomass to electricity, hydrogen, and biomass-derived molecules such as glycerol and alcohols, as well as converting 5-hydroxymethylfurfural to corresponding valuable chemicals, is reviewed.

Chapter 15: Reactors, Fundamentals, and Engineering Aspects for Photocatalytic and Photoelectrochemical Systems. Novel catalyst materials and new reactions have always been the hotspots in the research endeavors. As the places for the ultimate step, the reactors and associated parameters should also be given attention. This chapter provides insightful discussions on the key factors for practical approach of photocatalysis and PEC in solar fuel production. In particular, (i) fundamental rationales and mechanisms, (ii) the design and setup of photoreactors, and (iii) engineering aspects of photocatalytic and PEC systems with potential scalability are provided.

Chapter 16: Prospects of Solar Fuels. By several dream reactions, solar energy can be captured and stored in terms of solar fuels. This chapter first concludes the available processes that have been demonstrated to be feasible, and then attention is paid to explore the feasibility of the processes. Tremendous efforts are required from research, commercialization, and policies to achieve the solar fuel production at a large scale.

2

Artificial Photosynthesis and Solar Fuels

Jun Ke

Wuhan Institute of Technology, School of Chemistry and Environmental Engineering, Liufang Campus, No.206, Guanggu 1st road, Donghu New & High Technology Development Zone, Wuhan 430205, P.R. China

2.1 Introduction of Solar Fuels

In the Earth, all chemicals, in particular energy system, including fuels and their combustion products, are enclosed in a substance cycle. It is well known that fossil fuels including coal, petroleum, and natural gas mainly stem from the evolution of ancient animals and plants under the stratum for tens of thousands of years. When these fossil fuels are combusted, massive chemical energies can be converted into thermal energies, which can then be used to make liquid water vaporize for driving the electric generators to produce various available energy forms, such as electric power. Furthermore, during the combustion of fossil fuels, the intrinsic chemical reactions are oxidation of hydrocarbons, accompanying with outputs of thermal energy. Finally, the hydrocarbons and their derivatives are converted into carbon dioxide and water along with SO_x, NO_x, and cokes under imperfect combustion. It is known that when massive CO_2 and water are released into the air, the natural plants and alga can utilize CO_2 to produce hydrocarbons as own constituents and release dioxygen in the presence of sunlight and water, being defined as photosynthesis. Subsequently, these plants and alga enter food chains and finally become fossil fuels. Until now, carbon as energy carrier realizes the global cyclic process.

However, with the rapid development of industrial activities, energy consumption demands sharply increase, which greatly destroys the balance of global substance cycle [1]. Subsequently, the releasing amount of CO_2 significantly raises accompanying with other pollutants, resulting in a series of global pollutions, such as global warming and ozone depletion [2]. To overcome the coming energy crisis and environmental issues, chemists attempt to make the substance cycle rebalance by means of various promising solar-driven techniques, such as photocatalysis, CO_2 storage and utilization, water splitting, and N_2 fixation [3]. Meanwhile, these desirable techniques often have sustainable, clean, and benign metrics, which are beneficial to supporting the future sustainability of human being.

Solar-to-Chemical Conversion: Photocatalytic and Photoelectrochemcial Processes, First Edition.
Edited by Hongqi Sun.

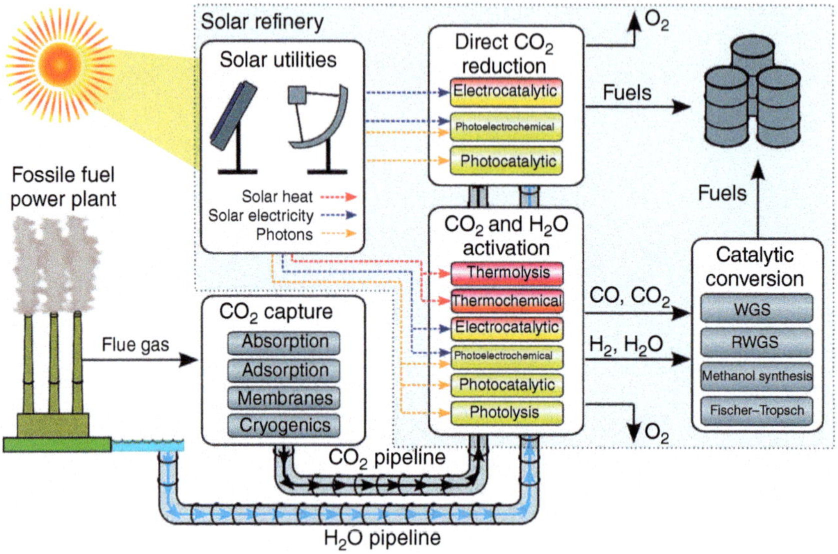

Figure 2.1 Schematic of solar fuel feedstocks (CO_2, H_2O, and solar energy) and production path on-site and/or transported to the solar refinery [4].

Figure 2.1 illustrates a blueprint of sustainable fuel production and consumption, where conventional power plant still consumes fossil fuel and produces CO_2 and water [4]. The released CO_2 can be captured by using absorption or adsorption techniques and react with H_2O to produce carbon monoxide (CO) and H_2 via thermochemical reactions that are triggered by indirect solar heat or solar-powered electric energy. Subsequently, CO and H_2 can be further utilized to transform into hydrocarbon fuels by various thermal catalytic conversions. These findings display a direction of CO_2 utilization and fuel productions while the solar energy utilization is still low in this process in spite of reducing CO_2 releasing. Inspired by natural photosynthesis, driving transformation of CO_2 with H_2O into fuels and O_2 under benign conditions is more desirable, where direct sunlight or solar-source electricity is the main energy source, as present in Figure 2.1. Nevertheless, it is demonstrated that the reaction is non-thermodynamic and extremely low rate under spontaneous condition. Therefore, to achieve the considerable efficiency of natural photosynthesis and commercialization, catalysts have to be introduced to accelerate the reaction rate, as similar as the chlorophyll, which is named by artificial photosynthesis.

2.2 Photosynthesis

2.2.1 Natural Photosynthesis

Photosynthesis is a chemical process that occurs in photoautotrophs (organisms that make their own food), in which light energy is converted into sugars and other organic compounds. It consists of a series of chemical reactions that require carbon dioxide and water to begin [5]. The light energy that hits the photoautotrophs is absorbed and drives these chemical reactions to produce carbohydrates and oxygen as a by-product. The following equation is the basis of photosynthesis:

$$6CO_2 + 6H_2O + \text{light energy} \rightarrow C_6H_{12}O_6 + 6O_2 \quad (2.1)$$

In past several decades, the rough photosynthesis paths have been reported by scientists based on excessive studies. In the plants, the chloroplast in leaf can absorb the sunlight and trigger the above reaction proceeding in the presence of a series of bioenzymes, such as chlorophyll II. Meanwhile, CO_2 from the air is captured by the leaf through diffusion process and brought to the chloroplast. Finally, the chlorophyll II excited by incident light catalyzes water split and releases hydrogen and oxygen. After that, the oxygen atoms are formed dioxygen and escape from chloroplast to air. Meantime, the hydrogen atoms react with CO_2 molecules into hydrocarbons, such as glucose. Until now, the plants fulfill the great photosynthesis process, which creates the suitable ecosystem for advanced living things, in which chemical fuels and O_2 can be further consumed to provide energy for aerobic organisms.

By systematically analyzing the process of natural photosynthesis, it can be found that CO_2 and water are the initial reactants and chlorophyll II as a catalyst is necessary to accelerate the reaction under sunlight irradiation. Generally speaking, CO_2 and water are abundant and cheap in the Earth, which can be obtained easily. Nevertheless, the reaction between CO_2 and water cannot proceed in thermodynamic aspect [6]. Therefore, for chemists, it is necessary to develop an efficient catalyst for triggering the photosynthesis reaction, which helps that the O_2 and chemical fuels can be produced in a cheap and sustainable way in a factory. Based on this inspiration, simulating natural photosynthesis process is a relentless pursuit for scientists, which is called as artificial photosynthesis [7].

2.2.2 Artificial Photosynthesis

Scientists around the world have been trying to replicate the natural reactions that occur during photosynthesis and have come across the science of artificial photosynthesis. The term artificial photosynthesis is used to refer to any mechanism made to capture light and store energy from the sun in chemical bonds of a solar fuel [8]. In general, as shown in Figure 2.2, the artificial photosynthesis includes three main steps: (i) The first step in artificial photosynthesis is for the reactants coming together. The reactants include sunlight along with water and carbon dioxide that is available in the atmosphere. (ii) These reactants then go through the whole process of photosynthesis artificially. Scientists have been known to use artificial leaves to split water, producing both oxygen and hydrogen. They are now creating artificial leaves using ruthenium and manganese complexes to mimic the natural process of photosynthesis (redox reaction of nicotinamide adenine dinucleotide phosphate ($NADPH/NADP^+$)). (iii) The products of the reactions are created as soon as water is split, producing both oxygen and hydrogen. The hydrogen is then either used directly as a fuel or a reductant for carbon dioxide to produce organic fuels [10].

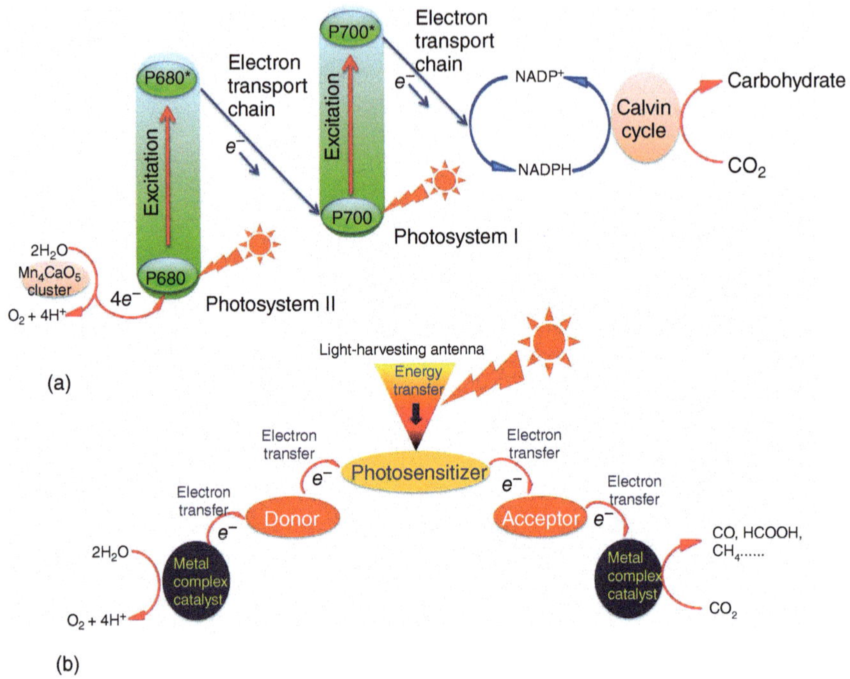

Figure 2.2 Schematic diagrams of (a) natural photosynthesis and (b) artificial photosynthesis based on molecular systems. Source: Liu et al. [9].

2.3 Principles of Photocatalysis

The term "photocatalysis" is often used in papers on catalytic reaction excited by simulated or natural light irradiation, where a catalyst has been used as a reaction center [11]. To simulate the efficient natural photosynthesis reaction, the use of a catalyst is inevitable in the view of physiochemical field. Because biological scientists find that in leave cells the enzyme is the key role in prompting the non-thermodynamic reaction, i.e. carbon dioxides reacting with water to produce dioxygen and hydrocarbons. Therefore, to accelerate this reaction under certain conditions, a photocatalyst often is used as similar in many industrial fields, such as sulfuric acid products, synthetic ammonia, and syngas. As shown in Figure 2.3, in general, there are many active sites existing on the surface of catalyst, which can act as reaction site and accelerate the transformation of substrates to products under certain reaction conditions [12]. Similarly, the catalytic reaction process is triggered at the active sites on the surface of photocatalyst under light irradiation. It is reported that the electrons and holes can be produced in the photocatalysts, which further participate the redox reaction [13].

In photocatalysis, many photocatalysts have been developed and investigated in the past decades [14]. Here, taking excessively studied semiconductor photocatalyst as an example, the whole photocatalytic process is described. Semiconductor is a kind of material with electrical conductivity between conductor (such as metals)

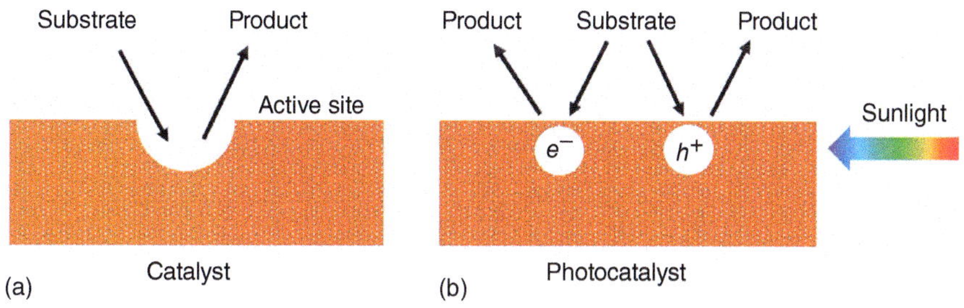

Figure 2.3 (a) Traditional catalytic processes on the surface of catalysts. (b) Classic photocatalytic paths over photocatalysts.

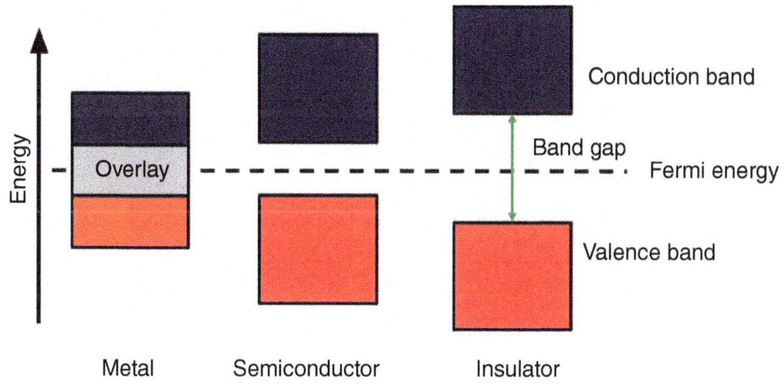

Figure 2.4 A band-gap diagram showing the different sizes of band gaps for conductors, semiconductors, and insulators.

and insulator (such as ceramic) [15]. The conductivity of a semiconductor usually increases with the increase of the temperature, which is opposite to that of a metal. The unique electronic property of a semiconductor is characterized by its valence band (VB) and conduction band (CB). The VB of a semiconductor is formed by the interaction of the highest occupied molecular orbital (HOMO), while the CB is formed by the interaction of the lowest unoccupied molecular orbital (LUMO). There is no electron state between the top of the VB and the bottom of CB. The energy range between CB and VB is called forbidden band gap (also called energy gap or bandgap), which is usually denoted as E_g, as shown in Figure 2.4.

The band structure, including the band gap and the positions of VB and CB, is one of the important properties for a semiconductor photocatalyst, because it determines the light absorption property as well as the redox capability of a semiconductor [16]. As shown in Figure 2.5, the photocatalytic reaction initiates from the generation of electron–hole pairs upon light irradiation. When a semiconductor photocatalyst absorbs photons with energy equal to or greater than its E_g, the electrons in VB will be excited to CB, leaving the holes in VB. These photogenerated electron–hole pairs may further be involved in the following three possible processes: (i) successfully migrate to the surface of semiconductor, (ii) be captured by the defect sites in bulk and/or on the surface region of semiconductor, and (iii) recombine and release the

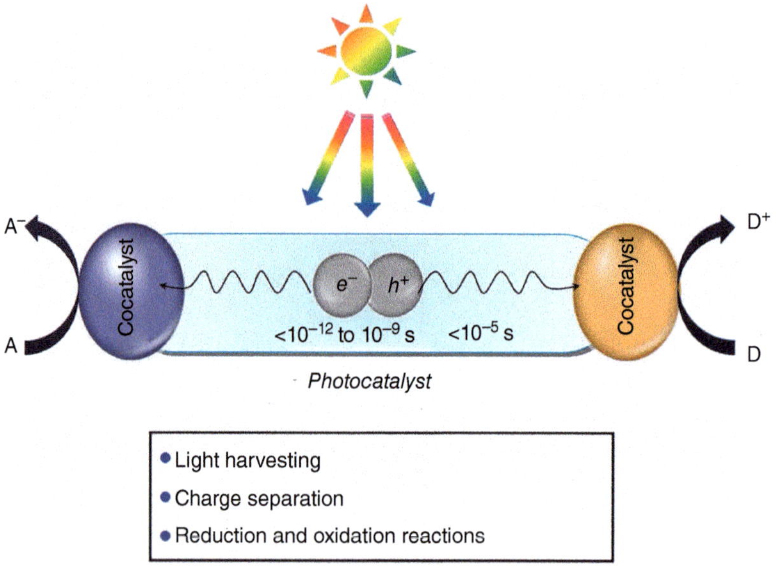

Figure 2.5 Proposed mechanism of photocatalytic reactions on semiconductor photocatalysts. Source: Ma et al. [16].

energy in the form of heat or photon. The last two processes are generally viewed as deactivation processes because the photogenerated electrons and holes do not contribute to the photocatalytic reaction. Only the photogenerated charges that reached to the surface of semiconductor could be available for photocatalytic reactions. The defect sites in the bulk and on the surface of semiconductor may serve as the recombination centers for the photogenerated electrons and holes, which will decrease the efficiency of the photocatalytic reaction.

Efficient charge separation is the most important factor that determines the photocatalytic activities. Various strategies could be applied for improving charge separation efficiency [17]. For example, preparation of semiconductor photocatalysts at high temperatures may lead to high crystallinity that diminishes the formation of charge recombination defect sites. Construction of various kinds of nanostructures such as nanowires (belts) and nanosheets may also facilitate charge transportation and promote charge separation efficiency. As compared to nanoparticle, one-dimensional nanostructures exhibit better photocatalytic activity because they have better charge mobility and can reduce the charge recombination. Furthermore, creation of "junctions" with built-in electric fields or chemical potential differences is also an effective strategy for improving charge separation efficiency [18]. The surface catalytic reaction is a successive step of charge separation. In principle, a photocatalytic reaction consists of two half-reactions, reduction reaction and oxidation reaction. The electrons in CB may initiate the reduction reaction, and the reduction capability is determined by the position of CB; the holes in the VB involve the oxidation reaction, and the oxidation capability is determined by the position of VB. For the water splitting reaction, the position of CB of a semiconductor photocatalyst should be more negative than the redox potential of H^+/H_2 (0 V vs. normal hydrogen electrode [NHE], pH = 7), while the energy level of VB should be more positive than the redox potential of O_2/H_2O (1.23 V vs. NHE, pH = 7). Sometimes, some particular surface sites of a semiconductor can act as the active centers themselves, especially for oxidation reaction on the surface of metal oxide semiconductors. However, in most cases, efficient photocatalytic reactions proceed only after loading noble metal and oxide cocatalysts on semiconductors.

2.4 Products of Artificial Photosynthesis

2.4.1 Hydrocarbons

Hydrocarbons as important products of photosynthesis provide almost all energy for living things that take advantage of chemical energy to maintain activities by oxidizing organics. In photosynthesis system, carbon dioxide is the mainly carbon source and transformed into hydrocarbons. Inspired by the nature, scientists attempt to hydrogenate CO_2 to produce hydrocarbons as substitutes to traditional fossil fuels. In general, CO_2 is one of the most thermodynamically stable carbon compounds. The photocatalytic reduction of carbon dioxide to hydrocarbons requires to consume large amounts of energy for dissociation of the C=O bond and formation of C—H bonds. Since the carbon atoms in CO_2 are in the highest oxidation state, this process may occur only with the participation of reducing agents providing a certain amount of electrons. On the other hand, the corresponding protons participate oxidative reactions, such as water oxidation. Meanwhile, hydrogen in water can further give rise to CO_2 reduction for formation of C–H. However, the reduction of CO_2 by water to give organic compounds such as methanol or methane entails a reaction with a high positive Gibbs free energy change: $CO_2 + 2H_2O \rightarrow CH_3OH + 1.5O_2$ ($\Delta G_0 = 702.2\,kJ\,mol^{-1}$); $CO_2 + 2H_2O \rightarrow CH_4 + 2O_2$ ($\Delta G_0 = 818.3\,kJ\,mol^{-1}$). Therefore, to overcome these energy obstacles requires external energy source such as thermal energy, plasma, electrics, and solar irradiation. For example, in 1973, Inoue et al. reported that HCHO and CH_3OH were produced by the reduction of CO_2 with H_2O in aqueous suspension systems under Xe lamp irradiation and electric field involving a variety of semiconductor photocatalysts such as TiO_2, ZnO, and WO_3 [19]. Meanwhile, trace formic acid and methane were detected over specific photocatalysts (SiC, GaP) by using special quantitative analysis methods. In contrast, in the absence of light source, no reduced products were detected in these systems, which strongly demonstrate that light excitation triggers the reduction reaction between CO_2 with H_2O in room temperature and normal pressure over the semiconductor photocatalysts. As shown in Figure 2.6, it is demonstrated that there are many reaction paths in reduction of CO_2 under external field excitation, leading to the diversity of products so that selectivity is another important parameter to determine the performance of CO_2 reduction photocatalysts. The possible reactions that can occur in the reduction of CO_2 in the aqueous medium, under band-gap illumination, ranging from one- to eight-electron transfer, with thermodynamic

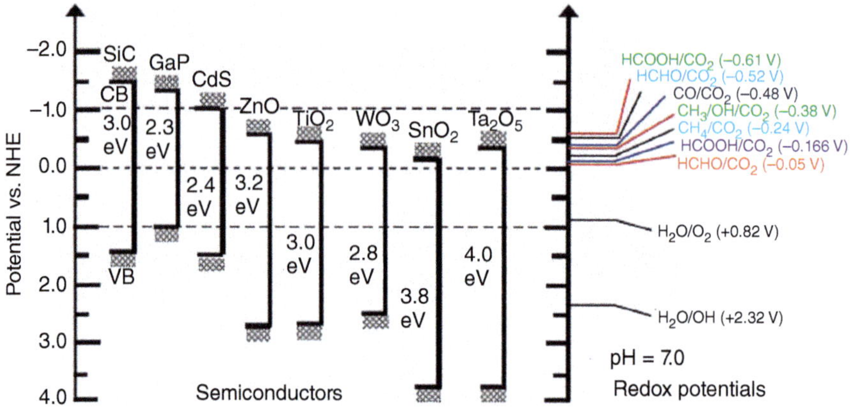

Figure 2.6 A schematic illustration of the energy correlation between semiconductor catalysts and redox couples in water. CB and VB denote a conduction band and a valence band, respectively. Source: Tu et al. [20].

potentials vs. NHE, are mentioned below:

$$CO_2 + e^- \rightarrow CO_2^{-} - 2.0\,V \tag{2.2}$$

$$CO_2 + 2H^+ + 2e^- \rightarrow HCOOH - 0.61\,V \tag{2.3}$$

$$CO_2 + 4H^+ + 4e^- \rightarrow HCHO + H_2O - 0.52\,V \tag{2.4}$$

$$CO_2 + 2H^+ + 2e^- \rightarrow CO + H_2O - 0.48\,V \tag{2.5}$$

$$CO_2 + 6H^+ + 6e^- \rightarrow CH_3OH + H_2O - 0.38\,V \tag{2.6}$$

$$CO_2 + 8H^+ + 8e^- \rightarrow CH_4 + H_2O - 0.24\,V \tag{2.7}$$

To improve the selectivity of expected product, Anpo and coworkers reported a series of studies on photocatalytic reduction of CO_2 with H_2O for producing methanol and methane with a high selectivity by designing various nanostructured photocatalysts, such as Cu/TiO_2 [21], TiO_2/Y-zeolites [22], Ti/FSM-16 [23], and TiO_2/glass [24] in Figure 2.7. The mechanism investigation results inferred that the local structure of TiO_2 tested by X-ray absorption near-edge structure (XANES) on the support has significant influences in product selectivity. In this section, different products and reaction mechanism are respectively discussed in detail.

2.4.1.1 Methane (CH_4)

As a high value-added fuel, CH_4, a natural gas, is widely used in residential living and industrial activities. Reduction of CO_2 to methane named by the methanation process is of substantial importance in industrial field as the following equation: $CO_2 + 2H_2 \rightarrow CH_4 + 2H_2O$. In general, this reaction is proceeded at high temperatures and pressures in the presence of metal catalysts, for example, Ru, Mo, and Ni.

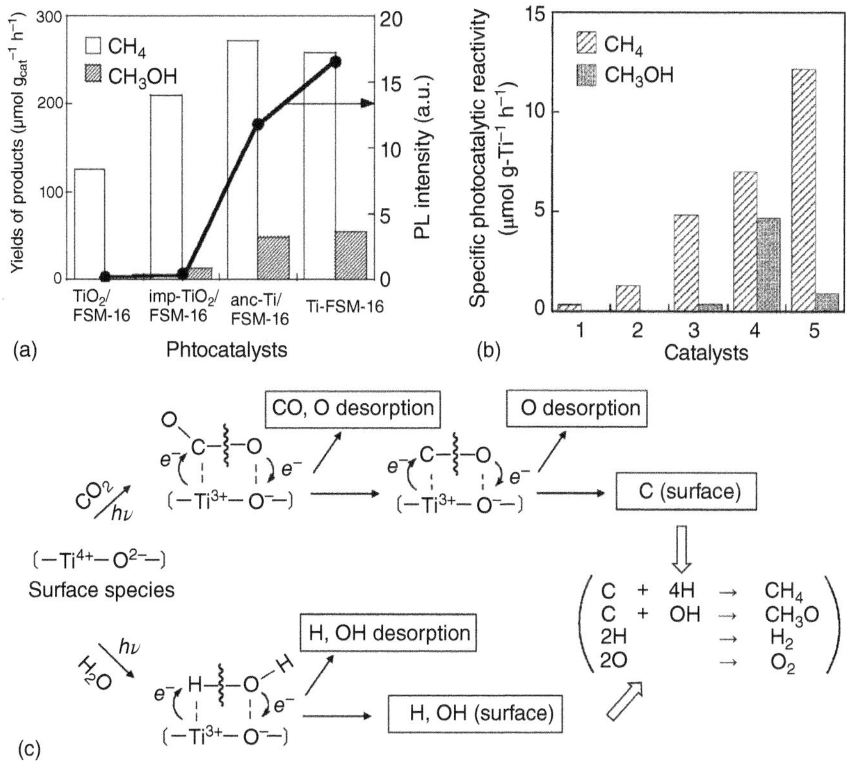

Figure 2.7 (a) Yields of the products produced during the photocatalytic reduction of CO_2 with H_2O and photoluminescence of various Ti/FSM-16 photocatalysts. (b) Product distribution of CO_2 photoreduction over 1-TiO_2, 2-10 wt% imp-TiO_2/Y-zeolite, 3-1.0 wt% imp-TiO_2/Y-zeolite, 4-ex-TiO_2/Y-zeolite, and 5-Pt-loaded ex-TiO_2/Y-zeolite. (c) Schematic mechanisms of the photocatalytic CO_2 reduction with H_2O on TiO_2. Source: (a) Ikeuea et al. [23]; (b) Anpo et al. [22]; (c) From Anpo et al. [21]. © 1995 Elsevier.

However, inspired by natural photosynthesis, chemists expect to produce CH_4 fuel under mild and environment-friendly conditions. In 1986, Willner and coworker reported an exciting result, in which gaseous CH_4 was successfully synthesized with a 0.0025% of quantum yield by using a $Ru(bpz)_3^{2+}$ as sensitizer and Ru metal colloid as catalyst under visible-light irradiation [25]. In Figure 2.8, the reaction mechanism displayed that $Ru^*(bpz)_3^{2+}$ produced from the light-excited $Ru(bpz)_3^{2+}$ can be reduced by triethanolamine (TEOA) as electron donor into $Ru(bpz)_3^{+}$ that can pair with $Ru(bpz)_3^{2+}$ with a redox potential of −0.86 V vs. Saturated Calomel Electrode (SCE). On the other hand, it is demonstrated that the redox potential of CH_4/CO_2 is equal to −0.24 V vs. NHE, which is less negative than that of $Ru(bpz)_3^{+}/Ru(bpz)_3^{2+}$. As a result, CO_2 molecular can be thermodynamically reduced into CH_4 molecular by $Ru(bpz)_3^{+}$ over Re metal catalyst in acidic media according to the Eq. (2.7).

However, the chemical yield of CH_4 from this expensive organic/inorganic photocatalytic system is low and unsustainable. In 1995, Kamber and coworkers took advantage of popular photocatalyst TiO_2 to reduce gaseous CO_2 into CH_4 under

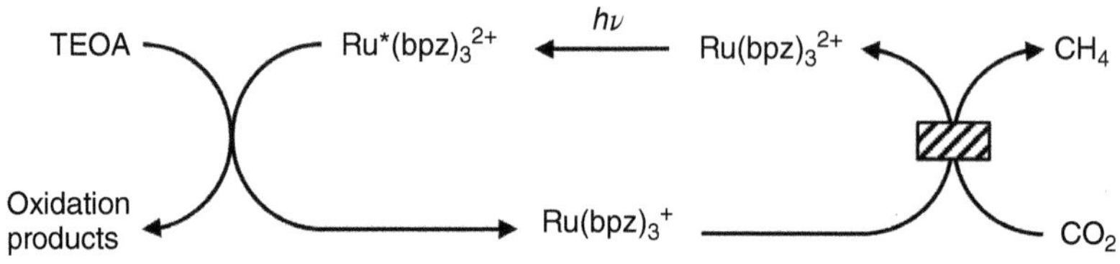

Figure 2.8 Schematic cycle for the photosensitized reduction of CO_2 to CH_4. Source: Maidan and Willner [25].

ultraviolet (UV) illumination at moderate temperature while the chemical yield and selectivity is not satisfying due to the nonspecific reaction over semiconductor photocatalysts [26]. The chemical yield of CO and H_2 is 27 and 14 times higher than that of CH_4, which indicated CH_4 seemed the by-products and the formation reactions of CO and H_2 are the dominant reduction reaction.

Considering the commercial implication in future, the selectivity to CH_4 is a key challenge in transforming CO_2 into CH_4. Recently, Tasbihi et al. reported Pt/TiO_2/mesoporous SiO_2 photocatalyst for significant enhancement of selectivity to CH_4 [27]. The main motivation of constructing this system is to elucidate that the introduced platinum (Pt) not only increases the activity of TiO_2 catalysts for CO_2 reduction but also modifies the selectivity toward highly reduced products. It is reported that compared with other noble cocatalysts, such as palladium, gold, silver, and rhodium, platinum has shown exceptional selectivity to methane in CO_2 photoreduction, and, therefore, the Pt nanoparticles are to be expected to improve the selectivity of catalytic processes [28]. To further reduce the cost of photocatalysts for CH_4 production, transition metal elements are investigated to substitute noble metal as well as maintain high selectivity. For example, Zhao and coworkers prepared a surface La-modified TiO_2 photocatalyst through a facile sol–gel route [29]. It was found that the La species was deposited on the surface of TiO_2 in the form of La_2O_3. More importantly, small part of La atoms replaced surface Ti atom forming a Ti—O—La bond, resulting in the generation of oxygen vacancies (OVs) and Ti^{3+}, which synergistically contributed to the enhanced CO_2 adsorption, water vapor activation, and charge separation. Finally, the selectivity to CH_4 in La-modified TiO_2 reached to ~80% compared with ~20% selectivity of TiO_2. Meanwhile, Ye and coworkers created a rich of oxygen vacancies on the (001) facet of Bi_2MoO_6 (BMO), which significantly improved the special CO_2 adsorption in a bidentate carbonate mode, thereby facilitating hydrogenation of intermediate *CO to form methane in a high selectivity of ~88% [30], as present in Figure 2.9. Meanwhile, a broader density functional theory (DFT) study was performed to further explore the possible CO_2 conversion pathways and intermediate species initiated by these two different CO_2 adsorption configurations in Figure 2.9d,e. For the defect-free BMO, only CO was produced in a rate of 0.12 $\mu mol\,g^{-1}\,h^{-1}$ with total consumed electron number (TCEN) of 0.24 $\mu mol\,g^{-1}\,h^{-1}$ in Figure 2.9g, while for BMO-OVs, the introduced OVs served as the natural active sites, leading to remarkable production of CH_4 with rate of 2.01 $\mu mol\,g^{-1}\,h^{-1}$ as the main product, CO of 0.27 $\mu mol\,g^{-1}\,h^{-1}$, and TCEN of 16.62 $\mu mol\,g^{-1}\,h^{-1}$, respectively.

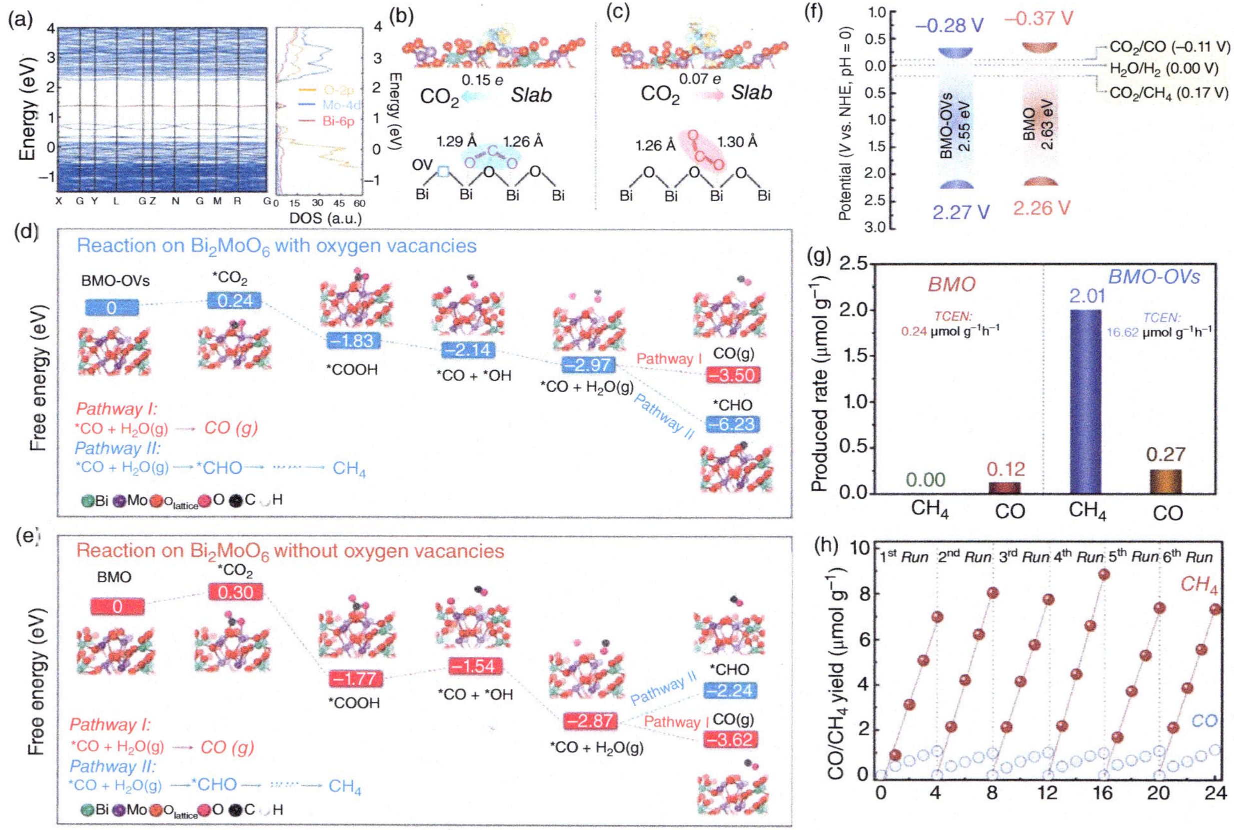

Figure 2.9 (a) Calculated band structures and total density of states of BMO-OVs. Absorptive formats of CO_2 on BMO-OVs (b) and BMO (c). CO_2 reduction pathways on BMO-OVs (d) and BMO (e). (f) The diagram of band positions. (g) CH_4 production and the TCEN value for CO_2 photoreduction over the BMO-OVs and BMO during four hours visible-light irradiation. (h) Typical time course of CO/CH_4 generated over BMO-OVs. Source: Yang et al. [30].

Moreover, new layered photocatalytic systems have been built to provide more active sites; therefore, high reaction rate can be achieved. For example, Wang and coworkers reported in situ prepared ultrathin SiC/reduced graphene oxide (RGO) nanosheets heterojunction for high yield and selectivity of CH_4 under visible-light irradiation [31]. The result demonstrated that ultrathin SiC nanosheets is beneficial for suppressing charge recombination and more energetic electrons can transfer to RGO for promoting CO_2 activation. However it is found that the amount of RGO added needed to be optimized due to the competitive reaction of CO/CO_2 (−0.52 V vs. NHE) with CH_4/CO_2 (−0.24 V vs. NHE). Excessive RGO in SiC/RGO will dilute the energetic electron density on RGO, which is beneficial for proceeding of two-electron CO/CO_2 reaction and suppresses the eight-electron process of CH_4/CO_2 despite the redox potential of CH_4/CO_2 being lower than that of CO/CO_2. Under the optimal conditions, the yield and selectivity of CH_4 reached to 6.72 $\mu mol\ h^{-1}\ g^{-1}$ and ~75%, respectively.

2.4.1.2 Methanol (CH_3OH)

Methanol, also known as wood alcohol, is considered an alternative to conventional transportation fuel due to its lower production costs, improved safety, and increased energy security. In general, methanol is cheaper than other fossil fuels with lower flammability risk. Also, methanol can be produced from a variety of carbon-based feedstocks. Unlike in the United States with abandoned blending methanol into gasoline, China allows around 12% of methanol used in fuel. As a fuel substitute, methanol does not have to compete for food consumption, such as ethanol production. In industrial production, methanol is primarily produced from the methane component of natural gas, while natural gas is an unsustainable source. Based on this requirement, chemists expect to produce methanol through greener, milder, and more efficient way. In 1980s, Halmann and coworkers have attempted to reduce CO_2 into methanol and other organic fuels over several semiconductor catalysts, such as $SrTiO_3$, WO_3, and TiO_2, with the photoelectron-assisted conditions, while the conversion efficiency was very low and time consuming [32]. After that, because of industrial production of methanol, the photoreduction of CO_2 to methanol is stagnant. In the twenty-first century, global warming is considered as a severe environmental problem, so ways of reducing CO_2 have been widely studied, including photoreduction. Wu [33] and Gondal and coworkers [34] tried to improve the photocatalytic conversion of CO_2 into methanol over traditional TiO_2 and modified ones by using novel light source, laser, and photoreactor, optical fiber, which to some degree enhanced the CO_2 conversion efficiencies and selectivity. However, the conversion rate was still at micromole level per photocatalyst weight hour, and by-product H_2 accounted for large proportion of reductive products.

With the rapid development of nanotechnology and nanocatalyst, morphologies, surficial properties, and heterojunction have been extensively investigated in the past, which extremely promote the design and fabrication of highly efficient semiconductor-based photocatalysts for CO_2 photoreduction. Zhang and coworkers synthesized Ag-based plasmonic photocatalysts with different constitutions and shapes for enhanced reduction of CO_2 to methanol, where Ag nanoparticles can accumulate photogenerated electrons and benefit for reducing the potential barriers of CO_2 reduction [35]. Under visible-light irradiation, with the assistance of anisotropic AgCl/Ag nanoparticles, the yield rate of methanol reached as high as 188 μmol g_{cat}^{-1}, and the activity of the photocatalysts has no obvious loss in the recycling reactions. Recently, Liu et al. prepared facet-dependent Cu_2O that was coupled with reduced graphene, which was used to investigate the effects of Cu_2O-exposed facet on the photoreduction of CO_2 to methanol [36]. The rhombic dodecahedral Cu_2O/rGO exhibited the highest methanol yield (355.3 μmol g_{cat}^{-1}), which was ca. 4.1–80.8 times superior to cubic octahedral Cu_2O/rGO and CuO/rGO after 20 hours of visible-light illumination.

To further improve the yield and selectivity of methanol, Sun and coworkers prepared a single-unit-cell Bi_2WO_6 layers that exhibited enhanced CO_2 photoreduction because the density of states at the CB from the surface atomic layer is significantly increased upon reducing the 3D bulk Bi_2WO_6 to the atomically thin Bi_2WO_6 layers [37]. The reduced dimension is beneficial for rapid movement of the photogenerated electrons and holes to the surface and thereby improves the solar CO_2 reduction. What's more, the increased surface charge density can effectively facilitate the two-dimensional (2D) conductivity, leading to the increase of CO_2 adsorption amount. Therefore, over single-unit-cell Bi_2WO_6, the predominant product of CO_2 photoreduction is methanol, and yield of CH_3OH reaches to 75 μmol h^{-1} g^{-1} under a 300 W Xe lamp irradiation, which is 125 times higher than that of bulk Bi_2WO_6 (0.6 μmol h^{-1} g^{-1}). Meanwhile, Mi and coworkers revealed the potential of III-nitride semiconductor nanostructures in solar-powered reduction of CO_2 into hydrocarbon fuels [38]. It was demonstrated that CO_2 molecules can be spontaneously activated on the clean nonpolar surfaces of wurtzite metal nitrides, InGaN, based on the results of ab initio calculations, in which the photoreduction conversion rate of CO_2 into methanol is ~0.5 mmol g_{cat}^{-1} h^{-1} under visible-light illumination (>400 nm). Moreover, it was found that upon incorporating a small amount of Mg dopant into InGaN, the photocatalytic performance of CO_2 reduction can be drastically enhanced. It is well known that the separation and transport of photogenerated electrons and holes are largely governed by the presence of surface band bending, for instance, upward and downward band bending. For Mg-doped InGaN nanowires, the introduction of Mg can make the surface potential of InGaN nanowires upward (up to 2 eV), leading to the near-flat surface band structure, which can apparently improve the photocatalytic performance in CO_2 reduction.

In addition, in 2018, Gong and his group reported 0D/2D materials with polymeric C_3N_4 nanosheets and CdSe quantum dots to enhance the separation and reduce the diffusion length of charge carriers [39]. The rapid outflow of carriers also restrains self-corrosion and consequently enhances the stability. Owing to quantum confinement effects of the QDs, the energy of the electrons could be adjusted to a level that inhibits the hydrogen evolution reaction (HER) (the main competitive reaction to the photocatalytic CO_2 reduction reaction) and improves the selectivity and activity for CH_3OH production from the photocatalytic CO_2 reduction reaction. With an optimal CdSe particle size of 2.2 nm, high selectivity (73%) and activity (186.4 mmol h^{-1} g^{-1}) for CH_3OH production with an apparent quantum efficiency

(AQE) of 0.91% are achieved, higher than most multicomponent CO_2 reduction reaction photocatalysts for CH_3OH production (generally 0.02–75 mmol h^{-1} g^{-1}).

Besides, inspired by biocatalytic reaction, based on high selectivity of enzyme, a series of enzyme-based inorganic/organic systems have been developed in the past. For example, Baeg and coworkers attached isatin–porphyrin to chemical-converted graphene for constructing a newly developed photocatalyst that was used sequentially to couple with enzymes in the presence of nicotinamide cofactor (NADH) under visible-light irradiation [40]. In this system, the selectivity of MeOH reached to almost 100%, while the yield (~5 μM h^{-1}) is lower than that of efficient semiconductor-based photocatalysts. Besides, in 2017, Park and coworkers further took advantage of enzyme-cascade system to improve the transferring of photoinduced electrons into a multienzyme for biocatalyzed reduction of CO_2 to methanol in a photoelectrocatalytic system [41]. In this tandem photoelectronic cell with an integrated enzyme cascade (TPIEC) system, hematite, and bismuth ferrite were used as photoanode and photocathode, respectively, and NADH and three different enzymes (formate dehydrogenase, formaldehyde dehydrogenase, and alcohol dehydrogenase) as specific reaction enzyme catalyzed CO_2 into formate, formaldehyde, and finally methanol, as shown in Figure 2.10. The average rate of methanol formation per reaction volume (~220 μM h^{-1}) is much higher than the pure photocatalytic system mentioned above.

2.4.1.3 Formaldehyde (HCHO)

As described in Eq. (2.3), HCHO is one of products that can be theoretically and experimentally formed, involving four-electron reactions, while more elementary reactions happen during the final formation of HCHO [42]. Meanwhile, HCHO is also detected as an intermediate for fulfilling next reactions and producing CO and other advanced hydrocarbons. For example, Ojha et al. found that HCHO and COO– functioned as important intermediates to form CO over Cu-modified S-doped g-C_3N_4 thin sheets through in situ diffuse reflectance infrared Fourier transform spectroscopy (DRIFTS) [43]. Meanwhile, it is found that the existing forms of Cu in S-doped g-C_3N_4 system include Cu^0, Cu^+, and Cu^{2+} three states, which have obvious effects in the adsorbed species and reaction pathways. Owing to the higher absorbance value on the Cu^0-decorated sheets, the bicarbonate species as an intermediate tended to promote CH_4 production. Furthermore, Song and coworkers explore the selectivity and photocatalytic activity of Cu- or Ni-doped TiO_2 catalysts by means of a hydrolysis method for the reduction of CO_2 under irradiation from Hg lamps in a CO_2/NaOH aqueous solution [44]. The results display that CO, HCHO, CH_3OH, and CH_4 can be yielded simultaneously in Cu/TiO_2 and Ni/TiO_2 system, while methane is identified as the major product. At the optimum value of 1%, the CO_2 conversion ability and light utilization ability reach to 29.43 μmol g^{-1} and 244.72 μmol g_{cat}^{-1}, respectively, which is almost seven times that of pure TiO_2 after 24 hours UV illumination. In Ni-doped TiO_2, the maximum value of the CO_2 conversion ability and light utilization ability, amounting to 43.97 and 460.30 μmol g_{cat}^{-1}, respectively, was observed after 24 hours UV irradiation. The CO_2 conversion ability can be indirectly defined as the sum of the mole number

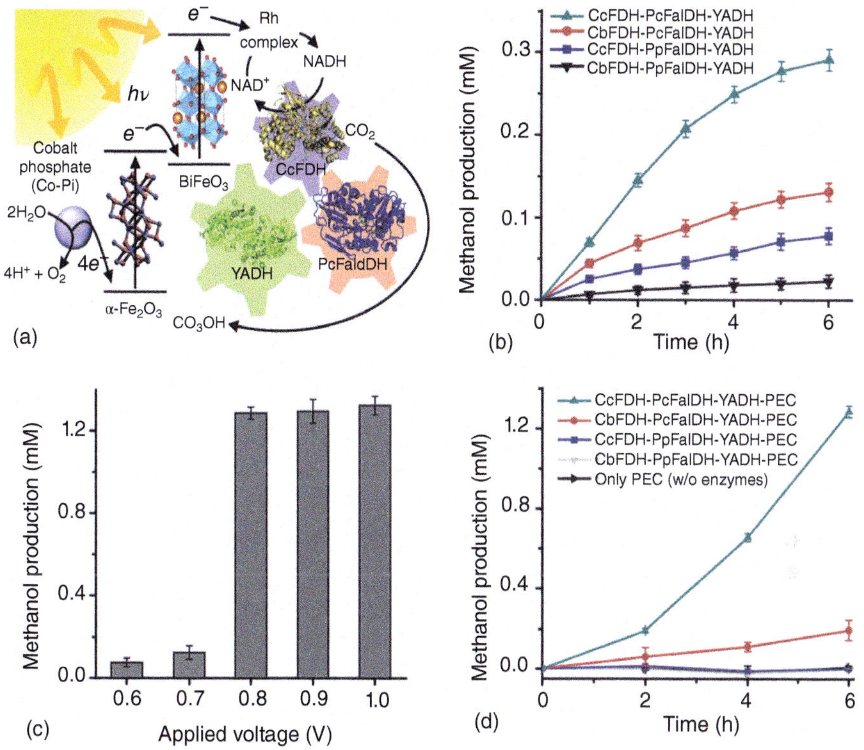

Figure 2.10 (a) Schematic illustration of the solar-assisted production of methanol from CO_2 and water through a three-enzyme cascade. (b) Methanol production as a function of time in multienzymatic systems. (c) Photoelectrochemical production of methanol by TPIEC under various applied potentials under visible-light illumination. (d) Methanol production of the TPIEC system in various multienzymatic systems with an external voltage of 0.8 V. Source: Kuk et al. [41].

of all independent carbon products, including CH_3OH, CO, HCHO, and CH_4. And meanwhile, the transformation of the photon is used to evaluate the light utilization ability, which is determined not only by the total quantity of carbon products produced but also by the H_2 yield. Therefore, the light utilization ability is the sum of the CO_2 conversion ability and the H_2 yield, which indicates that H_2 is the main product instead of reduced hydrocarbons. Recently, Garay-Rodríguez et al. reported a $Ba_3Li_2Ti_8O_{20}$/CuO composite for selective reduction of CO_2 to formaldehyde under visible-light irradiation [45]. The highest formaldehyde production is 11.6 $\mu mol\ g_{cat}{}^{-1}\ h^{-1}$, which is mainly attributed to lower particle size distribution and higher surface area compared with others, allowing an increase in the CO_2 adsorption.

On the other hand, owing to the ability to uptake and release electron, polyoxometalate (POM) can prolong the lifetime of the intermediate by directing the flow of the charge carriers, which enhances the efficiency of photocatalytic activity in CO_2 reduction [46]. For example, Barman et al. prepared a Mo-based polyoxometalate for selective formation of formaldehyde in photochemical carbon dioxide reduction

[47]. In this system, a maximum of 0.858 mmol of formaldehyde and 0.63 mmol of formic acid were obtained in the reactions with 0.01 μmol of catalyst loading. The overall maximum turnover number (TON) for photochemical carbon dioxide reduction is 546. Meanwhile, the results present that water as an electron donor releases O_2, electrons, and protons simultaneously through water oxidation process that is a pH-dependent reaction. A plausible mechanism of HCOOH formation is proposed, where CO_2 attached on Mo-based polyoxometalate is converted to CO_2 anion radical under the attack of the electron and subsequently is reduced to HCOO– and formic acid in the presence of large protons. Finally, formic acid is detached to vacate the active site to further produce formaldehyde.

To further improve the selectivity of HCHO in CO_2 photoreduction, the reaction parameters and reactors have been studied in the past, as similar as other chemical reactions. Recently, Brunetti et al. reported a continuous operating mode using C_3N_4–TiO_2 photocatalyst embedded in a dense Nafion matrix, where reaction pressure as a driving force was specifically and systematically investigated to determine reactor performance and product removal in photocatalytic CO_2 reduction [48]. The results showed that under the high feed pressure (5 bar), HCHO is the predominant production and MeOH is the secondary, where the carbon conversion efficiency achieved is 61 $\mu mol\ g_{cat}{}^{-1}\ h^{-1}$, better performing than other photocatalytic membrane reactors. Meanwhile, the effects of H_2O/CO_2 feed molar ratio and contact time on the membrane reactor performance were explored. Total converted carbon instead did not vary significantly with reaction pressure.

2.4.1.4 Formic Acid (HCOOH)

Formic acid, the two-electron reduction product of CO_2, has recently attracted attention as a storage source of H_2. Formic acid itself is an important chemical. It has been employed as a preservative and an insecticide and is also a useful acid, reducing agent, and source of carbon in synthetic chemical industries. Analyzing the reduction reaction of CO_2 to HCOOH, as shown in Eqs. (2.8, 2.10), it is found that this reaction involves multiple steps and multiple electrons so that the efficiency and yield are relatively low and the by-products undermine the selectivity of HCOOH through a series of competitive reactions. Therefore, unique photocatalysts or photoelectrocatalysts should be developed to meet the need of CO_2 conversion to HCOOH:

$$\text{Photocatalyst} + h\nu \rightarrow h^+ + e^- \tag{2.8}$$

$$H_2O + h^+ \rightarrow {}^{\cdot}OH + H^+ \tag{2.9}$$

$$CO_2 + 2H^+ + 2e^- \rightarrow HCOOH \tag{2.10}$$

Recently, bimetallic-graphene (Pt@Au@rGO) system has been reported by Kim and coworkers, in which the hybridized system is composed of gold nanoparticles (30 nm) as the core and ultrasmall platinum nanoparticles as the satellites wrapped by RGO [49]. The results demonstrate that the noble metal nanoparticles can efficiently improve the selectivity for HCOOH formation because noble metal can decrease the overpotential of hydrogen production from water splitting, therefore beneficial to the hydrogenation of CO_2, in particular Pt nanoparticles compared with other noble metals. On the other hand, it is reported that CB of g-C_3N_4 has enough high position, located at −1.23 V, which is more negative than the reduction potential of CO_2/CH_3OH (−0.38 V) and CO_2/HCOOH (−0.61 V). Therefore, it is often viewed as an effective photocatalyst for CO_2 photoreduction to methanol and formic acid. For example, Adekoya et al. constructed g-C_3N_4/(Cu/TiO_2) nanocomposites through coupling g-C_3N_4 with Cu-modified TiO_2 to suppress the rapid recombination of photoinduced charge carriers in pristine g-C_3N_4 [50]. As a result, the nanocomposite photocatalyst displayed enhanced photoreduction abilities of CO_2 to CH_3OH and HCOOH under UV and visible-light irradiation. Meanwhile, the yield of HCOOH (5069 μmol g_{cat}^{-1}) is much higher than that of CH_3OH (2574 μmol g_{cat}^{-1}), but the selectivity is not good. According to the analysis of HCOOH formation and related literatures, it seems that higher CB potential and lower overpotential hydrogen facilitate the yield of HCOOH, while it is observed that the by-product CH_3OH is higher, which is a strong competitive reaction due to the lower redox potential compared with that of HCOOH synthetic reaction [51]. Consequently, only introducing new component for enhanced charge separation and solar harvesting is hardly to achieve high selectivity of HCOOH. Therefore, novel photocatalysts with new reaction path of HCOOH should be considered and developed.

In 1992, Matsuoka et al. reported a oligo(*p*-phenylenes) organic molecular photocatalyst for CO_2 photoreduction to formic acid with small quantity of CO in a nonaqueous solvent under UV light illumination [52]. To further improve the selectivity of formic acid and efficiency, Tamaki and coworkers designed a series of Ru(II) supermolecular photocatalysts that exhibit high electivity and turnover frequency (TOF) under visible-light irradiation [53]. By means of combining photosensitizer units and catalyst units, a trinuclear complex displayed 0.061 of formic acid yield, 671 of TON, and 11.6 min^{-1} of TOF. Recently, Tamaki et al. prepared a novel sacrificial electron donor, 1,3-dimethyl-2-(*o*-hydroxyphenyl)-2,3-dihydro-1*H*-benzoimidazole, for efficient supermolecular photocatalyst, further increasing reduction efficiency of CO_2 to HCOOH, where the TON_{HCOOH} and TOF_{HCOOH} reached to 2766 and 44.9 min^{-1}, respectively, under visible-light irradiation [54].

Owing to the high cost and unrecyclable properties of homogeneous photocatalysts, chemical scientists pay more attention to combine solid semiconductor photocatalysts with liquid organic supermolecular photocatalysts together to overcome these obstacles as well as maintain high efficiency and selectivity. For instance, Suzuki et al. reported that a hybrid photocatalyst that consisted of Ru(II) complex and N-doped Ta_2O_5 anchored with an organic group was fabricated through a direct assembly route [55]. In Figure 2.11a, when the N-Ta_2O_5 was excited by suitable light source, the photogenerated electrons at the CB of N-Ta_2O_5 transferred to the Ru complex, in which the two-electron reduction of CO_2 to HCOOH can be easily triggered. Nevertheless, the solid semiconductor and organics should be assembled by a certain organic ligand to realize electron transfer process. To reach the goal, phosphonate and carboxylic groups were introduced to link the inorganic and organic components. The inorganic/organic system modified by

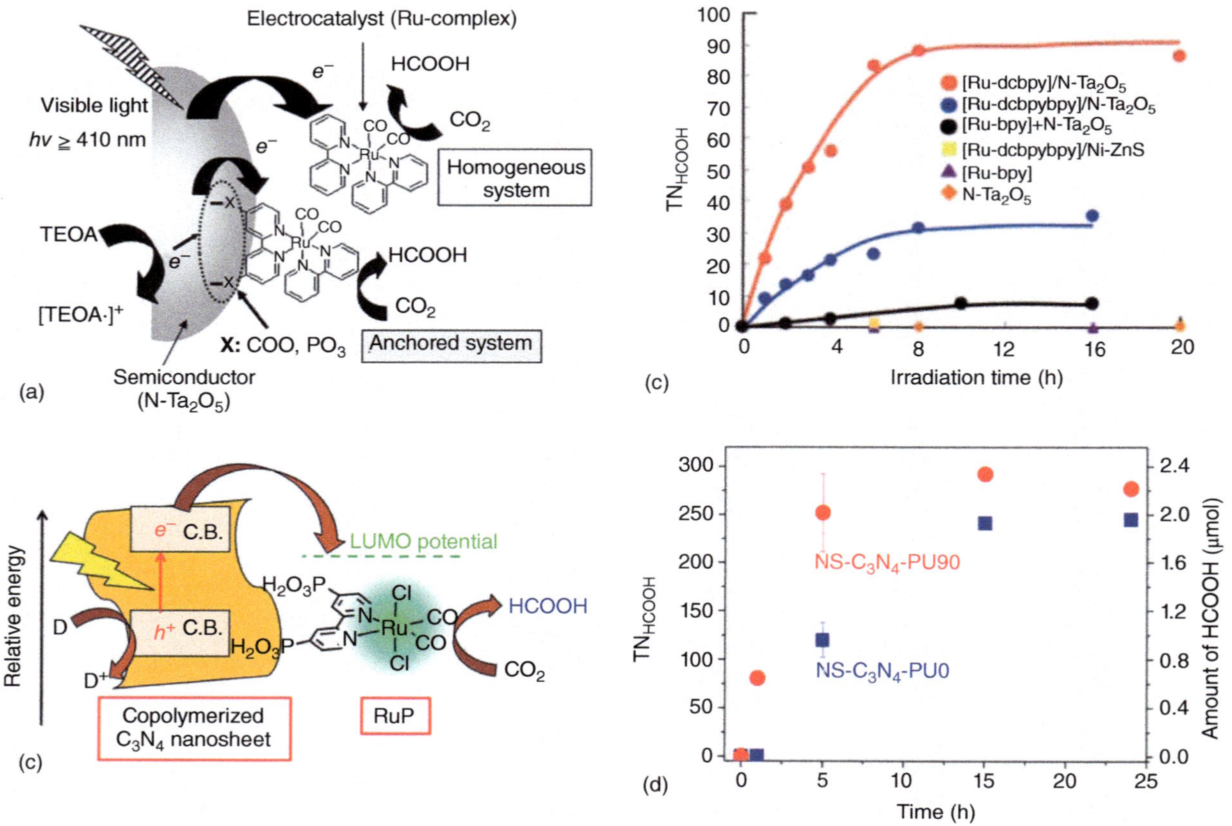

Figure 2.11 (a) Mechanism for the reduction of CO_2 by photocatalysis under visible light with a Ru complex and an N-Ta_2O_5 hybrid catalyst. (b) Turnover number for HCOOH formation from CO_2 over various molecular systems as a function of irradiation time. (c) Schematic mechanism of visible-light-driven photocatalytic CO_2 reduction over RuP/NS-C_3N_4 hybrids. (d) Time courses of photocatalytic CO_2 reduction using RuP/NS-C_3N_4-PU0 and RuP/NS-C_3N_4-PU90 under visible light ($\lambda > 400$ nm). Source: (a) Suzuki et al. [55]; (b) Sato et al. [55]; (c) Tsounis et al. [56]; (d) Tsounis et al. [56].

phosphonate or carboxylic groups exhibited excellent photoconversion activity of CO_2 to formic acid under visible-light irradiation with respect to the reaction rate and stability, as shown in Figure 2.11b. On the other hand, the rapid and efficient electron transfer can be achieved by adjusting the position of CB of semiconductor. For example, recently, Maeda and coworkers developed a Ru(II) complex/C_3N_4 hybrid photocatalyst for improving the photocatalytic reduction efficiency of CO_2 to HCOOH (selectivity >98%) under longer wavelength visible ($\lambda > 500$ nm) through a copolymerization preparation method, as shown in Figure 2.11c–d [56]. For the copolymerized C_3N_4 nanosheets fabricated by combing urea with phenylurea in high temperature, the absorption edge can be extended to 650 nm, which can significantly enhance photocatalytic activity and selectivity of modified Ru(II) complex/C_3N_4 system to HCOOH under longer-wavelength light source irradiation.

2.4.1.5 C2 Hydrocarbons

From the reaction procedure of CO_2 photoreduction, the reductive electrons mainly react with the adsorbed carbon species into C1 products because these reactions can be more easily driven thermodynamically. On the other hand, more valuable hydrocarbons including ethane and ethanol are more desirable such that it is inevitable to proceed multistep reactions involving more complicated reaction mechanism. It is well known that a multistep reaction means decaying of reaction efficiency and increasing of by-products. Therefore, selectivity is a key parameter for the production of C2 or higher hydrocarbons.

In general, there are two possible ways of CO_2 reduction to C_2H_6, which is from hydrogenation of ethylene intermediate or dimerization of two CH_3 adsorbates. Recently, Choi and his group introduced a thin Nafion layer on Pd-deposited TiO_2 nanoparticles that markedly enhances the photosynthetic conversion of CO_2 to hydrocarbons (mainly CH_4 and C_2H_6) in an aqueous suspension without any sacrificial electron donor under UV and solar irradiation conditions [57]. In this system, the Nafion layer functioned as a conductive and fixed layer to facilitate diffusion of protons onto the CO_2 reduction sites and immobilize intermediates of CO_2 reduction, which could help the serial electron transfers from the intermediate to the final product. The hybridized catalyst displays the CH_4 and C_2H_6 production of 123 and 12 $\mu mol\,g_{cat}^{-1}$, which apparently inhibited H_2 generation. Besides, Kulandaivalu et al. constructed a CQDs/Cu_2O nanocomposite photocatalyst with good optical, physical, and chemical properties that was able to selectively reduce CO_2 to C_2H_6 under visible-light irradiation [58]. The improved photoactivity is the accumulated C_2H_6 of 101.83 $\mu mol\,g_{cat}^{-1}$ for six hours under continuous visible-light illumination, while the pure Cu_2O photocatalyst produces C_2H_6 of 66.73 $\mu mol\,g_{cat}^{-1}$ for the same reaction time. The findings present that no methane or ethylene is detected over bare Cu_2O and CQDs/Cu_2O hybrids, indicating that the dimerization of CH_3 adsorbates is favorable in this system. The proposed step number of C_2H_6 production is eight, where both the photoinduced electrons and holes take part in the CO_2 photoreduction and 3.5 molecules of O_2 for each C_2H_6 molecular is released from water splitting process but no hydrogen is detected.

Apart from ethane, ethanol is also reported in the photoreduction of CO_2 in the past [59]. Recently, Liu et al. reported visible-light-responsive photocatalyst $BiVO_4$ for selective formation of ethanol under the condition of high-intensity visible-light irradiation [60]. It is known that the proton in water cannot capture the photogenerated electrons on $BiVO_4$ to produce H_2. In contrast, once CO_2 molecule is adsorbed and changed to CO_3^{2-}, the photogenerated electrons can react with CO_3^{2-} likely and give rise to methanol after protonation. Furthermore, under intense irradiation, a large number of C1 intermediate species are anchored on the surface of $BiVO_4$, which will facilitate dimerization of C1 intermediate species to form ethanol. Based on this strategy, Peng and coworkers took advantage of graphitic carbon nitride with high CB position for achieving selectively photocatalytic reduction of CO_2 under visible-light irradiation [61]. It was found that the starting materials of graphitic C_3N_4 have significant influence on the activity and selectivity of CO_2 reduction. Once graphitic carbon nitride was fabricated by using urea (denoted as u-g-C_3N_4) that possessed large surface area with mesoporous nanostructure, the reduced products of CO_2 were a mixture of CH_3OH and C_2H_5OH in the presence of NaOH (1.0 M), where the yields were 6.28 and 4.51 μmol g_{cat}^{-1} h^{-1}, respectively. Compared with u-g-C_3N_4, the sample prepared by using melamine (denoted as m-g-C_3N_4) mainly catalyzed CO_2 to C_2H_5OH under the same conditions, where the yield of C_2H_5OH was 3.64 μmol g_{cat}^{-1} h^{-1}. Meanwhile, high concentration O_2 and trace H_2 were detected in two samples, inferring that photoexcited holes at VB of C_3N_4 can be efficiently formed from water splitting and the corresponding electrons took part in the conversion of CO_2 with H^+ into alcohols. Moreover, by comparing u-g-C_3N_4 with m-g-C_3N_4, it shows that larger surface area, smaller crystal size, and lower crystallinity may be the main reasons for the apparent difference in selective formation of CH_3OH and C_2H_5OH in the present system. Owing to the nonporous structure of m-g-C_3N_4, the fast exchange of the formed $\cdot OCH_3$ or CH_3OH may be suppressed, thereby in turn benefiting the dimerization to produce C_2H_5OH in the present CO_2/NaOH system.

In addition, Pastrana-Martínez et al. prepared a graphene derivative–TiO_2 composites for photocatalytic water reduction of CO_2 into renewable fuels under UV–vis light irradiation [62]. Meanwhile, the findings present that the pH of solution has significant influence toward selective ethanol formation. The prepared GO/TiO_2 composite exhibited superior photocatalytic activity for EtOH production (144.7 μmol g_{cat}^{-1} h^{-1}) at pH 11.0 and for MeOH production (47.0 μmol g_{cat}^{-1} h^{-1}) at pH 4.0. It is found that the yield of C2 hydrocarbons is much lower than that of C1 products due to the demand of more electrons and complicated reaction mechanism, which cannot be varied. Therefore, it is more feasible that the long-chain hydrocarbons are formed as secondary products over photocatalysts that can realize efficient C1–C2 conversion in the future.

2.4.1.6 Other Hydrocarbons

Apart from these hydrocarbons produced in CO_2 photoreduction with high frequency, some complicated hydrocarbons with more than two carbon numbers have been reported recently. Fusco et al. reported firstly the formation of C2 hydrocarbons, acetic acid, through CO_2 photoreduction over TiO_2@PEI-grafted-MWCNTs hybrids under UV–Vis light irradiation [63]. Besides, Park et al. decorated Cu nanoparticles and CdS quantum dots on the surface of TiO_2 nanotubes, forming a ternary nanostructured photocatalyst that is capable of converting CO_2 and H_2O into C1–C3 hydrocarbons, including CH_4, C_2H_6, C_3H_6, and C_3H_8 (Figure 2.12b–e) [64]. In this system, CdS quantum dots are responsible of harvesting solar light and forming hot electrons that will rapidly move to Cu nanoparticles through TiO_2 nanotubes. At Cu nanoparticles, the concentrated electrons can react with CO_2 molecular and produce C_xH_y under visible-light excitation (above 420 nm), as shown in Figure 2.12a. In contrast, over five hours irradiation, free H_2 molecular that is more easily formed than CO_2 reduction products was not detected, which inferred that competitive H_2 evolution reaction has been suppressed efficiently in this CdS/(Cu–TNTs) hybrid system. The photocatalytic reduction of CO_2 over the CdS/(Cu–TNTs) hybrid is initiated likely through a one-electron reduction to form $\cdot CO_2^-$, reacting in turn with a H atom (H·) to produce hydrocarbons (Figure 2.12f). CO_2 hydrogenation leading to hydrocarbon formation appears to be a less likely pathway because of the absence of H_2 detected in the headspace of the photolysis reactor during five hours of irradiation with visible light. However, the yield and efficiency are sharply decreased owing to involvement of more electrons, formation of longer C–C chains, and production of more by-products, which severely undermine the proportion of expected products, such as C3 or C4 hydrocarbons. Therefore, to increase the formation possibility and efficiency of major hydrocarbons, electrocatalytic CO_2 reduction is paid attention through introducing more strong hot electrons to react with adsorbed CO_2 molecular over a variety of electrocatalysts, leading to longer carbon chain hydrocarbons [65].

2.4.2 Carbon Monoxide (CO)

Apart from the hydrocarbons, other carbon-containing fuels such as CO also are produced as main product in CO_2-involving photocatalysis [66]. From the view of redox potential, CO formation reaction (−0.48 eV vs. NHE) is inferior to the production of CH_4 (−0.38 eV vs. NHE) and CH_3OH (−0.24 eV vs. NHE), while the formation reactions of CH_4 and CH_3OH involve six- and eight-electron reaction process, in which a series of elemental reaction are unclear and slow. In contrast, carbon monoxide is simpler than CH_4, CH_3OH, and other hydrocarbons, which means fewer needs of reductive electrons. Therefore, the CO reduction reaction is thermodynamically triggered, and CO is a preferred product in CO_2 photoreduction. For instance, Miyauchi and coworkers reported CuO-decorated Nb_3O_8 nanosheets for photocatalytic CO_2 reduced into CO, and simultaneously the reaction pathway over this system had been deeply investigated through electron spin resonance (ESR) and isotope-labeled experiments [67]. The results indicate that amorphous copper oxide nanoclusters can work as efficient electrocatalysts grafted onto the surface of niobate nanosheets for the reduction of carbon dioxide to carbon monoxide. Furthermore, the photocatalytic activity and reaction pathway of Cu(II)-grafted Nb_3O_8 nanosheets were investigated using ESR analysis and isotope-labeled molecules

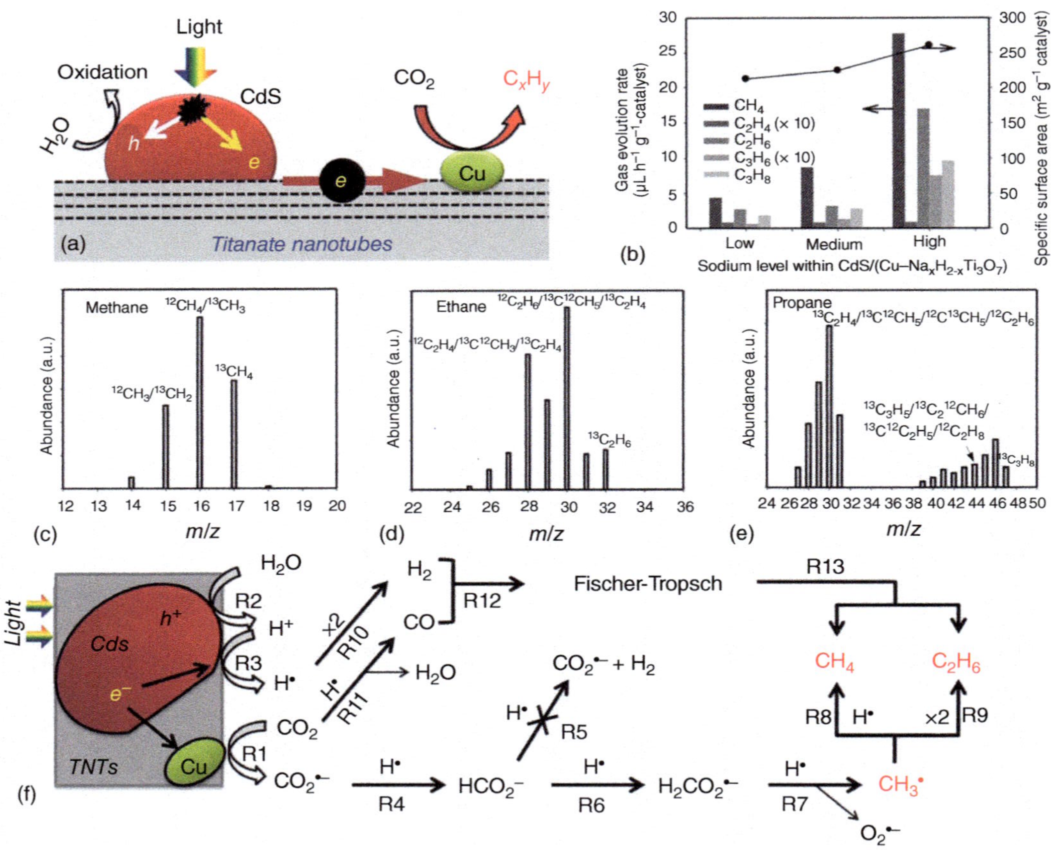

Figure 2.12 (a) Scheme of the photocatalytic CO_2 reduction over CdS/(Cu−$Na_xH_{2-x}Ti_3O_7$) irradiated by light. (b) Gas evolution rates of C1−C3 hydrocarbons on CdS/(Cu−$Na_xH_{2-x}Ti_3O_7$) as well as the specific surface area, where the Na/Ti ratios were 0.093 (low), 0.143 (medium), and 0.507 (high). (c−e) Mass spectra of the formed hydrocarbons (methane, ethane, and propane) with the labeling of ^{13}C. (f) Proposed elementary reaction mechanisms of photocatalytic CO_2 conversion into hydrocarbons. Source: Park et al. [64].

($H_2{}^{18}O$ and $^{13}CO_2$). The results of the labeling experiments demonstrated that under UV irradiation, electrons are extracted from water to produce oxygen ($^{18}O_2$) and then reduce CO_2 to produce ^{13}CO. ESR analysis confirmed that excited holes in the VB of Nb_3O_8 nanosheets react with water and that excited electrons in the CB of Nb_3O_8 nanosheets are injected into the Cu(II) nanoclusters through the interface and are involved in the reduction of CO_2 into CO. The Cu(II) nanocluster-grafted Nb_3O_8 nanosheets are composed of nontoxic and abundant elements and can be facilely synthesized by a wet chemical method. The nanocluster grafting technique described here can be applied for the surface activation of various semiconductor light harvesters, such as metal oxide and/or metal chalcogenides, and is expected to aid in the development of efficient CO_2 photoreduction systems. Furthermore, Tahir et al. synthesized Ag nanoparticles/TiO_2 nanowires core–shell heterojunction for efficient photoreduction of CO_2 to CO in the presence of hydrogen [68]. Ag-NPs coated over TiO_2 NWs exhibited strong absorption of visible light due to localized surface plasmon resonance (LSPR) excitation, trapped electrons, and hindered charge recombination rate. The synergistic effect of Ag-NPs coated over TiO_2 NWs for CO_2 conversion was evaluated in a gas-phase system under UV and visible-light irradiation. The plasmonic Ag-NPs/TiO_2 NWs demonstrated excellent photoactivity in the reduction of CO_2 into CO, CH_4, and CH_3OH under visible-light irradiation. The results show that 3 wt% Ag-NPs-loaded TiO_2 NWs was found to be the most active, giving the highest CO evolution of 983 mmol g^{-1} h^{-1} at selectivity 98%. This amount of CO produced was 23 times more than the TiO_2 NWs and 109 times larger than the yield of CO produced over the pure TiO_2. More importantly, the quantum yield was substantially enhanced for CO evolution. The LSPR excitation and synergic effect of Ag-NPs that can effectively accelerate the charge separation were proposed to be responsible for the enhancement of photocatalytic activity. The photostability of Ag-NPs/TiO_2 NWs evidenced in cyclic runs for selective CO production under visible light, yet photoactivity declined over the irradiation time under UV light.

Recently, it is well demonstrated that layered nanomaterials with excellent conductivity, e.g. graphene and C_3N_4, can facilitate to accelerate the charge transportation, which is a promising component to efficiently improve the separation efficiency of photogenerated electrons and holes in traditional semiconducting photocatalysts [69]. For instance, Ye and coworkers constructed a smooth carrier channel in the basal plane of organic polymeric C_3N_4 photocatalyst by enriching defects with short and strong C–N chains through a glycine linker so that the photoinduced carrier transfer is apparently enhanced. Based on this, the photoreduction ability of CO is increased by 29.2 times under a simulated sunlight irradiation, as shown in Figure 2.13 [70]. Compared with complicated hydrocarbons, H_2 has strong competition ability to CO due to the lower potential. Therefore, H_2 often consumes more photoexcited electrons and suppresses the CO production. Although H_2 is cleaner energy storage form, carbon element is not contained in the energy cycle, which is undesirable to control CO_2 level in atmosphere. In other words, as a competitive reaction, two-electron H_2 evolution reaction from water splitting should be suppressed deliberately. In 2011, Kudo and coworkers

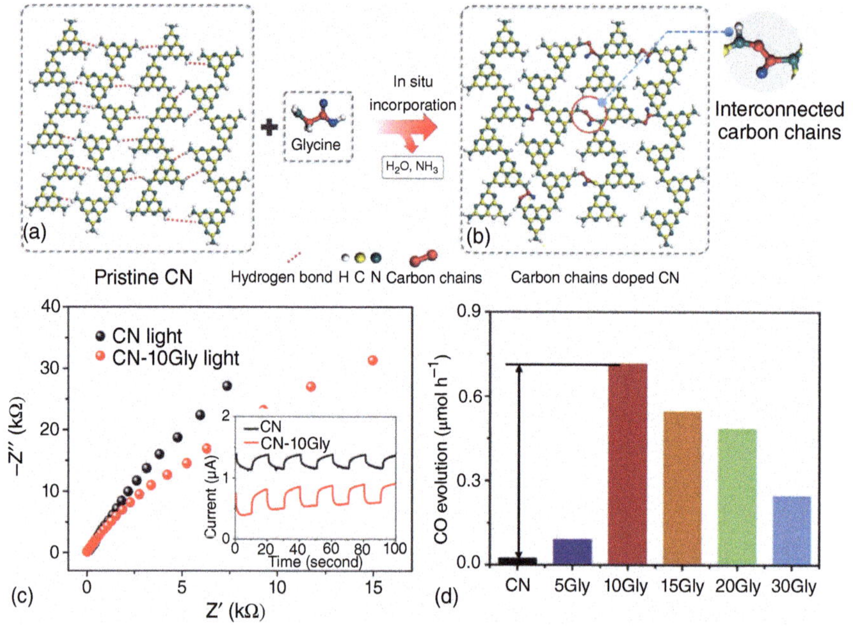

Figure 2.13 Structures of CN before (a) and after (b) doping with carbon chains. (c) Electrochemical impedance spectroscopy (EIS) Nyquist plots under irradiation condition. Inset: Periodic on/off photocurrent response under visible-light irradiation. (d) Comparison of the photocatalytic CO_2 reduction rate of the samples with different amount of glycine, respectively. Source: Ren et al. [70].

reported a series of Ag-modified $ALa_4Ti_4O_{15}$ (A = Ca, Sr, and Ba) photocatalysts that have 3.79–3.85 eV of band gaps and layered perovskite structures, displaying photocatalytic activities of CO_2 reduction to produce CO and HCOOH without any sacrificial reagents [71]. It is reported that d^0-type metal oxide photocatalysts can catalyze water splitting into H_2 and O_2 because of high CB positions and wide band gaps, which are expected to be active for CO_2 photoreduction at a suitable reaction sites on the surface of photocatalysts. Among these photocatalysts, Ag-loaded $BaLa_4Ti_4O_{15}$ showed the best photocatalytic activity. The Ag as a cocatalyst prepared by the liquid-phase chemical reduction method was loaded as fine particles with the size smaller than 10 nm on the edge of the $BaLa_4Ti_4O_{15}$. On the optimized $Ag/BaLa_4Ti_4O_{15}$ photocatalyst, CO was the main reduction product rather than H_2 even in an aqueous medium. Furthermore, evolution of O_2 in a stoichiometric ratio ($H_2 + CO{:}O_2 = 2:1$ in a molar ratio) indicated that water was consumed as a reducing reagent (an electron donor) for the CO_2 reduction, which demonstrated that CO_2 reduction accompanied with water oxidation was achieved using the $Ag/BaLa_4Ti_4O_{15}$ photocatalyst. After that, Teramura and coworkers designed a new three-component heterojunction for efficiently hampering the H_2 evolution and increasing the selectivity of CO. [72] The results indicated that Zn-doped Ga_2O_3 exhibited significant restraint on H_2 releasing from overall water splitting. Based on this result, they further deposited Ag onto the Zn–Ga_2O_3 to increase the production efficiency of CO because introduction of Ag as a cocatalyst could enhance the harvesting efficiency of sunlight and collect more photogenerated electrons for CO_2 reduction. In the case of $Ag/Zn/Ga_2O_3$, the selectivity toward CO evolution is higher than toward H_2 in the presence of $NaHCO_3$ solution, where the yields of CO and H_2 reached to 800 and 60 μmol g_{cat}^{-1}, respectively, over seven hours under UV irradiation. At the same time, the O_2 evolution was observed, which inferred that overall water splitting happened in this system while H_2 releasing was suppressed. These results indicate that the Ag-modified Zn-doped Ga_2O_3 realizes selective conversion of CO_2 and H_2O to CO and O_2 under UV irradiation.

On the other hand, it is reported that CO can originate from the secondary photolysis of unstable reduced products, such as HCO_2H. For example, Frei and coworker investigated the reaction mechanism of CO_2 photoreduction over Ti silicalite molecular sieve in the presence of methanol as electron donor by means of in situ FTIR [73]. It was found that HCO_2H, CO, and HCO_2CH_3 as reduced products were detected, in which mass proportion of CO was the highest. The formation of the products was studied through the infrared analysis of experiments with isotope-labeled reactants, such as $C^{18}O_2$, $^{13}CO_2$, and $^{13}CH_3OH$. The results inferred that the produced CO is derived from secondary photolysis of the reduced HCO_2H. In contrast, the formic acid is the primary two-electron reduction product of CO_2 at the ligand-to-metal charge transfer transition (LMCT) excited Ti centers. This means that since the complex hydrocarbons are the target products, the photolysis effect should be paid more attention for suppressing the formation of CO.

2.4.3 Dioxygen (O_2)

As discussed above, an ideal artificial photosynthesis system usually contains at least three different components: light harvesting, water oxidation, and CO_2 reduction [74]. To achieve CO_2 photoreduction accompanied with water oxidation, early single integrated systems are composed of semiconducting metal oxide with large band gaps absorbing UV light and metal or metal oxide cocatalysts, where considerable efforts are paid to functionalizing these materials and exploring various modifications [75]. Nevertheless, since the assembly of multifunctional units in an integrated device is extremely difficult, an alternative strategy is to divide the overall process into two half-reactions: water oxidation and CO_2 reduction. Once each half-reaction is well understood and optimized, the two reactions can be coupled in an integrated device. For example, it is demonstrated that combining two semiconductor materials in tandem (Z-scheme) mode can efficiently extend the light response into the visible region [76]. Recently, Arai et al. reported a photoelectrochemical system consisting of $SrTiO_3$ photoanode for water oxidation and InP photocathode for CO_2 reduction that produced O_2 and formate, respectively [77]. Besides, mononuclear $[Ru(bpy)_3]^{2+}$ is often combined with $S_2O_8{}^{2-}$ sacrificial acceptor to evaluate water oxidation photocatalysis systems in aqueous photosensitization system in the past decades [78]. For example, upon combining mild oxidant $[Ru(bpy)]_3{}^{3+}$ of well-defined potential (+1.26 V) with a Co_3O_4 nanotube in neutral aqueous solution, a hole can be efficiently transferred to the catalyst [79]. Subsequently, four holes transfer from $[Ru(bpy)_3]^{3+}$ to Co_3O_4 and react with two H_2O molecules to O_2 and $4H^+$, which indicates that Co_3O_4 is appropriate candidate for use in an artificial photosynthetic assembly.

On the water oxidation side, several groups proposed approaches to overcome challenges of efficiently combining molecular light absorbers with multi-electron water oxidation catalyst inspired by the tyrosine mediator design of nature's photosystem II. For instance, the Mallouk group reported a series of researches on coupling of Ir oxide nanocluster (IrO_x) with $[Ru(bpy)_3]^{2+}$ or a porphyrin visible-light absorber through a benzimidazole–phenol redox linking with both components covalently anchored on a TiO_2 surface [80]. Based on the results of transient optical spectroscopy, when the Ru chromophore absorbs a visible photon, an photoinduced electron injects into TiO_2 at an ultrafast speed, and subsequently an electron transfers from the benzimidazole–phenol mediator to the oxidized Ru complex for reducing the oxidized chromophore on a short time scale compared to the catalytic turnover of O_2 at the IrO_x particle. As a result, the competition between undesired transfer of electrons injected into TiO_2 back to the oxidized light absorber and hole injection into the IrO_x catalyst is shifted in favor of catalysis, thus improving the quantum efficiency of water oxidation by a factor of three. The approach offers substantial room for further efficiency improvement because the relative position of the light absorber and the catalyst with the attached mediator is not yet molecularly defined in the present system.

Recently, earth-abundant inorganic oxide catalysts have been widely used in water oxidation, accompanied with CO_2 photoreduction. In the past decades, Ir oxide nanoclusters have been viewed as efficient catalysts for water oxidation, which can be used to assemble directly various fashion systems for closing the photosynthetic cycle on the nanoscale. For example, Kim et al. prepared an all-inorganic polynuclear unit consisting of an oxo-bridged binuclear $ZrOCo^{II}$ group with iridium oxide nanocluster assembled on SBA-15 silica mesopore surface, which could produce CO and O_2 simultaneously [81]. Apart from Ir-based cocatalyst, other noble metals also are applied into the CO_2 photoreduction and water oxidation by deposition on semiconducting materials. For example, Fan et al. fabricated a binary functional photocatalyst by loading Au@Pd nanoparticles onto oxygen vacancy-rich TiO_2 through surfactant-free deposition–reduction method [82]. The obtained V_O-rich TiO_2 contributes more protons to water oxidation process, and meanwhile, the metallic Au@Pd sites are beneficial for CO_2 activation and proton supply. By varying the ratio of Au@Pd nanoparticles and V_O concentration, CH_4 selectivity of 96% can be obtained with minimal H_2 production. However, the Ir oxide or other noble metals are not a viable component for a scalable artificial photosystem because of the scarcity of this element. Therefore, earth-abundant metal oxides are developed as robust alternatives. For example, Yu et al. Cl-doped Cu_2O nanorods for photocatalytic CO_2 reduction conjugated with H_2O oxidation under visible-light irradiation [83]. Owing to the introduction of Cl atoms, the band structure of Cu_2O is optimized, leading to a more positive VB position for water oxidation, which promotes CO_2 activation and separation and transfer efficiency of

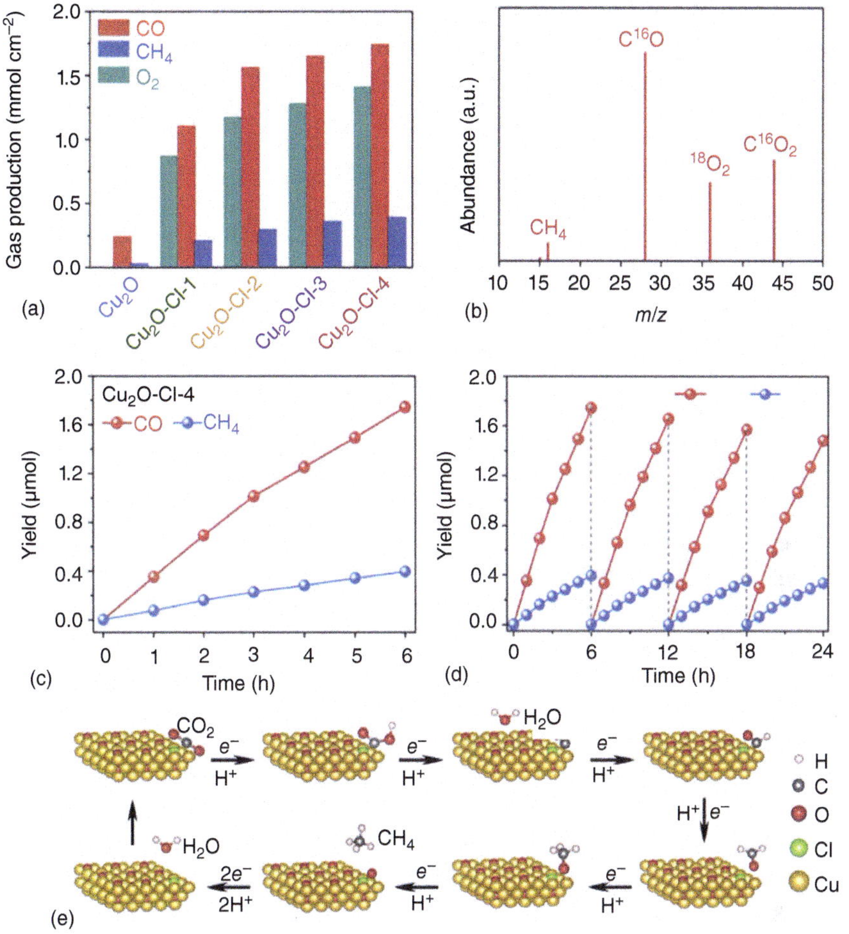

Figure 2.14 (a) Gas production under six hours visible-light irradiation of pure Cu_2O and Cl-doped Cu_2O samples. (b) Mass spectrum from the ^{18}O-labeling H_2O isotope experiments of Cu_2O–Cl-4. (c) Time-dependent product evolution over Cu_2O–Cl-4 under visible-light irradiation. (d) Stability test of Cu_2O–Cl-4. (e) Proposed reaction pathways of CO_2RR to CO and CH_4 on Cl-doped Cu_2O. Source: Yu et al. [83].

photoinduced charge carriers. As a result, in Figure 2.14, the Cl-doped Cu_2O composite exhibits excellent photocatalytic CO_2 reduction performance accompanied by favorable water oxidation ability under visible-light irradiation. DFT calculations demonstrate that the doping of Cl atoms facilitates to transform CO_2 into the intermediates of *COOH, *CO, and *CH_3O, which could enhance production of reductive CO and CH_4 and oxidative O_2. The CO, CH_4, and O_2 amount were improved to 1.74 μmol cm^{-2}, 0.39 μmol cm^{-2}, and 1.32 μmol cm^{-2}, respectively, for the Cu_2O–Cl-4, as shown in Figure 2.14a. Furthermore, it is found that the Cl-doped Cu_2O composite displays stronger affinity toward the *CO intermediate, which tends to be protonated and ultimately produce CH_4, leading to higher selectivity of CH_4 than that of pure Cu_2O.

Based on early reports of electro- or light-driven Co and Mn oxide catalysts, the findings display that these catalysts have the TOFs per metal center in the range 10^{-4} to 10^{-2} O_2 s^{-1} at overpotentials in the 30–400 mV range depending on the structure, pH, and temperature conditions, which are at least 2 orders of magnitude lower than TOF of IrO_2 clusters at similar overpotential [84]. In the past several years, therefore, intense efforts have been paid to enhance the TOF per projected area of the catalyst by optimizing the Co-, Mn-, and Ni-based metal oxide. For example, amorphous Co species (a-Co-E) synthesized through thermal treatment of a Co-ethylene diamine tetraacetic acid complex were combined with Bi_2WO_6, and the product exhibited an impressively higher photocatalytic O_2 evolution rate than that of the CoO_x-loaded counterparts [85]. The findings suggested that the photocatalytic O_2 production rate of the Bi_2WO_6/a-Co-E hybrid is apparently enhanced to 352.3 μmol g^{-1} h^{-1} with no obvious decrease in six hours, which is 3.6 times higher than that of Bi_2WO_6/CoO_x, testifying the excellent capability of a-Co-E to function as the cocatalyst over CoO_x. Graphitic carbon nitride and cubic cobalt manganese spinel (c-$CoMn_2O_4$) as light transducer and water oxidation cocatalyst were used to assemble effective water oxidation system [86]. Owing to acceleration of interface charge transfer rate and decreasing of the excessive energy barrier for O–O formation, the oxygen evolution rate of the g-$C_3N_4/CoMn_2O_4$, 18.3 μmol h^{-1}, is four times higher than that of pristine g-C_3N_4. To further provide more active sites for water oxidation, novel mesoporous MOFs are utilized as a host material to realize the strategy [87]. Co-based POM possessing stable water oxidation on central Co^{III} sites is coupled with MIL-100(Fe), which not only immobilizes homogeneous Co-POM cluster but also contributes more surface areas [88]. The results reveal that a 1.72-fold increasing of oxygen yield and TOF of 9.2×10^{-3} s^{-1} are achieved for the Co-POM/MIL-100(Fe) hybrids.

2.5 Perspective

The energy shortage and environmental-related problems are becoming the worsening global crisis and the great challenges for human in the twenty-first century. The artificial photosynthesis process involved in transformation of carbon dioxide and water to value-added chemical fuels is a promising strategy to rebuild the global energy consumption and elemental balance, and the potential rewards are enormous. Up to now, the main goal of scientists is to synthesize photoactive materials that are able to chemically couple these light-driven redox reactions together and to achieve conversion efficiency and selectivity that outperforms nature's photosynthesis. However, photoconversion of carbon dioxide with water contains thermodynamically uphill, multi-electron, multi-hole, and multi-proton processes occurring on a multicomponent photocatalyst, where many challenges are presented in the fields of catalysis, energy science, semiconductor physics, and engineering. To develop effective artificial photosynthesis and conversion, several key considerations must be balanced, including the following: (i) A deep understanding of processes that occur on the surface of photocatalysts during artificial photosynthesis processes, e.g. adsorption/desorption of gaseous reactants, products, and intermediates, as well as the role of adsorbed water. (ii) The potential great improvement of efficiency in carbon dioxide photoreduction can result from significantly increasing the lifetime of the charge-separated state. The time scale of this electron–hole recombination is two to 3 orders of magnitude faster than other electron transfer processes. Therefore, any process that inhibits electron–hole recombination would greatly increase the efficiency and rates of carbon dioxide reduction and water oxidation. For example, electrical conductivity and diffusion length of the photocatalyst should be as high as possible to minimize recombination of electron–hole pairs. Ultrathin nanostructures may also facilitate the charge carrier transport to surface reaction sites to participate reductive chemistry. Spatial separation of photoexcited electrons and holes can also be reduced by the electron–hole recombination, which was achieved by cocatalysts, Z-scheme, or heterostructures of coupling two semiconductors with properly aligned band structures. (iii) The kinetics of photocatalytic carbon dioxide reduction is also dependent upon light absorption by the photocatalyst. Many optical techniques are potentially useful for harvesting light for improvement of the efficiency including structuring for multiple light scattering to increase the effective optical path length, and upconversion to transform non-absorbed infrared light to absorbed visible light. (iv) Large surface area and porosity are required to maximize the adsorption, transport, and desorption of reactants, intermediates, and products. (v) As defects such as oxygen vacancies control most of the chemistry at many metal oxide surfaces, oxygen vacancies are believed to act as a very important role on electron trapping and activating CO_2. The oxygen vacancies can be created either by doping with other anions or cations or by thermal treatments of stoichiometric photocatalysts, and these defects can be detected with in situ electron paramagnetic resonance spectroscopy, UV photoemission spectroscopy, metastable impact electron spectroscopy, and so on. (vi) A high-efficiency process is also necessary to be founded on photoactive materials made of earth-abundant, nontoxic, light-stable, scalable, and low-cost materials.

In addition, the efficiency of photocatalytic reduction of carbon dioxide could be deactivated after irradiation for long time. The deactivating phenomenon in the photocatalytic reduction of carbon dioxide can be attributed to the following three reasons: (i) The adsorption or accumulation of intermediate products on the semiconductor surface could occupy photocatalytic reaction centers and also hinder the adsorption of carbon dioxide or water, which could lead to the deactivation of semiconductors under the continuous irradiation. (ii) The desorption of hydrocarbon production affects the continuous adsorption of reactants that take part in the photocatalytic reaction. (iii) The surface contamination of semiconductors shields the light absorption, resulting in the reduction of the generation of electron–hole pairs. Therefore, it is necessary to pay more attention to the study of semiconductor deactivation in future work.

Acknowledgments

This work is supported financially by the National Natural Science Foundation of China (21501138), the Natural Science Foundation of Hubei Province (2019CFB556), and the Science Research Foundation of Wuhan Institute of Technology (K201939).

References

1 (a) Oppenheimer, M. and Alley, R.B. (2005). *Clim. Change* 68: 257. (b) Leifeld, J. and Fuhrer, J. (2005). *Environ. Sci. Policy* 8: 410.

2 (a) Keller, F., Lee, R.P., and Meyer, B. (2020). *J. Cleaner Prod.* 250: 119484. (b) Kaur-Sidhu, M., Ravindra, K., Mor, S., and John, S. (2020). *Atmos. Pollut. Res.* 11: 252.

3 (a) Bachu, S., Bonijoly, D., Bradshaw, J. et al. (2007). *Int. J. Greenhouse Gas Control* 1: 430. (b) Rubin, E.S., Chen, C., and Rao, A.B. (2007). *Energy Policy* 35: 4444. (c) Davison, J. (2007). *Energy* 32: 1163.

4 Herron, J.A., Kim, J., Upadhye, A.A. et al. (2015). *Energy Environ. Sci.* 8: 126.

5 (a) McConnell, I., Li, G., and Brudvig, G.W. (2010). *Chem. Biol.* 17: 434. (b) Larkum, A.W. (2010). *Curr. Opin. Biotechnol.* 21: 271.

6 van Grondelle, R. and Boeker, E. (2017). *J. Phys. Chem. B* 121: 7229.

7 (a) Dogutan, D.K. and Nocera, D.G. (2019). *Acc. Chem. Res.* 52: 3143. (b) Bohne, C., Pan, Q., Ceroni, P. et al. (2015). *Faraday Discuss.* 185: 187. (c) Barber, J. (2009). *Chem. Soc. Rev.* 38: 185.

8 Dhakshinamoorthy, A., Navalon, S., Corma, A., and Garcia, H. (2012). *Energy Environ. Sci.* 5: 9217.

9 Liu, X., Inagaki, S., and Gong, J. (2016). *Angew. Chem. Int. Ed.* 55: 14924.

10 (a) Hamdy, M.S., Amrollahi, R., Sinev, I. et al. (2014). *J. Am. Chem. Soc.* 136: 594. (b) Zhao, G., Huang, X., Wang, X., and Wang, X. (2017). *J. Mater. Chem. A* 5: 21625.

11 Ke, J., Liu, J., Sun, H. et al. (2017). *Appl. Catal., B* 200: 47.

12 Hou, W. and Cronin, S. (2013). *Adv. Funct. Mater.* 23: 1612.

13 (a) Ke, J., Duan, X., Luo, S.L. et al. (2017). *Chem. Eng. J.* 313: 1447. (b) Luo, S., Ke, J., Yuan, M. et al. (2018). *Appl. Catal., B* 221: 215.

14 (a) Watanabe, K., Menzel, D., Nilius, N., and Freund, H. (2006). *Chem. Rev.* 106: 4301. (b) Banerjee, S., Pillai, S., Falaras, P. et al. (2014). *J. Phys. Chem. Lett.* 5: 2543. (c) Tahir, M. and Amin, N. (2013). *Renewable Sustainable Energy Rev.* 25: 560.

15 Ke, J., Adnan Younis, M., Kong, Y. et al. (2018). *Nano-Micro Lett.* 10: 69.

16 Ma, Y., Wang, X., Jia, Y. et al. (2014). *Chem. Rev.* 114: 9987.

17 (a) Zou, X., Dong, Y., Li, S. et al. (2018). *Sol. Energy* 169: 392. (b) Liu, J., Zhang, J., Wang, D. et al. (2019). *ACS Sustainable Chem. Eng.* 7: 12428. (c) Zhang, Z., Wang, S., Bao, M. et al. (2019). *J. Colloid Interface Sci.* 555: 342.

18 (a) Ke, J., Zhao, C., Zhou, H. et al. (2019). *Sustainable Mater.Technol.* 19: e00088. (b) Zou, X., Yuan, C., Dong, Y. et al. (2020). *Chem. Eng. J.* 379: 122380.

19 Inoue, T., Fujishima, A., Konishi, S., and Honda, K. (1979). *Nature* 277: 637.

20 Tu, W., Zhou, Y., and Zou, Z. (2014). *Adv. Mater.* 26: 4607.

21 Anpo, M., Yamashita, H., Ichihashi, Y., and Ehara, S. (1995). *J. Electroanal. Chem.* 396: 21.

22 Anpo, M., Yamashita, H., Ichihashi, Y. et al. (1997). *J. Phys. Chem. B* 101: 2632.

23 Ikeuea, K., Mukaia, H., Yamashitaa, H. et al. (2001). *J. Synchrotron Radiat.* 8: 640.

24 Anpo, M. and Chiba, K. (1992). *J. Mol. Catal.* 74: 207.

25 Maidan, R. and Willner, I. (1986). *J. Am. Chem. Soc.* 108: 8100.

26 Saladin, F., Forss, L., and Kamber, I. (1995). *J. Chem. Soc., Chem. Commun.* 5: 533.

27 Tasbihi, M., Fresno, F., Simon, U. et al. (2018). *Appl. Catal., B* 239: 68.

28 Ran, J., Jaroniec, M., and Qiao, S.Z. (2018). *Adv. Mater.* 30: 1704649.

29 Liu, Y., Zhou, S., Li, J. et al. (2015). *Appl. Catal., B* 168–169: 125.

30 Yang, X., Wang, S., Yang, N. et al. (2019). *Appl. Catal., B* 259: 118088.

31 Han, C., Lei, Y., Wang, B., and Wang, Y. (2018). *ChemSusChem* 11: 4237.

32 Aurian-Blajeni, B., Halmann, M., and Manassen, J. (1980). *Sol. Energy* 25: 165.

33 Wu, J.C.S., Lin, H.M., and Lai, C.L. (2005). *Appl. Catal., A* 296: 194.

34 Yahaya, A.H., Gondal, M.A., and Hameed, A. (2004). *Chem. Phys. Lett.* 400: 206.

35 An, C., Wang, J., Jiang, W. et al. (2012). *Nanoscale* 4: 5646.

36 Liu, S., Lu, J., Pu, Y., and Fan, H. (2019). *J. CO_2 Util.* 33: 171.

37 Liang, L., Lei, F., Gao, S. et al. (2015). *Angew. Chem. Int. Ed.* 54: 13971.

38 AlOtaibi, B., Kong, X., Vanka, S. et al. (2016). *ACS Energy Lett.* 1: 246.

39 Li, A., Wang, T., Li, C. et al. (2019). *Angew. Chem. Int. Ed.* 58: 3804.

40 Yadav, R.K., Oh, G.H., Park, N.J. et al. (2014). *J. Am. Chem. Soc.* 136: 16728.

41 Kuk, S.K., Singh, R.K., Nam, D.H. et al. (2017). *Angew. Chem. Int. Ed.* 56: 3827.

42 Mora-Hernandez, J.M., Huerta-Flores, A.M., and Torres-Martínez, L.M. (2018). *J. CO_2 Util.* 27: 179.

43 Ojha, N., Bajpai, A., and Kumar, S. (2019). *Catal. Sci. Technol.* 9: 4598.

44 He, Z., Jiang, L., Han, J. et al. (2014). *Asian J. Chem.* 26: 4759.

45 Garay-Rodríguez, L.F., Torres-Martínez, L.M., and Moctezuma, E. (2018). *J. Photochem. Photobiol., A* 361: 25.

46 (a) Khenkin, A.M., Efremenko, I., Weiner, L. et al. (2010). *Chem. Eur. J.* 16: 1356. (b) Chen, D., Sahasrabudhe, A., Wang, P. et al. (2013). *Dalton Trans.* 42: 10587.

47 Barman, S., Sreejith, S.S., Garai, S. et al. (2019). *ChemPhotoChem* 3: 93.

48 Brunetti, A., Pomilla, F., Marcì, G. et al. (2019). *Appl. Catal., B* 255: 117779.

49 Kumar, D., Lee, S.B., Park, C.H., and Kim, C.S. (2018). *Chem. Commun.* 54: 1571.

50 Adekoya, D.O., Tahir, M., and Amin, N.A.S. (2017). *J. CO_2 Util.* 18: 261.

51 Ohno, T., Murakami, N., Koyanagi, T., and Yang, Y. (2014). *J. CO_2 Util.* 6: 17.

52 Matsuoka, S., Kohzuki, T., Pac, C. et al. (1992). *J. Phys. Chem.* 96: 4437.

53 Tamaki, Y., Morimoto, T., Koike, K., and Ishitani, O. (2012). *Proc. Natl. Acad. Sci. U.S.A.* 109: 15673.

54 Tamaki, Y., Koike, K., and Ishitani, O. (2015). *Chem. Sci.* 6: 7213.

55 (a) Suzuki, T.M., Tanaka, H., Morikawa, T. et al. (2011). *Chem. Commun.* 47: 8673. (b) Sato, S., Morikawa, T., Saeki, S. et al. (2010). *Angew. Chem. Int. Ed.* 49: 5101.

56 Tsounis, C., Kuriki, R., Shibata, K. et al. (2018). *ACS Sustainable Chem. Eng.* 6: 15333.

57 Kim, W., Seok, T., and Choi, W. (2012). *Energy Environ. Sci.* 5: 6066.

58 Kulandaivalu, T., Abdul Rashid, S., Sabli, N., and Tan, T. (2019). *Diamond Relat. Mater.* 91: 64.

59 (a) Seeharaj, P., Kongmun, P., Paiplod, P. et al. (2019). *Ultrason. Sonochem.* 58: 104657. (b) Dai, W., Xu, H., Yu, J. et al. (2015). *Appl. Surf. Sci.* 356: 173.

60 Liu, Y., Huang, B., Dai, Y. et al. (2009). *Catal. Commun.* 11: 210.

61 Mao, J., Peng, T., Zhang, X. et al. (2013). *Catal. Sci. Technol.* 3: 1253.

62 Pastrana-Martínez, L.M., Silva, A.M.T., Fonseca, N.N.C. et al. (2016). *Top. Catal.* 59: 1279.

63 Fusco, C., Casiello, M., Catucci, L. et al. (2018). *Materials* 11: 307.

64 Park, H., Ou, H.H., Colussi, A.J., and Hoffmann, M.R. (2015). *J. Phys. Chem. A* 119: 4658.

65 (a) Kim, D., Kley, C.S., Li, Y., and Yang, P. (2017). *Proc. Natl. Acad. Sci. U.S.A.* 114: 10560. (b) Zhong, S., Yang, X., Cao, Z. et al. (2018). *Chem. Commun.* 54: 11324. (c) Calvinho, K.U.D., Laursen, A.B., Yap, K.M.K. et al. (2018). *Energy Environ. Sci.* 11: 2550.

66 (a) Iguchi, S., Teramura, K., Hosokawa, S., and Tanaka, T. (2016). *Appl. Catal., A* 521: 160. (b) Iguchi, S., Teramura, K., Hosokawa, S., and Tanaka, T. (2016). *Catal. Sci. Technol.* 6: 4978.

67 Yin, G., Nishikawa, M., Nosaka, Y. et al. (2015). *ACS Nano* 9: 2111.

68 Tahir, M., Tahir, B., Amin, N., and Zakaria, Z.Y. (2017). *J. CO_2 Util.* 18: 250.

69 (a) Zou, X., Dong, Y., Li, S. et al. (2018). *J. Taiwan Inst. Chem. Eng.* 93: 158. (b) Yu, L.L., Qin, J.Z., Zhao, W.J. et al. (2020). *Int. J. Photoenergy* 2020: 1.

70 Li, Y., Ren, J., Ouyang, S. et al. (2019). *Appl. Catal., B* 259: 118027.

71 Iizuka, K., Wato, T., Miseki, Y. et al. (2011). *J. Am. Chem. Soc.* 133: 20863.

72 Teramura, K., Wang, Z., Hosokawa, S. et al. (2014). *Chem. Eur. J.* 20: 9906.

73 Ulagappan, N. and Frei, H. (2000). *J. Phys. Chem. A* 104: 7834.

74 Kim, W., McClure, B.A., Edri, E., and Frei, H. (2016). *Chem. Soc. Rev.* 45: 3221.

75 (a) Wang, Y., Liu, M., Chen, W. et al. (2019). *J. Alloys Compd.* 786: 149. (b) Huang, C., Guo, R., Pan, W. et al. (2018). *J. CO_2 Util.* 26: 487. (c) Ke, J., Zhou, H., Liu, J. et al. (2019). *J. Colloid Interface Sci.* 555: 413.

76 Zhou, H., Wen, Z., Liu, J. et al. (2019). *Appl. Catal., B* 242: 76.

77 Arai, T., Sato, S., Kajino, T., and Morikawa, T. (2013). *Energy Environ. Sci.* 6: 1274.

78 (a) Bazzan, I., Volpe, A., Dolbecq, A. et al. (2017). *Catal. Today* 290: 39. (b) Song, F., More, R., Schilling, M. et al. (2017). *J. Am. Chem. Soc.* 139: 14198.

79 Helveg, S., Kisielowski, C.F., Jinschek, J.R. et al. (2015). *Micron* 68: 176.

80 Zhao, Y., Swierk, J.R., Megiatto, J.D. Jr., et al. (2012). *Proc. Natl. Acad. Sci. U.S.A.* 109: 15612.

81 Kim, W., Yuan, G., McClure, B.A., and Frei, H. (2014). *J. Am. Chem. Soc.* 136: 11034.

82 Fan, J., Cheng, L., Liu, Y. et al. (2019). *J. Catal.* 378: 164.

83 Yu, L., Ba, X., Qiu, M. et al. (2019). *Nano Energy* 60: 576.

84 (a) Zhang, Y., Wu, C., Jiang, H. et al. (2018). *Adv. Mater.* 30: 1707522. (b) Shrestha, S. and Dutta, P.K. (2018). *ACS Omega* 3: 11972.

85 Zhang, H., Guo, C., Ren, J. et al. (2019). *Chem. Commun.* 55: 14050.

86 Zhang, L., Yang, C., Xie, Z., and Wang, X. (2018). *Appl. Catal., B* 224: 886.

87 Paille, G., Gomez-Mingot, M., Roch-Marchal, C. et al. (2018). *J. Am. Chem. Soc.* 140: 3613.

88 Shah, W., Waseem, A., Nadeem, M.A., and Kögerler, P. (2018). *Appl. Catal., A* 567: 132.

3

Natural and Artificial Photosynthesis

Dimitrios A. Pantazis

Max-Planck-Institut für Kohlenforschung, Department of Molecular Theory and Spectroscopy, Kaiser-Wilhelm-Platz 1, Mülheim an der Ruhr 45470, Germany

3.1 Introduction

Conversion of sunlight into storable chemicals is a central and urgent challenge for modern science and technology. Its implications are not simply technological and economical: the outcome of this scientific endeavor can have historical significance, shaping the future of society and civilization on a global scale. The dire consequences of climate destabilization effected by the mounting use of fossil fuels are already amply manifested in extreme events and record-breaking average temperatures year after year. Political responses against the inexorably evolving climate crisis and international coordination so far fall short of even meeting conservative stated targets, for example, on CO_2 emissions. It is hoped that successful developments on a global scale [1] in the science and technology of capturing solar energy and storing it in chemical bonds, that is, the production of solar fuels, will aid in counterbalancing the use of coal, oil, and gas, potentially placing limits to future consequences of climate change.

It is an often-repeated statement that the energy contained in just one hour of sunlight reaching the earth would be sufficient to cover a year's worth of our current global energy use. Regardless of the numerical preciseness of this statement, it gives an idea of the vast potential of sunlight as a renewable energy source. Harnessing solar energy can take several different forms. Conversion of sunlight to electricity and heat is already achievable, and application of relevant technologies should be distributed and maximized as much as possible. However, this type of solar energy utilization does not and cannot alone substitute for the use of fossil fuels [2], which account for the vast majority of global energy consumption. That is why the conversion of sunlight into fuels represents the ultimate goal. In attempting to reach it, we wish to replicate with our own technical means a crucial biological process, the process of natural photosynthesis.

Solar-to-Chemical Conversion: Photocatalytic and Photoelectrochemcial Processes, First Edition.
Edited by Hongqi Sun.

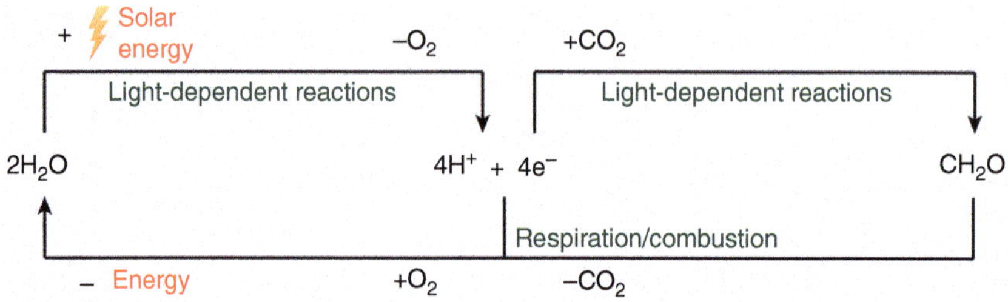

Figure 3.1 The light-dependent reactions of photosynthesis produce protons and electrons used in the conversion of CO_2 to carbohydrates (CH_2O) and other organic compounds, while oxidation of these organic compounds (biomass, food, fuels) by respiration or combustion releases the stored solar energy to power the metabolism of non-photosynthetic organisms and drive human technologies.

Photosynthetic organisms appeared very early in the history of life on our planet. They use the energy of sunlight to drive their metabolism, producing reduced compounds that can be afterward oxidized to release the stored energy. Of particular interest to us is oxygenic photosynthesis, which is believed to have emerged around 2.5 billion years ago and marked the use of water as electron donor, with dioxygen as a by-product. Oxygenic photosynthetic organisms today are plants, algae, and cyanobacteria. They oxidize water and reduce CO_2 to carbohydrates. The splitting of water into dioxygen and hydrogen equivalents was a crucial turning point in the evolution of life [3]. Oxygenic photosynthesis is essentially the only source of atmospheric O_2 on our planet. Surplus of photosynthetic biomass, i.e. organic products that were not oxidized soon after their formation (for example, through combustion or through respiration when used as food by other organisms; see Figure 3.1) but locked in the earth, led eventually to a surplus of dioxygen in the atmosphere. This generated the current atmospheric levels of O_2 and the ozone layer, with all the implications for the emergence of complex multicellular life forms that exploit high-energy oxidative metabolic pathways. Certain quantities of biomass that were not recycled into the food chain or oxidized by other means but instead buried in the earth happened to be transformed under special conditions and in geological time scales into coal, petroleum, and natural gas. Humans have always relied on and used the products of photosynthesis directly as food and fuel, while during the last two centuries we have also been using the indirect products of ancient photosynthesis, the fossil fuels.

In the present chapter the fundamental aspects, processes, and biomolecular systems of natural photosynthesis [4–8] are reviewed, with a brief discussion of corresponding artificial approaches that relate to solar fuels [9–18]. Artificial photosynthesis is an umbrella term that encompasses any kind of effort dedicated to replicating the fundamental conversion performed by natural photosynthesis. Here only selected biomimetic examples will be mentioned that draw direct inspiration from the natural system. A large variety of approaches for solar-to-chemical conversion will be extensively covered in subsequent chapters of this book.

3.2 Overview of Natural Photosynthesis

The molecular processes of natural photosynthesis can be distinguished into light-dependent and light-independent processes (occasionally referred to, somewhat misleadingly, as the "light" and "dark" reactions) [4]. In oxygenic photosynthesis by plants, algae, and cyanobacteria, the light-dependent reactions involve splitting of water and transfer of reducing equivalents to nicotinamide adenine dinucleotide phosphate ($NADP^+$) for the production of NADPH, coupled with chemiosmotic production of adenosine triphosphate (ATP), while the light-independent processes involve the carbon fixation reactions (CO_2 reduction). The overall transformation effected by the light-dependent photosynthetic reactions can be summarized as

$$2H_2O + 2NADP^+ + 3ADP + 3Pi \rightarrow O_2 + 2NADPH + 3ATP \qquad (3.1)$$

Figure 3.2 depicts the main enzymatic components of oxygenic photosynthesis. These are arranged in membranes, which for plants are the thylakoid membranes within chloroplasts. The process begins at the photosystem II (PS-II). Light is harvested by many chlorophyll molecules arranged in "antennae," proteins attached to the photosystems and evolved to combine light absorption with efficient excitation energy transfer at the charge separation centers. These contain specially arranged sets of closely coupled chlorophyll and pheophytin molecules where excitation induces electron transfer from one chromophore to another. Charge separation in PS-II results in creation of the strongest oxidant known in biology (estimated reduction potential of c. +1.2 eV) [19], at a group of chlorophylls designated as P680. This charge separation drives the oxidation of water into dioxygen, protons, and electrons. Four-electron water oxidation takes place at an active site called the oxygen-evolving complex (OEC) that contains an evolutionary unique Mn_4Ca cluster. PS-II couples this process with the two-electron reduction of a plastoquinone

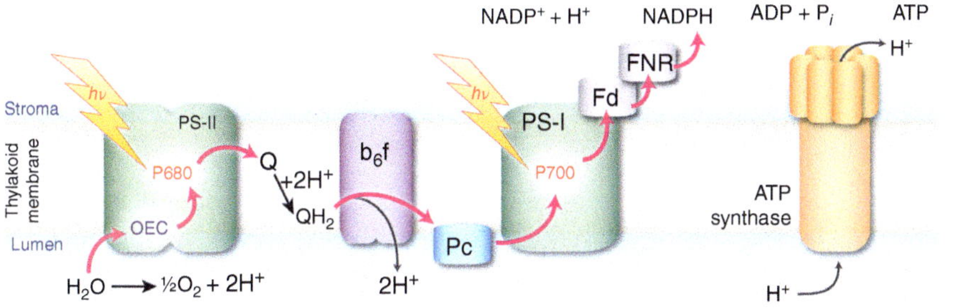

Figure 3.2 Schematic membrane arrangement of enzymes involved in the light-dependent reactions of oxygenic photosynthesis. Red arrows indicate the flow of electrons from the primary electron donor H_2O to the final electron acceptor $NADP^+$ aided by photoexcitation in photosystems II and I. Electron flow is tied to proton translocation across the membrane. The proton gradient created by accumulation of protons in the interior of the membrane powers the synthesis of ATP.

molecule:

$$2H_2O + 4h\nu \rightarrow O_2 + 4H^+ + 4e^- \tag{3.2}$$

$$Q + 2H^+ + 2e^- \rightarrow QH_2 \tag{3.3}$$

The reduced plastoquinone functions as a mobile electron carrier, transferring reducing equivalents from PS-II to the cytochrome b_6f complex. From there, another mobile carrier, the blue copper protein plastocyanin (PC), delivers electrons to photosystem I (PS-I). A second light absorption/charge separation event at PS-I, at a chlorophyll reaction center known as P700, promotes the electrons through chlorophylls (A_0), quinones (A_1), and a series of Fe–S centers (denoted F_X, F_A, and F_B) to ferredoxin (Fd) and finally to ferredoxin–NADP reductase (FNR), which catalyzes the formation of NADPH. NADPH functions as the carrier of hydrogen to be used in CO_2 fixation. The flow of electrons depicted in Figure 3.2 is coupled to generation of a proton gradient across the membrane that drives the synthesis of energy-rich ATP molecules by ATP synthase.

Photosynthesis can be viewed as consisting of four "working modules" [20]: the first module corresponds to a light-harvesting setup that funnels excitation energy to the second module, the charge-separating reaction center. Charge separation drives the transfer of electrons from an oxidative module (in oxygenic photosynthesis, the OEC) to a reducing module. In natural photosynthesis the Calvin–Benson cycle is used to reduce CO_2 to carbohydrates [4]. This part of natural photosynthesis, i.e. the light-independent CO_2 fixation reactions, will be discussed in a later section of this chapter. From the perspective of bioinspired artificial photosynthesis, the light-dependent reactions discussed above are a sufficient blueprint. Further mimicking the complete network of biological processes that culminates in synthesis of carbohydrates is not a strict requirement, although its study is certain to yield valuable insights and inspiration. The successful solar-driven generation of reducing equivalents by artificial means would already represent a complete achievement: the "reducing module" endpoint of an artificial photosynthetic system could as well involve the production of H_2 as solar fuel itself, or the utilization of reducing equivalents for other chemical transformations that are technologically important but do not directly mimic the CO_2 fixation reactions of natural photosynthesis. For this reason, emphasis will be placed in the following on the finer details of the light-dependent processes outlined above.

3.3 Light Harvesting and Excitation Energy Transfer

Light harvesting in natural photosynthesis is accomplished by pigment–protein complexes known as antenna complexes. These are associated with the photosystems, and their role is to collect as much as possible of the available light and funnel the excitation energy to them. The photosynthetic pigments are mostly bacteriochlorophylls, chlorophylls, and carotenoids (carotenes and xanthophylls). Figure 3.3 shows some of the chlorophylls utilized by photosynthetic organisms.

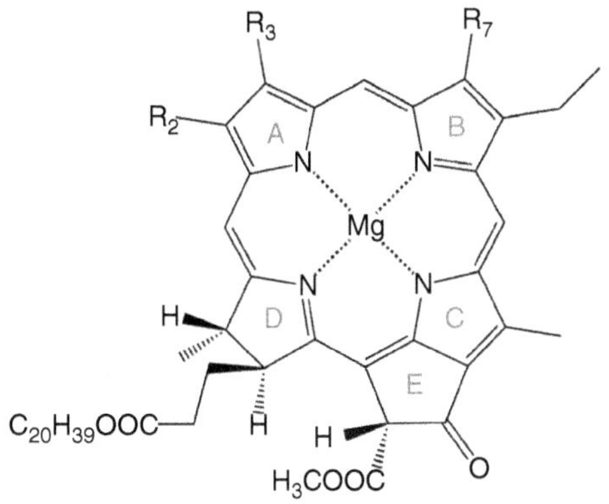

	R_2	R_3	R_7
Chl *a*	CH_3	C_2H_3	CH_3
Chl *b*	CH_3	C_2H_3	CHO
Chl *d*	CH_3	CHO	CH_3
Chl *f*	CHO	C_2H_3	CH_3

Figure 3.3 Selected types of chlorophyll molecules encountered in natural photosynthesis.

They differ in substituents on ring A at positions 2 and 3 and on ring B at position 7 of the chlorin system. Chlorin differs from the porphin tetrapyrrole in having a reduced D ring, while in bacteriochlorins the B ring is also reduced. The principal characteristic of the electronic structure of chlorophyll-type pigments is the conjugated π orbital system that is delocalized over the aromatic rings. Differences in substituents such as those shown in Figure 3.3 create slight variations in electronic structure that contribute to distinct optical properties by affecting the nature and energies of π–π* excitations, the lowest of which falls within the red or near-infrared (NIR) region. Thus, among the four types of chlorophyll shown in Figure 3.3, with respect to the most common chlorophyll *a* (Chl *a*) that has an absorption maximum (λ_{max}) at ca. 662 nm *in vitro*, Chl *b* is blueshifted (λ_{max} = 644 nm) and Chl *d* is redshifted (λ_{max} = 697 nm) [4]. Chl *f* was discovered only recently [21] and is the most redshifted chlorophyll known in oxygenic photosynthesis (λ_{max} = 707 nm). It has been suggested that its presence in key positions within PS-I and PS-II enables charge separation to occur with far-red light [22].

The absorption profile of photosynthetic pigments is a crucial determinant of the ability of organisms to utilize the wavelengths of sunlight available within their particular ecological environment. Besides the availability of distinct molecular species, fine-tuning the optical properties of individual (bacterio)chlorophylls can be achieved by (i) variable axial ligation at the Mg center, (ii) out-of-plane distortions of the (bacterio)chlorin ring, and (iii) the structured electrostatic environment of the protein matrix. These effects refer to the role of the protein in optimizing locally the properties of individual pigments. However, the truly amazing role of the protein in antenna complexes operates at the much larger scale of the precise and stable arrangement of arrays of pigment molecules (tens to hundreds of them) in terms of relative orientations and distances and in the control of their optical properties at the system level. The protein scaffold thus controls and optimizes the overall light-harvesting profile of the antenna complex, maximizes the effective pigment concentration while avoiding concentration quenching, and ensures efficient and directional energy transfer to the charge separation site. At the same time, antenna complexes incorporate mechanisms of photoprotection, which become

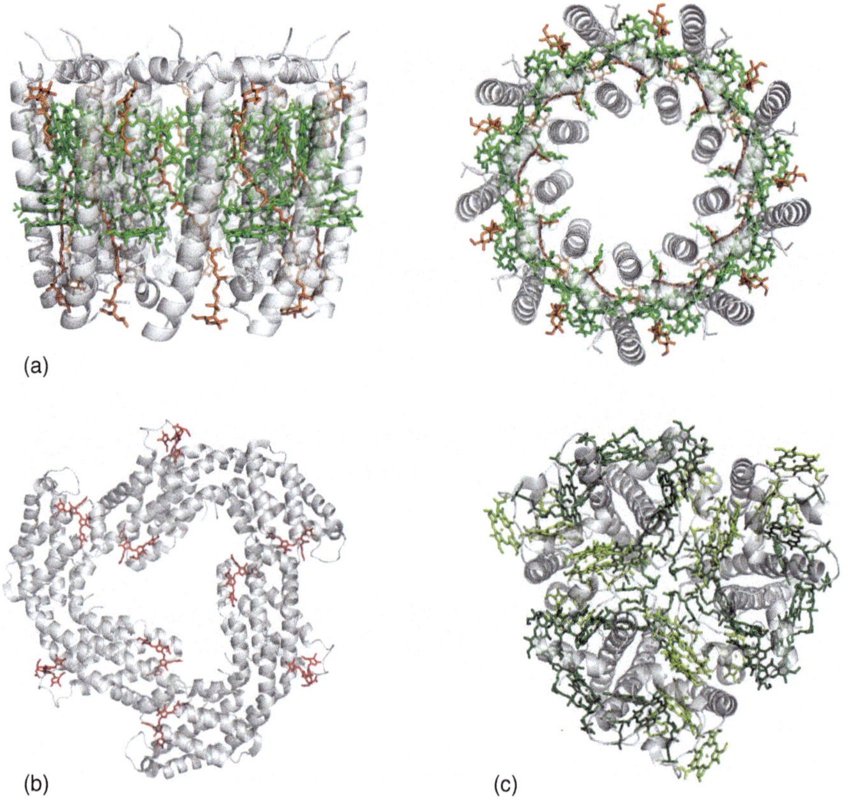

Figure 3.4 (a) Side and top view of the light-harvesting complex LH2 from the purple bacterium *R. acidophila* 10050 (PDB: 1KZU [24]), showing bacteriochlorophyll *a* pigments in green and carotenoids (rhodopin glucoside) in orange. (b) Rod structure of *c*-phycocyanin from the phycobilisome light-harvesting antenna of cyanobacterium *T. vulcanus* (PDB: 3O18 [25]). (c) Light-harvesting complex II (LHCII) from pea (PDB: 2BHW [26]), showing Chl *a* pigments in dark green and Chl *b* in light green.

particularly important under conditions of high light intensity. Photoprotection refers principally to eliminating chlorophyll triplet states that can persist long enough to react with molecular oxygen and is typically achieved by incorporating carotenoids within the protein matrix. These can play a dual role as light harvesters at wavelengths not covered by chlorophylls, i.e. green and blue light, and as chlorophyll triplet-state quenchers. There is a great diversity of antenna complexes in biology [23], characterized by variability in pigment composition and organization as well as in the mode of association with their respective reaction center complexes. Figure 3.4 depicts selected examples of light-harvesting complexes ranging from bacteria to higher plants.

Efficient unidirectional singlet energy transfer among chromophores relies on adequate spectral overlap between the emission of the donor pigment and the absorption of the acceptor pigment and on successful avoidance of alternative excited-state decay pathways. Suitable energy gradients that involve sacrificing some of the excitation energy ensure rapid transfer along the productive direction. The principal theory of excitation energy transfer is that of Förster resonance energy transfer (FRET) [27]. This relates the rates of excitation energy transfer to the dipole–dipole interactions between the donor and acceptor chromophores. The transfer rate according to FRET is determined by the relative orientations of the local electronic transition dipoles and the distance between them, and the theory formally applies to pigments that are spatially well separated, or equivalently when the electronic coupling between pigments is negligible. These conditions are rarely fully met in actual biological antenna complexes; therefore FRET most often provides a rough approximation at best. In this case a more appropriate approach is based on extensions of the Redfield theory [28] that treat the strong excitonic coupling non-perturbatively and take into account interactions with the environment. The close proximity of chromophores within the protein matrices implies strong electronic interaction between donor and acceptor pigments that results in shared excitonic energy levels that lie lower than those of the independent pigments. It is currently accepted that quantum coherence plays a key role in the remarkable efficiency of light harvesting in natural photosynthesis [29–31].

Considering biomimetic approaches to artificial versions of antenna complexes, certain challenges become immediately apparent. The pigments used in biological light harvesting are rather small molecules that are elaborately positioned and electronically fine-tuned by a "smart" protein matrix, resorting only to limited extent to covalent bonding. In this respect, it is tempting to consider the possible role of template-guided assembly of pigments as opposed to the conventional synthetic approach of covalently linking arrays of chromophores. It is remarkable that even when using several molecules of the same pigment, the protein matrix can modulate their absorption profile to produce a range of site energies with well-defined spatial distribution. This is important for expanding the spectral range and light-harvesting ability of the antenna beyond the intrinsic features of a given pigment, but its directed nature makes it also the basis of a crucial functionality: the creation of energy gradients within the antennae that enable efficient transfer of excitation energy to the reaction centers. It is likely that similar functionality could be built synthetically by utilizing ordering of distinct chromophores rather than manipulating the properties of a given pigment in a site-dependent manner. The high effective concentration of chromophores achieved in antenna complexes also appears hard to achieve in artificial analogs while avoiding concentration quenching (self-absorption) and seems to be possible in nature only because of the pigment organization imposed by the protein scaffold. Finally, the adaptability to changing light conditions, for example, by rerouting excitation energy transfer pathways, and the intrinsic photoprotection mechanisms of photosynthetic enzymes and antenna complexes are features that would be difficult, though not impossible [32], to replicate outside biology.

Artificial light-harvesting complexes [33, 34] can be conceived in the context of supramolecular chemistry as arrays of chromophores coupled through covalent linkages based on dendrimer architectures or assembled on scaffolds. Dendrimers [35, 36] lend themselves to V-shaped or circular arrangements of covalently bonded chromophores and can exhibit inherent directionality in excitation energy transfer at the molecular level. An important design concept in this area is the linkage of different but complementary chromophores to enable wide spectral coverage and to create energy gradients for efficient excitation energy transfer cascades. An additional requirement for a synthetic antenna complex is the successful interfacing with the synthetic reaction center. The coupling of the two modules should ensure efficient energy transfer from the light-harvesting system to the charge separation site while avoiding uncontrolled perturbation of the latter by the former.

Dendrimeric synthetic antenna complexes have been constructed using transition metal complexes of Ru or Os, bridged by oligopyridine-type ligands [37]. The choice of connecting ligand is important because it does not merely bring together mononuclear complexes into a closely packed ensemble, but it determines the overall nanoscale architecture of the dendrimeric structure and controls both the local electronic properties of the metal-based units and the electronic coupling between them [34]. Earth-abundant first-row transition metal complexes feature much less prominently in the field of light harvesting compared to 4d and 5d elements, but research efforts are currently directed toward changing this paradigm [38]. Multiporphyrin arrays, particularly utilizing zinc porphyrins, represent another common pattern in the context of bioinspired light-harvesting complexes [39–41]. On the other hand, light-harvesting dendrimers constructed using solely organic subunits have also been explored [42]. Figure 3.5 depicts two examples representative of ideas mentioned above. Another molecular approach worth mentioning is based on host–guest constructions, where molecules such as organic dyes or transition metal complexes are accommodated within internal cavities of dendrimers [45, 46]. Research and development of artificial light-harvesting systems is an active field with immense potential and diversity that goes far beyond what can be covered in the present chapter, so the interested reader is referred to the primary literature for more details.

3.4 Charge Separation and Electron Transfer

Referring back to Figure 3.2 that depicts the main components involved in oxygenic photosynthesis, it is useful to translate that scheme into a corresponding energy/electron flow diagram, the so-called Z-scheme of photosynthesis shown in Figure 3.6. The essential features are the utilization of two charge separation events

Figure 3.5 Examples of synthetic approaches to multi-chromophore arrays: (a) a nine-porphyrin array unit comprising a central free-base porphyrin core that acts as final acceptor and is surrounded by eight energy-donating zinc porphyrins [43]. Source: Choi et al. [41]. (b) Dendrimer consisting of a terrylenediimide (TDI) core with four attached perylenemonoimides (PMI) and eight peripheral naphthalenemonoimides (NMI) [44]. Source: Balzani et al. [34].

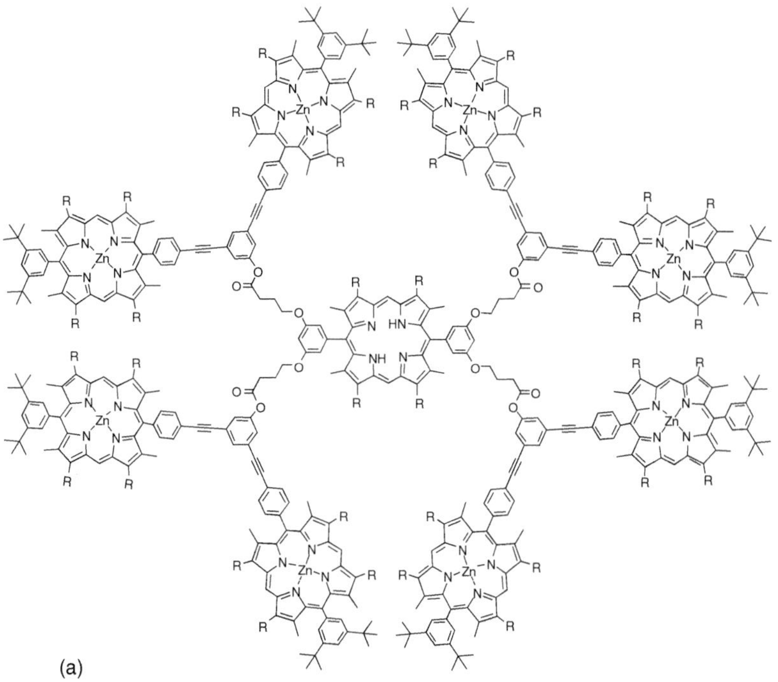

(a)

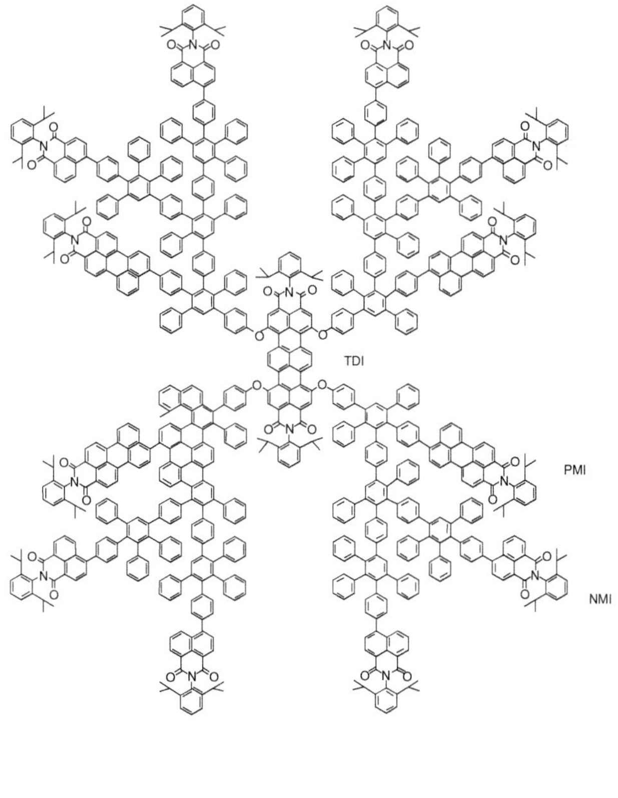

(b)

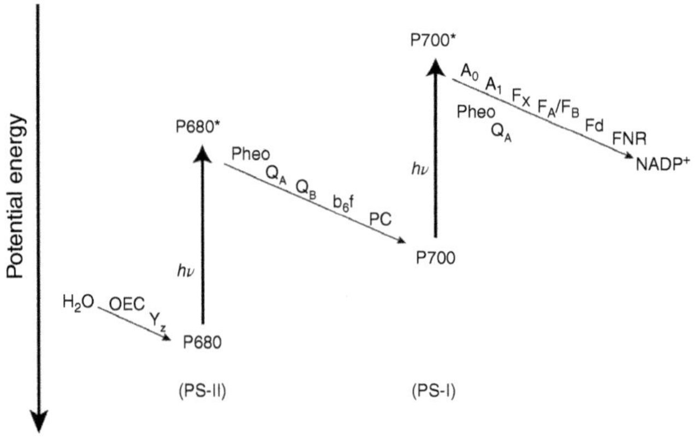

Figure 3.6 Simplified Z-scheme of natural oxygenic photosynthesis, showing how two photons are used per electron flowing from the terminal donor (H_2O) to the terminal acceptor ($NADP^+$) of the light-dependent reactions.

(at P680 of PS-II and P700 of PS-I) with distinct potentials and the closely spaced electron transfer cascades that contribute to stabilization of charge separation and unidirectionality of electron transfer. It is noted that purple bacteria utilize only a type II reaction center that lacks the water oxidizing ability of PS-II, whereas green sulfur bacteria utilize only type I centers, related to PS-I. A one-step process is in principle harder to apply to water splitting because of constraints placed on the reduction potential of the excited reaction center chromophore: it should be more positive than the water oxidation potential yet more negative than the hydrogen evolution potential. This creates limitations regarding the minimum excitation energy required to drive a single-step water splitting process. By contrast, the two-step process embodied in the Z-scheme of oxygenic photosynthesis relaxes these constraints by utilizing two photons per electron transferred from water to the final electron acceptor and hence being able to use sunlight of lower energy that what would have been necessary otherwise. In the context of artificial photosynthesis, an implementation of the Z-scheme (for instance, in multi-junction photovoltaic devices) would similarly offer higher flexibility in the choice of materials and redox linkers/mediators, requiring only that the excited-state potential of the reaction center at the oxidative side be lower than that of the reaction center at the reducing side.

Both photosystems have homodimeric structures and exhibit high similarity in the proteins and cofactors comprising their core regions, suggestive of their common evolutionary origin. In the following we will focus on the enzyme responsible for water oxidation, PS-II (see Figure 3.7). Crystallographic structures of PS-II are mostly available from thermophilic cyanobacteria such as *Thermosynechococcus elongatus* and *Thermosynechococcus vulcanus*. Conventional X-ray diffraction (XRD) studies, which first yielded a PS-II crystallographic model in 2001 [47] and make use of synchrotron X-ray radiation, have more recently been supplanted by approaches that utilize X-ray free-electron laser (XFEL) femtosecond pulses [48, 49].

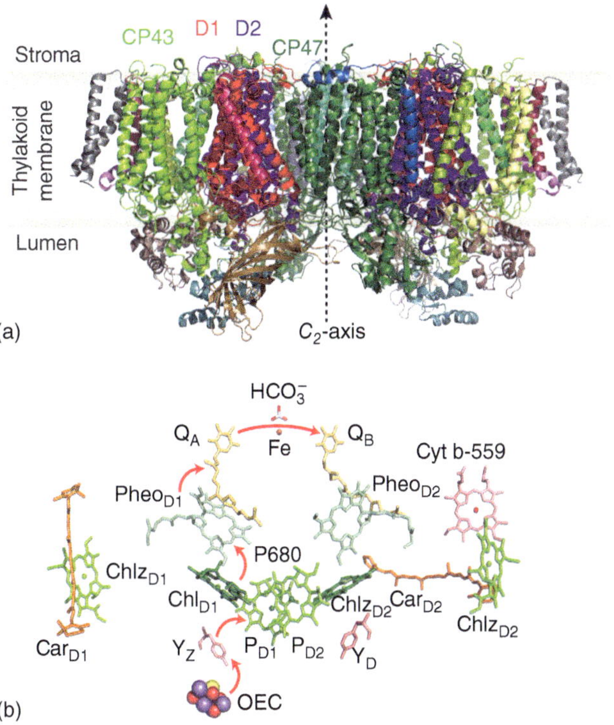

Figure 3.7 Photosystem II from *T. vulcanus* (PDB ID: 3WU2, a) and major redox-active components within a PS-II monomer involved in the main-pathway electron transfer indicated with red arrows (b).

Through a long series of XRD studies [50–56], the highest-resolution cyanobacterial PS-II crystallographic models currently stand at 1.9 Å [55] and 1.87/1.85 Å [56]. Presently available XFEL models have still not achieved comparable resolution, but they have opened the way for probing intermediate states of the water oxidation cycle [57–63]. Higher-plant PS-II structures that resolve internal cofactors have so far been reported from cryo-electron microscopy at comparatively lower resolution [64, 65].

Each monomer of cyanobacterial PS-II consists of 20 protein chains, a large number of organic cofactors such as chlorophylls, pheophytins, carotenoids, lipids, and quinones, and a number of inorganic cofactors that include hemes, a nonheme iron, calcium and chloride ions, and the tetramanganese–calcium cofactor (Mn_4CaO_x) of the OEC. Four transmembrane proteins comprise the core of the enzyme and host all major nonprotein cofactors; these are denoted D1, D2, CP43, and CP47. The PS-II core proteins are surrounded by additional proteins that may be either permanently or transiently attached [64, 66–70]. In contrast to the highly conserved core proteins, extrinsic or auxiliary proteins have larger variation between different species. The redox-active cofactors that participate in electron transfer from water to plastoquinone Q_B are arranged in quasi-symmetric branches (Figure 3.7). Importantly, only one branch is considered to be active in electron transfer, while the other is mostly involved in protective and regulatory functions [71]. The core chlorophylls of PS-II are four Chl *a* molecules denoted P_{D1}, P_{D2}, Chl_{D1}, and Chl_{D2}. These are assigned to the charge separation apparatus of PS-II, called P680 because of the absorption maximum at 680 nm observed for the cationic radical. Transfer of excitation energy to this set of chromophores results in a charge-separated state. Our knowledge of the possible nature of initial excited states and localization of charge separation among the core chromophores under various conditions is incomplete [72], but it is accepted that within picoseconds charge separation stabilizes in a state that can be described as [$P680^{•+}$ $Pheo_{D1}{}^{•-}$] [73]. The electron hole is principally localized on chlorophyll P_{D1} [74], as opposed to being equally distributed in the $P_{D1}P_{D2}$ pair, presumably as a combined result of the pigment conformation and the effect of the protein matrix [75–77]. Electron transfer on the acceptor side subsequently occurs within hundreds of picoseconds from $Pheo_{D1}{}^{•-}$ to the non-exchangeable plastoquinone Q_A. Electron transfer from $Q_A{}^-$ to the final electron acceptor plastoquinone Q_B is much slower compared with the previous electron transfer steps, up to a millisecond for the second reduction that creates the plastoquinol (Q_BH_2) that will be released to carry the two electrons further down the chain as depicted in Figures 3.2 and 3.6. The nonheme iron and its (bi)carbonate ligand mediate and modulate electron transfer between the Q_A and Q_B sites without Fe participating directly in redox chemistry [78–81].

The charge separation site of PS-II is interfaced with the water oxidation site via a redox-active tyrosine (D1-Tyr161) known as Y_Z. While the enzyme is poised in the [$P680^{•+}$ $Q_A{}^{•-}$] charge-separated state, Y_Z is oxidized by $P680^{•+}$ within tens of nanoseconds. The redox-active Y_Z is tightly hydrogen-bonded to the imidazole side group of histidine D1-His190 [82], which in turn is hydrogen-bonded to the conserved [83] asparagine D1-Asn298. Formation of the tyrosyl radical is thought to be coupled to proton shift from the phenolic proton of Y_Z to His190 and possibly to further proton translocation from His190 to Asn298 [84–86]. The tyrosyl radical $Y_Z{}^•$ is reduced directly by the Mn_4CaO_x cluster of the OEC in the micro- to millisecond time scale. Successive oxidations of the OEC by the $Y_Z{}^•$ radical formed after each light-driven charge separation event lead to accumulation of electron holes (oxidizing equivalents) at the manganese cluster. Four holes are stored at the OEC before it can catalyze the four-electron oxidation of water into dioxygen. The details of the catalytic cycle of the OEC will be discussed in the next section of this chapter.

Another redox-active tyrosine (D2-Tyr160, Y_D) is found in a position homologous to Y_Z (see Figure 3.7), but that branch does not contain a water oxidation site. Y_D presumably participates in regulatory and protective mechanisms of PS-II, such as influencing the charge distribution among the chlorophylls of $P680^{•+}$ or resetting the OEC to its resting state at night [75, 87–92]. Like Y_Z, the Y_D tyrosine is hydrogen-bonded to a histidine residue (D2-His189), but otherwise it is located in a hydrophobic region as opposed to the water-rich environment of Y_Z and displays slower redox kinetics compared with Y_Z [93–96]. A single water molecule present within a phenylalanine-rich cavity adjacent to Y_D and which can occupy either a proximal or a distal position with respect to the phenolic side chain is suggested to regulate the redox behavior of Y_D. [97, 98]

The central design principle of the electron transfer cascade is that the thermodynamic properties of each redox-active component and the kinetics of electron transfer contribute to stabilization of charge-separated states, ensuring high quantum yield [99] and directionality of electron transfer. The fast increase in the distance between electron and hole suppresses recombination reactions, but at the same time the multiple steps involved in the process lead to a decrease in free energy differences, reducing the total efficiency. PS-II successfully couples processes that occur in time scales spanning several orders of magnitude, but it is important to note that under normal operating conditions the enzyme has a lifetime of less than half an hour. Damage originates principally in formation of triplet-state chlorophyll, whose reaction with triplet dioxygen creates highly reactive, hence damaging singlet dioxygen [100]. The functionality of PS-II is restored through highly efficient repair mechanisms [101–103].

Research into artificial molecular charge-separating systems has a long history [39, 104–108]. The central challenge in artificial constructs is to stabilize the charge-separated state long enough that it can perform redox reactions. For the charge-separated state to be kinetically competent, it has been realized early on that species comprising at least three components, i.e. triads instead of simple electron donor–acceptor dyads, are required. A representative example of such a system is the molecular carotenoid–porphyrin–fullerene ($C–P–C_{60}$) triads [39, 109]. In this case light excitation leads first to formation of an excited singlet state localized on the central light-absorbing porphyrin dye ($C–P^*–C_{60}$). The initial excited state then relaxes to a charge-separated $C–P^{\bullet+}–C_{60}{}^{\bullet-}$ state. Charge recombination between the porphyrin and the fullerene is outcompeted by efficient hole transfer to the carotene, leading to the $C^{\bullet+}–P–C_{60}{}^{\bullet-}$ state with a quantum yield of 95% [109]. The spatial separation of charges in this state contributes to lifetimes in the scale of tens to hundreds of nanoseconds in solution or microseconds in a glass matrix [109, 110]. Even more complicated molecular constructs have been reported that incorporate their own antenna systems and photoprotection units [111, 112]. The use of components based on transition metal ions, particularly ruthenium photosensitizers that can be directly linked to manganese-based oxidation catalysts, also has a long history and is an active field of research [113–116]. A thorough overview of many additional molecular systems for photoinduced electron transfer is provided in the review by El-Khouly et al. [12] The challenges in this field, at least in terms of molecular systems discussed in the present chapter, remain the achievement of robustness, kinetic competence of charge-separated states, and coupling of the one-electron chemistry with accumulation of oxidizing equivalents so that concerted multi-electron transformations can be achieved.

3.5 Water Oxidation

Light harvesting, excitation energy transfer, and charge separation are functions shared by all types of photosynthetic organisms. What is special about oxygenic photosynthesis is the use of water as the ultimate electron donor by PS-II.

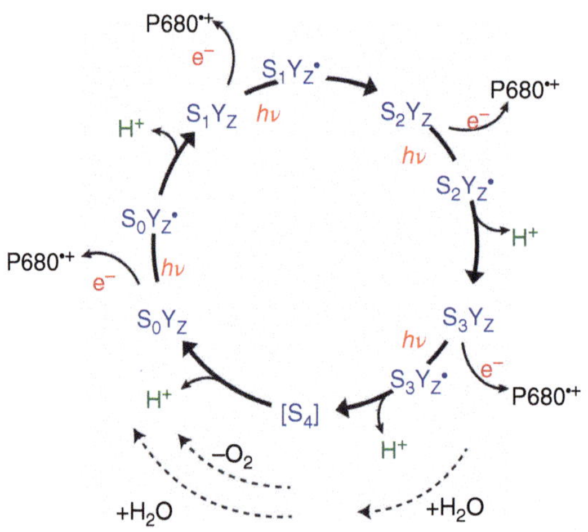

Figure 3.8 The cycle of intermediate oxidation states of the oxygen-evolving complex. Water is assumed to bind at the S_3 state of the cluster and upon reconstitution of the S_0 state. The S_4 state is a postulated but unobserved transient intermediate that decays spontaneously to S_0 with release of dioxygen.

Water oxidation takes place at an active site, the OEC, that harbors an inorganic oxo-bridged Mn_4Ca cluster. The cluster is readily assembled from Mn^{2+} and Ca^{2+} in solution through a process known as photoassembly [66, 117]. The OEC is successively oxidized by the Y_Z tyrosyl radical (D1-Tyr161), storing up to four oxidizing equivalents before releasing dioxygen. The storage/catalytic cycle of the OEC is described by the Kok–Joliot cycle of S_i states (i = 0–4), with S_0 being the lowest oxidation state of the OEC and S_4 the highest state that evolves dioxygen (Figure 3.8). Among those states S_1 is the resting state of PS-II, i.e. the state to which the enzyme reverts if left in the dark. Y_Z and the Mn cluster are in close spatial and electronic contact. Along each $S_i \rightarrow S_{i+1}$ transition, the intermediates $S_iY_Z^{\bullet}$ formed when $P680^{\bullet+}$ oxidizes Y_Z have a finite lifetime at low temperature and can be studied by electron paramagnetic resonance (EPR) spectroscopy [86, 118–125]. Studies of these intermediates provide information about the tyrosyl radical itself, its interaction with the manganese cluster, the spin state of the cluster, and changes in hydrogen bonding and protonation occurring during the S-state transition. Storing the four oxidizing equivalents before performing the four-electron water oxidation provides a low-energy pathway for oxidation of water to dioxygen and avoids formation of dangerous reactive intermediates that would result from partial oxidation of the substrate. An important feature of the OEC is that electrons and protons are removed in an alternate fashion along the S-state cycle [126–128]. This creates a redox-leveling effect, which means that the four successive oxidative steps can take place within a narrow range of potential.

The geometric structure of the OEC has been the subject of speculation for a very long time [129, 130], ever since EPR studies on the S_2 state in 1981 established the presence of four antiferromagnetically interacting Mn ions giving rise to a multiline $g = 2$ EPR signal arising from a cluster with total spin state $S = 1/2$ [131]. Extended X-ray absorption fine structure (EXAFS) studies provided increasingly detailed and accurate information about the metal–metal distances within the cluster over the next decades [130, 132–139], but no unique three-dimensional reconstruction of

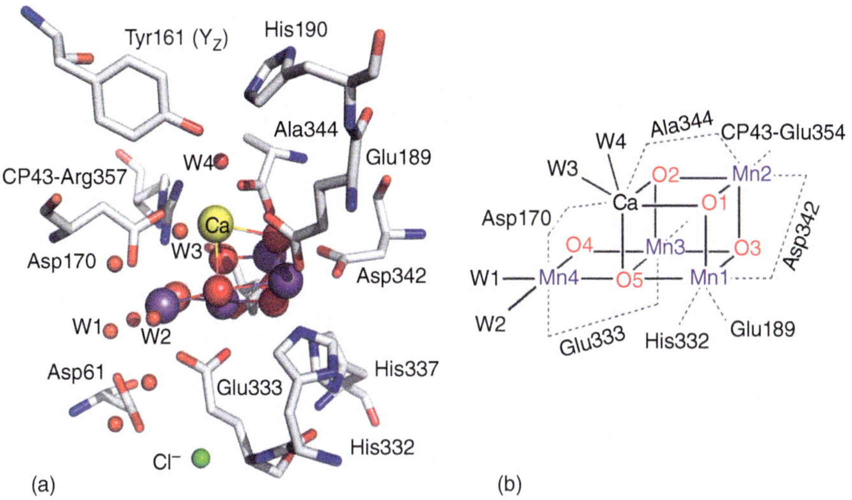

Figure 3.9 The Mn_4CaO_5 cluster and its protein pocket in the dark-stable S_1 state as revealed by protein crystallography (PDB ID: 3WU2, a), and a scheme showing the commonly used labeling of the ions comprising the inorganic core.

this information could be achieved without additional input from crystallography [140, 141]. The appearance of the first XRD structure of PS-II in 2001 [47] and the development of crystallographic models over the following years culminated in an atomic-resolution model of the OEC core in 2011 [55]. A particular challenge for crystallography was the control of X-ray radiation damage that led to reduction of the Mn ions [136, 142, 143] and compromised the quality and reliability of structural information contained in the fitted structural models [144–146]. This problem was addressed to large extent [147, 148] by the use of XFEL approaches [57], although certain structural details remain debatable [56, 149–151].

Our present view of the OEC cluster in the dark-stable S_1 state is depicted in Figure 3.9. It is an asymmetric Mn_4CaO_5 cluster, where three Mn ions and a Ca ion form a Mn_3CaO_4 cubane unit, while a fourth Mn ion is attached externally to this cubane both by coordination to an oxo bridge of the cubane and by a fifth oxo bridge (note that the protonation state of the bridges cannot be inferred from crystallographic models). The inorganic core is mostly ligated by carboxylates provided by the D1 and CP43 proteins: D1-Asp170, D1-Glu189, D1-Glu333, D1-Asp342, CP43-Glu354, and the C-terminal D1-Ala344. There is a single nitrogen-donor ligand, D1-His332, coordinated to Mn1. Four water-derived ligands, i.e. H_2O or OH, are identified in the crystallographic models; two of them are attached to Mn4 (W1 and W2), and two are attached to calcium (W3 and W4).

The second coordination sphere of the OEC contains the redox-active tyrosine and its hydrogen-bonded histidine partner D1-His190. The tyrosine hydrogen-bonds directly with one of the Ca-bound waters and hence is in close interaction with the cluster. Additional residues such as D1-His337, CP43-Arg357, D1-Asp61, and D2-Lys317, as well as a functionally required chloride ion, interact with the inorganic core and its ligands mostly via hydrogen bonds. These residues play important roles in regulating properties of the cluster and its ligands [152–157], such as the magnetic interaction between specific pairs of Mn ions and local pK_a values of various groups, and may influence or directly participate in proton translocation. An important additional aspect of the local environment of the OEC is the system of water channels and hydrogen-bonding networks that surround it. These channels and networks are crucial for connecting the active site of water oxidation to the solvent-exposed surface of the protein and play critical roles in substrate delivery, proton transfer, and product release [158–173].

There is a strong but complex connection between the geometric and electronic structure of the OEC. This connection is key for deciphering the structure of the other S-states and, eventually, for understanding the mechanism of biological water oxidation [7]. In the following, some of the currently most well-supported ideas about the geometric and electronic structure of the other S-states will be presented, with the caveat that there exist significant open questions and ambiguities about many of the specifics [7, 8].

A central question concerns the oxidation states of the Mn ions, their distribution within the cluster, and how they change along the S_i–S_{i+1} transitions. Important information on the electronic structure of the cluster can be obtained from magnetic resonance methods, as well as from XAS and XES [128, 174–183]. Structural interpretations of such data can in turn be achieved by spectroscopy-oriented quantum chemical methods [146, 150, 179, 184–193] that have been extensively benchmarked for high-valent manganese systems [150, 179, 186, 194–198] and additionally incorporate geometric information from EXAFS and crystallographic models. The dominant view is that the Mn oxidation states evolve from Mn(III)$_3$Mn(IV) in the S_0 state to Mn(III)$_2$Mn(IV)$_2$ in S_1, Mn(III)Mn(IV)$_3$ in S_2, and Mn(IV)$_4$ in S_3. This assignment is called the "high oxidation state scheme" [199] as opposed to the low oxidation state hypothesis [200–205] that assigns two more electrons to the Mn ions, with oxidation states ranging from Mn(III)$_3$Mn(II) in S_0 to Mn(III)$_2$Mn(IV)$_2$ in S_3. EPR spectroscopy has helped to identify spin states of all observable intermediates ($S = 1/2$ for S_0 [206–211], $S = 0$ for S_1 with a low-lying $S = 1$ state [212–217], two forms of S_2 with $S = 1/2$ and $S \geq 5/2$ [218–225], and $S = 3$ for S_3 [226–228]) but cannot uniquely assign absolute oxidation states. ^{55}Mn electron nuclear double resonance (ENDOR) studies of the S_2 state first demonstrated the "3+1" Y-shaped configuration of the cluster [176, 229], while ^{55}Mn hyperfine coupling parameters supported the Mn(III)Mn(IV)$_3$ oxidation state assignment for S_2 [176, 230, 231] and confirmed the absence of Mn(II) in the S_0 state [230, 232]. Electron–electron double resonance (ELDOR) detected nuclear magnetic resonance experiments (EDNMR) of the S_3 state [228] demonstrated that all Mn ions are similar and isotropic, consistent with the Mn(IV)$_4$ oxidation state assignment. X-ray absorption and emission spectroscopies broadly agree with the results of magnetic resonance spectroscopies regarding oxidation states and localization of oxidation events. X-ray absorption near-edge spectroscopy (XANES) shows a shift of the Mn K-edge to higher energies with each S-state transition, consistent with successive Mn-based oxidation [181, 233, 234] and a change in coordination in the $S_2 \rightarrow S_3$ transition [235]. XES studies observe changes in the Kβ', K$\beta_{1,3}$, and Kα lines. Recent time-resolved studies of $K\beta_{1,3}$ emission spectra of the OEC at room temperature that included comparisons with reference compounds support the high oxidation state assignment described above as well as Mn(III)–Mn(IV) oxidation in the $S_2 \rightarrow S_3$ transition [182], while room-temperature $K\alpha$ XES studies that similarly compared data on the OEC with those on synthetic compounds in different oxidation states confirmed that the OEC reaches the $Mn(IV)_4$ oxidation level in the S_3 state [183]. The same assignment of oxidation states is supported by independent studies of photoactivation of PS-II microcrystals, which measure the number of flash-driven electron removals required for assembly of an active manganese cofactor from Mn(II) and the Mn-free enzyme [236].

The direct interpretation of crystallographic models, supported by quantum chemical calculations, indicates that in the S_1 state the terminal Mn1 and Mn4 ions are present as Mn(III), with their Jahn–Teller axes aligned almost collinearly along O5. The precise protonation of the model has not been definitively assigned, with the protonation state of O5 (O^{2-} or OH^-) and W2 (H_2O or OH^-) remaining uncertain [145, 146, 150, 151, 237, 238]. The possibility of crystallographically unresolved structural heterogeneity in the S_1 state is also discussed [149, 239–241], which would not be unlikely given the spectroscopic heterogeneity reported both in the S_1 state and in the $S_1Y_Z^{\bullet}$ intermediate [118, 119, 242–246]. The preceding S_0 state has one more Mn(III) ion compared with S_1, and this has been assigned to Mn3, making Mn2 the only Mn(IV) ion of the cluster in S_0. The most likely protonation state assignment involves a hydroxy for O5, provided this bridge is unprotonated in S_1 [150, 247, 248], while a protonated O4 bridge in S_0 [249] is less likely according to spectroscopy [248].

A widely accepted structural/electronic model for the S_2 state posits the presence of two valence (redox) isomers, i.e. two geometrically similar forms with different distribution of oxidation states among the Mn ions [188, 250, 251] (Figure 3.10). This is based on quantum chemical calculations of exchange coupling constants, spin states, and ^{55}Mn hyperfine coupling parameters that first proposed explicit connections between modified crystallographic models and electronic structure data from magnetic resonance spectroscopies [188]. The two valence isomers differ in the position of the unique Mn(III) ion of the S_2 state, either at Mn1 ("open cubane" isomer $\mathbf{S_2}^{\mathbf{A}}$) or at Mn4 ("closed cubane" isomer, $\mathbf{S_2}^{\mathbf{B}}$). The different valence distribution has two important consequences: (i) the connectivity within the cluster is slightly different, as the central O5 bridge is more tightly bound to the Mn(IV) ion in each case rather than to the Mn(III) that exhibits a clear Jahn–Teller elongation axis in the O5 direction, and (ii) the exchange coupling topology is different in each isomer, resulting in different total spin states and related spectroscopic properties. Thus, the S_2^A isomer has a spin $S = 1/2$ ground state, whereas the $\mathbf{S_2}^{\mathbf{B}}$ isomer has a spin $S = 5/2$ ground state. These correspond exactly to the two observed EPR signals of the S_2 state at g = 2 and g = 4.1, respectively. Additionally, the computed ^{55}Mn hyperfine coupling constants for the $\mathbf{S_2}^{\mathbf{A}}$ model agree very well with the experimental constants measured for the $S = 1/2$ signal [146, 188]. Finally, the two quantum chemically derived valence isomers are almost isoenergetic and interconvertible over a low barrier, in direct analogy with the two EPR signals being interconvertible and having

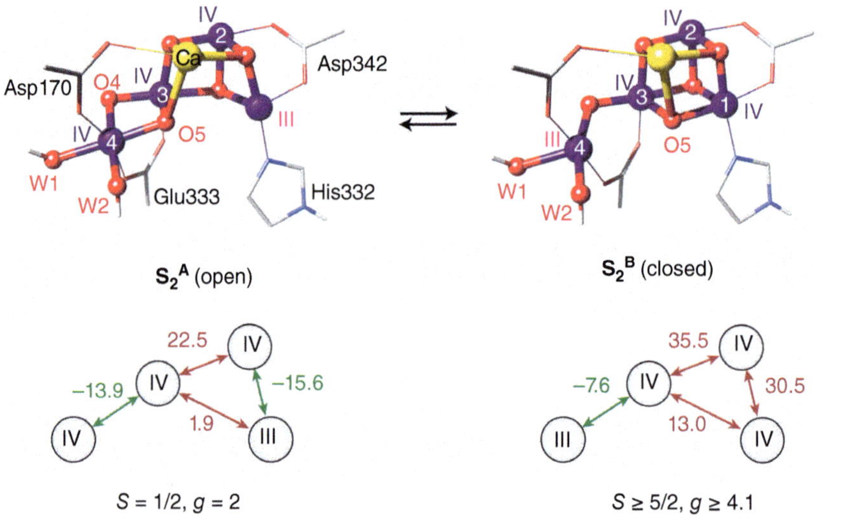

Figure 3.10 Proposed models for the inorganic core in the S_2 state of the OEC, with the first coordination sphere mostly omitted for clarity. The different magnetic topologies of the two valence isomers, as expressed through the pairwise exchange coupling constants J_{ij} (values shown in cm^{-1}; $J < 0$ is antiferromagnetic coupling), lead to different total spin states and g values for the corresponding EPR signals.

a small energy separation [252]. Although other possibilities are discussed in the literature [253, 254], no other interpretation satisfies all of the above experimental constraints.

The $S_2 \rightarrow S_3$ transition is the most complex among the observable transitions of the catalytic cycle and the subject of active research. The nature of the S_3 state itself remains contentious [255], particularly after recent XFEL models of PS-II were interpreted in terms of mutually incompatible valence states and structural forms [58, 61, 63]. A plausible scenario that is well supported by quantum chemical calculations on realistic models of the OEC and is maximally consistent with spectroscopic data on the electronic structure of the cluster is presented in Figure 3.11. It suggests that the presence of two valence isomers in the S_2 state plays a functional role as part of a gating mechanism [7, 191, 256]. The essential features of this mechanism are as follows: (i) formation of the tyrosyl radical, i.e. the $S_2Y_Z^\bullet$ intermediate, causes a reorientation of the dipole moment of the OEC toward Asp61 [257], which can act as proton acceptor [51, 152, 181, 258–260]; (ii) deprotonation of the cluster is required for the OEC to progress past the $S_2Y_Z^\bullet$ form [191]; and (iii) the deprotonated $\mathbf{S_2}^\mathbf{A}$ isomer cannot progress to the S_3 state, but the deprotonated $\mathbf{S_2}^\mathbf{B}$ is predicted to be so unstable in the presence of $Y_Z^\bullet$ that the cluster spontaneously reduces the tyrosyl radical before completion of a Mn1 → Mn4 O5 bridge shift to yield an all-Mn(IV) S_3 state species, $\mathbf{S_3}^\mathbf{B}$ [191]. This has an unusual five-coordinate Mn(IV) ion at Mn4 that correlates with the high total spin of $S = 6$, the unusually high local zero-field splitting of the five-coordinate Mn(IV) ion [261], and the ability to absorb in the NIR [191]. These properties can explain a whole list of observations regarding the S_3 state

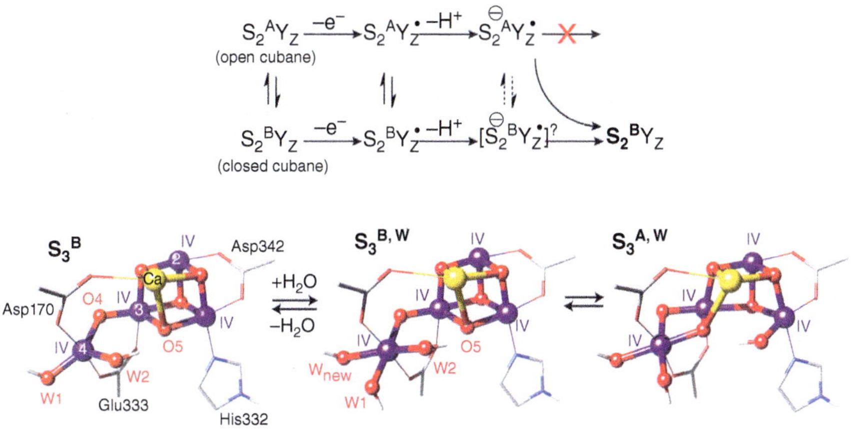

Figure 3.11 The $S_2 \rightarrow S_3$ transition according to Retegan et al. [191] and possible isovalent $Mn(IV)_4$ components of the S_3 state; the superscript "W" indicates binding of an additional water ligand.

that would otherwise be incomprehensible. These include the presence of different populations that give or do not give signals in the EPR [228, 242] and that absorb or do not absorb in the NIR [262–265]. This species can subsequently bind water either internally via the calcium ion [250] or externally [191, 266] through a channel associated with methanol and ammonia interaction with the OEC [157, 169, 170, 267–270] to give rise to additional isomers, all assignable to the S_3 state and all featuring four Mn(IV) ions [191, 255, 271, 272].

Alternative ideas for the S_3 state include formation of an oxyl radical as opposed to Mn-centered oxidation [273, 274] and onset of O—O bond formation as a peroxo or superoxo unit [58, 275–277]. These ideas are consistent to some extent with at least one of the available XFEL crystallographic models of the S_3 state, but not with the bulk of spectroscopic information that requires Mn-based oxidation in the $S_2 \rightarrow S_3$ transition [255] or with the most widely accepted interpretations of substrate exchange kinetics [278–280].

The uncertainty about the composition and nature of the S_3 state translates into uncertainty about the nature of the subsequent steps that remain experimentally unresolved and include formation of the active species after the final light-driven oxidation, formation of the O—O bond, release of dioxygen, and reconstitution of the S_0 state. A well-known radical-based mechanism for O—O bond formation has been proposed by Siegbahn and is based on structure $\mathbf{S_3}^{\mathbf{A,W}}$ of Figure 3.11 [281–288]. It involves ligand-based oxidation in the S_4 state to yield a terminal oxyl radical at Mn1 that couples with the oxo bridge that connects Mn3, Mn4, and Ca to form the O—O bond (Figure 3.12a). The same type of oxyl–oxo coupling can take place starting with the $\mathbf{S_3}^{\mathbf{B,W}}$ isomer. An alternative to the above mechanism was proposed by Shoji et al. [290] and also assumes $\mathbf{S_3}^{\mathbf{A,W}}$ to be the active species, but involves initiation of O—O bond formation at the $S_3Y_Z^{\bullet}$ intermediate coupled with intramolecular proton transfer. This circumvents the need to invoke an actual S_4 intermediate because the O—O bond is created with concomitant Mn(IV) reduction

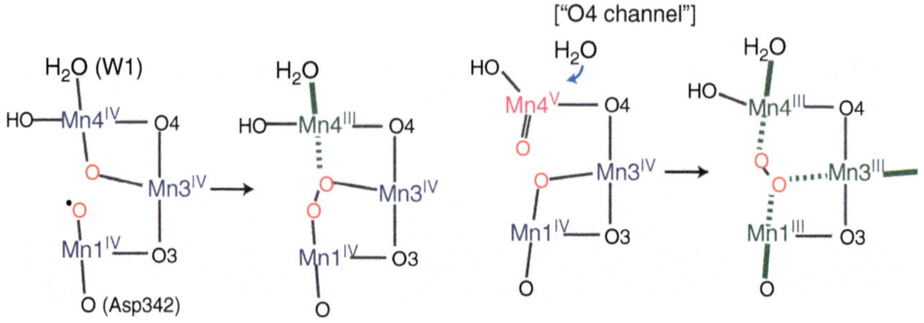

Figure 3.12 Two selected scenarios for the nature of the S_4 state and O–O bond formation from the computational literature: (a) formation of a Mn1(IV)-oxyl group in the S_4 state is followed by odd-electron radical oxyl–oxo coupling [285], and (b) formation of a five-coordinate high-spin Mn4(V)-oxo is followed by intramolecular nucleophilic coupling with concerted water binding [289]. Thick lines indicate direction of Jahn–Teller axes of Mn(III) ions.

to Mn(III) before the tyrosyl radical of the $S_3Y_Z^{\bullet}$ intermediate is reduced by the inorganic core [290]. It is noted that formation of a terminal oxyl radical [291–293] in the hypothetical S_4 state is connected to the high-spin octahedral Mn(IV) ion in the S_3 state, since formation of a genuine Mn(V) ion is unfavorable in such ligand field [289]. However, if water binding is not required for advancement of the OEC from the S_3 to the S_4 state, then the $\mathbf{S_3}^{\mathbf{B}}$ model of Figure 3.11 can also be a candidate for catalytic progression. In this case, Krewald et al. [289] demonstrated that Mn4 remains five-coordinate and forms a genuine Mn4(V)-oxo species, with two unpaired electrons localized on the high-spin Mn4(V) ion and no spin density on the equatorial oxo group [289]. This allows O—O bond formation to occur via genuine nucleophilic coupling that might occur synchronously with water binding to Mn4 (Figure 3.12b) [289], while the formation of three Jahn–Teller axes pointing simultaneously toward the newly formed O_2 unit would contribute to its irreversible expulsion from the active site. The mechanism of Krewald et al. provides access to thermodynamically favorable even-electron water oxidation [294, 295], potentially to a genuine single-step 4-electron transformation, and entirely avoids formation of potentially harmful radical intermediates [289]. Finally, it is worth mentioning an idea proposed by Zhang and Sun, as yet unsupported by quantum chemical calculations, according to which a redox isomerization in the highest state of the cycle could create a cluster with a highly oxidized Mn(VII) center and two terminal oxo groups that would couple to yield dioxygen [296]. Detailed discussions of these and other alternative hypotheses for the mechanism of biological O—O bond formation are available in recent literature [7, 255, 274–277, 287, 288, 290, 297–301].

It should be clear from the above that despite enormous strides, several aspects of the biological system, including its exact atomistic structure and crucial mechanistic details, remain incompletely understood for the later steps of the catalytic cycle. In the effort to better understand the natural water oxidation catalyst, a major target has been the synthesis of molecular mimics that reproduce structural and electronic properties of the OEC [302]. Following a long history in the development of oligonuclear manganese model complexes [303–305], the past decade has witnessed seminal achievements with the synthesis of manganese–calcium clusters that closely mimic the stoichiometry, metal oxidation states, and bonding topology of the OEC [306–318]. Landmark reports by the groups of Agapie [307] and Christou [308] established access to $Mn(IV)_3CaO_4$ cubanes, whose magnetic properties (ferromagnetic coupling to a total $S = 9/2$ state) mirror those of the cuboidal subunit of the OEC in specific states [198]. Zhang et al. [314] subsequently achieved the synthesis of a complex with a Mn_4CaO_4 core that reproduces the arrangement of metal ions of the OEC and has oxidation states equivalent to the S_1 state of the OEC, $Mn(III)_2Mn(IV)_2$. Moreover, the complex can be oxidized and produces spectroscopic signatures similar to those of the natural system [237, 314]. Extensions of this work include variants of the original complex with exchangeable solvent molecules [316].

Molecular biomimetic complexes are indispensable for elucidating structure–property correlations of relevance to the OEC, and they are a valuable source of insight into how specific geometric or electronic features of a polynuclear manganese cluster affect its overall properties and function [319–323]. At the same time, it should be acknowledged that structural mimics of the OEC have not been linked so far to appreciable water oxidizing activity. Water oxidation has been known for heterogeneous manganese oxides [324–328], but as far as molecular systems are concerned, manganese complexes reported to catalyze oxygen evolution are typically not direct mimics of the OEC, while their performance lags far behind noble metal molecular or solid-state catalysts [329–346]. There is undoubtedly vast unexplored potential for the development of biomimetic manganese-based molecular water oxidation catalysts. However, our current understanding of the biological system strongly indicates that its catalytic ability is not simply encoded in the structure of the inorganic cluster of the OEC, but depends critically on the protein matrix that both fine-tunes the properties of the cluster and performs crucial functions in terms of managing proton-coupled electron transfer and regulating the flow of substrate and product. Therefore, it is conceivable that any small-molecule mimic of the OEC, although useful as structural and electronic analog of the biological active site, is destined to fail as a practical water oxidation catalyst because it will not be able to reproduce the functionality that is taken care of by the PS-II enzyme as a whole. Promising approaches that would be arguably more suitable for large-scale realization of artificial photosynthesis are discussed in subsequent chapters.

3.6 Carbon Fixation

Capture of CO_2 and reduction to products such as CO, HCOOC, H_2CO, CH_3OH, or CH_4 are a principal target for artificial photosynthesis in the quest for solar fuels, as discussed in detail elsewhere in this book. A strictly biomimetic approach does not seem ideal in this case compared to photocatalytic and (photo)electrochemical reduction of CO_2. However, there are still lessons to be learned from biology, and here we will briefly cover the CO_2 fixation process in natural photosynthesis,

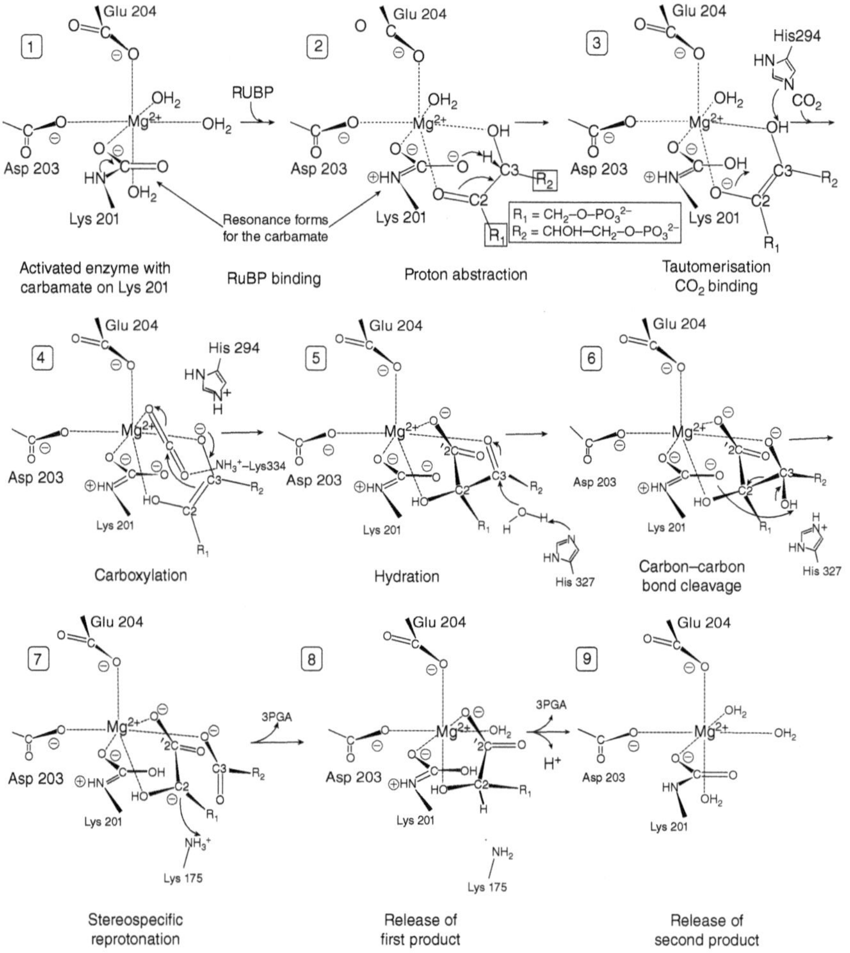

Figure 3.13 Reaction mechanism of RuBisCO proposed by Taylor and Andersson. Source: Taylor and Andersson [351].

which is carried out by the enzyme RuBisCO (ribulose-1,5-bisphosphate carboxylase/oxygenase) [4]. RuBisCO is considered the most abundant protein on earth [347, 348] and uses CO_2 to convert the five-carbon molecule ribulose-1,5-bisphosphate (RuBP) to two three-carbon 3-phosphoglycerate (3PGA) molecules, one of which incorporates the CO_2-derived carbon atom. This carboxylation reaction provides the substrate for subsequent reactions in the Calvin–Benson cycle that phosphorylate and reduce 3PGA using ATP and NADPH to produce glyceraldehyde 3-phosphate (G3P), the precursor molecule of glucose and other carbohydrates. In most photosynthetic organisms RuBisCO is present as a complex composed of eight copies of large (L) proteins and eight copies of small (S) proteins, L_8S_8 [349]. The active form of the enzyme is generated by carbamylation of an active site lysine residue [350] via reaction with CO_2 and subsequent binding of Mg^{2+} at the carbamate. This is the form that binds RuBP, at the Mg^{2+} ion [351]. The carboxylation reaction is thought to proceed by initial creation of the enediol form of RuBP, which reacts with CO_2 and is hydrated before C—C bond cleavage and release of the two 3PGA molecules (Figure 3.13).

RuBisCO evolved at a time when the atmosphere of our planet was much richer in CO_2 and did not contain much O_2. The oxygenation of the atmosphere posed a serious challenge for RuBisCO because O_2 is a competitive substrate to CO_2. Binding of O_2 by RuBisCO leads to an alternative reaction pathway that results in oxygenation of RuBP. This is an unproductive pathway (photorespiration) that leads to creation of 2-phosphoglycolate (2PGA) and eventual loss of previously fixed CO_2. Although several adaptations at the cellular or metabolic level exist in biology to deal with this problem, evolution has not come up with a "solution" at the molecular level, i.e. with restriction of oxygenase activity by adaptation of the enzyme itself. The inability of RuBisCO to discriminate strongly between CO_2 and O_2 is considered to be the reason for the utilization of very large quantities of the enzyme by photosynthetic organisms and is viewed as the primary reason for the low overall efficiency of natural photosynthesis [352–355].

3.7 Conclusions

Natural photosynthesis can show us how evolution solved the problem of converting solar to chemical energy to serve the biological needs of living organisms. The fundamental components of natural photosynthesis are conceptually the same as in any conceivable practical realization of artificial photosynthesis: light harvesting, charge separation, water oxidation, and CO_2 fixation. Many of the specifics of natural photosynthesis serve as blueprints and provide inspiration for the development of synthetic systems that might be conceived as "artificial leaves" [356, 357]. The operating principles of the OEC and its smart protein matrix are preeminent examples in this respect. However, there are other aspects of natural photosynthesis that are not ideal templates to be imitated in technological applications, such as nature's utilization of CO_2 to produce biomass. Research in natural photosynthesis and on the multiple questions that remain open, such as the details of water oxidation, will continue in tandem with efforts to develop artificial systems. It is hoped that insights from the former will fertilize the latter, because even if the future of artificial photosynthesis is not strictly biomimetic, it is inevitable that design principles will be shared.

References

1 Faunce, T.A., Lubitz, W., Rutherford, A.W. et al. (2013). *Energy Environ. Sci.* 6: 695–698.

2 Armaroli, N. and Balzani, V. (2016). *Chem. Eur. J.* 22: 32–57.

3 Lane, N. (2016). *Oxygen: The Molecule that Made the World*, 400. Oxford: Oxford University Press.

4 Blankenship, R.E. (2014). *Molecular Mechanisms of Photosynthesis*, 2e, 312. Chichester: Wiley.

5 Shevela, D. and Björn, L.O. (2017). *Govindjee, Photosynthesis: Solar Energy for Life*, 204. Singapore: World Scientific.

6 Krewald, V., Retegan, M., and Pantazis, D.A. (2016). *Top. Curr. Chem.* 371: 23–48.

7 Pantazis, D.A. (2018). *ACS Catal.* 8: 9477–9507.

8 Junge, W. (2019). *Q. Rev. Biophys.* 52: e1.

9 Wydrzynski, T.J. and Hillier, W. (2012). *Molecular Solar Fuels*, 553. Cambridge: The Royal Society of Chemistry.

10 Collings, A.F. and Critchley, C. (2005). *Artificial Photosynthesis: From Basic Biology to Industrial Application*, 313. Weinheim: Wiley-VCH.

11 Cogdell, R.J., Gardiner, A.T., Yukihira, N., and Hashimoto, H. (2018). *J. Photochem. Photobiol., A* 353: 645–653.

12 El-Khouly, M.E., El-Mohsnawy, E., and Fukuzumi, S. (2017). *J. Photochem. Photobiol., C* 31: 36–83.

13 Nocera, D.G. (2017). *Acc. Chem. Res.* 50: 616–619.

14 Cox, N., Pantazis, D.A., Neese, F., and Lubitz, W. (2015). *Interface Focus* 5: 20150009.

15 Kim, D., Sakimoto, K.K., Hong, D., and Yang, P. (2015). *Angew. Chem. Int. Ed.* 54: 3259–3266.

16 Barber, J. and Tran, P.D. (2013). *J. R. Soc. Interface* 10: 20120984.

17 Tachibana, Y., Vayssieres, L., and Durrant, J.R. (2012). *Nat. Photonics* 6: 511–518.

18 Lubitz, W., Reijerse, E.J., and Messinger, J. (2008). *Energy Environ. Sci.* 1: 15–31.

19 Rappaport, F. and Diner, B.A. (2008). *Coord. Chem. Rev.* 252: 259–272.

20 Brotosudarmo, T.H.P., Prihastyanti, M.N.U., Gardiner, A.T. et al. (2014). *Energy Procedia* 47: 283–289.

21 Chen, M., Schliep, M., Willows, R.D. et al. (2010). *Science* 329: 1318.

22 Nürnberg, D.J., Morton, J., Santabarbara, S. et al. (2018). *Science* 360: 1210.

23 Green, B. and Parson, W.W. (2003). *Light-Harvesting Antennas in Photosynthesis*, 516. Dordrecht: Springer.

24 Prince, S.M., Papiz, M.Z., Freer, A.A. et al. (1997). *J. Mol. Biol.* 268: 412–423.

25 David, L., Marx, A., and Adir, N. (2011). *J. Mol. Biol.* 405: 201–213.

26 Standfuss, J., Terwisscha van Scheltinga, A.C., Lamborghini, M., and Kühlbrandt, W. (2005). *EMBO J.* 24: 919–928.

27 Förster, T. (1948). *Ann. Phys.* 437: 55–75.

28 Redfield, A.G. (1965). *Advances in Magnetic and Optical Resonance*, vol. 1 (ed. J.S. Waugh), 1–32. Academic Press.

29 Panitchayangkoon, G., Hayes, D., Fransted, K.A. et al. (2010). *Proc. Natl. Acad. Sci. U.S.A.* 107: 12766.

30 Ishizaki, A. and Fleming, G.R. (2012). *Annu. Rev. Condens. Matter Phys.* 3: 333–361.

31 Fassioli, F., Dinshaw, R., Arpin, P.C., and Scholes, G.D. (2014). *J. R. Soc. Interface* 11: 20130901.

32 Straight, S.D., Kodis, G., Terazono, Y. et al. (2008). *Nat. Nanotechnol.* 3: 280–283.

33 Harriman, A. (2015). *Chem. Commun.* 51: 11745–11756.

34 Balzani, V., Credi, A., and Venturi, M. (2008). *ChemSusChem* 1: 26–58.

35 Newkome, G.R., Moorefield, C.N., and Vögtle, F. (2001). *Dendrimers and Dendrons*, 635. Weinheim: Wiley-VCH.

36 Balzani, V., Ceroni, P., Maestri, M., and Vicinelli, V. (2003). *Curr. Opin. Chem. Biol.* 7: 657–665.

37 Balzani, V., Campagna, S., Denti, G. et al. (1998). *Acc. Chem. Res.* 31: 26–34.

38 McCusker, J.K. (2019). *Science* 363: 484.

39 Gust, D., Moore, T.A., and Moore, A.L. (2001). *Acc. Chem. Res.* 34: 40–48.

40 Holten, D., Bocian, D.F., and Lindsey, J.S. (2002). *Acc. Chem. Res.* 35: 57–69.

41 Choi, M.-S., Yamazaki, T., Yamazaki, I., and Aida, T. (2004). *Angew. Chem. Int. Ed.* 43: 150–158.

42 Adronov, A., Gilat, S.L., Fréchet, J.M.J. et al. (2000). *J. Am. Chem. Soc.* 122: 1175–1185.

43 Ching Mak, C., Pomeranc, D., Sanders, J.K.M. et al. (1999). *Chem. Commun.*: 1083–1084.

44 Cotlet, M., Vosch, T., Habuchi, S. et al. (2005). *J. Am. Chem. Soc.* 127: 9760–9768.

45 Balzani, V., Bergamini, G., Ceroni, P., and Vögtle, F. (2007). *Coord. Chem. Rev.* 251: 525–535.

46 Hahn, U., Gorka, M., Vögtle, F. et al. (2002). *Angew. Chem. Int. Ed.* 41: 3595–3598.

47 Zouni, A., Witt, H.T., Kern, J. et al. (2001). *Nature* 409: 739–743.

48 Neutze, R., Wouts, R., van der Spoel, D. et al. (2000). *Nature* 406: 752–757.

49 Chapman, H.N., Fromme, P., Barty, A. et al. (2011). *Nature* 470: 73–77.

50 Kamiya, N. and Shen, J.-R. (2003). *Proc. Natl. Acad. Sci. U.S.A.* 100: 98–103.

51 Ferreira, K.N., Iverson, T.M., Maghlaoui, K. et al. (2004). *Science* 303: 1831–1838.

52 Biesiadka, J., Loll, B., Kern, J. et al. (2004). *Phys. Chem. Chem. Phys.* 6: 4733–4736.

53 Loll, B., Kern, J., Saenger, W. et al. (2005). *Nature* 438: 1040–1044.

54 Guskov, A., Kern, J., Gabdulkhakov, A. et al. (2009). *Nat. Struct. Mol. Biol.* 16: 334–342.

55 Umena, Y., Kawakami, K., Shen, J.-R., and Kamiya, N. (2011). *Nature* 473: 55–60.

56 Tanaka, A., Fukushima, Y., and Kamiya, N. (2017). *J. Am. Chem. Soc.* 139: 1718–1721.

57 Suga, M., Akita, F., Hirata, K. et al. (2015). *Nature* 517: 99–103.

58 Suga, M., Akita, F., Sugahara, M. et al. (2017). *Nature* 543: 131–135.

59 Kern, J., Alonso-Mori, R., Hellmich, J. et al. (2012). *Proc. Natl. Acad. Sci. U.S.A.* 109: 9721–9726.

60 Kupitz, C., Basu, S., Grotjohann, I. et al. (2014). *Nature* 513: 261–265.

61 Young, I.D., Ibrahim, M., Chatterjee, R. et al. (2016). *Nature* 540: 453–457.

62 Kern, J., Tran, R., Alonso-Mori, R. et al. (2014). *Nat. Commun.* 5: 4371.

63 Kern, J., Chatterjee, R., Young, I.D. et al. (2018). *Nature* 563: 421–425.

64 Wei, X., Su, X., Cao, P. et al. (2016). *Nature* 534: 69–74.

65 Su, X., Ma, J., Wei, X. et al. (2017). *Science* 357: 815.

66 Becker, K., Cormann, K.U., and Nowaczyk, M.M. (2011). *J. Photochem. Photobiol., B* 104: 204–211.

67 Shi, L.-X., Hall, M., Funk, C., and Schröder, W.P. (2012). *Biochim. Biophys. Acta, Bioenerg.* 1817: 13–25.

68 Fagerlund, R.D. and Eaton-Rye, J.J. (2011). *J. Photochem. Photobiol., B* 104: 191–203.

69 Bricker, T.M., Roose, J.L., Fagerlund, R.D. et al. (2012). *Biochim. Biophys. Acta, Bioenerg.* 1817: 121–142.

70 Pagliano, C., Saracco, G., and Barber, J. (2013). *Photosynth. Res.* 116: 167–188.

71 Rutherford, A.W., Osyczka, A., and Rappaport, F. (2012). *FEBS Lett.* 586: 603–616.

72 Mokvist, F., Sjöholm, J., Mamedov, F., and Styring, S. (2014). *Biochemistry* 53: 4228–4238.

73 Cardona, T., Sedoud, A., Cox, N., and Rutherford, A.W. (2012). *Biochim. Biophys. Acta, Bioenerg.* 1817: 26–43.

74 Diner, B.A. and Rappaport, F. (2002). *Annu. Rev. Plant Biol.* 53: 551–580.

75 Saito, K., Ishida, T., Sugiura, M. et al. (2011). *J. Am. Chem. Soc.* 133: 14379–14388.

76 Narzi, D., Bovi, D., De Gaetano, P., and Guidoni, L. (2015). *J. Am. Chem. Soc.* 138: 257–264.

77 Suomivuori, C.-M., Winter, N.O.C., Hättig, C. et al. (2016). *Theory Comput.* 12: 2644–2651.

78 Brinkert, K., De Causmaecker, S., Krieger-Liszkay, A. et al. (2016). *Proc. Natl. Acad. Sci. U.S.A.* 113: 12144–12149.

79 Müh, F., Glöckner, C., Hellmich, J., and Zouni, A. (2012). *Biochim. Biophys. Acta, Bioenerg.* 1817: 44–65.

80 Müh, F. and Zouni, A. (2013). *Photosynth. Res.* 116: 295–314.

81 Fletcher, S. (2015). *J. Solid State Electrochem.* 19: 241–250.

82 Saito, K., Shen, J.-R., Ishida, T., and Ishikita, H. (2011). *Biochemistry* 50: 9836–9844.

83 Kuroda, H., Kodama, N., Sun, X.-Y. et al. (2014). *Plant Cell Physiol.* 55: 1266–1275.

84 Kawashima, K., Saito, K., and Ishikita, H. (2018). *Biochemistry* 57: 4997–5004.

85 Chrysina, M., de Mendonça Silva, J.C., Zahariou, G. et al. (2019). *J. Phys. Chem. B* 123: 3068–3078.

86 Chrysina, M., Zahariou, G., Sanakis, Y. et al. (2011). *J. Photochem. Photobiol., B* 104: 72–79.

87 Vermaas, W.F.J., Renger, G., and Dohnt, G. (1984). *Biochim. Biophys. Acta, Bioenerg.* 764: 194–202.

88 Messinger, J. and Renger, G. (1993). *Biochemistry* 32: 9379–9386.

89 Faller, P., Debus, R.J., Brettel, K. et al. (2001). *Proc. Natl. Acad. Sci. U.S.A.* 98: 14368–14373.

90 Rutherford, A.W., Boussac, A., and Faller, P. (2004). *Biochim. Biophys. Acta, Bioenerg.* 1655: 222–230.

91 Diner, B.A., Bautista, J.A., Nixon, P.J. et al. (2004). *Phys. Chem. Chem. Phys.* 6: 4844–4850.

92 Jeans, C., Schilstra, M.J., Ray, N. et al. (2002). *Biochemistry* 41: 15754–15761.

93 Boussac, A. and Etienne, A.L. (1982). *Biochem. Biophys. Res. Commun.* 109: 1200–1205.

94 Styring, S., Sjöholm, J., and Mamedov, F. (2012). *Biochim. Biophys. Acta, Bioenerg.* 1817: 76–87.

95 Sjöholm, J., Mamedov, F., and Styring, S. (2014). *Biochemistry* 53: 5721–5723.

96 Ahmadova, N., Ho, F.M., Styring, S., and Mamedov, F. (2017). *Biochim. Biophys. Acta, Bioenerg.* 1858: 407–417.

97 Saito, K., Rutherford, A.W., and Ishikita, H. (2013). *Proc. Natl. Acad. Sci. U.S.A.* 110: 7690–7695.

98 Sirohiwal, A., Neese, F., and Pantazis, D.A. (2019). *J. Am. Chem. Soc.* 141: 3217–3231.

99 Romero, E., Novoderezhkin, V.I., and van Grondelle, R. (2017). *Nature* 543: 355–365.

100 Krieger-Liszkay, A., Fufezan, C., and Trebst, A. (2008). *Photosynth. Res.* 98: 551–564.

101 van Wijk, K.J., Nilsson, L.O., and Styring, S. (1994). *J. Biol. Chem.* 269: 28382–28392.

102 Nixon, P.J., Michoux, F., Yu, J. et al. (2010). *Ann. Bot.* 106: 1–16.

103 Jarvi, S., Suorsa, M., and Aro, E.M. (2015). *Biochim. Biophys. Acta* 1847: 900–909.

104 Meyer, T.J. (1989). *Acc. Chem. Res.* 22: 163–170.

105 Wasielewski, M.R. (1992). *Chem. Rev.* 92: 435–461.

106 Wasielewski, M.R. (2009). *Acc. Chem. Res.* 42: 1910–1921.

107 Redmore, N.P., Rubtsov, I.V., and Therien, M.J. (2003). *J. Am. Chem. Soc.* 125: 8769–8778.

108 Hammarström, L. and Styring, S. (2011). *Energy Environ. Sci.* 4: 2379–2388.

109 Kodis, G., Liddell, P.A., Moore, A.L. et al. (2004). *J. Phys. Org. Chem.* 17: 724–734.

110 Liddell, P.A., Kuciauskas, D., Sumida, J.P. et al. (1997). *J. Am. Chem. Soc.* 119: 1400–1405.

111 Gust, D., Moore, T.A., and Moore, A.L. (2009). *Acc. Chem. Res.* 42: 1890–1898.

112 Gust, D., Moore, T.A., and Moore, A.L. (2012). *Faraday Discuss.* 155: 9–26.

113 Sun, L.C., Hammarström, L., Åkermark, B., and Styring, S. (2001). *Chem. Soc. Rev.* 30: 36–49.

114 Karlsson, E.A., Lee, B.-L., Åkermark, T. et al. (2011). *Angew. Chem. Int. Ed.* 50: 11715–11718.

115 Kärkäs, M.D., Johnston, E.V., Verho, O., and Åkermark, B. (2014). *Acc. Chem. Res.* 47: 100–111.

116 Hammarström, L. (2015). *Acc. Chem. Res.* 48: 840–850.

117 Dasgupta, J., Ananyev, G.M., and Dismukes, G.C. (2008). *Coord. Chem. Rev.* 252: 347–360.

118 Petrouleas, V., Koulougliotis, D., and Ioannidis, N. (2005). *Biochemistry* 44: 6723–6728.

119 Havelius, K.G.V., Sjöholm, J., Ho, F. et al. (2010). *Appl. Magn. Reson.* 37: 151–176.

120 Ioannidis, N., Zahariou, G., and Petrouleas, V. (2006). *Biochemistry* 45: 6252–6259.

121 Zahariou, G., Chrysina, M., Petrouleas, V., and Ioannidis, N. (2014). *FEBS Lett.* 588: 1827–1831.

122 Zahariou, G. and Ioannidis, N. (2016). *Photosynth. Res.* 130: 417–426.

123 Havelius, K.G.V., Su, J.-H., Han, G. et al. (2011). *Biochim. Biophys. Acta, Bioenerg.* 1807: 11–21.

124 Cox, N., Ho, F.M., Pewnim, N. et al. (2009). *Biochim. Biophys. Acta, Bioenerg.* 1787: 882–889.

125 Peloquin, J.M., Campbell, K.A., and Britt, R.D. (1998). *J. Am. Chem. Soc.* 120: 6840–6841.

126 Dau, H. and Haumann, M. (2007). *Biochim. Biophys. Acta, Bioenerg.* 1767: 472–483.

127 Klauss, A., Haumann, M., and Dau, H. (2012). *Proc. Natl. Acad. Sci. U.S.A.* 109: 16035–16040.

128 Klauss, A., Haumann, M., and Dau, H. (2015). *J. Phys. Chem. B* 119: 2677–2689.

129 Wieghardt, K. (1989). *Angew. Chem. Int. Ed. Engl.* 28: 1153–1172.

130 Yachandra, V.K., Sauer, K., and Klein, M.P. (1996). *Chem. Rev.* 96: 2927–2950.

131 Dismukes, G.C. and Siderer, Y. (1981). *Proc. Natl. Acad. Sci. U.S.A.* 78: 274–278.

132 Yachandra, V.K., DeRose, V.J., Latimer, M.J. et al. (1993). *Science* 260: 675–679.

133 Sauer, K., Yano, J., and Yachandra, V.K. (2005). *Photosynth. Res.* 85: 73–86.

134 Yano, J., Kern, J., Sauer, K. et al. (2006). *Science* 314: 821–825.

135 Yano, J., Kern, J., Pushkar, Y. et al. (2008). *Philos. Trans. R. Soc. B* 363: 1139–1147.

136 Dau, H., Liebisch, P., and Haumann, M. (2004). *Phys. Chem. Chem. Phys.* 6: 4781–4792.

137 Haumann, M., Müller, C., Liebisch, P. et al. (2005). *Biochemistry* 44: 1894–1908.

138 Dau, H., Grundmeier, A., Loja, P., and Haumann, M. (2008). *Philos. Trans. R. Soc. B* 363: 1237–1243.

139 Glöckner, C., Kern, J., Broser, M. et al. (2013). *J. Biol. Chem.* 288: 22607–22620.

140 Grundmeier, A. and Dau, H. (2012). *Biochim. Biophys. Acta, Bioenerg.* 1817: 88–105.

141 Yano, J. and Yachandra, V. (2014). *Chem. Rev.* 114: 4175–4205.

142 Grabolle, M., Haumann, M., Müller, C. et al. (2006). *J. Biol. Chem.* 281: 4580–4588.

143 Yano, J., Kern, J., Irrgang, K.-D. et al. (2005). *Proc. Natl. Acad. Sci. U.S.A.* 102: 12047–12052.
144 Galstyan, A., Robertazzi, A., and Knapp, E.W. (2012). *J. Am. Chem. Soc.* 134: 7442–7449.
145 Luber, S., Rivalta, I., Umena, Y. et al. (2011). *Biochemistry* 50: 6308–6311.
146 Ames, W., Pantazis, D.A., Krewald, V. et al. (2011). *J. Am. Chem. Soc.* 133: 19743–19757.
147 Amin, M., Badawi, A., and Obayya, S.S. (2016). *Sci. Rep.* 6: 36492.
148 Amin, M., Askerka, M., Batista, V.S. et al. (2017). *J. Phys. Chem. B* 121: 9382–9388.
149 Shoji, M., Isobe, H., Yamanaka, S. et al. (2015). *Chem. Phys. Lett.* 623: 1–7.
150 Krewald, V., Retegan, M., Cox, N. et al. (2015). *Chem. Sci.* 6: 1676–1695.
151 Askerka, M., Vinyard, D.J., Wang, J. et al. (2015). *Biochemistry* 54: 1713–1716.
152 Rivalta, I., Amin, M., Luber, S. et al. (2011). *Biochemistry* 50: 6312–6315.
153 Vogt, L., Vinyard, D.J., Khan, S., and Brudvig, G.W. (2015). *Curr. Opin. Chem. Biol.* 25: 152–158.
154 Amin, M., Pokhrel, R., Brudvig, G.W. et al. (2016). *J. Phys. Chem. B* 120: 4243–4248.
155 Ghosh, I., Khan, S., Banerjee, G. et al. (2019). *J. Phys. Chem. B.*
156 Nakamura, S. and Noguchi, T. (2017). *J. Am. Chem. Soc.* 139: 9364–9375.
157 Lohmiller, T., Krewald, V., Pérez Navarro, M. et al. (2014). *Phys. Chem. Chem. Phys.* 16: 11877–11892.
158 Bondar, A.-N. and Dau, H. (2012). *Biochim. Biophys. Acta, Bioenerg.* 1817: 1177–1190.
159 Gabdulkhakov, A., Guskov, A., Broser, M. et al. (2009). *Structure* 17: 1223–1234.
160 Linke, K. and Ho, F.M. (2014). *Biochim. Biophys. Acta, Bioenerg.* 1837: 14–32.
161 Ho, F.M. (2012). *Molecular Solar Fuels* (eds. T.J. Wydrzynski and W. Hillier), 208–248. Cambridge: The Royal Society of Chemistry.
162 Ho, F.M. and Styring, S. (2008). *Biochim. Biophys. Acta, Bioenerg.* 1777: 140–153.
163 Murray, J. and Barber, J. (2008). *Photosynthesis. Energy from the Sun* (eds. J. Allen, E. Gantt, J. Golbeck and B. Osmond), 467–470. Springer Netherlands.
164 Vassiliev, S., Zaraiskaya, T., and Bruce, D. (2013). *Biochim. Biophys. Acta, Bioenerg.* 1827: 1148–1155.
165 Vassiliev, S., Comte, P., Mahboob, A., and Bruce, D. (2010). *Biochemistry* 49: 1873–1881.
166 Vassiliev, S., Zaraiskaya, T., and Bruce, D. (2012). *Biochim. Biophys. Acta, Bioenerg.* 1817: 1671–1678.
167 Debus, R.J. (2015). *Biochim. Biophys. Acta, Bioenerg.* 1847: 19–34.
168 Noguchi, T. (2015). *Biochim. Biophys. Acta, Bioenerg.* 1847: 35–45.
169 Retegan, M. and Pantazis, D.A. (2016). *Chem. Sci.* 7: 6463–6476.
170 Retegan, M. and Pantazis, D.A. (2017). *J. Am. Chem. Soc.* 139: 14340–14343.
171 Sakashita, N., Watanabe, H.C., Ikeda, T., and Ishikita, H. (2017). *Photosynth. Res.* 133: 75–85.

172 Sakashita, N., Watanabe, H.C., Ikeda, T. et al. (2017). *Biochemistry* 56: 3049–3057.

173 Kaur, D., Cai, X., Khaniya, U. et al. (2019). *Inorganics*: 7.

174 Haddy, A. (2007). *Photosynth. Res.* 92: 357–368.

175 Britt, R.D., Campbell, K.A., Peloquin, J.M. et al. (2004). *Biochim. Biophys. Acta* 1655: 158–171.

176 Peloquin, J.M., Campbell, K.A., Randall, D.W. et al. (2000). *J. Am. Chem. Soc.* 122: 10926–10942.

177 Lohmiller, T., Ames, W., Lubitz, W. et al. (2013). *Appl. Magn. Reson.* 44: 691–720.

178 Cox, N., Nalepa, A., Pandelia, M.-E. et al. (2015). *Methods in Enzymology*, vol. 563 (eds. Z.Q. Peter and W. Kurt), 211–249. Academic Press.

179 Krewald, V., Retegan, M., Neese, F. et al. (2016). *Inorg. Chem.* 55: 488–501.

180 Möbius, K., Lubitz, W., Cox, N., and Savitsky, A. (2018). *Magnetochemistry* 4: 50.

181 Dau, H. and Haumann, M. (2008). *Coord. Chem. Rev.* 252: 273–295.

182 Zaharieva, I., Chernev, P., Berggren, G. et al. (2016). *Biochemistry* 55: 4197–4211.

183 Schuth, N., Zaharieva, I., Chernev, P. et al. (2018). *Inorg. Chem.* 57: 10424–10430.

184 Pantazis, D.A., Orio, M., Petrenko, T. et al. (2009). *Chem. Eur. J.* 15: 5108–5123.

185 Pantazis, D.A., Orio, M., Petrenko, T. et al. (2009). *Phys. Chem. Chem. Phys.* 11: 6788–6798.

186 Schinzel, S. and Kaupp, M. (2009). *Can. J. Chem.* 87: 1521–1539.

187 Neese, F., Ames, W., Christian, G. et al. (2010). *Adv. Inorg. Chem.* 62: 301–349.

188 Pantazis, D.A., Ames, W., Cox, N. et al. (2012). *Angew. Chem. Int. Ed.* 51: 9935–9940.

189 Retegan, M., Cox, N., Pantazis, D.A., and Neese, F. (2014). *Inorg. Chem.* 53: 11785–11793.

190 Beckwith, M.A., Ames, W., Vila, F.D. et al. (2015). *J. Am. Chem. Soc.* 137: 12815–12834.

191 Retegan, M., Krewald, V., Mamedov, F. et al. (2016). *Chem. Sci.* 7: 72–84.

192 Orio, M., Pantazis, D.A., and Neese, F. (2009). *Photosynth. Res.* 102: 443–453.

193 Schinzel, S., Schraut, J., Arbuznikov, A.V. et al. (2010). *Chem. Eur. J.* 16: 10424–10438.

194 Schraut, J., Arbuznikov, A.V., Schinzel, S., and Kaupp, M. (2011). *ChemPhysChem* 12: 3170–3179.

195 Orio, M., Pantazis, D.A., Petrenko, T., and Neese, F. (2009). *Inorg. Chem.* 48: 7251–7260.

196 Pantazis, D.A., Krewald, V., Orio, M., and Neese, F. (2010). *Dalton Trans.* 39: 4959–4967.

197 Baffert, C., Orio, M., Pantazis, D.A. et al. (2009). *Inorg. Chem.* 48: 10281–10288.

198 Krewald, V., Neese, F., and Pantazis, D.A. (2013). *J. Am. Chem. Soc.* 135: 5726–5739.

199 Krewald, V., Neese, F., and Pantazis, D.A. (2015). *Isr. J. Chem.* 55: 1219–1232.

200 Jaszewski, A.R., Petrie, S., Pace, R.J., and Stranger, R. (2011). *Chem. Eur. J.* 17: 5699–5713.

201 Pace, R.J., Jin, L., and Stranger, R. (2012). *Dalton Trans.* 41: 11145–11160.

202 Chen, H., Case, D.A., and Dismukes, G.C. (2018). *J. Phys. Chem. B.*

203 Chen, H., Dismukes, G.C., and Case, D.A. (2018). *J. Phys. Chem. B* 122: 8654–8664.

204 Terrett, R., Petrie, S., Stranger, R., and Pace, R.J. (2016). *J. Inorg. Biochem.* 162: 178–189.

205 Petrie, S., Stranger, R., and Pace, R.J. (2018). *ChemPhysChem* 19: 3296–3309.

206 Åhrling, K.A., Peterson, S., and Styring, S. (1997). *Biochemistry* 36: 13148–13152.

207 Messinger, J., Nugent, J.H.A., and Evans, M.C.W. (1997). *Biochemistry* 36: 11055–11060.

208 Messinger, J., Robblee, J.H., Yu, W.O. et al. (1997). *J. Am. Chem. Soc.* 119: 11349–11350.

209 Åhrling, K.A., Peterson, S., and Styring, S. (1998). *Biochemistry* 37: 8115–8120.

210 Boussac, A., Kuhl, H., Ghibaudi, E. et al. (1999). *Biochemistry* 38: 11942–11948.

211 Deák, Z., Peterson, S., Geijer, P. et al. (1999). *Biochim. Biophys. Acta, Bioenerg.* 1412: 240–249.

212 Koulougliotis, D., Hirsh, D.J., and Brudvig, G.W. (1992). *J. Am. Chem. Soc.* 114: 8322–8323.

213 Dexheimer, S.L. and Klein, M.P. (1992). *J. Am. Chem. Soc.* 114: 2821–2826.

214 Yamauchi, T., Mino, H., Matsukawa, T. et al. (1997). *Biochemistry* 36: 7520–7526.

215 Campbell, K.A., Peloquin, J.M., Pham, D.P. et al. (1998). *J. Am. Chem. Soc.* 120: 447–448.

216 Campbell, K.A., Gregor, W., Pham, D.P. et al. (1998). *Biochemistry* 37: 5039–5045.

217 Matsukawa, T., Kawamori, A., and Mino, H. (1999). *Spectrochim. Acta, Part A* 55: 895–901.

218 Casey, J.L. and Sauer, K. (1984). *Biochim. Biophys. Acta, Bioenerg.* 767: 21–28.

219 De Paula, J.C. and Brudvig, G.W. (1985). *J. Am. Chem. Soc.* 107: 2643–2648.

220 Zimmermann, J.L. and Rutherford, A.W. (1984). *Biochim. Biophys. Acta, Bioenerg.* 767: 160–167.

221 Zimmermann, J.L. and Rutherford, A.W. (1986). *Biochemistry* 25: 4609–4615.

222 Cole, J., Yachandra, V.K., Guiles, R.D. et al. (1987). *Biochim. Biophys. Acta, Bioenerg.* 890: 395–398.

223 Kim, D.H., Britt, R.D., Klein, M.P., and Sauer, K. (1990). *J. Am. Chem. Soc.* 112: 9389–9391.

224 Horner, O., Rivière, E., Blondin, G. et al. (1998). *J. Am. Chem. Soc.* 120: 7924–7928.

225 Boussac, A. and Rutherford, A.W. (2000). *Biochim. Biophys. Acta, Bioenerg.* 1457: 145–156.

226 Sanakis, Y., Sarrou, J., Zahariou, G., and Petrouleas, V. (2008). *Photosynthesis: Energy from the Sun* (eds. J.F. Allen, E. Gantt, J.H. Golbeck and B. Osmond), 479–482. Dordrecht: Springer.

227 Boussac, A., Sugiura, M., Rutherford, A.W., and Dorlet, P. (2009). *J. Am. Chem. Soc.* 131: 5050–5051.

228 Cox, N., Retegan, M., Neese, F. et al. (2014). *Science* 345: 804–808.

229 Peloquin, J.M. and Britt, R.D. (2001). *Biochim. Biophys. Acta, Bioenerg.* 1503: 96–111.

230 Kulik, L.V., Epel, B., Lubitz, W., and Messinger, J. (2007). *J. Am. Chem. Soc.* 129: 13421–13435.

231 Charlot, M.-F., Boussac, A., and Blondin, G. (2005). *Biochim. Biophys. Acta, Bioenerg.* 1708: 120–132.

232 Kulik, L.V., Epel, B., Lubitz, W., and Messinger, J. (2005). *J. Am. Chem. Soc.* 127: 2392–2393.

233 Iuzzolino, L., Dittmer, J., Dörner, W. et al. (1998). *Biochemistry* 37: 17112–17119.

234 Dau, H., Iuzzolino, L., and Dittmer, J. (2001). *Biochim. Biophys. Acta, Bioenerg.* 1503: 24–39.

235 Dau, H., Liebisch, P., and Haumann, M. (2005). *Phys. Scr.* 2005: 844.

236 Cheah, M.H., Zhang, M., Shevela, D. et al. (2020). *Proc. Natl. Acad. Sci. U.S.A.* 117: 141–145.

237 Paul, S., Cox, N., and Pantazis, D.A. (2017). *Inorg. Chem.* 56: 3875–3888.

238 Nakamura, S. and Noguchi, T. (2016). *Proc. Natl. Acad. Sci. U.S.A.* 113: 12727–12732.

239 Kusunoki, M. (2011). *Photochem. Photobiol. B* 104: 100–110.

240 Shoji, M., Isobe, H., Tanaka, A. et al. (2018). *ChemPhotoChem* 2: 257–270.

241 Narzi, D., Mattioli, G., Bovi, D., and Guidoni, L. (2017). *Chem. Eur. J.* 23: 6969–6973.

242 Boussac, A., Rutherford, A.W., and Sugiura, M. (2015). *Biochim. Biophys. Acta, Bioenerg.* 1847: 576–586.

243 Nugent, J.H.A., Muhiuddin, I.P., and Evans, M.C.W. (2002). *Biochemistry* 41: 4117–4126.

244 Koulougliotis, D., Shen, J.-R., Ioannidis, N., and Petrouleas, V. (2003). *Biochemistry* 42: 3045–3053.

245 Koulougliotis, D., Teutloff, C., Sanakis, Y. et al. (2004). *Phys. Chem. Chem. Phys.* 6: 4859–4863.

246 Sioros, G., Koulougliotis, D., Karapanagos, G., and Petrouleas, V. (2007). *Biochemistry* 46: 210–217.

247 Pal, R., Negre, C.F.A., Vogt, L. et al. (2013). *Biochemistry* 52: 7703–7706.

248 Lohmiller, T., Krewald, V., Sedoud, A. et al. (2017). *J. Am. Chem. Soc.* 139: 14412–14424.

249 Saito, K., William Rutherford, A., and Ishikita, H. (2015). *Nat. Commun.* 6: 8488.

250 Bovi, D., Narzi, D., and Guidoni, L. (2013). *Angew. Chem. Int. Ed.* 52: 11744–11749.

251 Isobe, H., Shoji, M., Yamanaka, S. et al. (2012). *Dalton Trans.* 41: 13727–13740.
252 Vinyard, D.J., Khan, S., Askerka, M. et al. (2017). *J. Phys. Chem. B* 121: 1020–1025.
253 Corry, T.A. and O'Malley, P.J. (2019). *J. Phys. Chem. Lett.* 10: 5226–5230.
254 Pushkar, Y., Ravari, A.K., Jensen, S.C., and Palenik, M. (2019). *J. Phys. Chem. Lett.* 10: 5284–5291.
255 Pantazis, D.A. (2019). *Inorganics* 7: 55.
256 Narzi, D., Bovi, D., and Guidoni, L. (2014). *Proc. Natl. Acad. Sci. U.S.A.* 111: 8723–8728.
257 Retegan, M., Cox, N., Lubitz, W. et al. (2014). *Phys. Chem. Chem. Phys.* 16: 11901–11910.
258 Ishikita, H., Saenger, W., Loll, B. et al. (2006). *Biochemistry* 45: 2063–2071.
259 Debus, R.J. (2014). *Biochemistry* 53: 2941–2955.
260 Dilbeck, P.L., Hwang, H.J., Zaharieva, I. et al. (2012). *Biochemistry* 51: 1079–1091.
261 Gupta, R., Taguchi, T., Lassalle-Kaiser, B. et al. (2015). *Proc. Natl. Acad. Sci. U.S.A.* 112: 5319–5324.
262 Boussac, A., Sugiura, M., Kirilovsky, D., and Rutherford, A.W. (2005). *Plant Cell Physiol.* 46: 837–842.
263 Su, J.-H., Havelius, K.G.V., Ho, F.M. et al. (2007). *Biochemistry* 46: 10703–10712.
264 Ioannidis, N., Nugent, J.H.A., and Petrouleas, V. (2002). *Biochemistry* 41: 9589–9600.
265 Rappaport, F., Ishida, N., Sugiura, M., and Boussac, A. (2011). *Energy Environ. Sci.* 4: 2520–2524.
266 Capone, M., Narzi, D., Bovi, D., and Guidoni, L. (2016). *J. Phys. Chem. Lett.* 7: 592–596.
267 Pérez Navarro, M., Ames, W.M., Nilsson, H. et al. (2013). *Proc. Natl. Acad. Sci. U.S.A.* 110: 15561–15566.
268 Oyala, P.H., Stich, T.A., Debus, R.J., and Britt, R.D. (2015). *J. Am. Chem. Soc.* 137: 8829–8837.
269 Askerka, M., Vinyard, D.J., Brudvig, G.W., and Batista, V.S. (2015). *Biochemistry* 54: 5783–5786.
270 Guo, Y., He, L.-L., Zhao, D.-X. et al. (2016). *Phys. Chem. Chem. Phys.* 18: 31551–31565.
271 Capone, M., Bovi, D., Narzi, D., and Guidoni, L. (2015). *Biochemistry* 54: 6439–6442.
272 Shoji, M., Isobe, H., and Yamaguchi, K. (2015). *Chem. Phys. Lett.* 636: 172–179.
273 Suga, M., Akita, F., Yamashita, K. et al. (2019). *Science* 366: 334.
274 Pushkar, Y., Davis, K.M., and Palenik, M.C. (2018). *J. Phys. Chem. Lett.* 9: 3525–3531.
275 Corry, T.A. and O'Malley, P.J. (2018). *J. Phys. Chem. Lett.* 9: 6269–6274.
276 Isobe, H., Shoji, M., Shen, J.-R., and Yamaguchi, K. (2016). *Inorg. Chem.* 55: 502–511.
277 Isobe, H., Shoji, M., Suzuki, T. et al. (2019). *Theory Comput.* 15: 2375–2391.
278 Hillier, W. and Wydrzynski, T. (2004). *Phys. Chem. Chem. Phys.* 6: 4882–4889.

279 Hillier, W. and Wydrzynski, T. (2008). *Coord. Chem. Rev.* 252: 306–317.

280 Cox, N. and Messinger, J. (2013). *Biochim. Biophys. Acta, Bioenerg.* 1827: 1020–1030.

281 Siegbahn, P.E.M. (2008). *Chem. Eur. J.* 14: 8290–8302.

282 Siegbahn, P.E.M. (2009). *Acc. Chem. Res.* 42: 1871–1880.

283 Siegbahn, P.E.M. (2011). *J. Photochem. Photobiol., B* 104: 94–99.

284 Siegbahn, P.E.M. (2012). *Phys. Chem. Chem. Phys.* 14: 4849–4856.

285 Siegbahn, P.E.M. (2013). *Biochim. Biophys. Acta, Bioenerg.* 1827: 1003–1019.

286 Siegbahn, P.E.M. (2014). *Phys. Chem. Chem. Phys.* 16: 11893–11900.

287 Li, X. and Siegbahn, P.E.M. (2015). *Phys. Chem. Chem. Phys.* 17: 12168–12174.

288 Guo, Y., Li, H., He, L.-L. et al. (2017). *Phys. Chem. Chem. Phys.* 19: 13909–13923.

289 Krewald, V., Neese, F., and Pantazis, D.A. (2019). *J. Inorg. Biochem.* 199: 110797.

290 Shoji, M., Isobe, H., Shigeta, Y. et al. (2018). *Chem. Phys. Lett.* 698: 138–146.

291 Siegbahn, P.E.M. and Crabtree, R.H. (1999). *J. Am. Chem. Soc.* 121: 117–127.

292 K. Yamaguchi, Y. Takahara, T. Fueno, in *Applied Quantum Chemistry* (Eds.: V. H. Smith Jr., H. F. Scheafer III, K. Morokuma), D. Reidel, Boston, MA, 1986, pp. 155-184.

293 Lassalle-Kaiser, B., Hureau, C., Pantazis, D.A. et al. (2010). *Energy Environ. Sci.* 3: 924–938.

294 Krishtalik, L.I. (1986). *Biochim. Biophys. Acta, Bioenerg.* 849: 162–171.

295 Krishtalik, L.I. (1990). *Bioelectrochem. Bioenerg.* 23: 249–263.

296 Zhang, B. and Sun, L. (2018). *Dalton Trans.* 47: 14381–14387.

297 Najafpour, M.M., Heidari, S., Balaghi, S.E. et al. (2017). *Biochim. Biophys. Acta, Bioenerg.* 1858: 156–174.

298 Kawashima, K., Takaoka, T., Kimura, H. et al. (2018). *Nat. Commun.* 9: 1247.

299 Shoji, M., Isobe, H., Shigeta, Y. et al. (2018). *J. Phys. Chem. B* 122: 6491–6502.

300 Shoji, M., Isobe, H., and Yamaguchi, K. (2019). *Chem. Phys. Lett.* 714: 219–226.

301 Yamaguchi, K., Shoji, M., Isobe, H. et al. (2018). *Mol. Phys.* 116: 717–745.

302 Paul, S., Neese, F., and Pantazis, D.A. (2017). *Green Chem.* 19: 2309–2325.

303 Meelich, K., Zaleski, C.M., and Pecoraro, V.L. (2008). *Philos. Trans. R. Soc. B* 363: 1271–1281.

304 Mukhopadhyay, S., Mandal, S.K., Bhaduri, S., and Armstrong, W.H. (2004). *Chem. Rev.* 104: 3981–4026.

305 Mishra, A., Wernsdorfer, W., Abboud, K.A., and Christou, G. (2005). *Chem. Commun.*: 54–56.

306 Koumousi, E.S., Mukherjee, S., Beavers, C.M. et al. (2011). *Chem. Commun.* 47: 11128–11130.

307 Kanady, J.S., Tsui, E.Y., Day, M.W., and Agapie, T. (2011). *Science* 333: 733–736.

308 Mukherjee, S., Stull, J.A., Yano, J. et al. (2012). *Proc. Natl. Acad. Sci. U.S.A.* 109: 2257–2262.

309 Tsui, E.Y., Kanady, J.S., and Agapie, T. (2013). *Inorg. Chem.* 52: 13833–13848.

310 Kanady, J.S., Lin, P.-H., Carsch, K.M. et al. (2014). *J. Am. Chem. Soc.* 136: 14373–14376.

311 Han, Z., Horak, K.T., Lee, H.B., and Agapie, T. (2017). *J. Am. Chem. Soc.* 139: 9108–9111.

312 Lee, H.B., Tsui, E.Y., and Agapie, T. (2017). *Chem. Commun.* 53: 6832–6835.

313 Lee, H.B., Shiau, A.A., Oyala, P.H. et al. (2018). *J. Am. Chem. Soc.* 140: 17175–17187.

314 Zhang, C., Chen, C., Dong, H. et al. (2015). *Science* 348: 690–693.

315 Chen, C., Li, Y., Zhao, G. et al. (2017). *ChemSusChem* 10: 4403–4408.

316 Chen, C., Chen, Y., Yao, R. et al. (2019). *Angew. Chem. Int. Ed.* 58: 3939–3942.

317 Gerey, B., Gouré, E., Fortage, J. et al. (2016). *Coord. Chem. Rev.* 319: 1–24.

318 Li, Y., Yao, R., Chen, Y. et al. (2020). *Catalysts* 10: 185.

319 Tsui, E.Y. and Agapie, T. (2013). *Proc. Natl. Acad. Sci. U.S.A.* 110: 10084–10088.

320 Tsui, E.Y., Tran, R., Yano, J., and Agapie, T. (2013). *Nat. Chem.* 5: 293–299.

321 Krewald, V., Neese, F., and Pantazis, D.A. (2016). *Phys. Chem. Chem. Phys.* 18: 10739–10750.

322 Krewald, V. and Pantazis, D.A. (2016). *Dalton Trans.* 45: 18900–18908.

323 Romain, S., Rich, J., Sens, C. et al. (2011). *Inorg. Chem.* 50: 8427–8436.

324 Kurz, P. (2016). *Top. Curr. Chem.* 371: 49–72.

325 Frey, C.E., Wiechen, M., and Kurz, P. (2014). *Dalton Trans.* 43: 4370–4379.

326 Menezes, P.W., Indra, A., Littlewood, P. et al. (2014). *ChemSusChem.* 7: 2202–2211.

327 Najafpour, M.M., Abbasi Isaloo, M., Abasi, M., and Holynska, M. (2014). *New J. Chem.* 38: 852–858.

328 Wiechen, M., Najafpour, M.M., Allakhverdiev, S.I., and Spiccia, L. (2014). *Energy Environ. Sci.* 7: 2203–2212.

329 Najafpour, M.M., Renger, G., Hołyńska, M. et al. (2016). *Chem. Rev.* 116: 2886–2936.

330 Llobet, A. (2014). Molecular Water Oxidation Catalysis, Chichester: Wiley, p. 265.

331 Blakemore, J.D., Crabtree, R.H., and Brudvig, G.W. (2015). *Chem. Rev.* 115: 12974–13005.

332 Young, K.J., Brennan, B.J., Tagore, R., and Brudvig, G.W. (2015). *Acc. Chem. Res.* 48: 567–574.

333 Zhang, Q. and Guan, J. (2019). *ChemSusChem* 12: 3209–3235.

334 Kärkäs, M.D., Verho, O., Johnston, E.V., and Åkermark, B. (2014). *Chem. Rev.* 114: 11863–12001.

335 Arafa, W.A.A., Karkas, M.D., Lee, B.-L. et al. (2014). *Phys. Chem. Chem. Phys.*

336 Sala, X., Romero, I., Rodríguez, M. et al. (2009). *Angew. Chem. Int. Ed.* 48: 2842–2852.

337 Zhang, B. and Sun, L. (2019). *Chem. Soc. Rev.* 48: 2216–2264.

338 Parent, A.R. and Sakai, K. (2014). *ChemSusChem* 7: 2070–2080.

339 Liao, R.-Z., Kärkäs, M.D., Lee, B.-L. et al. (2015). *Inorg. Chem.* 54: 342–351.

340 Fukuzumi, S., Lee, Y.-M., and Nam, W. (2019). *Dalton Trans.* 48: 779–798.

341 Kärkäs, M.D. and Åkermark, B. (2016). *Dalton Trans.* 45: 14421–14461.

342 Garrido-Barros, P., Gimbert-Suriñach, C., Matheu, R. et al. (2017). *Chem. Soc. Rev.* 46: 6088–6098.

343 Ye, S., Ding, C., Liu, M. et al. (2019). *Adv. Mater.*: 1902069.

344 Thomsen, J.M., Huang, D.L., Crabtree, R.H., and Brudvig, G.W. (2015). *Dalton Trans.* 44: 12452–12472.

345 Matheu, R., Garrido-Barros, P., Gil-Sepulcre, M. et al. (2019). *Nat. Rev. Chem.* 3: 331–341.

346 Singh, A. and Spiccia, L. (2013). *Coord. Chem. Rev.* 257: 2607–2622.

347 Ellis, R.J. (1979). *Trends Biochem. Sci* 4: 241–244.

348 Raven, J. (2013). *New Phytol.* 198: 1–3.

349 Andersson, I. and Taylor, T.C. (2003). *Arch. Biochem. Biophys.* 414: 130–140.

350 Stec, B. (2012). *Proc. Natl. Acad. Sci. U.S.A.* 109: 18785.

351 Taylor, T.C. and Andersson, I. (1997). *J. Mol. Biol.* 265: 432–444.

352 Zhu, X.-G., Long, S.P., and Ort, D.R. (2010). *Annu. Rev. Plant Biol.* 61: 235–261.

353 Parry, M.A.J., Andralojc, P.J., Scales, J.C. et al. (2012). *J. Exp. Bot.* 64: 717–730.

354 Simkin, A.J., López-Calcagno, P.E., and Raines, C.A. (2019). *J. Exp. Bot.* 70: 1119–1140.

355 Brinkert, K. (2018). *Energy Conversion in Natural and Artificial Photosynthesis*, 127. Cham: Springer.

356 Nocera, D.G. (2012). *Acc. Chem. Res.* 45: 767–776.

357 Olmos, J.D.J. and Kargul, J. (2015). *Int. J. Biochem. Cell Biol.*, vol. 66, 37–44.

4

Photocatalytic Hydrogen Evolution

Amanj Kheradmand, Yuxiang Zhu, Shengshen Gu and Yijiao Jiang

Macquarie University, School of Engineering, Sydney 2109, Australia

4.1 Introduction

Impending depletion of fossil-based energy resources, increased worldwide demand for energy, and carbon dioxide (CO_2) emissions from burning fuels cause two serious challenges for our global sustainable development: energy and environment [1]. These challenges have stimulated research toward innovative utilization of renewable and environmentally benign energy resources [2]. Hydrogen (H_2) has one of the highest energy densities of around 120–140 kJ g^{-1}, which is about 2.4 times as much energy as natural gas [3]. Unlike natural gas, when H_2 is burned, there are no CO_2 or any other greenhouse gas emissions. The only by-product is water vapor. Therefore, H_2 is considered as a clean, nonpolluting, and carbon-free renewable energy carrier. H_2 itself can be used to power vehicles and heat buildings. On the other hand, large quantity of H_2 has been widely used as a reactant in the petroleum and chemical industries such as ammonia synthesis and methanol production.

The most common method to produce H_2 is steam reforming of natural gas ($CH_4 + H_2O \rightarrow CO + 3H_2$) at very high temperatures. In this process, an excess of water is often employed to recover hydrogen by use of carbon monoxide (CO) through the water–gas shift reaction ($CO + H_2O \rightarrow CO_2 + 3H_2$), resulting in significant CO_2 emissions. A simple method of H_2 production is to split water into H_2 and oxygen (O_2) using electricity. However, this technology requires large energy inputs. As the Earth is irradiated by sunlight with a huge amount of energy every day, a possible avenue to produce H_2 is to use the photocatalytic route for water splitting into H_2 by the renewable solar energy. Although the intrinsic idea of photocatalytic hydrogen generation using solar energy is appealing, the process suffers from very low efficiencies of photocatalytic materials. Therefore, the main challenge lies in developing potent photocatalysts that enable maximum harness solar energy to achieve high H_2 yields.

Since Fujishima and Honda discovered the photoelectrochemical water splitting into H_2 over the titanium dioxide (TiO_2) electrode in 1972 [4], TiO_2 has become the most widely studied photocatalyst because of its low cost, non-toxicity, and

Solar-to-Chemical Conversion: Photocatalytic and Photoelectrochemcial Processes, First Edition.
Edited by Hongqi Sun.

abundance. Despite these attributes, this wide-band-gap semiconductor has been hindered for the potential applications since it can make use of only 3–4% of solar radiation, i.e. the ultraviolet (UV) part. Considerable efforts have gone into developing visible-light-responsive TiO_2 photocatalysts by band-gap engineering such as doping with metal or nonmetal ions onto TiO_2, but their limitations in terms of low conversion efficiency and weak light absorption still remain evident.

Numerous semiconductors have been synthesized and used in the photocatalytic H_2 generation over the past decade. Nevertheless, most of them possess a wide band gap, which has very limited light absorption in visible and near-infrared (NIR) light region that accounts for about 96% of the whole solar spectrum. Since the first report by Wang et al. that graphitic carbon nitride (g-C_3N_4) exhibits visible-light photocatalytic activity in hydrogen generation [5], the development of photocatalysts has been shifted potentially from inorganic semiconductors to polymer conjugated materials. Extensive efforts have been attempted to improve the photocatalytic activity over g-C_3N_4. Morphology control and surface modification of the g-C_3N_4 nanostructure have been considered as effective strategies to enhance g-C_3N_4 photocatalytic activity, with larger specific surface areas and efficient mass diffusion during the photocatalytic hydrogen generation. Besides ordered mesoporous g-C_3N_4, tremendous efforts have been devoted to endow g-C_3N_4 with specific nanoarchitectures, such as nanosheets (NSs), nanospheres, nanorods, nanotubes (NTs), etc. However, the high recombination rate of the photogenerated electron–hole pairs and the low light absorbance still hinder their widespread applications. Similar to TiO_2-based photocatalysts, it is highly desirable to construct C_3N_4-based photocatalyst coupling with other semiconductors and cocatalysts in order to accelerate the charge separation rate and maximize the light harvesting.

The development of photocatalysts generally proceeds by trial and error in both theory and experimentation. Up to now, the exploitation of cost-effective photocatalysts is still slow due to the lack of knowledge concerning mechanistic understanding of materials structure and light-induced hydrogen generation process. For instance, how the water molecules adsorb on the surface and how they are split into hydrogen and oxygen remain a matter of debate. We have known that most semiconductor catalysts that demonstrate the photocatalytic activity for hydrogen generation from water are inefficient and require the use of sacrificial reducing reagents such as methanol and triethanolamine (TEA). These molecules can capture the photoinduced holes in the valence band and provide more electrons in the conduction band and thus drive the photoinduced H_2 evolution. What the actual roles of these sacrificial agents are not yet well understood. Mechanistic studies in these respects have been conducted. Adding of sacrificial agents can greatly improve the hydrogen generation performance by scavenging the photoinduced holes in the valence band, but the reaction pathway for the consumption of sacrificial agent is still lacking. Besides, some sacrificial agents might play more roles except consuming the holes. Advanced spectroscopic techniques have been employed to identify surface processes, including adsorption/desorption of sacrificial reagents and intermediate capture, as well as the role of adsorbed water. Knowledge of the reaction pathways is essential, as it can guide the design of targeted photocatalysts with active and selective sites tuned toward the favored pathway and thus greatly improve overall efficiency. Understanding of these fundamentals would provide more scientific discoveries for researchers to rationally design better photocatalytic systems, leading to improved hydrogen generation efficiency.

4.2 Fundamentals of Photocatalytic H_2 Evolution

Generally, a photocatalytic system for hydrogen generation from water requires reactants, photocatalyst, photoreactor, and light supply [6]. The reactant can be water itself or even mixing with a sacrificial reagent. For efficient H_2 evolution, sufficient interactions between the reactants, photocatalyst, and light are required [7]. Photocatalysts are working under light irradiation that can be UV, visible, and NIR light. The process of water splitting into H_2 over photocatalysts is shown in Figure 4.1. The photocatalyst absorbs photon energy to promote the transfer of electrons from the valence band to the conduction band. The energy band gap is based on the difference between the energy of the valence band and the conduction band.

In general, the photocatalytic process involves three steps: the excitation, bulk diffusion, and surface transfer of charge carriers. Photocatalysis starts with irradiation of light with energy higher than or equal to its band gap of a semiconductor. When photons reach the surface of the valence band, the electron–hole pairs generate (Eq. (4.1)). The process of photoinduced charge generation is very rapid and was reported on the order of femtoseconds [9]. These electrons migrate from the valence band, leaving holes, to the conduction band. The photogenerated electrons participate in the reduction process (Eq. (4.2)), while the holes are consumed in the oxidation process (Eq. (4.3)) [10]. When excited charges are generated, separation/migration and charge recombination are two significant competitive processes inside the semiconductor photocatalyst that largely impact photocatalytic reaction efficiency. The generated electrons are very unstable, and they tend to go back to the previous position. A tendency of an electron to reduce its energy and hole to increase its energy exist. This phenomenon is called recombination rate. If electron–hole carriers recombine, there is no redox reaction, just releasing energy. Otherwise, these can contribute to the oxidation and reduction reactions via being captured by species to complete reactions. There are many strategies to avoid recombination such as loading with cocatalysts [11], using sacrificial agents [12], and coupling with another semiconductor with different band gaps [13].

The oxidation–reduction reactions occurring during photocatalytic hydrogen generation are outlined below:

$$2h\upsilon \rightarrow \quad 2e_{CB}^{-} + 2h_{VB}^{+} \tag{4.1}$$

$$2H^{+}_{(aq)} + 2e^{-}_{CB} \rightarrow \quad H_{2(g)} \tag{4.2}$$

$$H_2O_{(l)} + 2h_{VB}^{+} \rightarrow \quad \frac{1}{2}O_{2(g)} + 2H^{+}_{(aq)} \tag{4.3}$$

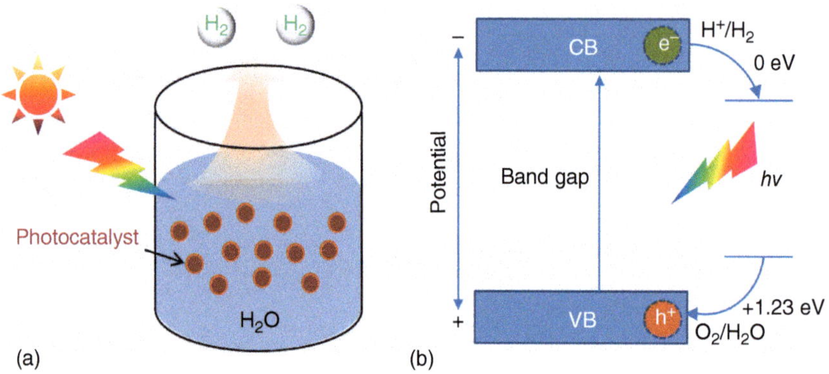

Figure 4.1 Schematic illustration of photocatalytic hydrogen generation over a semiconductor photocatalyst [8].

Overall reaction:

$$2h\upsilon + H_2O_{(l)} \rightarrow \frac{1}{2}O_{2(g)} + H_{2(g)} \tag{4.4}$$

For the overall water splitting, the electrochemical potential for the H_2 evolution reaction (HER) is at 0 V vs. normal hydrogen electrode (NHE) at pH = 0, and the electrochemical potential for the oxygen evolution reaction (OER) is at +1.23 V vs. NHE, leading to a theoretical minimum band gap of 1.23 eV that a semiconductor needs to perform this reaction [14]. Figure 4.2 shows the band structure of various semiconductors and the redox potentials of water splitting [15].

Semiconductors should meet the requirements to be suitable for photocatalytic H_2 generation. One of the most important properties is to possess an appropriate band gap, which determines the light absorption range. This feature defines the semiconductor ability of performing photoinduced electron transfer to the adsorbed species. For example, the lower the band gap, the wider the light absorption threshold. As well as the band gap, the conduction band and valence band positions of a semiconductor must be suitable for the desired redox reaction. For example, in the photocatalytic water splitting, the valence band must be appropriately positive to oxidize water for oxygen evolution, and the conduction band must be appropriately negative to be able to reduce water for H_2 generation. Considering that the ability of photocatalysts to maximum harness the sun's energy is very important, an efficient photocatalyst needs to absorb the sunlight efficiently. The major portion of the solar spectrum is visible and NIR light, which means that to harness the sunlight successfully, the semiconductor must have a band gap smaller than 3.0 eV. Another important feature of photocatalysts is chemical stability and photocorrosion resistance for preventing from their decomposition during the photoinduced reaction. The amount of active sites is also one of the important features that should be taken into account in photocatalysis. Being abundant and low cost are other properties of an appropriate semiconductor to enable its practical application.

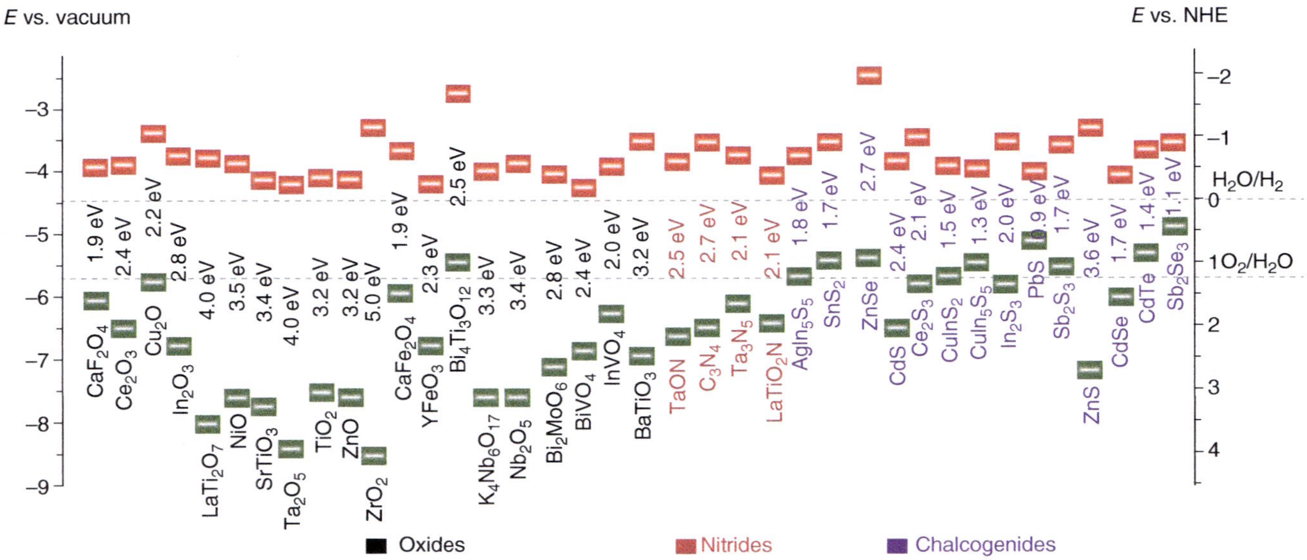

Figure 4.2 Band levels of various semiconductor photocatalysts. Source: Lu et al. [15].

4.3 Photocatalytic H_2 Evolution Under UV Light

4.3.1 Titanium Dioxide (TiO_2)-Based Semiconductors

Titanium dioxide (TiO_2) was the first semiconductor for photocatalytic water splitting since Fujishima and Honda discovered the photoelectrochemical water splitting using TiO_2 as the anode and Pt as the cathode in 1972 [4]. There are three polymorphs for TiO_2: brookite, anatase, and rutile. Due to the difficulties in synthesis, brookite is rarely studied as a photocatalyst. Rutile phase is thermodynamically stable compared with brookite and anatase. Anatase phase often exhibits superior photocatalytic activity. The crystal structures of both anatase and rutile are shown in Figure 4.3. The structures of both anatase and rutile can be regarded as the chains of TiO_6 octahedra. In each octahedral unit, Ti^{4+} ion is surrounded by an octahedron of six O^{2-}. Due to the difference in the distortion of each octahedron and their assembly patterns, the distances between the atoms are different in anatase and rutile phase of TiO_2. In more details, the Ti–Ti distances are larger in anatase phase, whereas the Ti–O distances are shorter compared with rutile phase of TiO_2 [16].

TiO_2 consists of Ti and O atoms and their electron configurations are [O]$1s^2 2s^2 2p^4$ and [Ti]$1s^2 2s^2 2p^6 3s^2 3p^6 4s^2 3d^2$ [17]. The hybridization between the Ti 3d orbital and O 2p orbital results in the formation of discrete energy levels, namely, the conduction band and valence band, between which is the void band gap. The valence band of TiO_2 mainly consists of O 2p states, and the conduction band primarily is contributed by Ti 3d states. Being chemically stable, benign to humans and the environment, photostable, easy to produce, and cheap are some properties of TiO_2 [18, 19].

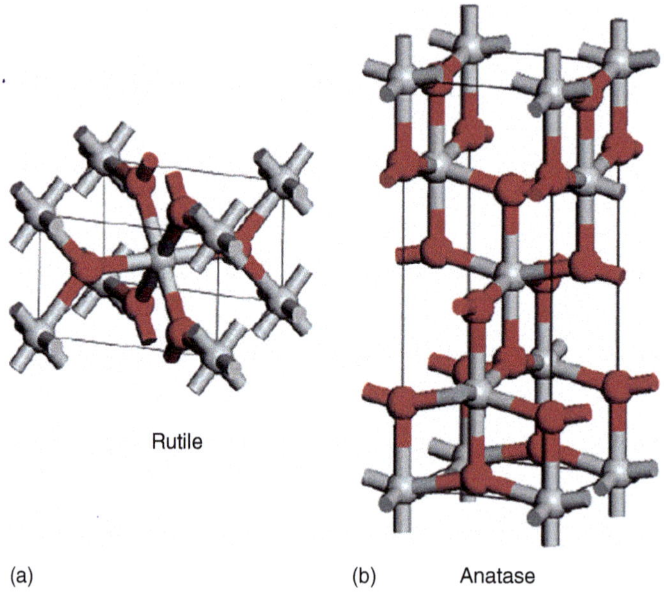

Figure 4.3 Schematic cells for rutile phase (a) and anatase phase (b) of TiO_2. Source: Thompson and Yates [16].

The capability of TiO_2 to absorb light depends on the existence of the partially filled valence band and almost empty conduction band. Anatase has an indirect band gap of ~3.2 eV, whereas rutile possesses a direct band gap of ~3.0 eV. The positions of valence band and the conduction band of TiO_2 are +2.53 and −0.52 eV, respectively. The importance of TiO_2 as a photocatalyst is due to its capability to oxidize a great number of organic compounds into harmless CO_2 and H_2O and to split water into O_2 and H_2 [20]. This property lies in its appropriate electronic band structure.

TiO_2-based photocatalysts have demonstrated great potentials to the application in photocatalytic hydrogen generation. However, the wide applications of TiO_2 photocatalysis in industry are rather limited due to the wide band gap of TiO_2, which restricts its light absorption only within the UV wavelengths, which comprises only 4% of solar radiation. The main portion of the solar spectrum is contributed by visible light and NIR light. On the other hand, a large number of surface defects may act as the recombination center for the electron–hole pairs, leading to a shorter lifetime of the charge carriers and thus a low quantum efficiency. Numerous strategies to improve the efficiency of TiO_2 have been applied so far. Metal and nonmetal doping of TiO_2 can narrow the band gap of TiO_2 and thus extend the absorption edge to the visible-light region due to the creation of new energy levels below the conduction band or above the valence band of TiO_2.

Metal deposition into a TiO_2 structure reduces the recombination rate of electron–hole pairs due to their higher work function compared with bare TiO_2, which leads to better photocatalytic activity [21]. Among the most investigated metals, Pt, with the highest work function, has been often used as a cocatalyst for H_2 generation thanks to its effective electron withdrawing from the TiO_2 conduction band and low overpotential toward the reduction of protons to produce H_2. Pt was loaded on TiO_2 via impregnation method for photocatalyst H_2 generation under UV light [22]. The results show that particle agglomeration was reduced by doping Pt on TiO_2, which leads to a better distribution of e^-/h^+, limiting charge recombination. The light absorption increases after doping Pt, resulting in an improved photocatalyst activity. The effect of the amount of Pt on photocatalytic H_2 generation was studied. One weight percent doping of Pt on TiO_2 exhibits the best H_2 generation performance. Si et al. synthesized Au nanorod TiO_2 nano-dumbbells [23]. Scanning electron microscopy (SEM) and transmission electron microscopy (TEM) images reveal that the morphology of TiO_2 nanoparticles remains without change after the loading of Au nanorods. The activity of photocatalyst under UV light is 11 595.6 $\mu mol\,g^{-1}\,h^{-1}$, which is much higher than its activity under visible light, 157.9 $\mu mol\,g^{-1}\,h^{-1}$. This wide difference is probably due to electron transfer direction, transfer efficiency, and photogenerated electron density. Ouyang et al. fabricated Ru/TiO_2 [24]. The band gap of Ru/TiO_2 decreased from 3.1 to 2.8 eV when the mass content of Ru was increased from 0 to 5 wt%. The photocatalytic activity under UV illumination enhances in the presence of Ru up to a loading of 3 wt%, which could be probably due to the lowest charge recombination for 3-Ru/TiO_2. Quantum efficiency value for 3Ru/TiO_2 was 0.6%. Zn/TiO_2 was fabricated via sol–gel autoignition method [25]. SEM images show that Zn is uniformly dispersed in the TiO_2 without changing the anatase phase of TiO_2. Up to a specific amount

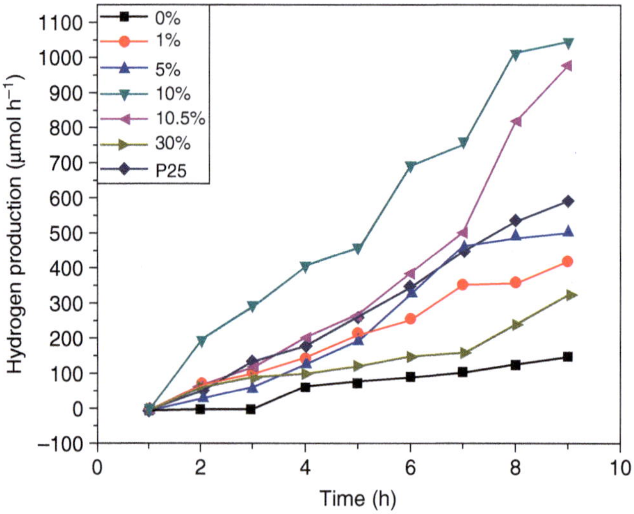

Figure 4.4 Photocatalytic H_2 evolution of pure TiO_2, Degussa P25, and TiO_2 with different amounts of Zn. Source: Al-Mayman et al. [25].

of Zn doping, the band gap was decreased and then increased. Photocatalytic H_2 generation shown in Figure 4.4 enhances with the increase of Zn content up to 10 wt%, after which decreased, which could be related to the decrease of the TiO_2 surface active sites and larger particle size.

Recent researches also reported that bimetallic catalysts showed a better performance compared with their monometallic counterparts due to the facile electron transfer from the conduction band of TiO_2 to the deposited metal nanoparticles, in which the Schottky energy barrier with suitable height is formed. As shown in Figure 4.5, Wang et al. reported that a remarkable enhancement of photocatalytic H_2 evolution was achieved on the alloyed Au–Pt/TiO_2 nanocomposites [26]. In comparison with the bare TiO_2, stronger metal-support interaction between the alloyed structures and TiO_2 as well as higher electron population on the Au–Pt/TiO_2 photocatalysts may contribute to the promoted photocatalytic activity.

Bimetallic Au–Pd particles immobilized on P25 TiO_2 were studied for photocatalytic hydrogen generation using a wide range of alcohol aqueous solution as shown in Figure 4.6 [27]. The highest activity was achieved on the Au-core/Pd-shell structure with the Au–Pd molar ratio of 25/75. It was reported that Au–Pt supported on commercial P25 TiO_2 upon prereduction under H_2 at 500 °C dramatically enhances H_2 production compared with the monometallic Au/TiO_2 and Pt/TiO_2 NPs [28]. It was suggested that the improved activity is the result of a combination of positive aspects including efficient charge separation process and weakened metal–H bonds on bimetallic catalysts. Jiang et al. reported that the bimetallic Ag–Pt deposited TiO_2 hollow spheres exhibited enhanced activity relative to their monometallic counterparts, which can be attributed to the formation of Schottky energy barriers with suitable height between the bimetallic nanoparticles and TiO_2 [29].

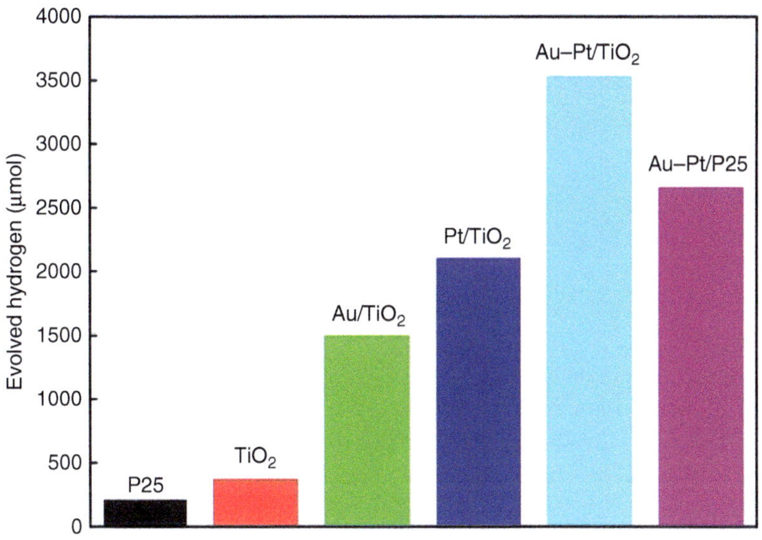

Figure 4.5 Hydrogen evolution on the bare and metallic TiO_2 photocatalysts using the benchmark Degussa P25 TiO_2 and its bimetallic materials as the references. Source: Wang et al. [26].

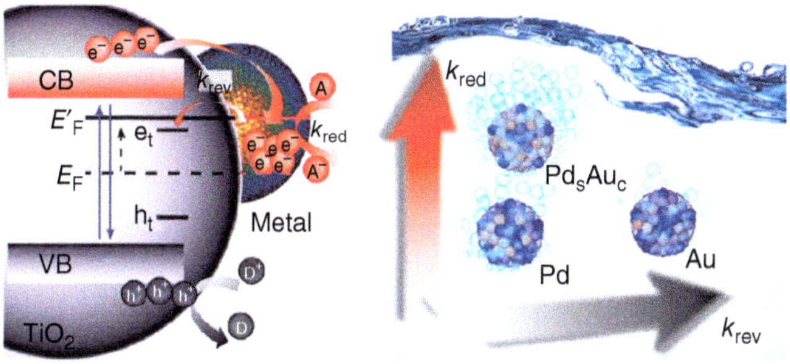

Figure 4.6 Photocatalytic hydrogen evolution on TiO_2-supported Au–Pd nanoparticles using a range of alcohols. Source: Su et al. [27].

Recent research results indicated that doping with nonmetallic elements, especially nitrogen (N), sulfur (S), and carbon (C), could greatly shift the absorption edge of TiO_2 to the visible region [30]. Asahi et al. presented that doping TiO_2 with nitrogen and other anionic species by substituting the O atoms would be an effective way to narrow the band gap of TiO_2 for better use of visible-light energy [31]. Based on the density of states (DOS) calculations, they found that the mixing of N 2p states with O 2p states would cause a significant decrease of the band gap. The hybridization between N 2p states and O 2p states leads to the formation of an intermediate energy level just above the valence band top of TiO_2, which caused the band-gap narrowing effect and redshift of the absorption edge [32]. As a result of this, the mechanism of the photoexcitation for N-doped TiO_2 is probably direct electron excitation from the

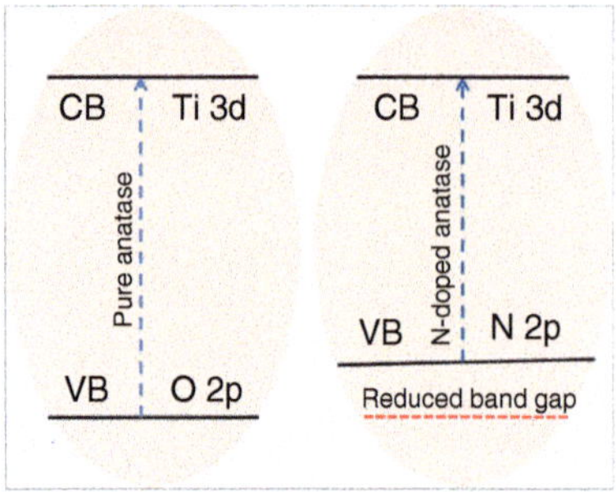

Figure 4.7 Schematic representation of the band structure of pure and N-doped anatase TiO_2. Note that the energies are not in scale. Source: Di Valentin et al. [32].

N 2p states to the conduction band of TiO_2. The intermediate energy level formation in the N-doped TiO_2 is illustrated in Figure 4.7 [16].

Coupling with other semiconductor is another strategy to increase the activity of TiO_2, which creates direct Z-scheme or conventional heterojunction type. Meng et al. fabricated TiO_2/CdS hybrid hierarchical photocatalyst [33]. The photocatalytic activity of TiO_2/CdS has a remarkable enhancement compared with the pure TiO_2 and CdS. This enhancement could be due to the increase of light harvesting and efficient electron–hole separation on the interface of two photocatalysts. Cu_2O/TiO_2 was fabricated by a facile ethanol reduction method [34]. These two semiconductors create p–n heterojunction, which improves its activity because of the improvement of visible-light absorption, limited charge recombination, faster interfacial charge transfer, and more surface reactive sites. The photocatalytic activity of 2.5 wt%-Cu_2O/TiO_2 was investigated with different sacrificial reagents such as methanol, anhydrous ethanol, ethylene glycol, and glycerol (Figure 4.8). It can be concluded that the higher proportion of carbon atom number and hydroxyl group number leads to lower photocatalytic H_2 generation rate. This proportion for ethylene glycol, methanol, and glycerol are 1 and for ethanol is 2.

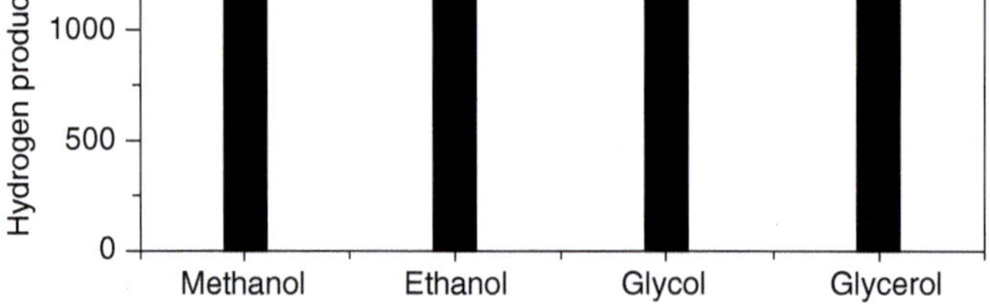

Figure 4.8 The photocatalyst H_2 generation rate of 2.5 wt%-Cu_2O/TiO_2 with different sacrificial reagents. Source: Li et al. [34].

Besides, coupling TiO_2 materials with other narrow-band-gap semiconductors or metal clusters possessing discrete energy levels can significantly extend the absorption edge of the TiO_2-based catalysts into the longer-wavelength region, thus enhancing the photocatalytic properties [27]. Controlling porosity for ordered macroporous and mesoporous structures is extensively explored in TiO_2 photocatalysis because of having the abundance of active sites, high surface area, and easily accessible channels [35, 36]. There are several methods that have been used to synthesize mesoporous TiO_2 such as sol–gel, hydrothermal, solvothermal, and template methods.

4.3.2 Other Types of UV-Responsive Photocatalysts

Wu et al. fabricated ZnO-dotted porous ZnS cluster microspheres for photocatalytic H_2 evolution under UV light [37]. Theoretical calculation and the experiments show that with the dotting of ZnO, the photogenerated electrons/holes migrate more efficiently and also more active sites are accessible for the photocatalytic reaction. The activity in photocatalytic hydrogen generation was 367 μmol h^{-1} without using any cocatalyst, and the activity shows no difference upon adding 1 wt% Pt, indicating the existence of the active sites in ZnO-dotted porous ZnS cluster microspheres. Figure 4.9 depicts this photocatalyst forming a type II heterojunction and also its schematic structure.

$SrTiO_3$ nanoparticles were synthesized through three different methods such as the solid-state reaction, the polymerized complex method, and the milling assistant method [38]. The photocatalyst activity shows that the sample that was fabricated with the polymerized complex method had higher activity in comparison with the other two methods. These differences come from the different uniformity of components, particle size, and particle aggregation extent for different methods. CdS nanoparticles were fabricated via precipitation method using different zeolites ZSM-5, H-Y, and H-β as templates [39]. These samples named CdS-Z, CdS-Y, and

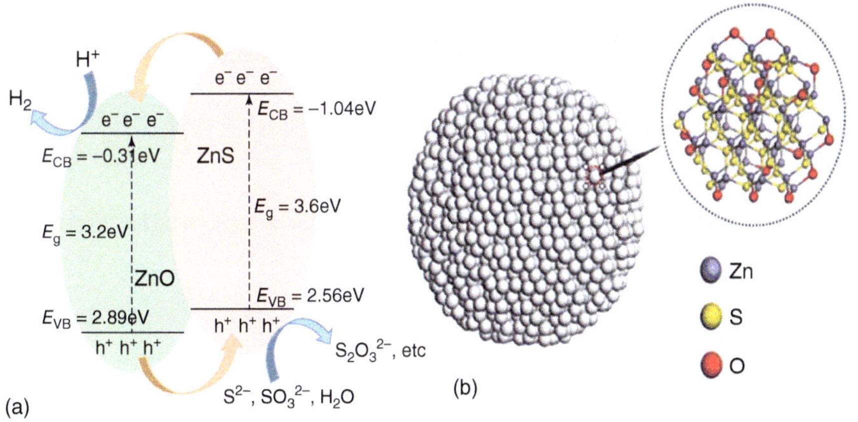

Figure 4.9 (a) The energy band structure of ZnO/ZnS heterojunction. (b) The graphic structure of ZnO-dotted ZnS. Source: Wu et al. [37].

CdS-B. CdS-Z showed a blueshift from UV–visible (UV–vis) absorption spectra analysis and had a higher band-gap energy compared with CdS-Y and CdS-B. There is a good correlation between band gap and particle size, i.e. the smaller particle size, the higher band gap. The highest photocatalyst H_2 generation rate was for CdS-Y, which could be ascribed to its smallest particle size and its highest surface area.

4.4 Photocatalytic H_2 Evolution Under Visible Light

4.4.1 Carbon Nitride (C_3N_4)-Based Semiconductor

Carbonic nitride (C_3N_4) polymers first discovered in 1834 [40]. Melem, melam, melamine, and melon were considered as heptazine- and triazine-based compounds. A yellow, amorphous, and insoluble material, i.e. melon, was thereafter discovered. In 1922, C_3N_4 was firstly introduced by Franklin, who identified C_3N_4 as the final product upon calcination of melon and proposed its structure. However, none has paid attention to the above melon-based C_3N_4. It is regarded as unconfirmed species for a long time because of its insolubility in numerous solvents and chemical inertness [41]. To date, studies on its preparation and characterization are still underway. In 2006, Goettmann et al. used g-C_3N_4 in catalytic activation of benzene [42]. In 2009, Wang et al. found out that g-C_3N_4 can split water for H_2 and O_2 production [5]. Because of several features like being metal-free, non-toxic, low cost, and chemically stable, C_3N_4 has attracted universal attention. Generally, C_3N_4 has been synthesized through thermal condensation from nitrogen-rich precursors like urea [43], cyanamide [44], melamine [45], and 3-amino-1,2,4-triazole [46]. The composition, crystallinity, and structure of C_3N_4 have strong influences on the photocatalytic properties of C_3N_4 materials [47].

C_3N_4 possesses seven phases, namely, α-C_3N_4, β-C_3N_4, pseudocubic C_3N_4, g-*h*-triazine, cubic C_3N_4, g-*o*-triazine, and g-*h*-heptazine. Among all the allotropes of C_3N_4, tri-*s*-triazine-based g-C_3N_4 is the most stable and energetically favorable allotrope at the ambient atmosphere as revealed by density functional theory (DFT) calculations (Figure 4.10). Thus, tri-*s*-triazine has been widely identified as the basic unit to form g-C_3N_4. Polymeric g-C_3N_4 is a metal-free p-type semiconductor. The band gap of bare g-C_3N_4 is about 2.7 eV, which is tunable by the amount of nitrogen in the C_3N_4 structure. The narrow band gap of g-C_3N_4 leads to an onset of visible-light absorption at around 450 nm. g-C_3N_4 has the required band positions of the conduction band at −1.3 V and the valence band at 1.4 V vs. NHE for photocatalytic water splitting at pH = 7. Despite the mentioned fascinating properties of C_3N_4, some disadvantages like low surface area, high charge recombination rate, and limited visible-light absorption hinder its practical application.

Similar to TiO_2-based photocatalysts, numerous strategies have been attempted to overcome these shortcomings such as metal doping [48], nonmetal doping [49], surface modification with other semiconductors [50], morphology control [51], and coupling with another semiconductor [52]. Design of g-C_3N_4 nanostructure is a promising method to improve photocatalytic H_2 generation. Various structures

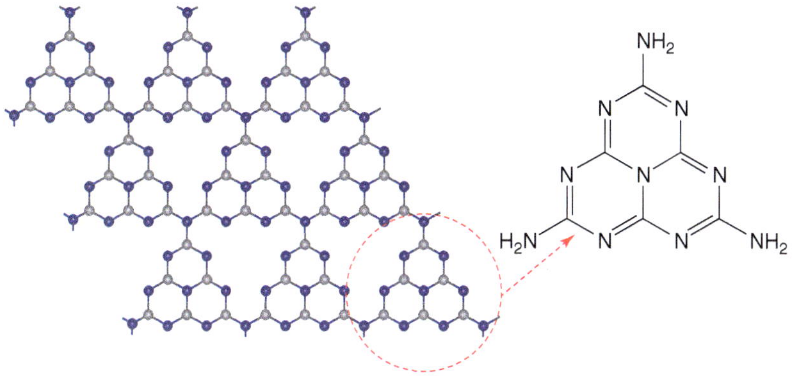

Figure 4.10 Tri-*s*-triazine-based structure of *g*-C_3N_4. The C and N atoms are indicated by gray and blue balls [8].

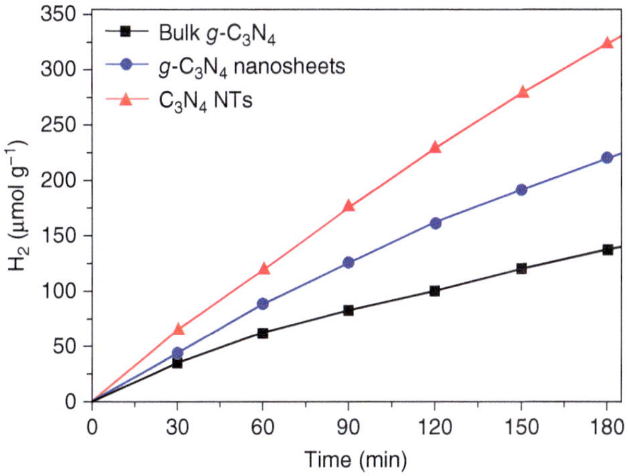

Figure 4.11 Photocatalytic H_2 production over the bulk C_3N_4, C_3N_4 NSs, and C_3N_4 NTs under visible-light irradiation. Source: Zhu et al. [54].

such as bulk g-C_3N_4, g-C_3N_4 nanosheets (C_3N_4 NSs), g-C_3N_4 nanotubes (C_3N_4 NTs), and mesoporous g-C_3N_4 have been synthesized [53, 54]. Recent studies have shown the superior activities of C_3N_4 NTs under visible-light irradiation. Zhu et al. prepared bulk C_3N_4, C_3N_4 NSs, and C_3N_4 NTs for photocatalytic H_2 generation [54]. Figure 4.11 shows that C_3N_4 NTs have higher activity compared with bulk C_3N_4 and C_3N_4 NSs, which could be ascribed to more active sites, higher photogenerated carrier transfer efficiency, and better mass transfer.

Mesoporous g-C_3N_4 (MCN) has gained attention in recent years. Several methods for synthesis of MCN such as the soft and hard templating approaches have been developed. Guo et al. fabricated bulk g-C_3N_4 and MCN for photocatalytic H_2 evolution under visible-light irradiation [55]. MCN shows higher activity than bulk g-C_3N_4, which could be attributed to porous structure, more active sites, and high surface area. Although morphology control of g-C_3N_4 provides more surface area

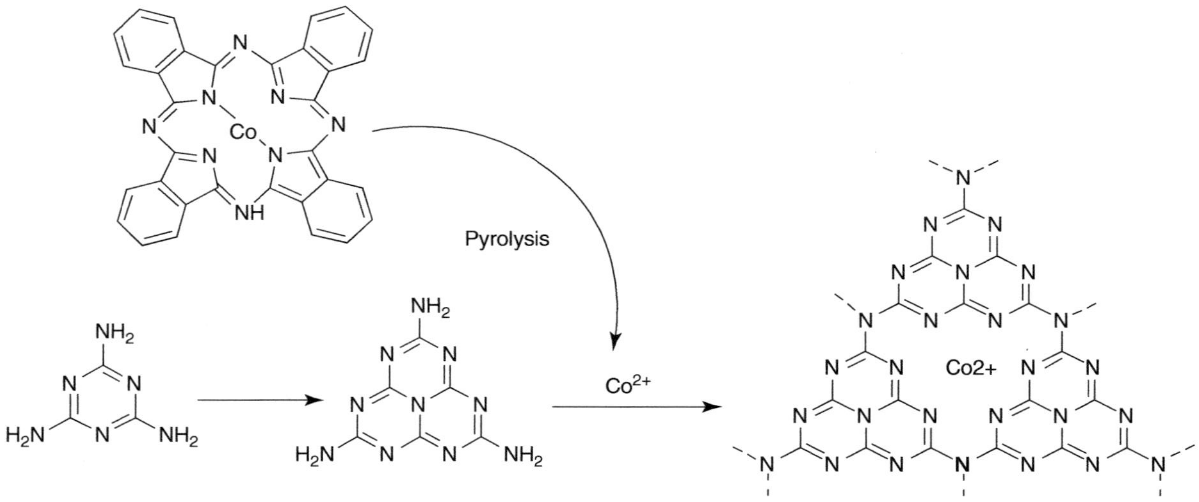

Figure 4.12 Graphic design of preparation of cobalt-doped g-C_3N_4. Source: Chen et al. [57].

and active sites, it still suffers from weak visible-light absorption and a high recombination rate. These drawbacks have restricted the application of g-C_3N_4-based photocatalysts.

One method to overcome these problems is the surface modification of g-C_3N_4 with metal and metal oxide semiconductors. Pt–Pd bimetallic alloy nanoparticles were decorated on C_3N_4 NSs via chemical deposition precipitation [56]. The morphology results show that Pt–Pd nanoparticles were well dispersed on the surface of C_3N_4 NSs. Photocatalytic H_2 generation reveals enhancement for PtPd/C_3N_4 NSs compared with the bare C_3N_4 NSs, which is probably because of higher light absorption, lower recombination rate, and improved charge separation. Cobalt was doped into g-C_3N_4 using one-step thermal polycondensation [57]. Figure 4.12 shows a schematic structure of Co-doped g-C_3N_4. The photocatalyst activity of Co/g-C_3N_4 was three times higher than that of the bare g-C_3N_4. This enhancement could be attributed to the higher surface area, lower band gap, and lower recombination rate of photogenerated electron–hole pairs.

Non-metal and metal oxide decoration on g-C_3N_4 is another strategy to enhance the photocatalyst activity of C_3N_4. Xu et al. fabricated sulfur-doped MCN (mpgCNS) using thiourea as a precursor and SiO_2 nanoparticles as the hard template for photocatalytic H_2 evolution [58]. They used an in situ method for sulfur doping. The TEM images show that mpgCNS has pores with a diameter of 10–20 nm, aligning well with the size of the SiO_2 template, and is consistent with the pore size distribution profile. Sulfur doping in MCN enhances the light absorption in both the UV and visible-light regions. mpgCNS exhibits 30 times higher activity than that of the bare g-C_3N_4 for photocatalytic water splitting as a result of stronger and extended light absorption in the visible-light region induced by sulfur doping, faster charge transfer, and efficient mass diffusion in the mesoporous structure. Sun et al. synthesized K/C_3N_4 NSs through hydrothermal treatment [59]. The high-resolution TEM (HRTEM) analysis shows that the K/C_3N_4 sample comprises 6–7 layers of g-C_3N_4. After introducing K into the structure of C_3N_4 NSs, the band gap decreases, leading to better light absorption. From the band-gap analysis and X-ray photoelectron spectroscopy (XPS) studies, the position of the valence band and the conduction band was determined. The results show the upshift of the conduction band of K/C_3N_4, which makes electrons more reductive and enhances their photoreduction ability, leading to an increase of photocatalytic H_2 production. Boron-doped g-C_3N_4 was prepared using a microwave heating method for photocatalytic H_2 evolution [60]. X-ray diffraction (XRD) and XPS analysis confirm the successful doping of B into the g-C_3N_4 structure. Electrochemical impedance spectroscopy (EIS), photoluminescence (PL), and photocurrent results demonstrate that upon B doping on g-C_3N_4, the charge transfer and the photogenerated electron–hole separation become more efficient and the lifetime of photogenerated electrons increases (Figure 4.13).

Another effective method to increase the photoactivity of C_3N_4 is coupling with another semiconductor with appropriate band potentials with C_3N_4. Upon hybridization of C_3N_4 with another semiconductor, the band bending is created at the interface of the heterojunction due to the potential difference between the two components. Based on the band positions of the semiconductors, the electronic

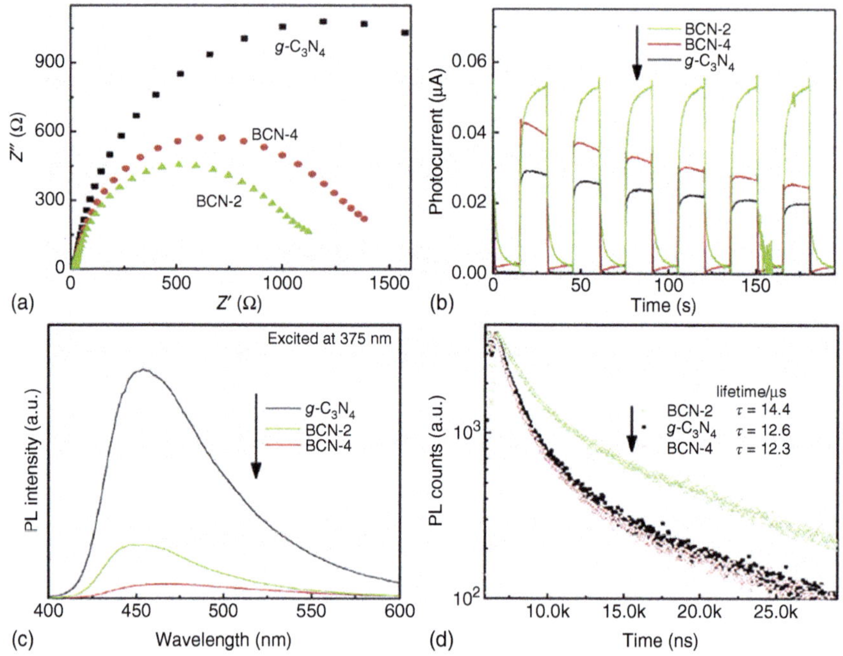

Figure 4.13 (a) EIS, (b) photocurrent response, and steady-state (c) and transient (d) PL spectra of g-C_3N4 and B-doped g-C_3N_4 samples. Source: Chen et al. [60].

structures of the formed heterojunction are categorized into three different types as depicted in Figure 4.14. In a type I heterojunction, the valence band position of semiconductor B is lower than that of semiconductor A, while the conduction band is higher. Since the electrons will move up and down to reach energy balance, they will transfer and accumulate on semiconductor A. Such arrangements are also called straddling gap. In a type II alignment, both the conduction band and valence band positions of semiconductor B lie higher than those of semiconductor A. The photogenerated electrons will transfer from the conduction band of semiconductor B to the conduction band of semiconductor A, and holes transfer from valence band (A) to valence band (B) simultaneously. The overpotentials between the two components induce upward or downward band bending, resulting in charge migration in the opposite direction. Therefore, type II (staggered gap) provides the optimum band positions for efficient electron–hole pair separation resulting in enhanced electron lifetimes, thus improving photocatalytic activity. Type III heterojunction shows the same charge carrier transfer as in type II – only that the band-edge potentials are even further set off. This alignment of band gap and potentials is defined as a broken gap.

Zhu et al. prepared type I and II heterojunction of CoOx nanoparticles in C_3N_4 NTs [61]. CoO and Co_3O_4 were decorated on C_3N_4 NTs through a facile one-pot method but under different calcination atmospheres, i.e. under vacuum and air condition, respectively. SEM-EDS and TEM images show that CoO and Co_3O_4 were distributed on the surface of C_3N_4 NTs. Figure 4.9a,b shows the photocatalytic H_2 generation

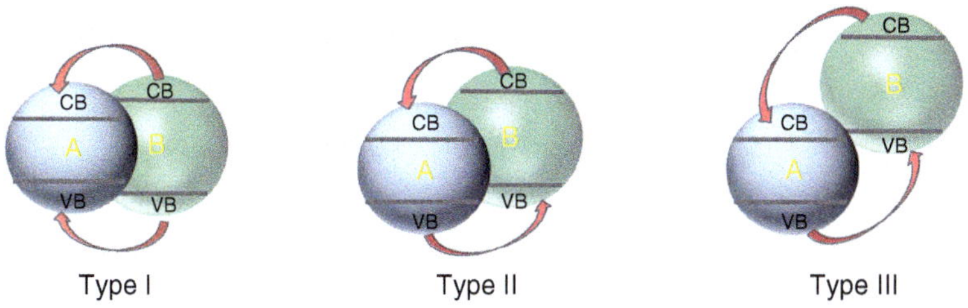

Figure 4.14 Schematic illustration of the charge transfer for the three types of heterojunctions [8].

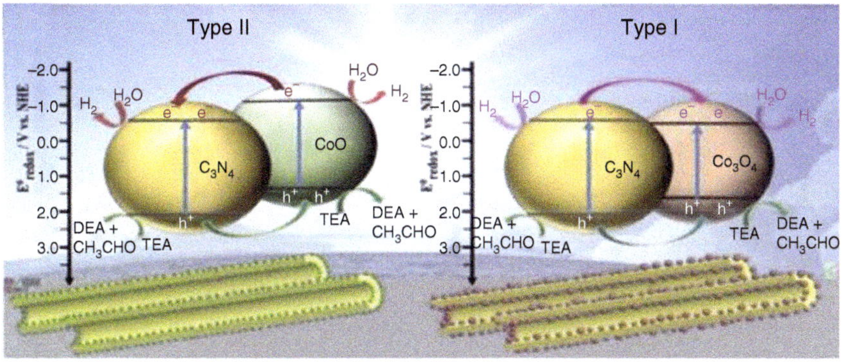

Figure 4.15 Graphic illustration of the proposed mechanism for cobalt oxide/C_3N_4 NT heterojunction for photocatalytic H_2 evolution. Source: Zhu et al. [61].

for different samples. CoO/C_3N_4 NT samples had higher activity in comparison with Co_3O_4/C_3N_4 NT samples, which could be because of finer CoO NPs and also more negative conduction band of CoO. With having band-gap results from the UV–vis diffuse reflectance spectroscopy (DRS), the conduction band was obtained (Figure 4.15). CoO and Co_3O_4 on C_3N_4 NTs form type II and I heterojunction, respectively.

Ag_2CrO_4 nanoparticles-decorated C_3N_4 NTs were fabricated for photocatalytic H_2 evolution using methanol as a sacrificial agent under visible-light irradiation ($\lambda \geq 420$ nm) [62]. Z-scheme formation of Ag_2CrO_4/C_3N_4 NTs could prevent photogenerated electron–hole pair recombination and high separation of photogenerated carriers, as a result of improving H_2 generation. Kailasam et al. fabricated tungsten oxide on MCN using Pt as a cocatalyst for photocatalytic H_2 generation [63]. WO_3/MCN composites were fabricated via impregnation method and showed a very high photocatalytic H_2 evolution under visible-light irradiation, because of the high surface area and a synergetic effect between MCN and WO_3, which led to increased charge separation as shown in Figure 4.16. Upon modification with WO_3, the photoinduced holes in the valence band of g-C_3N_4 and the photogenerated electrons in the conduction band of WO_3 recombine at the interface, leaving the electrons in the conduction band of g-C_3N_4 and holes in the valence band of

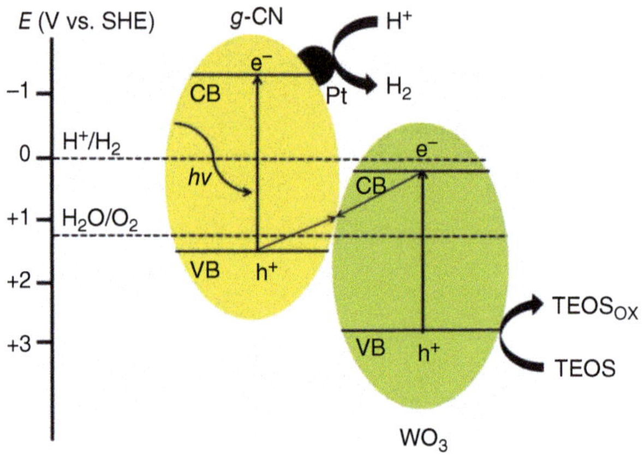

Figure 4.16 Schematic diagram of the photocatalytic mechanism in WO_3/MCN. Source: Kailasam et al. [63].

WO_3. Thus, the overall charge separation was improved and thus led to a higher photocatalytic activity.

4.4.2 Other Types of Visible-Light-Responsive Photocatalysts

Polyaniline (PANI) and polypyrrole (PPy), conducting polymers, were used to sensitize Ta_3N_5 for improved photocatalytic activity [64]. Adding these two polymers to Ta_3N_5 enhances the charge transfer efficiency and reduces the photogenerated electron and hole recombination rate of Ta_3N_5, leading to the increase of electron–hole separation efficiency and the improvement of its photocatalytic activity. These polymers that cover Ta_3N_5 particles simplified the migration of the photogenerated electrons and holes, which prevent self-photocorrosion of the Ta_3N_5. Due to the presence of the protonated nitrogen ($-N^+$) state in PPy, PPy electric conductivity was lower than that of PANI, which dropped the sensitizing ability of PPy in comparison with PANI. Consequently, the photocatalytic H_2 generation of Ta_3N_5/PANI was higher than that of Ta_3N_5/PPy. Norouzi et al. used different types of biochar such as sewage sludge (SS), soft wood pellet (SWP), and rice husk (RH) at 700 °C and then loaded Fe_3O_4 on them via an impregnation method for photocatalytic H_2 generation under visible-light illumination [65]. The highest activity was related to Fe_3O_4-2.5%/RH. This enhancement could be because of its low band gap, high photocurrent density, and low charge transport resistance.

SiC nanowires were fabricated by simple carbothermal reduction and modified by acid oxidation on the surface [66]. The FTIR spectra of the SiC nanowires before and after surface modification show that no new band arises and the intensities of all bands increase after surface modification. The band gaps of modified and unmodified SiC nanowires are 2.35 and 2.27 eV, respectively. Figure 4.17 compares the photocatalytic activity of the modified and unmodified SiC nanowires through three noncontinuous experiments. The results display that the modified SiC nanowires

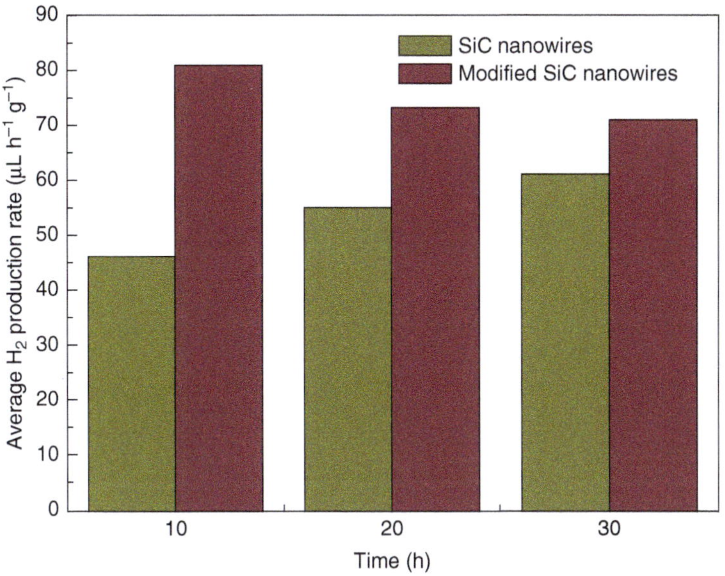

Figure 4.17 The average H_2 evolution rate of the SiC nanowires and modified SiC nanowires. Source: Hao et al. [66].

can absorb visible light and show excellent photocatalytic H_2 production compared with the unmodified one.

4.5 Photocatalytic H_2 Evolution Under Near-Infrared Light

Despite the above endeavors, the absorption of C_3N_4 NT-based catalysts with a longer-wavelength absorption of sunlight is still a challenge. Sensitization is one of the attractive techniques to extend the light absorption toward a higher wavelength range. Since the major part of solar energy is NIR light, it is important to construct NIR-responsive material for photocatalytic H_2 generation. Lanthanide upconversion nanoparticles (UCNPs), a kind of fluorescein, are generally composed of an inorganic host and lanthanide dopant ions acting as sensitizers and activators as shown in Figure 4.18. There are five basic mechanisms of lanthanide upconversion processes: excited-state absorption, energy transfer upconversion, cross-relaxation, cooperative sensitization upconversion, and photon avalanche. Basically, ideal host materials should be transparent and have low lattice phonon energies in order to maximize the radiative emission and minimize the non-radiative loss. Generally, fluorides exhibit high chemical stability and low phonon energies and thus are usually used as the host materials. By far, $NaYF_4$ has been recognized as the most popular host for lanthanide dopant ions for upconversion processes.

Zhu et al. prepared UCNPs supported on C_3N_4 NTs [67]. Lanthanide-doped UCNPs are able to convert NIR excitation to visible and even UV emissions via

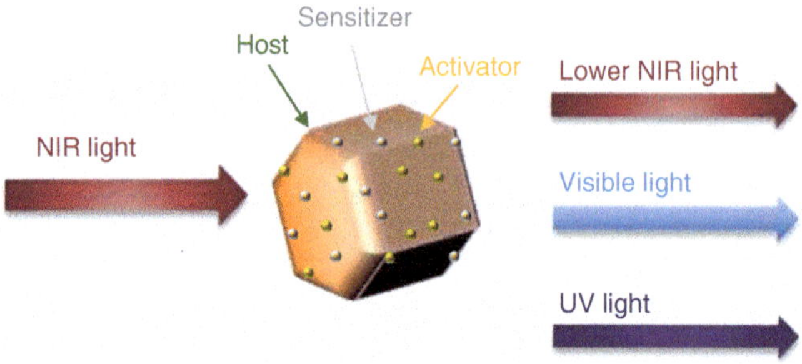

Figure 4.18 Illustration of lanthanide upconversion nanoparticles (UCNPs).

a multiphoton process. $NaYF_4$: Yb, Tm, Gd (NYFG) and $NaYF_4$: Yb, Tm (NYF) were decorated on C_3N_4 NTs separately to form the heterojunction structures. NYFG/C_3N_4 NTs exhibited the highest H_2 generation with an apparent quantum efficiency (AQE) of 0.80‰, about 1.4 times higher than that of NYF/C_3N_4 NTs under 980 nm laser irradiation. This enhanced photocatalytic activity is attributed to the synergistic effect, stronger interaction, higher emission intensity, and faster charge transfer between the two nanocomposites. As depicted in Figure 4.19, under NIR irradiation, the pumping of 980 nm light only excites the Yb^{3+} ions. Yb^{3+} ions act as the sensitizer, and three successive energy transfers from Yb^{3+} to Tm^{3+} ions populate its $^3H_5$, $^3F_{2,3}$, and $^1G_4$ levels. The cross-relaxation process between Tm^{3+} is responsible for populating the $^1D_2$ level. Gd^{3+} ions cannot absorb 980 nm photons directly. In the Yb^{3+}-Tm^{3+}-Gd^{3+} tridoped system, the excited Tm^{3+} ions in the high level can transfer energy to Gd^{3+} to promote its excitation. After the excitation of NYFG NPs, the C_3N_4 NTs is directly excited by the energy transfer from the high levels of Tm^{3+} and Yb^{3+} ions by the free resonance electron transfer process. Consequently, the photogenerated electrons are excited from the valence band to the conduction band of C_3N_4 NTs, leaving holes in the valence band. The separated electrons are trapped by H_2O to produce H_2, whereas the holes accumulated in the valence band will react with TEA to form diethylamine (DEA) and acetaldehyde.

Upconversion luminescence agents, Tm^{3+}, Yb^{3+}: $NaYF_4$ and Er^{3+}: $Y_3Al_5O_{12}$, were synthesized via hydrothermal and sol–gel method [68]. Tm^{3+}, Yb^{3+}: $NaYF_4$–Er^{3+}: $Y_3Al_5O_{12}$/MoS_2–$NaTaO_3$ nanocomposite was fabricated through the ultrasonic dispersion and liquid boiling method. Compared with MoS_2–$NaTaO_3$, Tm^{3+}, Yb^{3+}: $NaYF_4$–Er^{3+}: $Y_3Al_5O_{12}$/MoS_2–$NaTaO_3$ shows much higher activity, which could be due to the upconversion luminescence effect. Nitrogen (N)-doped carbon dots (CDs)/CdS hybrid photocatalyst was designed by a simple solvothermal method under visible/NIR light irradiation for photocatalytic hydrogen production [69]. Upconverted PL spectra show that N-CDs act as NIR light harvester to enhance the NIR light utilization of photocatalysts. N-CDs can upconvert NIR to visible light for further exciting CdS nanoparticles shown in Figure 4.20. The results show that no H_2 was detected for CdS photocatalyst under NIR light. With adding different amounts of N-CDs, photocatalytic H_2 generation occurred.

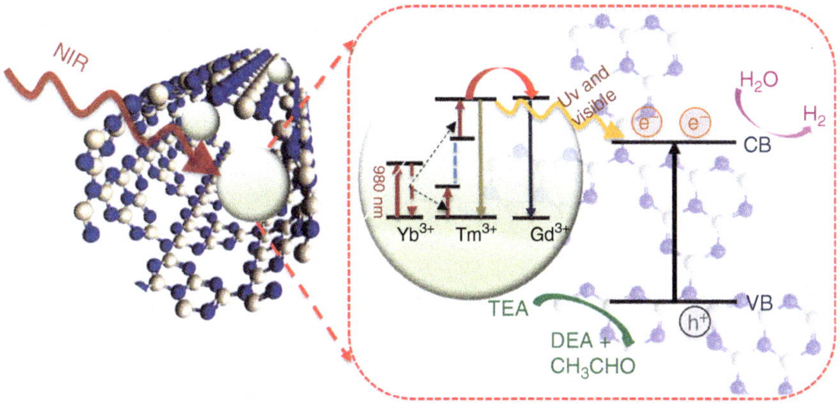

Figure 4.19 Illustrative diagrams of energy transfer among NYFG(15)/C_3N_4 NTs. Source: Zhu et al. [67].

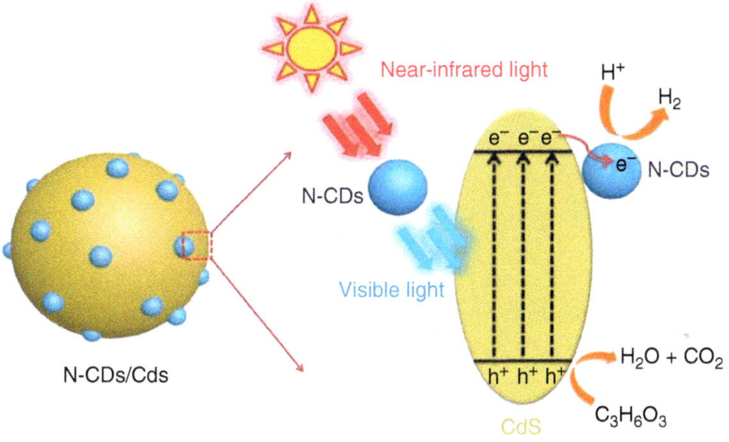

Figure 4.20 The proposed mechanism of the improved photocatalytic activity in the N-CDs/CdS photocatalyst. Source: Shi et al. [69].

Two-dimensional (2D) hybrid of black phosphorus (BP)/WS_2 was prepared as a metal-free NIR-driven photocatalyst [70]. The pure BP shows a wide range of absorption from UV to NIR region, while the pure WS_2 shows the main absorption in the visible-light region. The pure BP and WS_2 show poor activity, while a high amount of H_2 was released under three hours NIR light irradiation. This enhancement could be due to the low recombination rate of photogenerated electron–hole, an effective charge separation. FeS_2/TiO_2 was fabricated by wet chemical synthesis for improving the photocatalytic hydrogen evolution rate over an extensive range of absorption wavelengths, UV to NIR spectrum [71]. TEM image shows that the heterostructures were composed of TiO_2 and FeS_2. The absorption of TiO_2 shows that this material cannot absorb any wavelength over 500 nm. Since FeS_2 has a band gap of 0.95 eV, its absorption spectrum

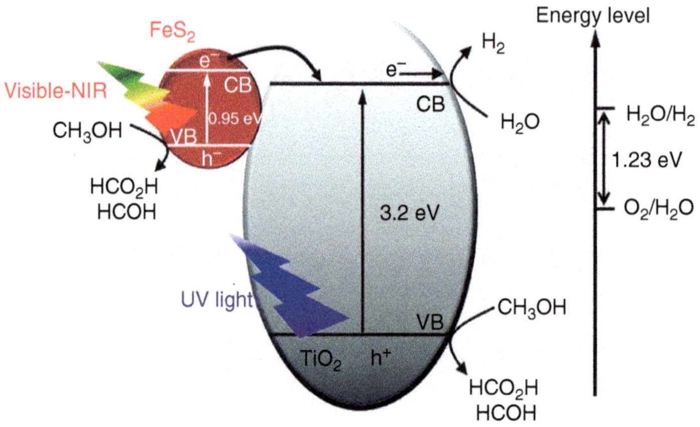

Figure 4.21 Graphic diagram of the FeS_2–TiO_2 heterostructures under ultraviolet, visible, and NIR light irradiation for photocatalytic H_2 evolution. Source: Kuo et al. [71].

covers from UV to NIR wavelength ranging from 300 to 1100 nm. As a result, FeS_2–TiO_2 heterostructure displays a wide absorption range from UV to NIR wavelength. The photocatalytic activity results present a high activity of FeS_2–TiO_2 and poor activity of either FeS_2 or TiO_2. The low activity of FeS_2 could be due to its band gap, which is less than the required band gap for water splitting. Also, TiO_2 shows low activity because of the large band gap, which allows just the absorption of UV light. Figure 4.21 shows the proposed mechanism for photocatalyst H_2 generation over FeS_2–TiO_2. The high activity of FeS_2–TiO_2 could be ascribed to the efficient absorption in the regions of visible and NIR light by FeS_2.

BP NSs were loaded on TiO_2 mesocrystals (TMC) using 3 wt% Pt as a cocatalyst [72]. Absorption ability of TMC was limited to UV light, while BP NS absorption covers from UV to NIR region. As a result, BP NS/TMC displays strong absorption in the visible and NIR regions. BP NS/Pt (3 wt%)/TMC shows the highest activity compared with BP NS and Pt (3 wt%)/TMC, representing the effective utilization of solar light because of wide absorption and the strong interaction between TMC and BP NS. It is found that a type II heterojunction is anticipated at the interface between TMC and BP NS. $CuInS_2$ (CIS) and Cu-In-Zn-S (CIZS) quantum dots (QDs) were synthesized by a one-pot aqueous method [73]. CIZS QD was used as light-harvesting material for photocatalytic H_2 evolution. Although it has a narrow band gap, this material shows outstanding energy conversion efficiency in the visible and NIR regions. CIZS QD has low recombination rate and high PL lifetime compared with CIS QDs, attributing to the reduction of defects because of the inclusion of Zn. This also leads to higher photocatalytic activity.

Jiang et al. fabricated amorphous TiO_{2-x} comprising Ti^{3+} from the dissolution of Ti^0 by water in an anaerobic environment [74]. XRD results show diffraction pattern that is attributed to amorphous phase, demonstrating the successful synthesis of amorphous TiO_{2-x}. XPS test verifies the presence of Ti^{3+} on the surface of the

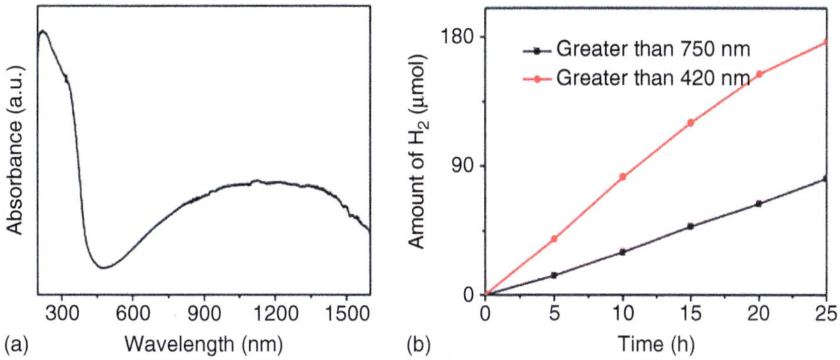

Figure 4.22 (a) The UV–vis near-infrared absorption spectrum. (b) Photocatalytic hydrogen evolution for amorphous TiO_2–x. Source: Jiang et al. [74].

fabricated amorphous TiO_2 to form amorphous TiO_{2-x}. Based on XRD analysis, it can be concluded that 120 hours of anaerobic hydrolysis is the optimum time to eliminate residual Ti^0 or Ti^{2+} and to produce the amorphous TiO_{2-x}. Figure 4.22a shows amorphous TiO_{2-x} is able to absorb wavelength in the range of 420–1500 nm, which offers a great potential for utilizing visible and NIR light. Figure 4.22b shows that the photocatalyst activity over wavelengths higher than 420 nm is better than its activity over wavelengths higher than 750 nm.

4.6 Roles of Sacrificial Reagents and Reaction Pathways

Upon excitation, most of the photoinduced electron–hole pairs will recombine within nanoseconds. Although the whole water splitting process can be achieved on the surface of rutile, the quantum yield efficiency is extremely low. Therefore, sacrificial agents (hole scavengers or electron donors) are often required to inhibit the recombination between the photoinduced charge carriers and improve the photocatalytic hydrogen production. The sacrificial agents react irreversibly with the photogenerated holes in the valence band, thus achieving higher photocatalytic hydrogen production. In this process, electron donors are added continuously to sustain H_2 production because they can sustainably consume the photoinduced holes in the valence band, thus providing more electrons in the conduction band. Organic compounds such as methanol, ethanol, ethylenediaminetetraacetic acid (EDTA), CN^-, lactic acid, and formaldehyde are commonly used as electron donors to enhance H_2 production.

In terms of the photocatalytic hydrogen generation from the methanol/water mixture, Miwa et al. proposed that the hydrogen generation from methanol/water mixture proceeded stepwise following the equations below [75]:

$$CH_3OH \xrightarrow[hv]{\text{Catalysts}} HCOH + H_2 \tag{4.5}$$

$$HCOH + H_2O \xrightarrow[hv]{\text{catalysts}} HCOOH + H_2 \text{ Global} \tag{4.6}$$

$$HCOOH \xrightarrow[hv]{\text{catalysts}} CO_2 + H_2 \tag{4.7}$$

Chen et al. also suggested that the hydrogen generation process from methanol/water mixture could be a one-step process, illustrated as follows [76]:

$$CH_3OH + H_2O \xrightarrow[hv]{\text{catalysts}} CO_2 + 3H_2 \tag{4.8}$$

According to the above proposed mechanisms, hydrogen molecules are produced in each step. It is reported that the methanol decomposition occurs easier compared with the water molecule splitting [75].

In 1983, Kawai and coworkers found that hydrogen could be produced if gas-phase methanol was used in the Pt/TiO_2 photocatalytic system, which indicated that hydrogen could be produced through the photocatalytic methanol reforming. If the mixture of water and methanol was injected into the photocatalytic system, enhanced hydrogen production was observed, revealing that the water molecules reacted with the adsorbed methanol on the surface of catalysts that contributed to the increased hydrogen evolution efficiency [77]. By using the in situ FTIR and time-resolved infrared (IR) spectroscopy, Li's group examined the photocatalytic hydrogen production process on Pt/TiO_2 from methanol. They found that the deposition of Pt would occupy some active sites for methanol adsorption [77]. During the photocatalytic hydrogen generation process from methanol, surface species CH_2O, CH_2OO, and HCOO were detected, which were derived from the photocatalytic methanol decomposition process. Recently, Yang's group employed a high-resolution STM technique to study the photocatalytic process of single methanol molecules on TiO_2(110) surface [78]. In this work, methanol photodissociation, photoinduced migration of formaldehyde, and formaldehyde photo-desorption were probed. This single molecular study on the methanol photocatalytic process successfully showed the surface chemistry process on how the methanol molecules could be split on the excited semiconductor nanocatalysts. Recently, in situ ^{13}C and ^{1}H NMR spectroscopy was applied to monitor the gas–liquid–solid photocatalytic H_2 generation process under real working conditions. As depicted in Figure 4.23, the surface-adsorbed methanol was firstly oxidized to formaldehyde via a two-electron oxidation pathway, followed by spontaneous hydrolysis and methanolysis to methanediol and methoxymethanol, rather than methyl formate and formic acid that have been previously reported in gaseous CH_3OH photocatalysis [25].

Lee and coworkers investigated the effect of CN^- as a hole scavenger on the photocatalytic H_2 production from water [79]. Pelizzetti et al. also reported high activity in photocatalytic O_2 evolution with the presence of CN^- [80]. Although cyanide is harmful to the environment, CN^- ions can be easily transformed into innocuous OCN^- ion during photocatalysis. Galińska and Walendziewski investigated the effect of various sacrificial reagents on the photocatalytic efficiency of water splitting

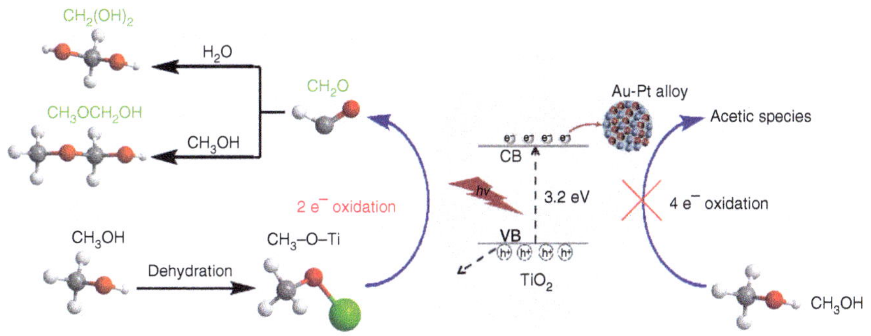

Figure 4.23 Methanol and its dissociative species underwent two-electron oxidation in photocatalytic H_2 production process over Au–Pt alloyed TiO_2 nanocomposites. Source: Al-Mayman et al. [25].

over Pt/TiO_2 [81]. Besides, Na_2S and EDTA were also employed as hole scavengers to achieve effective water splitting under illumination. Among the commonly used sacrificial agents, methanol was found to give the highest amount of H_2. Similar investigations were carried out by Nada et al. to study the influence of different electron donors on H_2 generation under UV irradiation using RuO_2/TiO_2 [82]. This study showed that EDTA achieved the highest H_2 evolution, followed by methanol, ethanol, and lactic acid. Several papers reported on the addition of carbonate salts to suppress backward reaction (recombination of H_2 and O_2 into H_2O). Sayama and Arakawa tested photocatalytic water splitting on Pt/TiO_2 with the addition of carbonate salts and observed significant enhancement of hydrogen and oxygen evolution stoichiometrically [83]. Particularly, addition of Na_2CO_3 while using Pt/TiO_2 was found to be effective for improving hydrogen and oxygen generation. The surface of the used Pt/TiO_2 catalyst was analyzed by IR spectroscopy, and various types of carbonate species (HCO_3^-, CO_3^-, $HCO\cdot_3$, and $C_2O_6^{2-}$) were found on the surface [84]. During the reaction, photogenerated holes are consumed by these carbonate species to form carbonate radicals, which is favorable for photogenerated electron–hole pair separation. The evolution of CO_2 and O_2 in this reaction may promote the desorption of O_2 from the surface of catalysts; therefore this would minimize the backward reaction of H_2 and O_2 to form H_2O. The desorbed CO_2 will be dissolved and then converted into HCO_3^-, followed by H_2 production.

On the other hand, other additives such as K_2CO_3, NaOH, NaCl, HCl, and Na_2SO_4 used in the photocatalytic H_2 production showed promising results [85]. Although the addition of carbonate salt can significantly enhance the photocatalytic H_2 generation, excess amount of carbonate salts could decrease the beneficial effect. This is because the adsorption of carbonate species onto the catalyst surface could reduce the light harvesting. Ye et al. explored the "phosphorylation" strategy for boosted hydrogen generation over the Pt-loaded carbon nitride photocatalysts [86]. They found that the addition of K_2HPO_4 agent in the reaction system could significantly enhance the photocatalytic activity for hydrogen generation, with an apparent quantum yield of 26.1% at 420 nm. No matter what sacrificial reagent and semiconductor photocatalysts they used, the promotion effect for hydrogen generation by K_2HPO_4

was obvious. The possibility that the originated hydrogen was from K_2HPO_4 was excluded by the isotopic analysis. Based on the linear sweep voltammetry studies and calculations, they attributed the enhanced production of hydrogen to the synergistic effect of the enhanced proton reduction and the improved hole oxidation.

4.7 Summary and Outlook

This chapter starts with the general introduction of photocatalytic hydrogen generation, followed by the basic photocatalytic process on the surface of TiO_2. TiO_2-based photocatalysts show the potential for the solar hydrogen generation by splitting water. However, despite their merits including low cost, high stability, and nontoxicity, the practical applications of TiO_2-based photocatalysts are hindered due to the large band gap and the high rate of electron–hole pair recombination. In order to overcome these two problems, various methods have been put forward, including doping of TiO_2 with non-metal and metal elements and sensitization of TiO_2 with dyes, metal clusters, and narrow-band-gap semiconductors. These methods can effectively extend the absorption edge of the TiO_2-based catalysts into the visible-light region, and thus a better use of solar energy could be expected. To inhibit the electron–hole recombination, cocatalysts are deposited on the surface to trap the photoinduced electrons/holes, thus prolonging the lifetime of the charge carriers. Besides, the deposited cocatalysts also provide the reactive sites to facilitate the redox reactions. Although the overall water splitting is feasible on the surface of TiO_2 catalysts in the presence of suitable cocatalysts, the quantum yield efficiency is extremely low. In such a case, sacrificial agents are often used to boost the efficiency. Some investigations have been conducted on hydrogen generation mechanism studies to achieve a better understanding of the reaction process and provide guidance for the design of more efficient catalysts in the future.

g-C_3N_4-based materials are widely studied for the photocatalytic H_2 generation under visible-light irradiation. A large number of approaches have been developed to improve the photocatalytic activity of bulk g-C_3N_4, including morphology control and surface modification. The crystal structure and electronic and optical properties are presented. In order to overcome the shortcomings of low visible-light response and fast electron–hole recombination, constructing heterojunction architectures of metal/C_3N_4 NTs, semiconductor/C_3N_4 NTs, and sensitizer/C_3N_4 NTs have been put forward. To further extend the absorption of visible-light response toward NIR light, sensitization is another typical strategy to efficiently harvest solar energy. Typically, an NIR-triggered C_3N_4-based heterojunction can be designed by incorporating UCNPs. Although numerous efforts have been put into the research on the photocatalytic hydrogen generation, there are still needs to prepare more efficient catalysts to achieve commercially acceptable quantum efficiency for solar H_2 generation. Besides, a better understanding of reaction pathways is essential on the photocatalytic hydrogen generation process, including the reaction mechanisms and the charge separation properties.

References

1 Dresselhaus, M.S. and Thomas, I.L. (2001). *Nature* 414: 332.
2 Potočnik, J. (2007). *Science* 315: 810.
3 Australian Government Chief Scientist, 2018, 59.
4 Fujishima, A. and Honda, K. (1972). *Nature* 238: 37.
5 Wang, X., Maeda, K., Thomas, A. et al. (2009). *Nat. Mater.* 8: 76.
6 Fajrina, N. and Tahir, M. (2019). *Int. J. Hydrogen Energy* 44: 540.
7 Chouhan, N., Ameta, R., Meena, R.K. et al. (2016). *Int. J. Hydrogen Energy* 41: 2298.
8 Y. Zhu, 2018. *Carbon Nitride Nanotube-based Photocatalysts for Solar to Chemical Energy Conversion.* Australia: University, Sydney.
9 Choi, W., Termin, A., and Hoffmann, M.R. (1994). *J. Phys. Chem.* 98: 13669.
10 Acar, C., Dincer, I., and Zamfirescu, C. (2014). *Int. J. Energy Res.* 38: 1903.
11 Yang, J., Wang, D., Han, H., and Li, C. (2013). *Acc. Chem. Res.* 46: 1900.
12 Ahmed, A.Y., Kandiel, T.A., Ivanova, I., and Bahnemann, D. (2014). *Appl. Surf. Sci.* 319: 44.
13 Serpone, N. and Emeline, A.V. (2012). *J. Phys. Chem. Lett.* 3: 673.
14 Roland, M. (2014). *Adv. Funct. Mater.* 24: 2420.
15 Lu, Q., Yu, Y., Ma, Q. et al. (2016). *Adv. Mater.* 28: 1917.
16 Thompson, T.L. and Yates, J.T. (2006). *Chem. Rev.* 106: 4428.
17 Chen, X. and Mao, S.S. (2007). *Chem. Rev.* 107: 2891.
18 Chen, C., Ma, W., and Zhao, J. (2010). *Chem. Soc. Rev.* 39: 4206.
19 Hernandez-Alonso, M.D., Fresno, F., Suarez, S., and Coronado, J.M. (2009). *Energy Environ. Sci.* 2: 1231.
20 Carp, O., Huisman, C.L., and Reller, A. (2004). *Prog. Solid State Chem.* 32: 33.
21 Ismail, A.A. (2012). *Appl. Catal., B* 117–118: 67.
22 Escobedo Salas, S., Serrano Rosales, B., and de Lasa, H. (2013). *Appl. Catal., B* 140–141: 523.
23 Si, Y., Cao, S., Wu, Z. et al. (2018). *Appl. Catal., B* 220: 471.
24 Ouyang, W., Muñoz-Batista, M.J., Kubacka, A. et al. (2018). *Appl. Catal., B* 238: 434.
25 Al-Mayman, S.I., Al-Johani, M.S., Mohamed, M.M. et al. (2017). *Int. J. Hydrogen Energy* 42: 5016.
26 Wang, F., Jiang, Y., Lawes, D.J. et al. (2015). *ACS Catal.* 5: 3924.
27 Su, R., Tiruvalam, R., Logsdail, A.J. et al. (2014). *ACS Nano* 8: 3490.
28 Gallo, A., Marelli, M., Psaro, R. et al. (2012). *Green Chem.* 14: 330.
29 Jiang, Z., Zhu, J., Liu, D. et al. (2014). *CrystEngComm* 16: 2384.
30 Ma, Y., Wang, X., Jia, Y. et al. (2014). *Chem. Rev.* 114: 9987.
31 Asahi, R., Morikawa, T., Ohwaki, T. et al. (2001). *Science* 293: 269.
32 Di Valentin, C., Pacchioni, G., and Selloni, A. (2004). *Phys. Rev. B* 70: 85116.
33 Meng, A., Zhu, B., Zhong, B. et al. (2017). *Appl. Surf. Sci.* 422: 518.
34 Li, Y., Wang, B., Liu, S. et al. (2015). *Appl. Surf. Sci.* 324: 736.
35 Petkovich, N.D. and Stein, A. (2013). *Chem. Soc. Rev.* 42: 3721.
36 Li, W., Wu, Z., Wang, J. et al. (2014). *Chem. Mater.* 26: 287.

37 Wu, A., Jing, L., Wang, J. et al. (2015). *Sci. Rep.* 5: 8858.
38 Liu, Y., Xie, L., Li, Y. et al. (2008). *J. Power Sources* 183: 701.
39 Sathish, M., Viswanathan, B., and Viswanath, R.P. (2006). *Int. J. Hydrogen Energy* 31: 891.
40 Ong, W.-J., Tan, L.-L., Ng, Y.H. et al. (2016). *Chem. Rev.* 116: 7159.
41 Wang, X., Blechert, S., and Antonietti, M. (2012). *ACS Catal.* 2: 1596.
42 Goettmann, F., Fischer, A., Antonietti, M., and Thomas, A. (2006). *Chem. Commun.*: 4530.
43 Lee, S.C., Lintang, H.O., and Yuliati, L. (2012). *Chem. Asian J.* 7: 2139.
44 Goettmann, F., Fischer, A., Antonietti, M., and Thomas, A. (2006). *Angew. Chem. Int. Ed.* 45: 4467.
45 Zhai, H.-S., Cao, L., and Xia, X.-H. (2013). *Chin. Chem. Lett.* 24: 103.
46 Mane, G.P., Talapaneni, S.N., Lakhi, K.S. et al. (2017). *Angew. Chem. Int. Ed.* (56): 8481.
47 Lakhi, K.S., Park, D.-H., Al-Bahily, K. et al. (2017). *Chem. Soc. Rev.* 46: 72.
48 Wang, Y., Wang, Y., Chen, Y. et al. (2015). *Mater. Lett.* 139: 70.
49 Guo, S., Zhu, Y., Yan, Y. et al. (2016). *Appl. Catal., B* 185: 315.
50 Li, K., Su, F.-Y., and Zhang, W.-D. (2016). *Appl. Surf. Sci.* 375: 110.
51 Tian, J., Liu, Q., Asiri, A.M. et al. (2015). *Sens. Actuators, B* 216: 453.
52 Liu, Y., Su, F.-Y., Yu, Y.-X., and Zhang, W.-D. (2016). *Int. J. Hydrogen Energy* 41: 7270.
53 Kheradmand, A., Zhu, Y., Zhang, W. et al. (2019). *Int. J. Hydrogen Energy* 44: 17930.
54 Zhu, Y., Marianov, A., Xu, H. et al. (2018). *ACS Appl. Mater. Interfaces* 10: 9468.
55 Guo, Y., Liu, Q., Li, Z. et al. (2018). *Appl. Catal., B* 221: 362.
56 Xiao, N., Li, S., Liu, S. et al. (2019). *Chin. J. Catal.* 40: 352.
57 Chen, P.W., Li, K., Yu, Y.X., and De Zhang, W. (2017). *Appl. Surf. Sci.* 392: 608.
58 Chem, J. M., 2012. *Journal of Materials Chemistry* 22: 15006.
59 Sun, S., Li, J., Cui, J. et al. (2019). *Int. J. Hydrogen Energy* 44: 778.
60 Chen, P., Xing, P., Chen, Z. et al. (2018). *Int. J. Hydrogen Energy* 43: 19984.
61 Zhu, Y., Wan, T., Wen, X. et al. (2019). *Appl. Catal., B* 244: 814.
62 Che, Y., Lu, B., Qi, Q. et al. (2018). *Sci. Rep.* 8: 16504.
63 Kailasam, K., Fischer, A., Zhang, G., Zhang, J., and Thomas, A. (2015). *Chem. Sus. Chem.* 1404.
64 Dao, V.-D., Le Chi, N.T.P., Van Thuan, D. et al. (2019). *J. Alloys Compd.* 775: 942.
65 Norouzi, O., Kheradmand, A., Jiang, Y. et al. *Int. J. Hydrogen Energy* 2019 https://doi.org/10.1016/j.ijhydene.2019.09.119.
66 Hao, J.-Y., Wang, Y.-Y., Tong, X.-L. et al. (2012). *Int. J. Hydrogen Energy* 37: 15038.
67 Zhu, Y., Zheng, X., Lu, Y. et al. (2019). *Nanoscale* https://doi.org/10.1039/C9NR05276C.
68 Chen, Y., Lu, C., Tang, L. et al. (2016). *Sol. Energy Mater. Sol. Cells* 149: 128.
69 Shi, W., Guo, F., Li, M. et al. (2019). *Sep. Purif. Technol.* 212: 142.
70 Zhu, M., Zhai, C., Fujitsuka, M., and Majima, T. (2018). *Appl. Catal., B* 221: 645.
71 Kuo, T.-R., Liao, H.-J., Chen, Y.-T. et al. (2018). *Green Chem.* 20: 1640.

72 Elbanna, O., Zhu, M., Fujitsuka, M., and Majima, T. (2019). *ACS Catal.* 9: 3618.
73 Liu, X.-Y., Zhang, G., Chen, H. et al. (2018). *Nano Res.* 11: 1379.
74 Jiang, J., Tang, X., Zhou, S. et al. (2016). *Green Chem.* 18: 2056.
75 Miwa, T., Kaneco, S., Katsumata, H. et al. (2010). *Int. J. Hydrogen Energy* 35: 6554.
76 Chen, J., Ollis, D.F., Rulkens, W.H., and Bruning, H. (1999). *Water Res.* 33: 661.
77 Kawai, M., Naito, S., Tamaru, K., and Kawai, T. (1983). *Chem. Phys. Lett.* 98: 377.
78 Wei, D., Jin, X., Huang, C. et al. (2015). *J. Phys. Chem. C* 119: 17748.
79 Lee, S.G., Lee, S., and Lee, H.-I. (2001). *Appl. Catal., A* 207: 173.
80 E. Pelizzetti, E. Borgarello, N. Serpone, M. Gratzel, in *Catal. Energy Scene* (Eds.: S. Kaliaguine and A. Mahay), Elsevier, 1984, pp. 327–334.
81 Galińska, A. and Walendziewski, J. (2005). *Energy Fuels* 19: 1143.
82 Nada, A.A., Barakat, M.H., Hamed, H.A. et al. (2005). *Int. J. Hydrogen Energy* 30: 687.
83 Sayama, K. and Arakawa, H. (1992). *J. Chem. Soc., Chem. Commun.*: 150.
84 Liu, H., Yuan, J., and Shangguan, W. (2006). *Energy Fuels* 20: 2289.
85 Ni, M., Leung, M.K.H., Leung, D.Y.C., and Sumathy, K. (2007). *Renewable Sustainable Energy Rev.* 11: 401.
86 Liu, G., Wang, T., Zhang, H. et al. (2015). *Angew. Chem. Int. Ed.* 54: 13561.

5

Photoelectrochemical Hydrogen Evolution

Zhiliang Wang and Lianzhou Wang

Nanomaterials Centre, School of Chemical Engineering and Australian Institute for Bioengineering and Nanotechnology, The University of Queensland, St Lucia, QLD 4072, Australia

5.1 Background of Photoelectrocatalytic Water Splitting

Powered by fossil fuel over two centuries, our society has made great achievements. However, as a cost, we face serious energy and environment crisis as well. The depletable fossil resource has endangered the society development with the energy shortage. The environment pollution caused by fossil fuel application has threatened human being's living condition. People have paid more and more attention to the renewable energy, including the solar energy, wind, tide, biomass, and so on. Compared with other candidates, solar energy is the most abundant and sustainable energy. The total solar power reaching the Earth's surface is about 120 000 TW, which is several thousand times the total consumption of our society (estimated to be 30 TW by 2030). This means that only 0.1% of the solar energy is enough to power our society. Since the solar energy can hardly be directly stored or used for our industry, we need to convert solar energy into a suitable energy carrier, and hydrogen (H_2) is the best choice [1, 2].

Hydrogen energy has the highest mass specific energy density (143 MJ kg^{-1}) and been the pursuit around the world [3] (https://www.energy.gov/eere/fuelcells/doe-technical-targets-hydrogen-production-photoelectrochemical-water-splitting; https://www.energy.gov/eere/fuelcells/hydrogen-production). For example, France has targeted on 20% energy demand contributed by H_2 energy in their perspective study of "Developing Hydrogen for the French Economy." Japan is ambitious to replace the traditional residential energy system with H_2 by 2030 according to their hydrogen strategy launched by the Ministry of Economy, Technology, and Industry (METI) on 2016. Hydrogen can be produced from natural gas reforming, biomass gasification, or water electrolysis (https://www.energy.gov/eere/fuelcells/hydrogen-production). In the viewpoint of technology readiness, natural gas reforming has come into commercial application, and water electrolysis is still at the R&D stages. However, due to the more and more serious issues of CO_2 footprint, solar-driven water splitting for H_2 production is always stated as the dreaming

Solar-to-Chemical Conversion: Photocatalytic and Photoelectrochemcial Processes, First Edition.
Edited by Hongqi Sun.

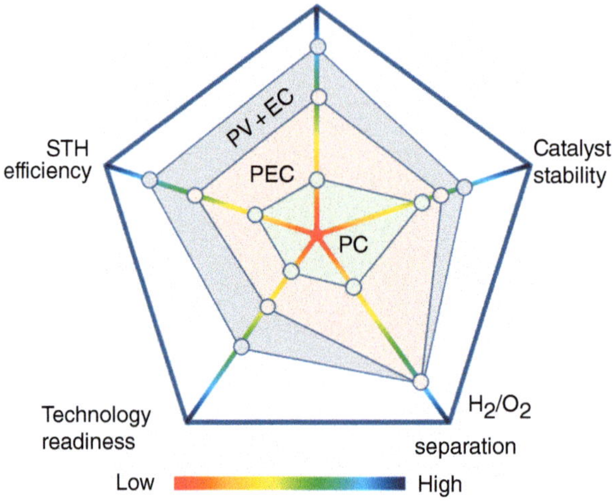

Figure 5.1 The schematic illustration of the features of H_2 production by PV + EC, PEC, and PC in the five aspects of cost, STH efficiency, technology readiness, H_2/O_2 separation, and catalyst stability.

technology in many governments' H_2 energy roadmap. A rough estimation of the technology indicates that 3.5×10^{13} l water is needed to supply H_2 for the world energy consumption in the form of H_2. This only accounts for 0.000 002% of the amount in the world.

Generally speaking, there are three ways to achieve solar-to-hydrogen (STH) energy conversion, which is the water splitting via the photocatalyst powder-based photocatalysis (PC), film-based photoelectrocatalysis (PEC), and photovoltaic-driven water electrocatalysis (PV + EC) [4, 5]. In terms of the cost, STH efficiency, stability, technology readiness, and so on, they have different features as shown in Figure 5.1. For the PV + EC process, it holds the highest technology readiness and STH efficiency since the Si photovoltaic (PV) has already been commercialized with an efficiency over 15%. It is very easy to apply the PV with the commercialized water electrolyzer for H_2 production. But the investment of PV + EC is very high due to the high price and energy penalty of PV panel. On the contrary, PC needs the lowest cost for the H_2 production system, but its STH efficiency is very low. Moreover, it also faces the challenges like H_2/O_2 separation, catalyst recycle, and so on. By comparison, the PEC process has the best balance between the cost and the STH efficiency. Due to the separated photoanode and photocathode, the H_2/O_2 separation is no longer an issue. Moreover, the semiconductors used here are similar to the case of PC, so the cost is not high. Meanwhile, sharing the same principle as the PV process, the STH efficiency is also higher than that of PC. Therefore, PEC-based hydrogen production has aroused worldwide attention.

As it is widely recognized for PEC process, the performance efficiency (η_{PEC}) is determined by three key criteria, including light absorption efficiency (η_{abs}), charge separation and transfer efficiency (η_{CST}), and surface reaction efficiency (η_{rea}) as described by Eq. (5.1): [6, 7]

$$\eta_{PEC} = \eta_{abs} \times \eta_{CST} \times \eta_{rea} \tag{5.1}$$

The η_{abs} indicates how many photons can be absorbed for charge generation. It is largely determined by the band gap of semiconductors. Meanwhile, it is also affected by other factors, e.g. nanostructure of the photoelectrode, film thickness, and even the electrode substrates. The η_{CST} reveals how many effective charges can be generated for surface reaction. It is a good merit to show the charge recombination in the photoelectrode. The η_{rea} shows the percentage of separated charges involved in surface reaction. It reveals the electrocatalytic capability of the semiconductor surface or loaded cocatalyst. By choosing suitable hole or electron scavenger, it is easy to investigate these three parameters. For example, H_2O_2 is widely used as the hole scavenger in photoanodes like hematite, Ta_3N_5, TiO_2, etc. [7–10] Na_2SO_3 is normally used in $BiVO_4$-based photoanode as the scavenger, [11–13] while $K_2S_2O_8$ has been applied as electron scavenger in CuO photocathode [14]. In the presence of scavenger, the surface reaction efficiency is regarded as unit (i.e. $\eta_{rea} = 100\%$) due to the favorable surface reaction kinetics with the scavenger reagent. So the charge separation efficiency and the surface water splitting reaction efficiency can be calculated accordingly.

Moreover, the three steps have a huge time scale difference as shown in Figure 5.2. For example, light absorption can be completed within 10^{-10} s by transferring energy from photons to "hot" electron–hole excitons. After 10^{-7}–10^{-3} seconds, the charge separation and transfer happened; meanwhile large amount of electron–hole pairs is recombined again during the cooling-down process. For the surface reaction, it is even slower at the time scale of ms ∼ s, as a result of which many generated electron–hole pairs are trapped by the recombination centers [7, 15]. It can be found that the rates of light harvest, separated photocharge generation, and surface reaction do not match with each other at all, which is the major loss of the solar conversion efficiency. How to balance the time scale between these steps is of great significance for achieving efficient PEC water splitting. Among the three steps, the charge separation and transfer is the keystone to bridge the charge generation and consumption. Therefore, optimizing this step plays a critical significance in the PEC solar hydrogen production. In the following part, we will discuss about how to accelerate the charge separation and transfer in photoelectrode and address on how to balance the sluggish surface reaction with the photocharge generations.

5.2 Mechanism of Charge Separation and Transfer

The primary driving force for charge separation and transfer comes from the band bending. Lots of experts have developed rigorous reviews to explain the principle of band bending in detail, including Yates [16], Koval [17], Morrison [18], Nozik, [19, 20] and so on. Herein, we will take the simplest model to provide a qualitative explanation based on semiconductor–electrolyte interface (SEI). When

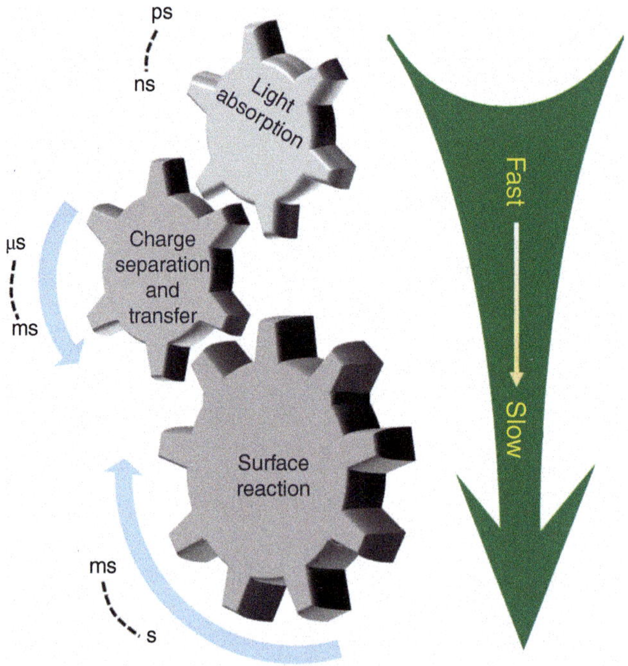

Figure 5.2 The schematic illustration of three steps in PEC water splitting process. Source: From Wang and Wang [7]. © 2018 Elsevier.

the semiconductor contacts with the electrolyte, the work function (Φ) difference between the semiconductor and the electrolyte solution will lead to charge exchange at the SEI. Taking the p-type semiconductor for an example, it normally has higher work function than the electrolyte. Therefore, the electrons will pass from the electrolyte to the semiconductor (or in other words, holes inject into the electrolyte), resulting in a negatively charged layer adjacent to the electrolyte as shown in Figure 5.3a–d. In this situation, an electric field directed from the bulk toward the surface is built up gradually to prevent this charge exchange. When the Fermi level of the semiconductor equals to that of the electrolyte solution, it reaches a new balance at the SEI, and the region with the built-in field is called space charge layer (Figure 5.3c,d). Since the charge density in the solution is much larger than that in the semiconductor, the space charge layer majorly exists in the semiconductor surface as shown in Figure 5.3c. When the SEI is stimulated under illumination, photogenerated electrons and holes will be produced in the space charge layer (Figure 5.3e,f) and are further swiped toward the surface for reaction (electrons) or bulk for collection (holes), with below hypotheses applied:

(i) Perfect interface without surface states.
(ii) Completely ionized acceptors (p-type semiconductor) with homogeneous distribution at the density of N_a within the space charge layer.
(iii) There is no net charge beyond the space charge layer.

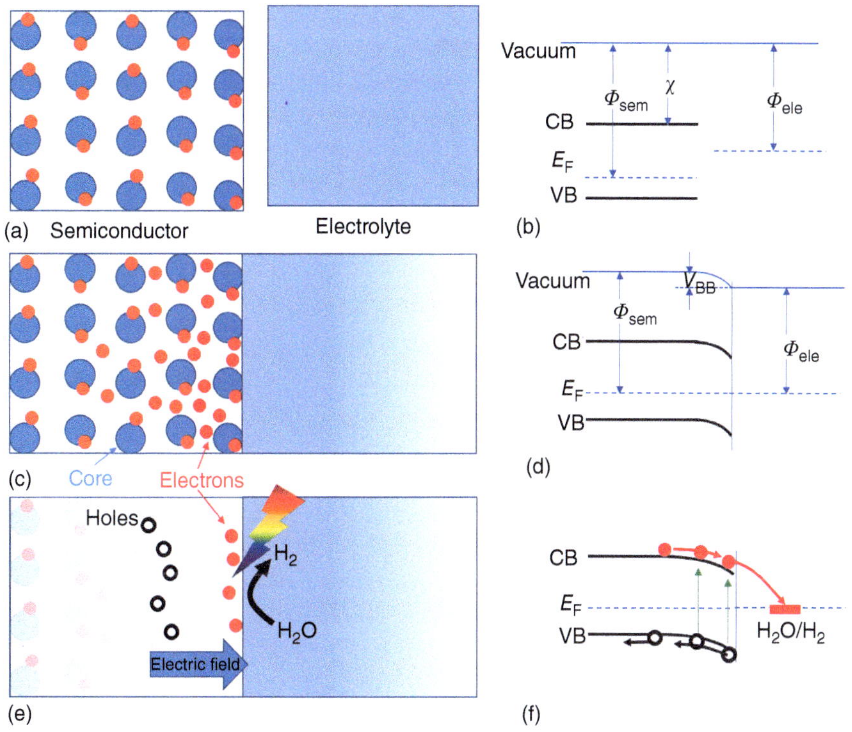

Figure 5.3 The illustration of semiconductor–electrolyte interface and the corresponding band bending. (a, b) The scheme and energy level of semiconductor and electrolyte in vacuum. The electrons are uniformly distributed around the core of atoms (blue dot). (c, d) The schematic illustration of electron distribution and the energy band bending when SEI is built. The electrons (red dot) are accumulated at the SEI interface. (e, f) The schematic illustration of electron–hole separation and transfer under illumination.

It has been able to derive the built-in electric field (E_{bi}) and the width of space charge layer (D) expressions as follows: [16]

$$E(z) = \frac{eN_a}{\varepsilon_0 \varepsilon_r}(D - z) \quad (0 \leq z \leq D) \tag{5.2}$$

$$D = \left[\frac{2\varepsilon_0 \varepsilon_r V_{BB}(0)}{eN_a}\right]^{1/2} \tag{5.3}$$

wherein e is the elemental charge, ε_0 and ε_r are the dielectric constant of vacuum and the relative one of the semiconductors, z is the depth from the surface, and $V_{BB}(0)$ is the band bending at the surface ($z = 0$).

This model provides a simple but useful understanding of the photogenerated charge separation in the space charge layer and the charge transfer at the SEI. According to Eq. (5.2), we can conclude that extending the space charge layer (D) or improving the carrier concentration (N_a) can improve the built-in electric field, therefore enhancing the charge separation in the semiconductor.

However, for a complete PEC process, it is also very important to consider the photocharge collection by the conductive substrates (e.g. holes in p-type semiconductor) so as to close the circuit for a complete water splitting (i.e. electrons for water reduction and holes for water oxidation). Otherwise, there will be serious back-recombination between the uncollected holes and the electrons [21]. Therefore, it requires that the photoelectrode has a good conductivity for charge transfer as well, whose significance has been underestimated.

Theoretically, there are two ways for charge transfer according to the driving force difference: drift in an electric field (ruled by Coulomb's law) and diffusion by concentration difference (ruled by Fick's law). In the bulk of semiconductor photoelectrode, it is normally treated as an electric-field-free region, which will lead to flat energy band in the bulk. In this case, the only driving force for charge transfer comes from diffusion caused by the carrier concentration difference between the bulk and surface region, and the charge transfer efficiency heavily depends on the carrier diffusion length. Table 5.1 summarized the diffusion length of the frequently applied semiconductors in PEC research [22]. For comparison, some typical PV materials are presented in the table, including CuZnSnSe (CZTSe), Si, and perovskite organic lead halides ($MAPbI_3$, $MAPbI_{3-x}Cl_x$). These PV semiconductors all show much longer diffusion length than the metal oxide used in PEC. Furthermore, comparing the diffusion length with the light penetration length, it reveals that only a small part of the absorbed light can be used for generated effective photocharges. When the photocharges generated deeper than their diffusion length, they have few chances to survive for the surface reactions. So we need to take the thickness of the film into consideration to balance the charge transfer and light harvest. However, this discussion above is under the assumption that the static potential is equal everywhere within the bulk, which is suitable for conductor (e.g. metals) but not accurate for semiconductor. Due to the low carrier concentration and low dielectric screen of the electric field, the potential drop on the semiconductor film is unavoidable, and the electric field can exist in the bulk region as well. Therefore, the energy band in the bulk will change slightly, providing some drift force for charge transfer. In this case, it is highly desirable to create a high-speed way (i.e. low resistance) for charge transfer. In the following part, we will briefly introduce how we can improve the charge transfer in semiconductor film for efficient PEC water splitting.

5.3 Strategy for Improving Charge Transfer

There are two kinds of semiconductor films for PEC application: the continuous film and the particulate film. For the continuous film, the electrode is normally fabricated in a bottom-up process (e.g. hydrothermal, chemical vapor deposition, physical vapor deposition, etc.) where the semiconductor film is epitaxially grown on the conductive substrate. For the particulate film, the electrode is fabricated from the well-prepared semiconductor particles via a post-film formation process (e.g. doctor blading, electrophoresis, participle transfer, etc.). The two kinds of films need different strategies to improve the charge transfer. In the case of continuous film, the

Table 5.1 The diffusion length and light penetration of some typical semiconductors.

Semiconductor	Diffusion length	Light penetration	References
WO_3	Hole ~700 nm	~7000 nm (λ = 440 nm)	[23, 24]
TiO_2	Hole ~120 nm	250 nm (λ = 380 nm)	[25]
Fe_2O_3	Hole 2–4 nm	118 nm (λ = 550 nm)	[26]
$BiVO_4$	Hole ~100 nm Electron ~10 nm	~500 nm (λ = 530 nm)	[27]
Ta_3N_5	Hole ~18 μm	N.A.	[28]
CdS	Hole ~1300 nm	~62 nm (λ = 500 nm)	[29]
CZTSe	Electron ~2000 nm		[30]
Si	Hole 168 μm	2–3 μm (λ = 650)	[23, 31]
$MAPbI_3$	Hole ~100 nm Electron ~120 nm	N.A.	[32]
$MAPbI_{3-x}Cl_x$	Hole ~1000 nm Electron ~1290 nm	N.A.	[32]

Source: From Xiao et al. [22]. © 2018 John Wiley & Sons.

crystal grains have intimate contact with each other, so the intrinsic conductivity of the semiconductor will dominate the charge transfer in the photoelectrode. For the particulate film, the particle connection is loosened and requires more effort to improve the charge transfer among the particles and at the interface of particle/substrate.

5.3.1 Improving the Charge Transfer in Continuous Film

For the photoelectrode with continuous film, the intrinsic conductivity of the semiconductor plays a key role in the charge transfer. Therefore, it is beneficial to develop strategies to improve the carrier concentration for increasing conductivity. The most widely applied method is elemental doping, which can drastically change the carrier concentration. For example, in hematite system, the dopants of Ti, Pt, Nb or Ta, etc. have been reported to improve the carrier concentration by one to two order of magnitudes [33, 34]. Similar phenomena were observed in W- or Mo-doped $BiVO_4$ [35], Ag-doped copper-based chalcogenides (e.g. CIGS, CGSe), and so on [36, 37]. Besides the foreign element doping, it is possible to create some self-doping by conducting the defect formation process.

Oxygen vacancies are one of the most intensively investigated dopants in metal oxide photoanodes. During the oxygen vacancy formation, there will produce extra electrons to compensate the charges follow Eq. (5.4): [38, 39]

$$2O_O \rightarrow O_2 + V_O'' + 2e^{\cdot} \tag{5.4}$$

In Fe_2O_3 photoanode, the influence of oxygen vacancies has been scrutinized by precisely controlling the oxygen vacancy concentration in the photoanode [39]. In Figure 5.4a, it is shown that the carrier concentration is gradually increased by prolonging the N_2 treatment of the hematite photoanode. However, it is found that the photocurrent improved first and then decreased when the carrier concentration goes

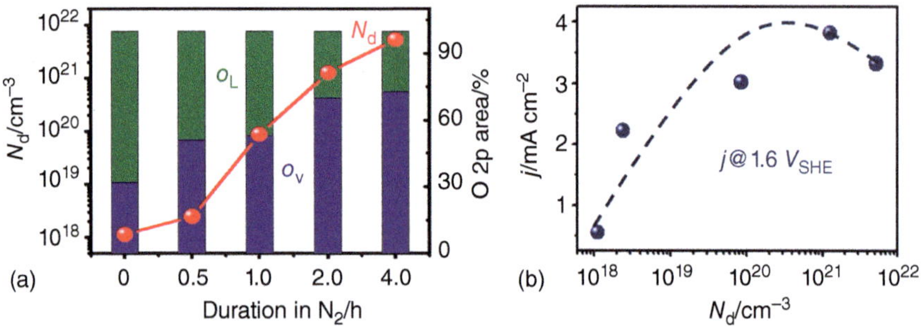

Figure 5.4 (a) The oxygen vacancy concentration change by adjusting the treatment duration in N_2. (b) The volcano relationship between photocurrent and oxygen vacancy concentration. Source: Wang et al. [39]. © 2019 Willey.

too high (Figure 5.4b). This is due to the oxygen vacancy-induced recombination in the photoelectrode. Even though the oxygen vacancies have improved the charge transfer, they also induce the recombination centers, which may drastically annihilate the photogenerated electrons and holes. This leads to the volcano-shape curves of the carrier concentration–photocurrent relationship.

Besides the doping strategies, there are also some reports on reducing film thickness to decrease the length of charge transfer pathway. As a trade-off, the light harvest will decrease for a thin film. Therefore, it needs a rational design of the microstructure of the electrode to increase the light penetration. Qiu et al. reported the deposition of ultrathin Fe_2O_3 film on the nanospike conductive substrate, which achieved excellent charge collection for the high PEC performance [40–42]. Similar strategies have been applied for $BiVO_4$ photoelectrode fabrication [43].

In addition, nanostructure engineering is also widely applied in improving the charge transfer in the photoelectrode. For the three-dimensional continuous film, the charge transfer through the film is relatively random, which will lead to low charge collection efficiency. Alternatively, when the semiconductor is fabricated into one-dimensional nanorod or two-dimensional (2D) nanoflake, the charge transfer will be restricted along some special direction due to the quantum confinement. This concept has been verified by ZnO where the ZnO nanorod photoelectrode always shows superior charge transfer than the ZnO film electrode.

5.3.2 Improving the Charge Transfer in Particulate Photoelectrodes

The particulate photoelectrode arouses researchers' interests since it allows to apply the highly crystallized semiconductor powder as the photoresponse layer for the photoelectrode [44]. Normally, it needs a high temperature and controlled atmosphere to prepare well semiconductor crystals, which is too harsh for the conductive layers (e.g. metal foil, fluorine-doped tin oxide) to keep good conductivity. For example, the well-crystallized $SrTiO_3$ is normally prepared over 1000 °C in the air. It is almost impossible to find any suitable substrates to process this in situ $SrTiO_3$ photoelectrode fabrication due to the very high temperature. The particulate

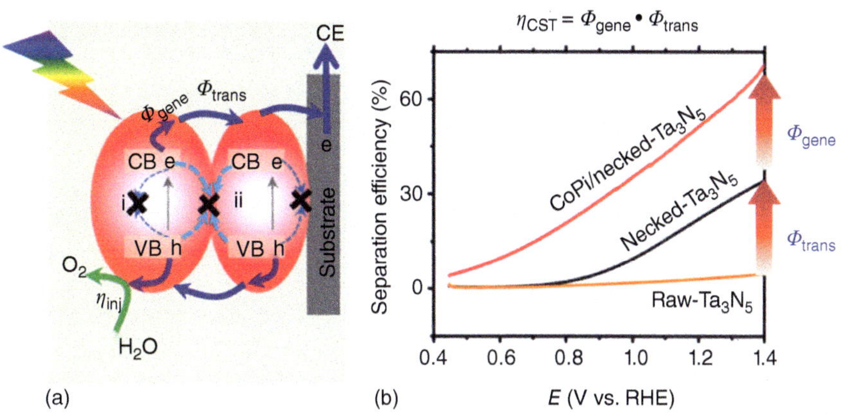

Figure 5.5 (a) The schematic charge transfer in particulate Ta_3N_5 photoanode. (b) The change of charge separation and transfer efficiency for Ta_3N_5 photoanode when the charge transfer and generation are improved gradually. Source: Wang et al. [45]. Licensed under CC BY 3.0 Unported.

photoelectrode is usually fabricated via a top-down process by immobilizing the well-synthesized semiconductor particles on the conductive layer. This fabrication process will create large amounts of interfaces, including the particle–particle interface and particle–substrate interface. How to strengthen the contact at these interfaces is the key concern from the particulate photoelectrode fabrication since the charge transfer at the interface will affect the performance of the photoelectrode significantly.

Necking treatment is reported to be an effective post-treatment strategy [45, 46]. Taking the Ta_3N_5 as example, Wang et al. reported that a Ta_3N_5 particulate film can be strengthened by coating with diluted $TaCl_5$ solution and treated at a relatively low temperature. The photocurrent of necked Ta_3N_5 is profoundly improved compared with the raw Ta_3N_5 photoelectrode [45]. In this research, they also proposed that the apparent η_{CST} is actually a production of charge generation efficiency in the semiconductor and the transfer efficiency at the interfaces, highlighting the significance of charge transfer in the PEC process as shown in Figure 5.5a. The necking treatment can improved the charge transfer efficiency profoundly, which leads to obvious η_{CST} increase as shown in Figure 5.5b. Moreover, passivating the surface recombination centers with CoPi to improve the change generation efficiency in the particles can improve the η_{CST} further. The effectiveness of necking treatment in improving the charge transfer has also been confirmed in other particulate electrode systems, e.g. TaON, $BaTaO_2N$, and so on [46, 47].

Moreover, Domen's group has developed a so-called particle transfer method to solve the contact issue at the particle–substrate interface [47, 48]. A monolayer of semiconductor crystals is intimately embedded with the conductive layer to ensure the facile charge transfer as well as to avoid the challenge from the particle stacking. A metal contact layer (e.g. Au, Ni) between the semiconductor particles and the conductor layer is applied to adjust the work function difference. The particles used in this method are usually at large crystal size (~μm), which makes the film performed like a single-crystal layer. Due to this reason, the charge recombination among different particles is eliminated, and the highly crystallized particles ensure the good charge transfer in a single crystal. This method has been applied for fabrication of many different photoelectrodes, e.g. $SrTiO_3$ [49], $BiVO_4$ [50], $LaTiO_2N$ [51], chalcogenide [52, 53], and so on.

Besides the abovementioned methods, it is reported that changing the atmosphere during photoelectrode fabrication can tailor the charge transfer as well. During the fabrication of GaN:ZnO photoanode, it is found that the GaN:ZnO photoanode prepared by traditional nitridation shows very poor particle connection and the charge transfer in the film is also very poor. By comparison, when there exists water vapor in the ammonia, it can enhance the particle connection in the film [54]. Benefiting from the moisture-assisted nitridation process, the GaN : ZnO photoanode can achieve a photocurrent of 4 mA cm^{-2}, which is far better than the one prepared from traditional method [55].

Above all, it can be concluded that the charge transfer is an essential criterion for fabricating effective photoelectrode. On the one hand, the semiconductor crystals need to have high quality to ensure a large charge diffusion length in the particles. On the other hand, the particles need to have intimate intercollection to ensure the generated charges can transfer through the thick films.

5.4 Strategy for Improving Electron–Hole Separation

An efficient photoelectrode requires not only good charge collection but also highly efficient electron–hole separation in the crystal. The charge separation efficiency is determined by the electric field intensity (E_{bi}) in the space charge layer as shown in Eq. (5.2). Therefore, the strategies to enhance the E_{bi}, such as formation of junction, crystal facet engineering, and improving carrier concentration, can effectively promote the charge separation. Alternatively, due to the competition of charge recombination, the ways to suppress the recombination centers can also result in significant improvement of charge separation. In the following sections (5.4.1–5.4.3), we will discuss them in detail.

5.4.1 Heterojunction Formation

Creating another junction besides the semiconductor–electrolyte junction has been verified to be an effective strategy for promoting the charge transfer in the photoelectrode. For example, the $BiVO_4/WO_3$ heterojunction has shown superior charge separation to the pure $BiVO_4$ or WO_3 [35]. Due to the Fermi level difference between the two contacted semiconductors, there exists another space charge layer at the interface, which will provide additional driving force. Especially, when the two semiconductors are p-type and n-type semiconductor, respectively, the electric field can be as strong as 10^4 V cm^{-1}, which can efficiently separate the photogenerated electron–hole pairs. Normally, people also call it as buried junction in photoelectrodes. This includes the p–n Si junction [56], Cu_2O-based junction (e.g. Cu_2O with TiO_2, Ga_2O_3, or ZnO) [57–59], and chalcogenide-based junction (e.g. CdS with CuZnSnS, CuInGaSe, CuGaSe, etc.) [29, 36, 37, 60–62]. Without the buried junction, the Cu_2O or chalcogenides mentioned above show very weak charge separation since the semiconductor–electrolyte junction cannot provide sufficient driving force for this process. Once the buried junction is built, the photocurrent is boosted. For example, the pure Cu_2O photocathode shows very limited photocurrent for water reduction. But Luo et al. found that the coating with Ga_2O_3 can lead to the formation of well-matched p–n junction with an open circuit over 1 V [58]. Therefore, they are able to fabricate high-performance Cu_2O-based photocathode with the solar hydrogen production efficiency over 3%. However, it should be noted that the match between different semiconductors is critical for the effective junction formations. The lattice constant and work function are the two most significant parameters that should be taken into consideration. For example, even though many different junctions have been reported between $BiVO_4$ and other metal oxides (e.g. Fe_2O_3, TiO_2, ZnO), WO_3 is always the best one matching with it.

For all the above junctions, they are localized junction with limited interaction distance. Beyond the space charge layer, it still faces the driving force issue due to the flat band in the bulk of semiconductors. Nowadays, people attempt to solve this problem by creating the gradient doping to the semiconductors [56, 63, 64]. The dopant concentration is believed to change the Fermi level in the semiconductor. With the gradient dopant concentration, the Fermi level will change continually and finally will lead to the formation of bulky band bending. This concept has been confirmed in W-doped $BiVO_4$ photoanode [56]. The W concentration is changed from 0 to 4 wt% in a depth profile, leading to a gradient junction formation in the $BiVO_4$. In hematite, the gradient structure has resulted in an unpredicted low onset potential of this PEC water splitting [64].

5.4.2 Crystal Facet Control

Besides the formation of junction, the exposed crystal facet is also significant in determining the charge separation. Due to the electron affiliation difference of different crystal facets, the photogenerated electrons and holes will accumulate on difference facets. Therefore, tuning the exposed facet during PEC process can significantly change the PEC performance. Taking $BiVO_4$ as example, it has been widely recognized that the (110) and (010) facets are the hole and electron accumulation facet, respectively, for $BiVO_4$ crystal [65–67]. When fabricating $BiVO_4$ electrode, exposing the hole-accumulated facet will ensure the efficient charge consumption during PEC process. Han et al. have achieved (001) facet-exposed $BiVO_4$ photoelectrode by the epitaxial growth on SnO_2 [68]. The photocurrent of the [001] oriented $BiVO_4$ photoanode was reported to be 16 times higher than the randomly oriented $BiVO_4$ photoanodes. Similar facet promoting effect is also reported in WO_3, TiO_2, and so on. Moreover, for some polarized semiconductors (e.g. ZnO, GaN), they usually have an axial of polarization, along whose direction it can generate a polarized electric field. By selectively controlling the exposed facets, it is possible to control the direction of the polarization in the photoelectrode and provide additional electric field for charge separation.

For a special case of selectively exposed facet, the 2D structure is also appealing in PEC charge separation. The reducing of one dimension can confine the charge transfer and avoid the potential bulk charge recombination [69–71]. For example, hematite has been reported to endure very serious bulk recombination due to the short hole diffusion length (2–4 nm). But when the hematite photoanode has the 2D nanostructure to facilitate the hole transfer, the charge separation is promoted, and the PEC response is improved profoundly [70]. Moreover, Butburee et al. have fabricated a porous 2D TiO_2 photoanode with ultrathin TiO_2 single crystal as shown in Figure 5.6a [71]. This porous 2D structure largely facilitates the charge separation. Moreover, the additional branches at the edge of the 2D flake facilitates the surface reaction greatly, which ensures a near-ideal PEC response of TiO_2 with very low onset potential and extremely steep photocurrent increase during the potential scan (Figure 5.6).

The 2D structure shows advantage in reducing charge diffusion length. As a trade-off, it also has a demerit of decreasing the width of space charge region, which is not good for charge separation. In order to compensate the driving force loss, 2D structured heterojunction is a good choice to combine the advantages of heterojunction and 2D structures. This concept was verified by a SnO_2/SnS_2 photoelectrode. The SnO_2 nanoflake was further deposited with SnS_2 to fabricate a type II heterojunction structure. The 2D SnO_2/SnS_2 heterostructure shows a much better PEC performance than the bare 2D SnS_2 photoanode due to the improved charge separation.

5.4.3 Surface Passivation

In order to provide more separated photogenerated charges, we need to not only improve the driving force for charge separation but also suppress the charge recombination. One major source of charge recombination comes from the surface trap states. Due to the adsorbed surface species, unsaturated coordination (e.g. dangling bonds), and defects, there exist many states that can trap photogenerated electrons and holes, leading to the charge recombination [72]. Fortunately, it is possible to passivate these trapping states by different methods.

The most widely used strategy is to deposit a passivation layer using TiO_2, Al_2O_3, Ga_2O_3, etc. [73, 74] With these large band-gap metal oxide, it is possible to depress the charge recombination in Fe_2O_3, Cu_2O, and Si photoelectrode. But considering the limited conductivity of these metal oxides, it requires the passivation layer to be thin enough to ensure the tunneling of photocharges.

Besides the passivation layer, it has been reported to realize surface passivation through chemical treatment. For example, it has reported that the UV light/ozone treatment is very efficient in removing the surface trap states in ZnO [75]. This passivation is explained with the removal of surface dangling –OH group when interacting with ozone. Similarly, using concentrated H_2O_2 treatment is also reported to realize surface passivation in TiO_2 and hematite photoanodes [8, 76]. Furthermore,

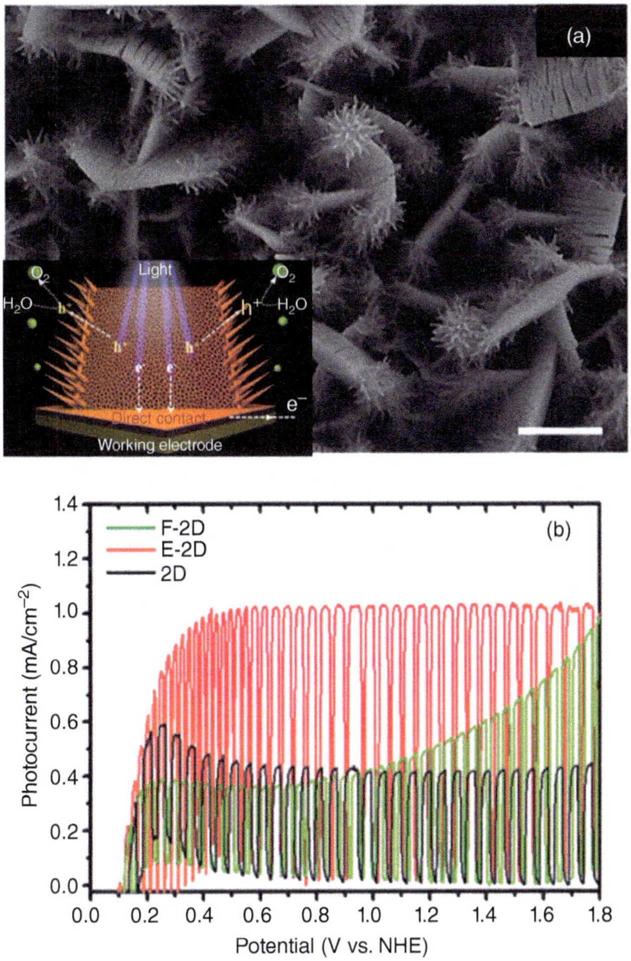

Figure 5.6 (a) The image and (b) the photoresponse (red curve) of branched 2D porous TiO_2 single-crystal nanosheet photoelectrode. Insert (a) illustrates the photocharge transfer during PEC process. Source: Butburee et al. [71]. Reproduced with the permission from Wiley.

treating the metal oxide or metal oxynitride in strong acid is able to remove the surface defect as well. For example, it is found that when treating hematite in HCl acid, the photocurrent of hematite photoelectrode will improve and the onset potential will decrease [8]. It is believed that the acid treatment can remove the surface defective layer, resulting in the suppressed charge recombination.

Interestingly, it also has reported that some surface states have close relationship with the surface reactions where they may work as intermediates during the reactions. Wang et al. reported that the surface state passivation strategies, like ALD Al_2O_3 coating and H_2O_2 treatment, actually significantly change this surface state distribution (C_{ss} in Figure 5.7a) [8]. But this distribution can be exactly predicted with the j–V curves (dj/dV) of the PEC response. The strong correlation between

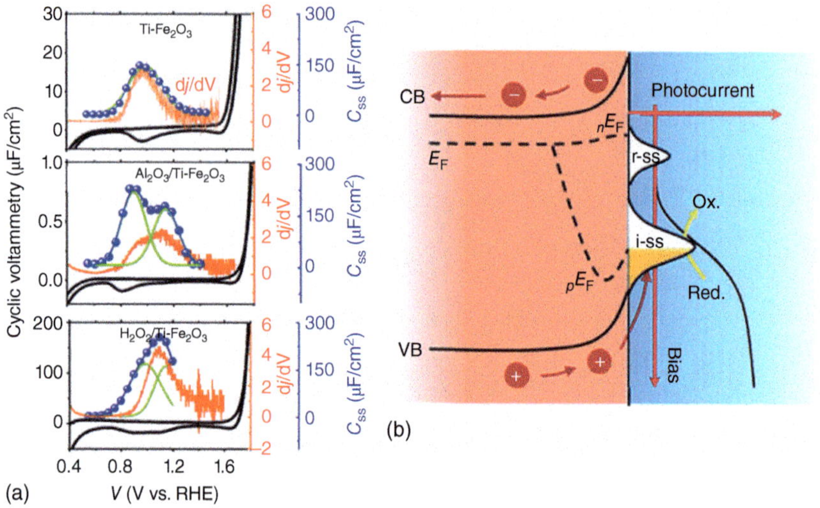

Figure 5.7 (a) The surface state distribution characterization by cyclic voltammetry (CV) and the surface state related capacitance (C_{ss}). The different hematite photoelectrodes are compared including Ti doping, Al_2O_3 passivation, and H_2O_2 treatment. (b) The schematic illustration of PEC reaction via the surface states as intermediates. Source: From Wang et al. [8]. © 2016 Royal Society of Chemistry.

C_{ss} and dj/dV suggests that the actual PEC process will be highly possibly happened via the surface states as reaction intermediates as shown in Figure 5.6b. However, in this research, they also pointed out that the surface states may have different intrinsic property (recombination center or reaction intermediate) due to their energy difference. These results suggest that it needs to be careful when dealing with the surface states.

5.5 Surface Cocatalyst Design

The surface reaction is the last step for solar conversion, and it heavily depends on the surface property of the semiconductors. For most semiconductors, the surface shows very poor catalytic capability for reaction due to the high overpotential for surface reaction [7]. In order to improve the surface reaction kinetics, the cocatalyst is an essential part for photoelectrode design. For example, due to the hydrophobic surface property of Ta_3N_5 photoanode, the raw electrode shows a very poor PEC response. However, after the loading of hybrid cocatalyst, e.g. NiOOH/ferrihydrite (Fh) and Ir and Co molecular catalyst on Ta_3N_5 photoanode, its photocurrent could approach the theoretical 12.6 mA cm^{-2} under 100 mW cm^{-2} simulated sunlight [77].

Actually, due to the complexity of surface water splitting reaction, the cocatalyst design is still a trial-and-error process in order to match the cocatalysts with the semiconductors efficiently. Basically, the electrocatalysts for hydrogen/oxygen evolution reaction (HER/OER) are the potential candidates as cocatalysts in PEC process, for example, the Ni-, Co-, and Fe-based oxide, hydroxide, layered double

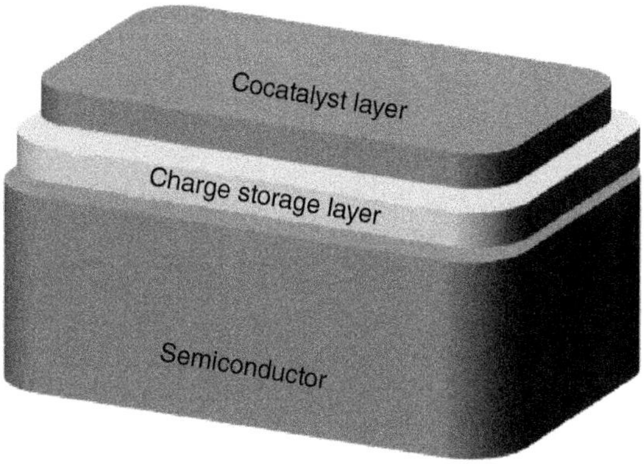

Figure 5.8 The schematic illustration of cocatalyst design strategy based on a charge storage layer between the semiconductor and the effective cocatalyst layer.

hydroxide (LDH), and phosphide are the most frequently used catalysts in water oxidation, and Pt and No-Mo alloy for water reduction.

There are two ways to evaluate the effectiveness of a cocatalyst: (i) based on the overpotential of current at dark (j_{dark}) and light (j_{light}) and (ii) based on the surface reaction efficiency (η_{rea}). Due to the catalytic feature, the overpotential of j_{dark} and j_{light} will dramatically decrease once loading the cocatalysts. It should be mentioned that the decreased overpotential of j_{light} does not guarantee the effect of cocatalyst since the passivation of surface trap states can also lead to lower overpotential for photoresponse. Therefore, it is important to investigate cocatalyst based on both the j_{dark} and j_{light}. Another empirical value, η_{rea}, is a good figure of merit to investigate the surface cocatalyst efficiency. By taking scavenge in the electrolyte (j_{scav}), it is assumed an unit surface reaction efficiency (100%). Then the surface reaction efficiency for water splitting reaction can be derived by a ratio, $(j_{light}-j_{dark})/j_{scav}$.

After many years of intensive investigations of cocatalyst–semiconductor relationship, people have achieved some fundamental understanding. The researchers from Boettcher's group have shown the potential drop on different cocatalysts. They found that the ionic permeable cocatalyst (e.g. NiOOH) can form an adaptive Schottky junction with the semiconductor, resulting in more efficient semiconductor–catalyst charge transfer compared with the case in dense cocatalyst (e.g. IrO_2) where a buried junction with constant Schottky barrier is formed [78]. Additionally, in order to match with the sluggish surface reaction, Li's group has proposed a cocatalyst design strategy: adding another charge storage layer between semiconductor and the surface catalyst as indicated in Figure 5.8 [79]. The charge storage layer can capture the vast amount of photogenerated charges first and further pass the charges to the more active cocatalyst. With Ta_3N_5 as an example, ferrihydrite was found to be a good hole storage layer to fast capture the photogenerated holes from Ta_3N_5 matrix and transfer the holes to the surface cocatalyst (e.g. Co_3O_4, Ir-based molecular catalyst) for further water oxidation. This design strategy is found to be very effective in achieving high photocurrent and long stability for Ta_3N_5 photoanode. Similar strategy is applied in designing efficient hematite photoanode.

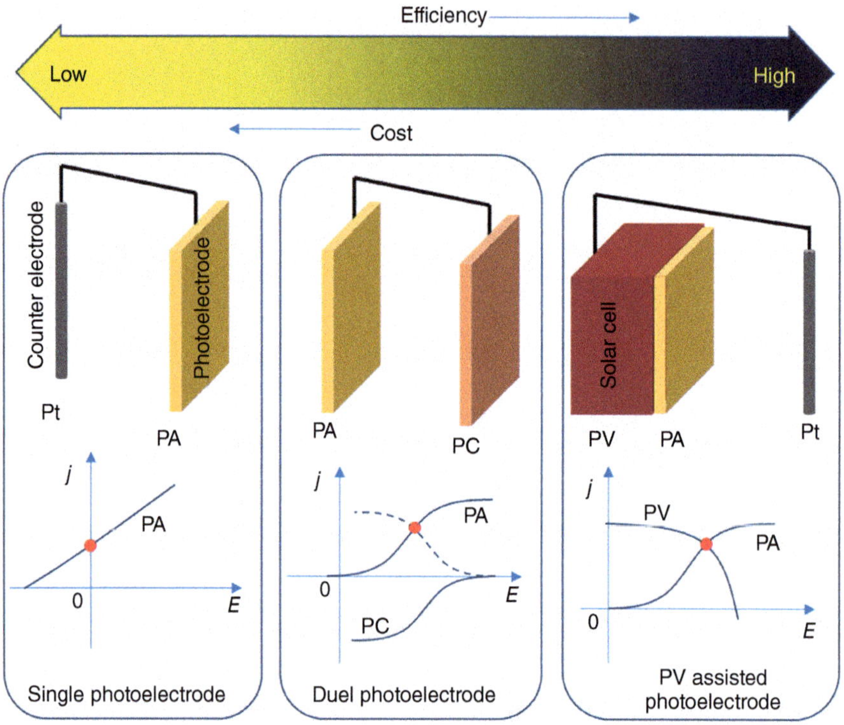

Figure 5.9 The schematic illustration of three different types of unbiased PEC water splitting systems. PA, PC, and PV represent photoanode, photocathode, and photovoltaic, respectively. The *j*–*E* curves represent their photoresponses in a three-electrode system. The intersections (red dots) indicate the unbiased operation points.

5.6 Unbiased PEC Water Splitting

The final goal of STH conversion is to achieve unbiased water splitting where spontaneous HER and OER occur without applying external bias. Therefore, in practice, we can observe significant photocurrent at 0 V bias in a two-electrode system. There are three configurations for achieving unbiased water splitting as shown in Figure 5.9: single photoelectrode, dual photoelectrode, and PV-assisted photoelectrode [80].

For the water splitting based on single photoelectrode, it requires the band gap to be large enough for HER and OER. The pioneer attempt based on TiO_2 single crystal has shown the feasibility of the single photoelectrode for unbiased STH conversion process [81]. Recently, Ebaid et al. have reported the success of unbiased over all water splitting based on III–V nitride ($In_{0.33}Ga_{0.67}N$) photoelectrodes [82].

Even though the single-photoelectrode system has a simple structure for the whole cell design, its efficiency is seriously limited by the band gap, and the candidate semiconductors are limited. In order to achieve higher STH efficiency, people normally adopt the dual-photoelectrode system, with the HER on photocathodes and OER on photoanodes. Under this configuration, it requires the intersection of the photocurrent–potential curves for both of the photoelectrodes. Therefore, it is

Table 5.2 The reported unbiased PEC water splitting system based on photoanode–photocathode dual-photoelectrode system.

Photoanode	Photocathode	j@0 V bias/STH efficiency	References
$BiVO_4/NiFeO_x$	$Cu_2O/Ga_2O_3/TiO_2/NiMo$	$2.45\ mA\ cm^{-2}/3\%$	[58]
$BaTaON/CoO_x$	$La_5Ti_2Cu_{1-x}Ag_xS_5O_7/Pt$	N.A./0.1%	[83]
$BiVO_4/FeOOH/$			
NiOOH	$(Ag,Cu)GaSe_2/CuGa_3Se_5/CdS/Pt$	$0.55\ mA\ cm^{-2}/0.67\%$	[84]
$BiVO_4/CoFeO_x$	$(CuGa_{1-y}In_y)_{1-x}Zn_{2x}S/CdS/TiO_2/Pt$	0.4 mA/1.1%	[85]
TiO_2	$p\text{-}Si/TiO_2/Fe_2O_3$	$0.15\ mA\ cm^{-2}/0.18\%$	[86]
$Fe_2O_3/NiFeO_x$	$a\text{-}Si/TiO_2/Pt$	$0.74\ mA\ cm^{-2}/0.91\%$	[87]
$BiVO_4/CoPi$	p-Si/Pt	$0.46\ mA\ cm^{-2}/0.57\%$	[88]
$BiVO_4/NiFeO_x$	$(ZnSe)_x(CuInGaSe)_{1-x}/CdS/Ti/Mo/Pt$	0.9 mA/0.91%	[89]

highly desired to develop photoanode with low onset potential and photocathode with a high onset potential. Taking the typical Cu_2O-$BiVO_4$ system as example, Pan et al. have fabricated effective Cu_2O photocathode with an onset potential over 0.9 V_{SHE}, which ensures the intersection photocurrent over 2 mA cm^{-2} under a tandem configuration when combining with the $BiVO_4$ photoanode [58]. The other dual-photoelectrode systems for unbiased water splitting are summarized in Table 5.2.

Besides these two configurations discussed above, the PV-combined PEC system represents a highly efficient STH system. For many PEC systems, only within 0.5 V external bias is needed to achieve the overall water splitting, which can be easily provided by the solar cell. For example, $BiVO_4$ photoanode is normally reported with an onset potential around 0.3–0.4 V_{SHE}. So it can be compensated with Si-based, perovskite-based, and dye-sensitized solar cells. Kim et al. have used Si solar cell to drive $BiVO_4 + Fe_2O_3$ dual photoanode to achieve an STH efficiency over 7.7% without applied bias [12]. Among all the three types of unbiased water splitting, it is much easier to achieve higher efficiencies by PV + PEC benefiting from the boosting of PV research. As a trade-off, it also has the highest cost [80]. If we aim at the practical application in the future, we need to pay more effort in improving the efficiency of single- and dual-photoelectrode systems.

5.7 Conclusion and Perspective

The solar hydrogen production has long been regarded as the promising route to power our society sustainably and environmentally. Even though the feasibility has been verified from time to time, the key issue is the STH efficiency in the present stages. Charge separation and transfer process is not only the most important but also the most challenging step in water splitting. Therefore, more attention should be paid on improving the charge separation and transfer in the photoelectrode. The defect engineering, crystal facet control, nanostructure fabrication, etc. have been confirmed to effectively improve the charge separation and transfer process. But this promotion effect is limited, and more efforts are appealed to achieve a groundbreaking progress in the following aspects:

(1) Designing well-crystallized semiconductors with a low concentration of recombination centers and a high charge mobility.
(2) Fabricating well-aligned interfaces of the semiconductor–substrate and semiconductor–semiconductor.
(3) Achieving well match between the semiconductor and cocatalyst.

By well addressing the above issues, it is possible to prepare photoelectrodes with low bulk charge recombination, efficient charge transfer, and outstanding surface reaction, which will guarantee a high STH conversion efficiency.

References

1 Abas, N., Kalair, A., and Khan, N. (2015). *Futures* 69: 31–49.
2 Perez, R. and Perez, M. (2009). *The IEA SHC Solar Update*, vol. 50: 2–3.
3 Barreto, L., Makihira, A., and Riahi, K. (2003). *Int. J. Hydrogen Energy* 28: 267–284.
4 Coridan, R.H., Nielander, A.C., Francis, S.A. et al. (2015). *Energy Environ. Sci.* 8: 2886–2901.
5 Chu, S., Li, W., Yan, Y. et al. (2017). *Nano Futures* 1: 022001.
6 Dotan, H., Sivula, K., Grätzel, M. et al. (2011). *Energy Environ. Sci.* 4: 958–964.
7 Wang, Z. and Wang, L. (2018). *Chin. J. Catal.* 39: 369–378.
8 Wang, Z., Fan, F., Wang, S. et al. (2016). *RSC Adv.* 6: 85582–85586.
9 Fu, G., Yan, S., Yu, T., and Zou, Z. (2015). *Appl. Phys. Lett.* 107: 171902.
10 Wang, G., Wang, H., Ling, Y. et al. (2011). *Nano Lett.* 11: 3026–3033.
11 Kim, T.W. and Choi, K.-S. (2014). *Science*: 343: 990–994.
12 Kim, J.H., Jo, Y., Kim, J.H. et al. (2015). *ACS Nano* 9: 11820–11829.
13 Wang, S., Chen, P., Yun, J.H. et al. (2017). *Angew. Chem. Int. Ed.* 56: 8500–8504.
14 Wang, Z., Zhang, L., Schülli, T. et al. (2019). *Angew. Chem. Int. Ed.* 131: 1042.
15 Chen, Z., Jaramillo, T.F., Deutsch, T.G. et al. (2010). *J. Mater. Res.* 25: 3–16.
16 Zhang, Z. and Yates, J.T. Jr., (2012). *Chem. Rev.* 112: 5520–5551.
17 Koval, C.A. and Howard, J.N. (1992). *Chem. Rev.* 92: 411–433.
18 Morrison, S.R. (1980). *Electrochemistry at Semiconductor and Oxidized Metal Electrodes*. Wiley.
19 Boudreaux, D., Williams, F., and Nozik, A. (1980). *J. Appl. Phys.* 51: 2158–2163.
20 Williams, F. and Nozik, A. (1984). *Nature* 312: 21.
21 Le Formal, F., Pendlebury, S.R., Cornuz, M. et al. (2014). *J. Am. Chem. Soc.* 136: 2564–2574.
22 Xiao, M., Wang, Z., Lyu, M. et al. (2018). *Adv. Mater.*: 1801369.

23 Pala, R.A., Leenheer, A.J., Lichterman, M. et al. (2014). *Energy Environ. Sci.* 7: 3424–3430.

24 Tacca, A., Meda, L., Marra, G. et al. (2012). *ChemPhysChem* 13: 3025–3034.

25 Salvador, P. (1984). *J. Appl. Phys.* 55: 2977–2985.

26 Joly, G., Williams, J.R., Chambers, S.A. et al. (2006). *J. Appl. Phys.* 99: 053521.

27 Seabold, J.A., Zhu, K., and Neale, N.R. (2014). *Phys. Chem. Chem. Phys.* 16: 1121–1131.

28 de Respinis, M., Fravventura, M., Abdi, F.F. et al. (2015). *Chem. Mater.* 27: 7091–7099.

29 Piekoszewski, J., Castaner, L., Loferski, J. et al. (1980). *J. Appl. Phys.* 51: 5375–5379.

30 Lee, Y.S., Gershon, T., Gunawan, O. et al. (2015). *Adv. Energy Mater.* 5: 1401372.

31 Jeong, I.-S., Kim, J.H., and Im, S. (2003). *Appl. Phys. Lett.* 83: 2946–2948.

32 Stranks, S.D., Eperon, G.E., Grancini, G. et al. (2013). *Science* 342: 341–344.

33 Wang, Z., Liu, G., Ding, C. et al. (2015). *J. Phys. Chem. C* 119: 19607–19612.

34 Bora, D.K., Braun, A., and Constable, E.C. (2013). *Energy Environ. Sci.* 6: 407–425.

35 Kim, J.H. and Lee, J.S. (2019). *Adv. Mater.* 31: 1806938.

36 Zhang, L., Minegishi, T., Kubota, J., and Domen, K. (2014). *Phys. Chem. Chem. Phys.* 16: 6167–6174.

37 Yokoyama, D., Minegishi, T., Maeda, K. et al. (2010). *Electrochem. Commun.* 12: 851–853.

38 Deml, M., Holder, A.M., O'Hayre, R.P. et al. (2015). *J. Phys. Chem. Lett.* 6: 1948–1953.

39 Wang, Z., Mao, X., Chen, P. et al. (2019). *Angew. Chem. Int. Ed.* 58: 17604.

40 Qiu, Y., Leung, S.-F., Zhang, Q. et al. (2014). *Nano Lett.* 14: 2123–2129.

41 Li, J., Qiu, Y., Wei, Z. et al. (2014). *Energy Environ. Sci.* 7: 3651–3658.

42 Hussain, S., Hussain, S., Waleed, A. et al. (2016). *ACS Appl. Mater. Interfaces* 8: 35315–35322.

43 Qiu, Y., Liu, W., Chen, W. et al. (2016). *Sci. Adv.* 2: e1501764.

44 Wang, Q. and Domen, K. (2019). *Chem. Rev.*

45 Wang, Z., Qi, Y., Ding, C. et al. (2016). *Chem. Sci.* 7: 4391–4399.

46 Abe, R., Takata, T., Sugihara, H., and Domen, K. (2005). *Chem. Lett.* 34: 1162–1163.

47 Ueda, K., Minegishi, T., Clune, J. et al. (2015). *J. Am. Chem. Soc.* 137: 2227–2230.

48 Hisatomi, T., Kubota, J., and Domen, K. (2014). *Chem. Soc. Rev.* 43: 7520–7535.

49 Ham, Y., Minegishi, T., Hisatomi, T., and Domen, K. (2016). *Chem. Commun.* 52: 5011–5014.

50 Kuang, Y., Jia, Q., Ma, G. et al. (2017). *Nat. Energy* 2: 16191.

51 Minegishi, T., Nishimura, N., Kubota, J., and Domen, K. (2013). *Chem. Sci.* 4: 1120–1124.

52 Kumagai, H., Minegishi, T., Moriya, Y. et al. (2014). *J. Phys. Chem. C* 118: 16386–16392.

53 Liu, J., Hisatomi, T., Ma, G. et al. (2014). *Energy Environ. Sci.* 7: 2239–2242.

54 Wang, Z., Han, J., Li, Z. et al. (2016). *Adv. Energy Mater.* 6: 1600864.

55 Wang, Z., Zong, X., Gao, Y. et al. (2017). *ACS Appl. Mater. Interfaces* 9: 30696–30702.

56 Abdi, F.F., Han, L., Smets, A.H. et al. (2013). *Nat. Commun.* 4: 2195.

57 Luo, J., Steier, L., Son, M.-K. et al. (2016). *Nano Lett.* 16: 1848–1857.

58 Pan, L., Kim, J.H., Mayer, M.T. et al. (2018). *Nat. Catal.* 1: 412.

59 Paracchino, A., Mathews, N., Hisatomi, T. et al. (2012). *Energy Environ. Sci.* 5: 8673–8681.

60 Zhao, J., Minegishi, T., Zhang, L. et al. (2014). *Angew. Chem. Int. Ed.* 53: 11808–11812.

61 Yokoyama, D., Minegishi, T., Jimbo, K. et al. (2010). *Appl. Phys Express* 3: 101202.

62 Jiang, F., Harada, T., Kuang, Y. et al. (2015). *J. Am. Chem. Soc.* 137: 13691–13697.

63 Wang, F., Septina, W., Chemseddine, A. et al. (2017). *J. Am. Chem. Soc.* 139: 15094–15103.

64 Han, J., Zong, X., Wang, Z., and Li, C. (2014). *Phys. Chem. Chem. Phys.* 16: 23544–23548.

65 Li, R., Zhang, F., Wang, D. et al. (2013). *Nat. Commun.* 4: 1432.

66 Chen, R., Pang, S., An, H. et al. (2018). *Nat. Energy* 3: 655.

67 Zhu, J., Fan, F., Chen, R. et al. (2015). *Angew. Chem. Int. Ed.* 54: 9111–9114.

68 Han, H.S., Shin, S., Kim, D.H. et al. (2018). *Energy Environ. Sci.* 11: 1299–1306.

69 Wang, L. and Sasaki, T. (2014). *Chem. Rev.* 114: 9455–9486.

70 Peerakiatkhajohn, P., Yun, J.H., Chen, H. et al. (2016). *Adv. Mater.* 28: 6405–6410.

71 Butburee, T., Bai, Y., Wang, H. et al. (2018). *Adv. Mater.* 30: 1705666.

72 Klahr, B., Gimenez, S., Fabregat-Santiago, F. et al. (2012). *J. Am. Chem. Soc.* 134: 4294–4302.

73 Liu, R., Zheng, Z., Spurgeon, J., and Yang, X. (2014). *Energy Environ. Sci.* 7: 2504–2517.

74 Hisatomi, T., Le Formal, F., Cornuz, M. et al. (2011). *Energy Environ. Sci.* 4: 2512–2515.

75 Ju, S., Lee, K., Yoon, M.-H. et al. (2007). *Nanotechnology* 18: 155201.

76 Mukherjee, B., Wilson, W., and Subramanian, V.R. (2013). *Nanoscale* 5: 269–274.

77 Liu, G., Ye, S., Yan, P. et al. (2016). *Energy Environ. Sci.* 9: 1327–1334.

78 Nellist, M.R., Laskowski, F.A., Lin, F. et al. (2016). *Acc. Chem. Res.* 49: 733–740.

79 Liu, G., Shi, J., Zhang, F. et al. (2014). *Angew. Chem. Int. Ed.* 53: 7295–7299.

80 Sayama, K. and Miseki, Y. (2014). *Synthesiology* 7: 79–91.

81 Fujishima, A. and Honda, K. (1972). *Nature* 238: 37–38.

82 Ebaid, M., Priante, D., Liu, G. et al. (2017). *Nano Energy* 37: 158–167.

83 Hisatomi, T., Okamura, S., Liu, J. et al. (2015). *Energy Environ. Sci.* 8: 3354–3362.

84 Kim, J.H., Kaneko, H., Minegishi, T. et al. (2016). *ChemSusChem* 9: 61–66.

85 Hayashi, T., Niishiro, R., Ishihara, H. et al. (2018). *Sustainable Energy Fuels* 2: 2016–2024.

86 Kargar, A., Khamwannah, J., Liu, C.H. et al. (2016). *Nano Energy* 19: 289–296.
87 Jang, J.-W., Du, C., Ye, Y. et al. (2015). *Nat. Commun.* 6: 7447.
88 Xu, P., Feng, J., Fang, T. et al. (2016). *RSC Adv.* 6: 9905–9910.
89 Kaneko, H., Minegishi, T., Nakabayashi, M. et al. (2016). *Adv. Funct. Mater.* 26: 4570–4577.

6

Photocatalytic Oxygen Evolution

Huayang Zhang, Wenjie Tian and Shaobin Wang

School of Chemical Engineering and Advanced Materials, The University of Adelaide, North Terrace, Adelaide, SA 5005, Australia

6.1 Introduction

6.1.1 Configuration of Photocatalytic Water Oxidation

Since the discovery of the Honda–Fujishima effect [1], tremendous attention has been centered on solar-light-driven water splitting by adopting various inorganic semiconductors that possess strong photoinduced redox abilities. Since the oxidative half-reaction, namely, water oxidation, is a four-electron transfer and four-proton removal process to form oxygen from water molecules, it is generally considered as the rate-limiting step to achieve overall solar water splitting. Therefore, much attention has been given to water oxidation. Photocatalytic water oxidation is usually divided into two kinds of configurations, namely, homogeneous and heterogeneous systems, as shown in Figure 6.1 [2, 3]. In earlier studies, water oxidation activities have been tested in homogeneous phases [4]. Generally for homogeneous water oxidation photocatalysts, the most studied materials are the molecular metal clusters where the mononuclear or multinuclear metal centers are coordinated to external ligands [5]. Depending on the various coordination configurations, the light absorption, catalytic activity, and stability of the molecular cluster catalysts were highly controlled and tuned by their bridging ligand or bonding environment. Besides, sacrificial chemical oxidants such as $AgNO_3$ and $[Ru(bpy)_3]^{2+}$ (bpy = 2,2′-bipyridine) in combination with persulfate ($S_2O_8^{2-}$) are commonly used to oxidize the homogeneous catalyst to a state where it can oxidize water. Notably, the large number of sacrificial oxidants in the reaction medium can actively destroy the structure of the molecular complexes. The stability of the homogeneous water oxidation catalysts (WOCs) is always an important issue that affects their studies. Compared with the homogeneous system, particular catalyst materials for heterogeneous systems have arisen as attractive candidates for photocatalytic water oxidation, as they are more stable in terms of the chemical and catalytic stability. In addition, they have the merits of being simpler, cheaper, easier to be recycled, and more universal to be developed and used than the homogeneous system. Generally, in the heterogeneous system, an electron acceptor is still needed when using the single component.

Solar-to-Chemical Conversion: Photocatalytic and Photoelectrochemcial Processes, First Edition.
Edited by Hongqi Sun.

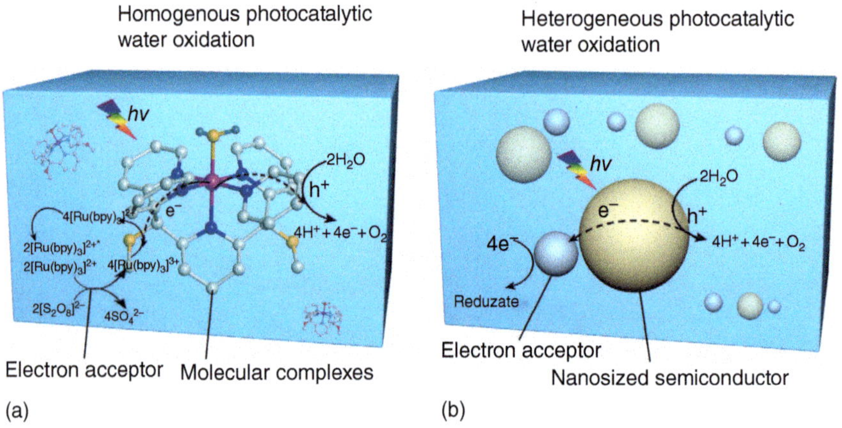

Figure 6.1 Schematic illustration of photocatalytic oxygen evolution systems: (a) homogeneous and (b) heterogeneous configuration.

6.1.2 Mechanism, Thermodynamics, and Kinetics Toward Efficient Oxygen Evolution

The photocatalytic oxygen evolution is a relatively slow reaction, as it concerns the transference of four electrons and multiple protons along with the formation of an O—O bond ($2H_2O \rightarrow 4H^+ + 4e^- + O_2$). To date, there exist two kinds of principles for the reaction pathway, i.e. water nucleophilic attack (WNA) and interaction of two metal–oxygen moieties (I2M) [6–8]. In the WNA reaction pathway, H_2O molecule functions as the nucleophile to attack the electrophilic high-valent single metal–oxo (M=O) site, which was preformed through the deprotonation of hydroxyl bound by metal center. This attack commences on the cleavage of the M=O π-bond and the accompanying formation of a metal hydroperoxide intermediate (M—OOH), which can consequently undergo oxidation to release O_2. It was recently suggested that water oxidation pathway of oxygen-evolving complex (OEC) in nature and most heterogeneous catalysts follow the WNA process [9]. In contrast, the I2M process requires the radical coupling of two metal–oxo sites that contain a vital radical nature, forming [M–O–O–M] intermediates that will undergo further oxidation and release O_2 more quickly. Notably, I2M is the proposed water oxidation mechanism for various homogeneous complex catalysts, as the formation of intermolecular O—O bond through association of two molecules is normally slow, serving as the rate-determining step in most cases.

The minimum potential for oxygen evolution reaction (OER) is $+1.23\,V_{RHE}$ (V vs. standard hydrogen electrode [RHE]). Coupled with the reduction half-reaction ($4H^+ + 4e^- \rightarrow 2H_2$), the least energy required for the overall water splitting reaction is 1.23 eV, meaning that the band-gap energy should be larger than 1.23 eV to straddle the HER and OER potentials [10, 11]. However, a larger band gap will result in inadequate light absorption. For heterogeneous WOCs, corresponding to the band-gap energy of 1.23 eV, the semiconductor material needs to absorb an equivalent light wavelength of 1100 nm, which is in the near-infrared (NIR) range of the sunlight spectrum [12]. In general, the band-gap energies of most photocatalysts are larger than 2 eV, meaning that they can only absorb light wavelengths of ≤620 nm [12]. Therefore, the high requirement of appropriate band-gap energy highly limits the catalyst selection for photocatalytic water oxidation [10, 12]. Besides, 237 kJ mol^{-1} standard Gibbs free energy is required for overall water splitting, as it is an uphill chemical process. It means that the energies of the photogenerated holes should be high enough to overcome the overpotentials related to the oxidative half-reaction, namely, η_{ox} [13]. In this case, external electron acceptors (sacrificial reagents such as $Na_2S_2O_8$ or $AgNO_3$) may need to be added in the photochemical system to overcome η_{ox} and actuate water oxidation [8].

Since Mn_4CaO_4 was identified as the OEC of photosystem II (PS-II) in the leaves of plants [14], most recent studies have been pointed to exploit various earth-abundant elements based on photocatalysts to mimic this function [14–16]. From the perspective of a practical application that desires a low cost of production, this chapter will concentrate on catalysts on the basis of low-cost and earth-abundant elements, which have been exploited to promote the efficiency in photocatalytic water oxidation. From homogeneous to heterogeneous catalysis, in this chapter, we attempt to outline the state-of-the-art development of molecular complexes and nanosized semiconductor materials for photoinduced water oxidation. Through referencing some classic examples, we tried to present some fundamental insights into how the photocatalytic water oxidation performance is enhanced. Particular attention was given to the different strategies to select suitable catalysts to optimize the activity and efficiency. Besides, the mechanisms about the unique catalytic properties of zero-dimensional (0D), one-dimensional (1D), and two-dimensional (2D) materials are compared, and low-dimensional (LD) metal-free semiconductors are also described briefly. At last, conclusive remarks will be summarized, and the key challenges and the future strategies for the management of catalysts for photocatalytic water oxidation will be discussed.

6.2 Homogeneous Photocatalytic Water Oxidation

6.2.1 Molecular Complexes and Polyoxometalates

In nature, the protein OEC near PS-II is responsible for water oxidation. To be specific, the Mn tetramer cluster ($CaMn_4O_5$) catalyzes the conversion of H_2O to O_2 by means of the positive charge equivalents [17]. Until very recently, the precise structure of the $CaMn_4O_5$ core has been revealed by a series of high-resolution characterization techniques [18, 19]. The admitted model indicates that $CaMn_4O_5$ consists of a Mn_3CaO_4 cubane motif and a "dangler" Mn linked via two bridging oxides, building a definite asymmetric Mn_4O_4 cluster (Figure 6.2). This cluster is further harbored in the coordination environment provided by the protein residues [20].

Attempts to emulate the structure and function of the multimetallic $CaMn_4O_5$ have yielded a handful of low nuclearity metal complexes containing organic ligand for water oxidation in homogeneous systems, as shown in Table 6.1 [19–21].

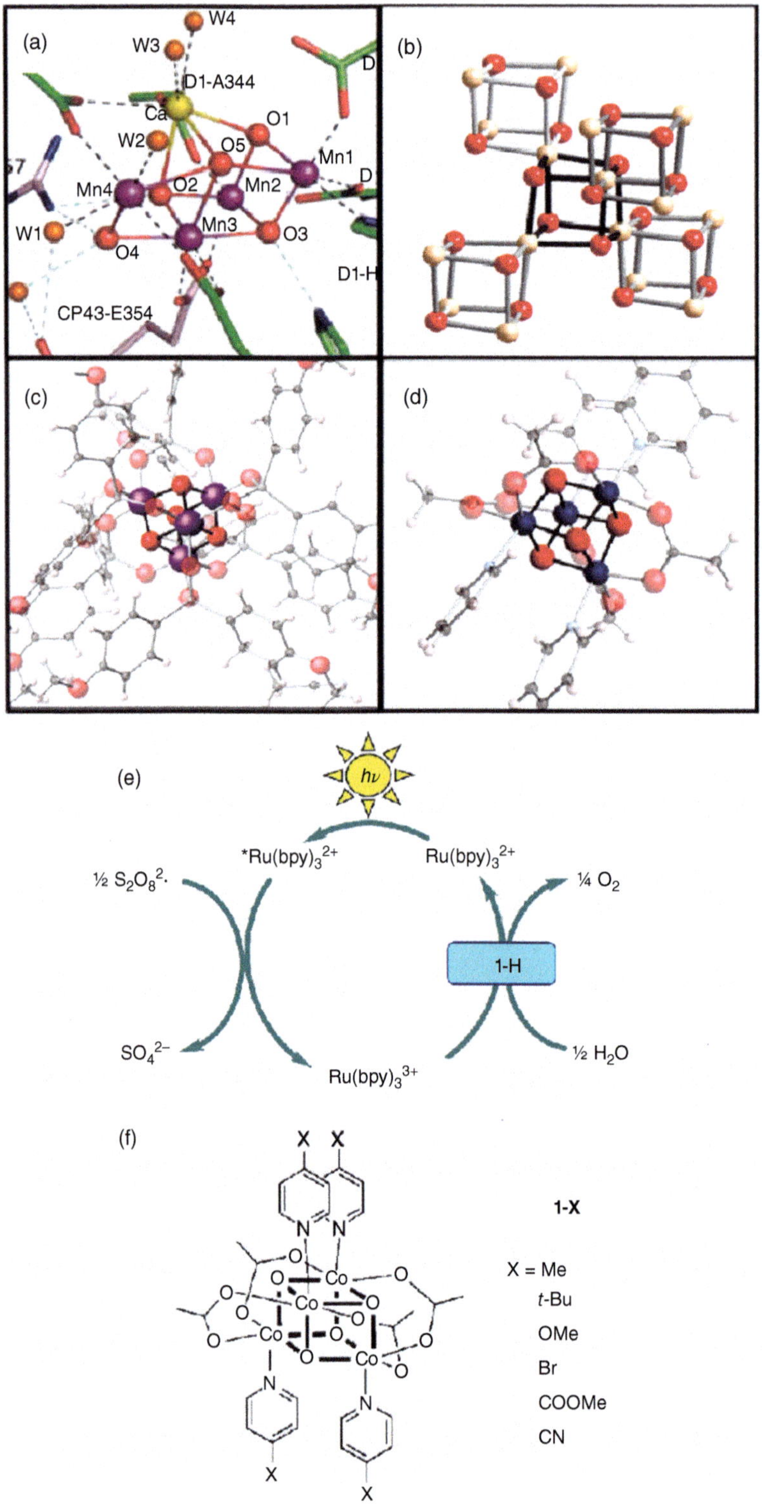
(a)
W4
W3
D1-A344
Ca
O1
W2
O5
Mn1
Mn4
O2
Mn2
Mn3
W1
O4
O3
D1-H
CP43-E354
(b)
(c)
(d)
(e)
hν
*Ru(bpy)$_3^{2+}$
Ru(bpy)$_3^{2+}$
½ $S_2O_8^{2-}$
¼ O_2
1-H
SO_4^{2-}
Ru(bpy)$_3^{3+}$
½ H_2O
(f)
1-X
X = Me
t-Bu
OMe
Br
COOMe
CN

Figure 6.2 Structures of (a) PS-II-OEC natural WOCs and (b) λ-MnO_2, (c) $Mn_4O_4L_6$ core, and (d) $Co_4O_4(Ac)_4(py)_4$ artificial WOCs. Source: From McCool et al. [27]. © 2011 American Chemical Society. (e) Cycle diagram for the electron/proton transfer during the photocatalytic water oxidation of 1-H in the $Ru(bpy)_3^{2+}/S_2O_8^{2-}$ system and (f) molecular structure of 1-X (X = H, Me, *t*-Bu, OMe, Br, COOMe, CN). Source: From Berardi et al. [29]. © 2012 American Chemical Society.

Early studies had proved that even the mononuclear molecular complexes such as single Fe cluster [22] and single Co cluster [23] are enough for water oxidation catalysis, which broke the dogma that at least two metal sites in one cluster monomer are needed for catalytic water oxidation [24, 25]. Until very recently, documented examples of molecular complexes containing higher nuclearity TM centers have been reported for water oxidation [26]. McCool et al. firstly described a well-structured $Co_4O_4(OAc)_4(py)_4$ cluster containing a cubical Co_4O_4 core and used it for photochemical water oxidation in the $Ru^{III}(bpy)_3^{3+}$ photo-oxidant system [27]. In situ observation of catalytic species had confirmed the inherent capacities of this core type, which provide a strong evidence that the tetranuclear oxo core is the crucial structural trait responsible for imitating the natural OEC of the natural photosynthetic enzyme (Figure 6.2b–d). Interestingly, the Co_4O_4 cubane core, rather than those "incomplete cubane" $Co_3O_{3,4}$ cores or "half-cubane" Co_2O_2 cores, was proved to be the smallest catalytic unit of Co-based molecular clusters by Smith et al. [28]. In their work, the impact of cluster nuclearity on the kinetics of water oxidation was examined among a series of cobalt–oxo clusters including cubanes, dimers, and trimers. In the typical $Ru(bpy)_3^{2+}/S_2O_8^{2-}$ photo-oxidant system, neither trimers nor dimers were found able to evolve O_2, despite owning the same ligand positions as their cubane counterparts. The key feature of Co_4O_4 cubanes for the intramolecular water oxidation pathway was proved to be the four one-electron redox metals, which can provide necessary terminal coordination sites for substrate aquo/oxo formation. Another promising case about the ligand–reactivity relationship stems from the inspection of the influence of ligand substitution on the cubane Co_4 cluster [29]. Diverse para-substituted pyridines X (X = H, Me, *t*-Bu, OMe, Br, COOMe, CN) were chosen as terminal ligands of the tetracobalt core, which were directly conjugated to each of the four cobalt sites (Figure 6.2e,f). The redox and kinetic properties of all photoinduced intermediates and electron transfer could be favored to different degrees according to various electron-rich pyridine ligands.

Compared with clusters containing organic ligand, a new class of metal–oxo cores surrounded by all-inorganic coordination, i.e. polyoxometalate (POM)-based cluster, have recently been proven as one kind of light-driven WOCs [30, 31]. POMs with well-defined nanostructures can undergo multi-electron redox transformations awarding themselves catalytic redox activity. Han et al. synthesized three POM clusters with various Ni nuclearity ($\{Ni_{12}\}$, $\{Ni_{13}\}$, and $\{Ni_{25}\}$ cores). The Ni-O configuration was constructed as $\{Ni_3O_3\}$ quasi-cubane or $\{Ni_4O_4\}$ cubane cores, which were then connected by the inorganic ligands and finally surrounded by

Table 6.1 Homogeneous catalysts for photocatalytic water oxidation.

Metal sites	Ligand in the coordination sphere	Light condition	Sacrificial agent	TON [a] or TOF [b]	References
Polynuclear clusters					
Co_4O_4	$(Ac)_4(py)_4$	$\lambda > 400$ nm	$[Ru(bpy)_3]^{2+} - Na_2S_2O_8$	TON:42	[27]
Co_4O_4	$(OAc)_4(py)_4$	UV/visible	$[Ru(bpy)_3]^{2+} - Na_2S_2O_8$	TOF: 3.4/s/μmol cata.	[28]
$Co^{III}{}_4(\mu\text{-}O)_4$	$(\mu\text{-}CH_3COO)_4(p\text{-}NC_5H_4X)_4]$, 1-X (X = H, Me, *t*-Bu, OMe, Br, COOMe, CN)	$\lambda > 400$ nm	$[Ru(bpy)_3]^{2+} - Na_2S_2O_8$	Total TON: 140	[29]
$\{Co_4(OH)_3(PO_4)\}_4$	$(AsW_9O_{34})_4$	$\lambda > 420$ nm	$[Ru(bpy)_3]^{2+} - Na_2S_2O_8$	TON: 71.5	[30]
$Co_4(H_2O)_2$	$(PW_9O_{34})_2$	$420 < \lambda < 470$ nm	$[Ru(bpy)_3]^{2+} - Na_2S_2O_8$	TON: 224 ± 11	[31]
$Ni_{25}(H_2O)_2OH)_{18}(CO_3)_2(PO_4)_6$	A-α-$\{SiW_9O_{34}\}$	$\lambda > 420$ nm	$[Ru(bpy)_3]^{2+} - Na_2S_2O_8$	TON: 204.5	[4]
Mononuclear complexes					
Co	(Me_6tren)	$\lambda > 420$ nm	$[Ru(bpy)_3]^{2+} - Na_2S_2O_8$	TON: 420	[23]
Fe	Tetra-amidomacrocyclic	Visible light	$[Ru(bpy)_3]^{2+} - Na_2S_2O_8$	TON: 220	[22]
Co	Salophen	Visible light	$[Ru(bpy)_3]^{2+} - Na_2S_2O_8$	TON: 17	[123]
Co	Porphyrin	$400 < \lambda < 800$ nm	$[Ru(bpy)_3]^{2+} - Na_2S_2O_8$	TON: 121.8	[124]
Co	Porphyrin	$400 < \lambda < 800$ nm	$[Ru(bpy)_3]^{2+} - Na_2S_2O_8$	TON: 570	[125]
Co	Triazole	$\lambda > 420$ nm	$[Ru(bpy)_3]^{2+} - Na_2S_2O_8$	TON: 90	[126]
Co	Polypyridine	$\lambda > 420 \pm 10$ nm	$[Ru(bpy)_3]^{2+} - Na_2S_2O_8$	TON: 51 ± 3	[127]

a) TON: turnover number for water oxidation, representing moles of oxygen produced by per mole catalyst.
b) TOF: turnover frequency for water oxidation, representing moles of oxygen produced by per mole catalyst per second.

the lacunary {A-α-SiW_9O_{34}} POM units [4]. Unlike the aforementioned report by Smith that proposed Co_4O_4 cubane as the smallest active unit in Co clusters, in this work, not only the {Ni_4O_4} cubane but also the {Ni_3O_3} quasi-cubane were claimed as the active units for visible-light-driven water oxidation in $Ru(bpy)_3^{2+}$/ $S_2O_8^{2-}$ photo-oxidant system. Particularly, the water oxidation efficiencies follow the order of {Ni_{12}} < {Ni_{13}} < {Ni_{25}}, which was initially explained from the perspective of band-gap structures and crystal structural features:

i) Increased HOMO potential gap between $Ru(bpy)_3$ and Ni_xO_y cores from {Ni_{12}} to {Ni_{25}} makes the clusters easily oxidized by $Ru(bpy)_3$. In that case, the charge transfer efficiencies between $Ru(bpy)_3$ and POMs might be improved.
ii) The increasing Ni nuclearity results in more exposed {Ni_xO_y} fragments for easy contact with {$Ru(bpy)_3$} and water molecules. Mn_4POM is the first manganese-containing POM that could undergo multiple electron transfers under visible-light irradiation, directing to photoinduced oxygen evolution. To imitate structure and activity of the PS-II-OEC, a set of hybrid ligands containing an inorganic POM platform and carboxylate bridges were used to stabilize a defective $Mn^{III}{}_3Mn^{IV}O_3$ cubane core (Figure 6.3) [20]. The interaction of organic and inorganic residues affords a stable and flexible coordination environment to support stepwise one-electron oxidation of the $Mn^{III}{}_3Mn^{IV}O_3$ core to obtain high-valent Mn^{IV} states that are responsible for photocatalytic water oxidation. For clusters with water-derived ligands such as oxo, hydroxy, or aqua ligands, studies had proved that they share a common structural requirement and reaction pathways feature with organic ligands coordinated clusters. Theoretical investigations revealed that the water adsorption could induce dimensionality change of ligand-free $Mn_4O_4^+$ clusters, involving a transformation from a 2D ring-like structure to a common cuboidal octa-hydroxy cluster. The cluster dimensionality crossover would facilitate hydrogen abstraction and possibly eventually light-induced oxygen evolution.

6.2.2 Mechanism Details and the Stability

It was confirmed that the WNA mechanism applies to the mononuclear and part of multinuclear complex catalysts. For example, a hexanuclear cobalt molecular cluster was recently reported as an efficient visible-light-driven WOC [32]. Systematic research reveals that the long-distance adjacent Co ions in this cluster enforce every Co nuclear function as active site independently, following the WNA mechanism during water oxidation. However, many studies had revealed that, for most multinuclear MCCs to effectively catalyze water oxidation, they usually contain at least two activated metal sites separated by an appropriate distance for promoting intramolecular O=O bond formation and following the I2M pathway [33].

Regardless of the vigorous activity in water oxidation, several reports found that a certain number of heteroatomic ligands were incapable of preventing the cluster cores from decomposition into catalytically active films or nanoparticles (NPs) [34]. Whether these molecular clusters or complexes maintain their molecular integrities

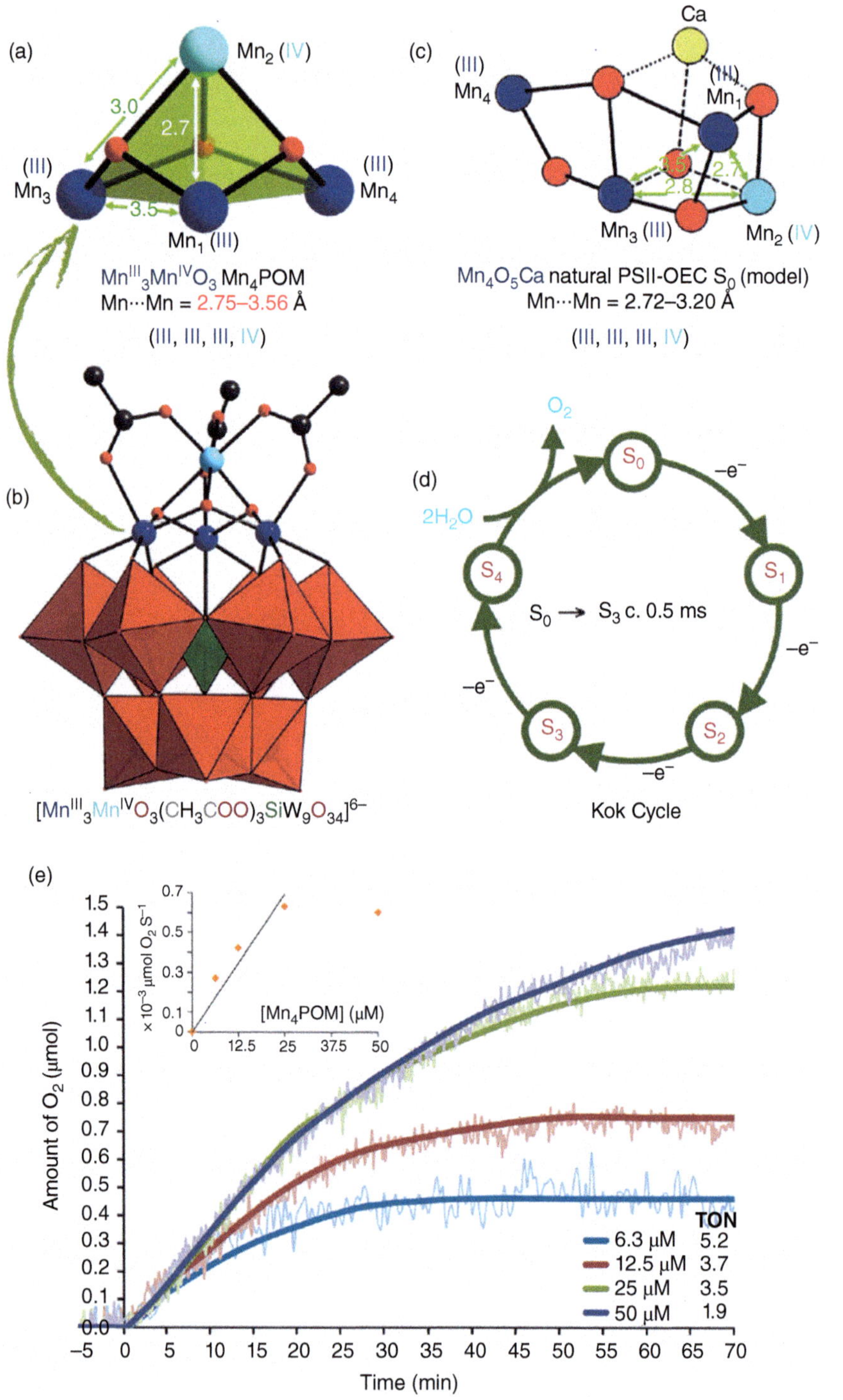

Figure 6.3 Ball-and-stick model of (a) Mn_4POM core, (b) polyhedral Mn_4POM, and (c) S_0 state of the natural OEC model. (d) Cycle diagram for the electron transfer within the $S_0 \rightarrow S_4$ Kok cycle of the natural PS-II-OEC. (e) Photocatalytic oxygen evolution of Mn_4POM within the $[Ru(bpy)_3]^{2+}/Na_2S_2O_8$ system. Source: From Al-Oweini et al. [20]. © 2014 John Wiley & Sons.

during catalysis or merely function as precursors of activated NPs is complicated and controversial [24]. A typical example of this challenge is the molecularly defined Co_4POM, which had been presumed by Hill et al. to be an extremely active homogeneous WOC in either chemical or photochemical water oxidation reaction [31, 35], exploited later to discover that the Co_4POM was deposited at higher Co_4-POM concentrations (500 μM) and lower overpotentials (<500 mV) as an amorphous CoO_x film on the electrode surface to behave as a heterogeneous WOC [36, 37]. Follow-up reports by Hill and his colleagues gave a clear evidence again that, within certain situations in which some specific chemical oxidants are employed, the observed oxygen evolution results from the homogeneous clusters rather than heterogeneous particles [38].

6.3 Heterogeneous Photocatalytic Water Oxidation

One point we need to declare is that the solubility of the homogeneous WOCs makes them very difficult to be recovered from the reaction media with intact structures, which will narrow their practical application. Currently, increasing studies are focusing on the development of heterogeneous semiconductor materials, and researches have extended from TiO_2 to ZnO, WO_3, Fe_2O_3, Co_3O_4, and others, as shown in Figure 6.4.

Unlike homogeneous catalysts, whose actives sites can immediately contact with the reaction medium to an extreme, for heterogeneous catalysts, the photoinduced holes within the semiconductors must travel through long distances to the surface to react with water. Unless separated effectively, the holes can easily recombine with electrons trapped on the surface or in the bulk of semiconductor materials, leading to a deficient OER. To reduce electron–hole recombination and accelerate hole diffusion kinetics, reducing the size of semiconductors to nanoscale is a promising approach because of the attractive attributes of nanosized materials including short carrier transfer distance, readily tunable band-gap structure, large surface area, and unique electronic and optical properties that can be adjusted by doping or dimensional modification [13, 39, 40]. A lot of recent researches concentrating on the

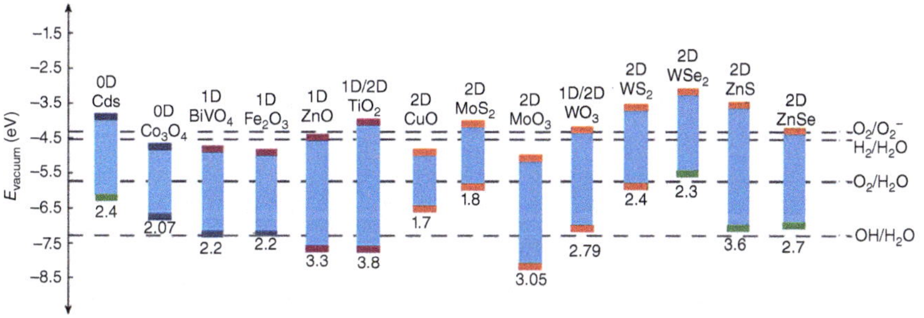

Figure 6.4 Illustration of the band positions of common semiconductor materials and the energy level toward water splitting.

management of catalyst size and dimensionality have been recorded with excellent water oxidation catalytic performance (Table 6.2). Here, we limit the scope of the nanosized materials to LD materials, which should have at least one dimension of $\leq$100 nm, and typical examples include 0D NPs, nanoclusters, and quantum dots (QDs); 1D nanotubes (NTs) and nanowires (NWs)/or nanorods (NRs); and 2D nanosheets and layers. Various categories of 0D/1D/2D-based hybrid materials, which could be inorganic, organic, metallic–organic, or inorganic–organic matters, are also included. With investigation of the unique properties and inherent advantages of 0D, 1D, and 2D materials, in the following sections, we retrospect how recent research works control and exploit their properties for better improving the photocatalytic water oxidation performance. We believe that understanding the interaction of nanosized photocatalysts with water could be essential for the bottom-up design of artificial water oxidation nanocatalysts.

6.3.1 Unique Properties of Nanosized Semiconductor System

6.3.1.1 Quantum Confinement

An essential feature of nanomaterials is the quantum confinement effect, which endows the materials with unique electronic, optical, and structural properties compared with their bulk counterparts [41–43]. Quantum confinement effects can be achieved when the NP size is close to the exciton Bohr radius. In that case, the energy level spacing of the particle will exceed kT (k is Boltzmann's constant and T is temperature), and the corresponding energy differences ($>kT$) will restrict the electron–hole mobility [39]. In 1974, Dingle et al. first confirmed the carriers confinement in 2D materials, where numerous restricted electron and hole states of rectangular potential wells were observed [44]. Compared with bulk materials, the quantum confinement effect will bring about modified properties in nanosized materials, two of which are of particular importance. The first one is the energy states changing from quasi-continuous states to discrete and well-separated states due to the reduced atom number in the specific dimension direction. Figure 6.5a–d shows the density of states (DOS) as a function of energy in 3D to 0D semiconductor materials, which provide novel opportunities to control charge diffusion directions and pathways. Another valuable property is depicted in Figure 6.5e. Depending on the characteristics of a semiconductor, once the specific dimension diameter decreases to a particular value, an increase (blueshift) in band-gap energy is observed. To be specific, the conduction band (CB) edge shifts toward a lower potential, while the valence band (VB) moves to more oxidizing position. According to Marcus–Gerischer theory, this change is expected to increase the thermodynamic driving force, which can accelerate interfacial charge transfer and enhance the water splitting rate [45, 46]. For example, since their CB edges are more positive than the hydrogen evolution potential, bulk WO_3, $BiVO_4$, and Fe_2O_3 are inefficient to produce hydrogen despite their strong visible-light absorption [11]. However, by reducing at least one dimension of these bulk materials to a certain level, the band-gap position will shift to cover the reduction potential, allowing for efficient water splitting on these semiconductors. Although the quantum confinement is

Table 6.2 Heterogeneous catalysts for photocatalytic water oxidation.

Low-dimensional semiconductors	Low-dimensional size (diameter/ thickness) (nm)	Light condition	Sacrificial agent	TOF(s^{-1})/O_2 yield ($\mu mol\ h^{-1}$)	References
0D semiconductors					
Co_3O_4 nanoclusters in SBA-15	7.6	Visible light	$[Ru(bpy)_3]^{2+} - Na_2S_2O_8$	$1140\ s^{-1}$	[65]
Co_3O_4 nanoclusters in KIT-6	~25	$\lambda > 420$ nm	$[Ru(bpy)_3]^{2+} - Na_2S_2O_8$	$4.05 \pm 0.10 \times 10^4\ s^{-1}/70.5 \pm 1.7$	[67]
Co_3O_4 quantum dots	3–4	$\lambda > 420$ nm	—	$\mu mol\ h^{-1}$	[72]
Co_3O_4 quantum dots	4.5	$\lambda > 420$ nm	$[Ru(bpy)_3]^{2+} - Na_2S_2O_8$	$2.7 \times 10^{-4}\ s^{-1}$	[73]
α-Fe_2O_3 quantum dots	3	$\lambda > 420$ nm	$AgNO_3$	$0.31\ \mu mol\ h^{-1}\ g^{-1}$	[70]
Co_3O_4 nanoparticles	3	$\lambda > 450$ nm	$[Ru(bpy)_3]^{2+} - Na_2S_2O_8$	$130\ \mu mol\ h^{-1}$	[69]
CoO nanoparticles	5–8	532 nm	—	$20\ ml\ h^{-1}$	[77]
Spinel CoO_x nanocubanes	<2	$\lambda > 400$ nm	$[Ru(bpy)_3]^{2+} - Na_2S_2O_8$	$0.023\ s^{-1}/1.3$ mol/mol c_o	[81]
α-Fe_2O_3	5.4	UV/visible light	$AgNO_3$	$1072\ \mu mol\ h^{-1}\ g^{-1}$	[83]
$BiVO_4$ nanocrystals	4.4	UV/visible light	$[Ru(bpy)_3]^{2+} - Na_2S_2O_8$	$4.32\ \mu mol\ h^{-1}$	[84]
1D semiconductors					
$BiVO_4$ quantum tube	5	Visible light	—	$0.11\ \mu mol\ h^{-1}$	[89]
2D semiconductors					
WO_{3-x} nanosheets	2.1	Visible light	$NaIO_3$	$1593\ \mu mol\ h^{-1}\ g^{-1}$	[58]
Zn/Ti, LDHs	—	Visible light	$AgNO_3$	$268.3\ \mu mol\ h^{-1}\ g^{-1}$	[104]
Zn/Ce, LDHs				$626.1\ \mu mol\ h^{-1}\ g^{-1}$	
Zn/Cr LDHs				$1073.3\ \mu mol\ h^{-1}\ g^{-1}$	
Co–Fe LDHs	—	400 nm < λ < 700 nm	$AgNO_3$	$15\ \mu mol\ h^{-1}$	[105]

(continued)

Table 6.2 (*Continued.*)

Low-dimensional semiconductors	Low-dimensional size (diameter/ thickness) (nm)	Light condition	Sacrificial agent	TOF(s^{-1})/O_2 yield ($\mu mol\ h^{-1}$)	References
(Cu/Ti) LDHs	—	400 nm < λ < 700 nm	$AgNO_3$	8.16 $\mu mol\ h^{-1}$	[103]
NiTi LDHs	~2.0	λ > 400 nm	$AgNO_3$	~2148 $\mu mol\ h^{-1}\ g^{-1}$	[106]
BiOCl nanolayer	~2.0	λ > 400 nm	$AgNO_3$	56.85 $\mu mol\ g^{-1}\ h^{-1}$	[107]
Ultrathin Fe_2O_3 nanosheets	~3.5	420 nm < λ < 700 nm	$[Ru(bpy)_3]^{2+} - Na_2S_2O_8$	70 $\mu mol\ g^{-1}\ h^{-1}$	[102]
LD semiconductor-based hybrids					
1D TiO_2 nanorods/0D $Co(OH)_2$ nanoclusters	—	λ > 400 nm	$AgNO_3$	11 $\mu mol\ h^{-1}$	[110]
1D WO_2/1D WO_3 nanorods	—	420 nm < λ < 700 nm	$AgNO_3$	220 $mmol\ h^{-1}\ g^{-1}$	[53]
1D Si/1D TiO_2 nanowires	—	Visible light	—	458 $\mu mol\ h^{-1}\ g^{-1}$	[112]
2D WO_3/2D $CoWO_4$ nanosheets	~6	λ > 400 nm	$[Ru(bpy)_3]^{2+} - Na_2S_2O_8$	1.6 $mmol\ h^{-1}\ g^{-1}$	[113]
0D carbon dots/ 0D $BiVO_4$	2–8	Visible light	—	0.51 $\mu mol\ h^{-1}$	[90]
0D Co_3O_4 QDs/ 2D g-C_3N_4 nanosheets	—	λ > 400 nm	$[Ru(bpy)_3]^{2+} - Na_2S_2O_8$	10 $\mu mol\ h^{-1}$	[128]
0D α-Fe_2O_3/ 2D rGO	—	UV/visible light	$AgNO_3$	752 $\mu M\ h^{-1}\ g^{-1}$	[121]

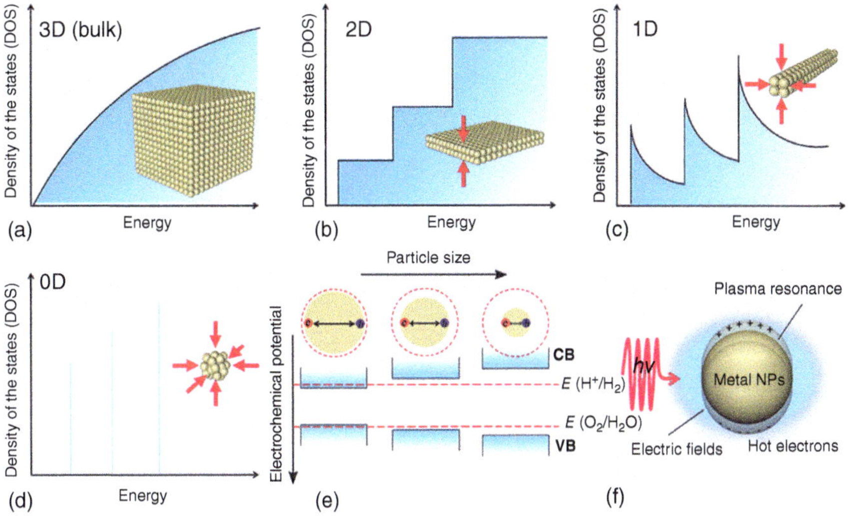

Figure 6.5 Comparison of electronic DOS of (a) 3D bulk semiconductor materials and LD semiconductor materials including (b) 2D, (c) 1D, and (d) 0D. The inset arrows on related models indicate the quantum confinement direction and dimensionalities. (e) Quantum size effect on band structure. (f) Schematic illustration of surface plasmon resonance excitation on metallic NPs.

common in nanomaterials, the upper limit of dimensionality or size to trigger this effect will exactly depend on the individual electronic structure of each semiconductor [47]. For example, Li's group found that even though the thickness was as thin as 1.0 nm for 2D TiO_2 nanosheets, the quantum confinement was not apparent as no obvious changes are observed in the band-gap structure and electronic property. As such, they inferred that the quantum effects of some specific LD semiconductors should only occur on some specific dimensionality [48]. Nevertheless, since the diffusion distance of minority charge carriers in most semiconductors is below several nanometers, it is still challenging to control the quantum confinement within this range [49]. We will discuss in detail the quantum confinement in water oxidation in below sections where applicable.

6.3.1.2 Localized Surface Plasmon Resonance (LSPR)

Apart from the quantum confinement effect, another beneficial feature of metal-based nano-semiconductors is the (localized) surface plasmon resonance (SPR/LSPR). Plasmon resonance can be defined as the collective resonant oscillation of free conduction electrons restricted in metallic nanomaterials under light irradiation, while SPR are oscillations restricted to the surfaces of nanomaterials and interact actively with light [50]. As shown in Figure 6.5f, irradiating metal NPs at their plasmon frequency can produce energy ("hot") electrons and intense electric fields on the particle surface, which can be further extracted and applied in catalytic reactions [51]. In the past decade, SPR has been widely employed on specific 0D noble metals, e.g. Au, Ag, and Pt in photocatalysis [52]. Besides, plasmonic semiconductors usually possess excellent light absorption due to stronger light

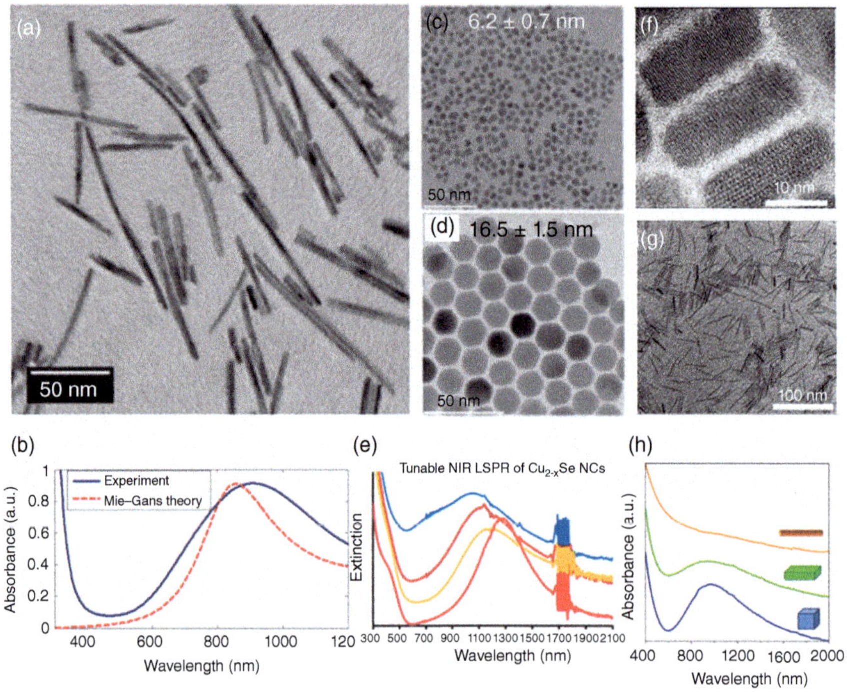

Figure 6.6 (a) TEM image of $WO_{2.83}$ NRs. (b) LSPR in 1D $WO_{2.83}$ NRs (experimental measurement and theoretical simulation). Source: Manthiram and Paul Alivisatos [55]. (c,d) Representative TEM images and (e) tunable LSPR of $Cu_{2-x}Se$ NCs with uniform size of 6.2 nm and 16.5 nm. Source: Liu et al. [57]. (f) HRTEM of $Cu_{2-x}Te$ nanoplates. (g) TEM images of $Cu_{2-x}Te$ nanorods. (h) UV–vis spectra nanoplate and nanorod. Source: Reproduced with permission. Li et al. [56]. Copyright 2013, American Chemical Society.

trapping. However, the SPR effect of these costly noble metal 0D materials is only beneficial to visible-light harvesting. Further studies discovered that LSPR can also be found in transition metal-based catalysts [50, 53]. It was reported that the LSPR effect could be employed to optimize the light absorption maximum in the NIR and mid-infrared ranges of non-noble metal oxide-based semiconductors [50, 54]. As shown in Figure 6.6a–e, through the formation of cations or oxygen vacancies, some nonstoichiometric self-doped semiconductors (e.g. $Cu_{2-x}Se$, $Cu_{2-x}Te$, WO_{3-x}) are identified as attractive LSPR hosts [55–57] and were applied for improving the photocatalytic water oxidation [58, 59]. It is also indicated that the plasma resonance frequency can be controlled by modifying the sizes (Figure 6.6d,e) or shapes (Figure 6.6f–h) of nanosized materials.

6.3.1.3 Surface Area and Exposed Facet-Enhanced Charge Transfer

As the catalytic reaction occurs on the surface, the surface area to volume proportion of catalysts plays a vital role in heterogeneous catalysis. The large surface area to volume ratio of nanosized materials is another promising feature, which makes the most of the catalyst surface atoms where reactions occur. For instance, a catalyst with a size of 100 nm has only around 1% atoms distributed on the surface, whereas this value increases to 10% and 90% for a catalyst of 10 and 1 nm size, respectively [16]. Furthermore, the larger specific surface area can promote the activity, selectivity, and faradaic efficiency of the catalytic reactions by improving the formation of edges and defects or by preferentially exposing highly active crystal facets. Besides, the more exposed surface atoms also make it easier to regulate nanosized material properties through elemental doping, surface modification, and defect engineering [60]. Taking doping as an example, most of the dopants are confined inside the structure of bulk materials; however, in nanosized materials, dopants are positioned close to the surface, acting as active catalytic sites. More importantly, doping can be adopted to adjust the electron density and enhance the light absorption of nanosized semiconductor materials.

In addition to the above three main superiorities, nanosized materials also own some other features that can be employed in photochemical energy conversions such as the reduced carrier collection pathways [45], enhanced light distribution [61], and multiple exciton generations [62]. All these advantages have been intensively studied for the water oxidation system. This will be discussed in below sections.

6.3.2 Zero-Dimensional Semiconductor Materials for Photocatalytic Water Oxidation

In general, 0D photocatalysts including metal nanoclusters or NPs and metal oxide QDs have been intensively investigated. For size-selected photocatalytic materials (<5 nm), most if not all of the metal sites are on the surface with the low-coordinated environment and favorable for the conversion of adsorbed species during water oxidation [63]. In this part, we summarize common 0D non-noble metal-based semiconductors for photocatalytic water oxidation.

6.3.2.1 0D Metal Complexes and Nanoclusters

In natural photosynthetic systems, water oxidation is catalyzed by a Mn_4CaO_4 cluster, which has stimulated the construction of transition metal clusters and complexes as effective catalysts for solar water oxidation at the 0D molecular level [64]. In a rather interesting study, Jiao and Frei reported that cobalt or manganese oxide nanoclusters loaded on mesoporous silica (SBA-15) can exhibit effective water oxidation under mild pH conditions [65, 66]. The Co_3O_4 and MnO_x nanoclusters on SBA-15 provided large quantities of surface metal atoms per projected area, while the mesoporous silica frameworks offered the nanoclusters with high dispersion stability and prevented the destruction of surface active sites. Photocatalytic water oxidation by these nanoclusters was performed using $Na_2S_2O_8$ as the electron acceptor. Due to the large geometrical surface area provided by such a nanostructure, increased TOF were obtained, which were over a thousand times higher than that of the bulk-sized counterparts. In the following work, Jiao's group employed various substances, including mesoporous silica KIT-6, SBA-15, and γ-Al_2O_3 NPs to support Co_3O_4 nanoclusters for fabricating a series of WOCs [67]. The results verified that the Co_3O_4 cluster with a smaller size resulted in a higher water oxidation activity. Compared with SBA-15, KIT-6 was determined to be able to afford more accessible loading channels, due to its 3D porous structure. More importantly, they concluded that the principal function of these mesoporous skeletons in photocatalytic water oxidation reaction is to physically disperse the Co_3O_4 nanoclusters rather than to participate in the reaction. Another typical example of an all-in-one molecular heterogeneous photocatalytic system was recently proposed by encapsulating the Co_4-POMs (the catalysts) in the pores of a porphyrinic MOF (the photosensitizer) through strong H-bonds [68]. The external Co–OH_2 sites could serve as the proton shuttling and redox catalytic centers of Co_4-POM, while the organic linkers of MOF could behave as light-harvesting units capable of transferring electrons to adjacent catalytic centers.

6.3.2.2 Metal Oxide Quantum Dots and Nanocrystals

Compared with the above in situ formed 0D metal complexes or clusters, the ex situ preparation allows for better control over surface approachability and dimensionality of the 0D materials such as QDs [69]. Because of the unique optical property, quantum effect, and relatively high photostability, semiconductor QDs have emerged as extraordinary 0D photocatalysts. Over the past decades, transition metal-based QDs such as Co_3O_4 and Fe_2O_3 had been developed [70, 71], while studies over the applications of these QDs for water splitting just started in recent years. Guo's group first tried to synthesize the Co_3O_4 QDs through a reverse micelle technique and applied them for wireless water oxidation [72]. The QDs exhibited a stable oxygen evolution (0.40 $\mu mol\,h^{-1}$) even in the absence of a sacrificial agent or cocatalyst. Mechanistic studies revealed that the band gap of Co_3O_4 QDs was increased by the quantum confinement effect and the shifted CB minimum was lower than the hydrogen reduction potential. In this case, the QDs were able to realize overall water splitting under visible-light irradiations. Following this success, Wang et al. synthesized α-Fe_2O_3 QDs (3 nm) using the same method and investigated them for wireless visible-light-driven water oxidation [70]. Also, by adding ammonium hydroxide into the benzyl alcohol route, Niederberger and coworkers successfully synthesized monodisperse Co_3O_4 QDs with sizes of 4.5 nm for excellent visible-light-driven water oxidation performance [73].

Although some QDs present considerable water oxidation performances, the presence of the stabilizing agent, an organic ligand, attached on the surface of QDs during synthesis could inevitably block the accessibility of surface active centers and then limit their extensive applications. In addition, the annealing or drying process for removing this ligand, in turn, would affect the dispersion of QDs. Fortunately, approaches to increase the dispersibility and to synthesize the ligand-free NPs have been achieved successfully [69, 74, 75]. Grzelczak et al. [69] reported the first example of ligand-free Co_3O_4 NPs with tunable sizes (Figure 6.7a–d). As shown in Figure 6.7e, at Co_3O_4 particle size of 3 nm, the surface area was elevated seven times higher than that of 40 nm size particles. With the reduced particle sizes, the absorption band edge displayed gradual blueshift toward shorter wavelengths, further revealing the apparent quantum size effect. The Co_3O_4 NPs with excellent colloidal stability were examined in the photocatalytic system with $Na_2S_2O_8$ as the sacrificial agent and $Ru[bpy]_3^{2+}$ as the photosensitizer. The oxygen evolution rates

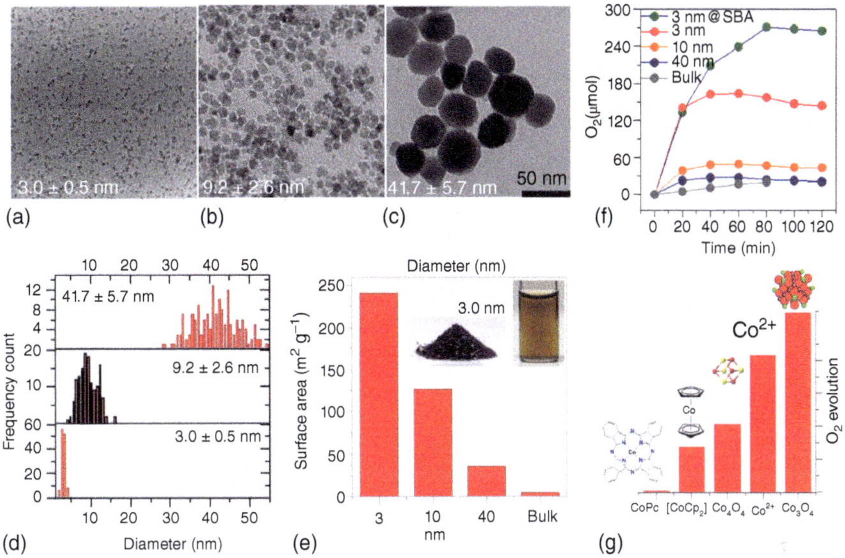

Figure 6.7 (a–c) TEM images and (d) size distributions of Co_3O_4 NPs with different diameters (3, 10, and 40 nm). (e) The relationship between particle diameter and surface area. Inset: Images of NPs in the form of powder or dispersed in water. (f) Visible-light-driven oxygen evolution of Co_3O_4 NPs with different diameters using $Ru(bpy)_3{}^{2+}$ as the sensitization system. (g) Visible-light-driven oxygen evolution from various cobalt compounds placed on SBA-15. Source: Reproduced with permission. Grzelczak et al. [69]. Copyright 2013, American Chemical Society.

and calculated turnover number (TON) per mole of cobalt gradually raised with reduced particle sizes, from 0.1 (bulk Co_3O_4) to 2.3 (3 nm Co_3O_4) (Figure 6.7f). The researchers also anchored the 3 nm Co_3O_4 onto SBA-15 and found that the OER rate was improved evidently, compared with the homogeneous dispersed 3 nm Co_3O_4. Besides, these heterogeneous Co_3O_4 NPs were more active than other homogeneous counterparts, such as CoPc, $CoCp_2$, Co^{2+} and Co_4O_4 (Figure 6.7g).

Later, Blakemore et al. [76] reported a pulsed laser ablation method to obtain surfactant-free, quantum-confined Co_3O_4 NPs (<5 nm) with an excellent water oxidation activity. Interestingly, the Co_3O_4 NP suspensions prepared in this work showed yellow color, which is completely different from that of bulk Co_3O_4 (black). This optical spectroscopic property adds more evidences to quantum confinement. Judging from the increase in direct and indirect band gaps as a function of the reciprocal squared particle diameter, this work precisely deduced that the maximum diameter to achieve quantum confinement effects of Co_3O_4 would be 5 nm. This work gives us an idea to rigorously control the experimental condition to employ the quantum confinement effects for water oxidation.

The above prediction about the quantum confinement of cobalt oxides can be partly supported by a recent work from Liao et al. [77] In this work, CoO nanocrystals were employed as a photocatalyst to split water into oxygen and hydrogen under visible-light irradiations in the absence of sacrificial reagents. Through diffuse reflectance spectroscopy, they determined that the band gaps

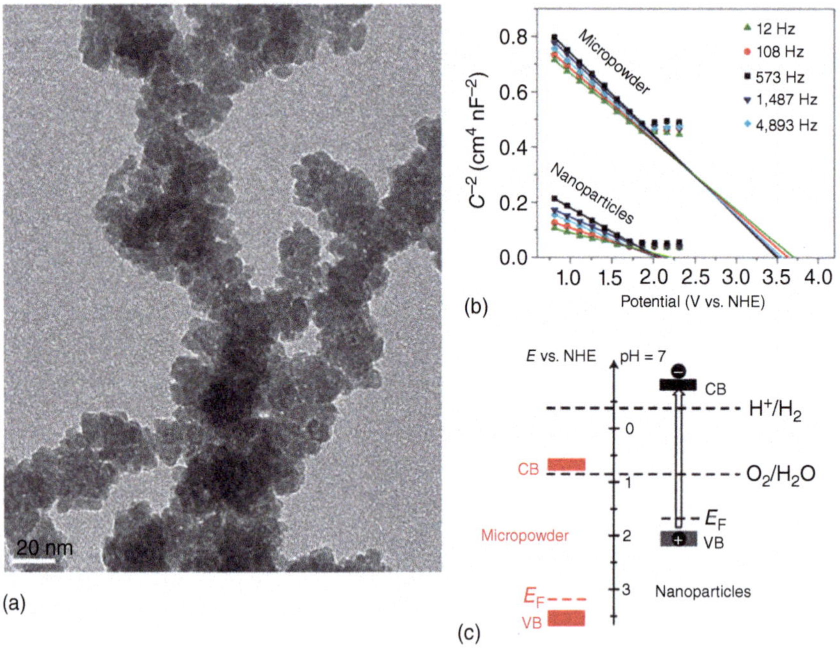

Figure 6.8 (a) TEM image and (b) Mott–Schottky plots for CoO NPs and micropowders. The flat band potentials are obtained via extrapolation method. (c) Band positions of CoO NPs and micropowders according to the band gaps and flat band potentials. Source: Reproduced with permission Liao et al. [77]. Copyright 2014, Nature Publishing Group.

of both CoO nanocrystals and bulk counterparts were very close to 2.6 eV. They further concluded that the comparatively large size of CoO NPs (5–8 nm, as shown in Figure 6.8a) led to the loss of the quantum confinement effect. However, Mott–Schottky plots confirmed that the nanoscale morphology could considerably alter the band-edge positions of a semiconductor, in spite of the absence of quantum confinement effect (Figure 6.8b,c). In this case, the researchers assumed that it is the band-edge position rather than the band gap resulted in the more excellent photocatalytic performance of CoO than their bulk counterparts.

The ligand-free single crystals with atomic-level accuracy allow for easier surface accessibility and less-affected geometric structures due to the absence of external coordination configuration [69]. Spinel-type AB_2O_4 NPs consisting of M_4O_4 cubane-like structure are the most representative NPs for photochemical water oxidation [78], and their catalytic performance is usually geometry site dependent [70, 71]. Taking spinel Co_3O_4 as example, two distinct cobalt sites exist in Co_3O_4: Co^{2+} tetrahedral sites ($Co^{2+}{}_{Td}$) and Co^{3+} octahedral sites ($Co^{3+}{}_{oh}$). The $Co^{3+}{}_{oh}$ and neighboring oxygen atoms are in a Co_4O_4 cubane composition, which has been recognized to be critical for water oxidation [79]. Wang and his coworkers had proposed that both the WNA and I2M pathway could occur on the distinct $Co^{3+}{}_{oh}$ sites. Direct coupling on adjacent oxo-bridged $Co^{3+}{}_{oh}$ OH binuclear site across a large number of cubic faces can facilitate the intramolecular O=O bonds formation, following the I2M water oxidation pathway [80]. Otherwise, the single $Co^{3+}{}_{oh}$

OH sites without the cooperation of the adjacent $Co^{3+}{}_{oh}$ sites will go through the catalytic cycle following the I2M water oxidation pathway. Through a phase transformation approach, Jiao et al. synthesized unsupported spinel CoO_x nanocubes with a size below 2 nm and an average cobalt oxidation state close to 3^+ [81]. Based on the EXAFS spectrum, the authors suggested a nanocube structure (~1.1 nm) consisting of 13 cobalt atoms, among which 12 cobalt atoms are on the surface and all the cobalt octahedra are interconnected with edge sharing. Interestingly, they claimed that the superior water oxidation catalytic activity of this type of nanocubanes compared with previous reported Co_3O_4 spinels was due to the high percentage of Co_4O_4 cubanes in a small cluster (8 Co_4O_4 with 13 Co atoms). The authors assumed that the WNA pathway that occurred at the single CoIIIOH site would be the principal water oxidation mechanism for this nanocubane catalyst.

Some other semiconductor NPs like CoSe [82], α-Fe_2O_3 [83], and $BiVO_4$ [84] were also employed as photocatalytic WOCs. For example, Osterloh's group first investigated the photocatalytic water oxidation using freely dispersed α-Fe_2O_3 nanocrystals [83]. In this work, hematite with different sizes including 5.4 nm α-Fe_2O_3 NPs, 44 nm ultrasonicated bulk-α-Fe_2O_3, and 120 nm bulk-type-α-Fe_2O_3 were achieved by hydrolysis of $FeCl_3 \cdot 6H_2O$. Beneficial to the decreased diffusion path of photogenerated holes, the suspended crystals of α-Fe_2O_3 showed a robust photocatalytic performance under irradiations with initial O_2 evolution rates of up to 1072 $\mu mol\ h^{-1}\ g^{-1}$ in the presence of $AgNO_3$.

6.3.3 One-Dimensional Semiconductor Materials for Photocatalytic Water Oxidation

Since carbon nanotubes (CNTs) were discovered by Lijima [85], extensive interests have been activated in studying 1D nanomaterials, including NWs, NRs, and NTs. Substantial breakthroughs have been achieved in applying 1D materials in various fields of modern techniques, for example, photocatalytic water splitting system. By preferential exposure of selected facets, 1D photocatalysts possess several advantages, for example, high accessible surface area and directed transfer channels [13], endowing them with a better ability for transfer of (photo)generated charges away from potential recombination sites. Attributed to their linear morphologies, 1D semiconductors are mainly assembled onto the photoanodes as the nanoarrays and widely applied in photoelectrochemical water oxidation [86]. Therefore, in this section, we mainly make a brief investigation about the implementation of 1D nanomaterials for photocatalytic water oxidation.

Electrons tend to migrate randomly and interact with scattering sites easily on 0D, 2D, or 3D semiconductor materials, bringing about a high possibility of charge recombination [45, 87]. Distinctly, the confined conductive channels of 1D semiconductors [47] can direct electrons to transfer along the 1D diffusion channel to external circuit and thus reduce electron–hole recombination to a great extent. 1D nanostructured semiconductors in the form of NWs, NRs, or NTs have been broadly reported. Xie's group first fabricated novel quantum tubes of $BiVO_4$, which had ultra-narrow diameters (of about 5 nm) and ultrathin wall thicknesses

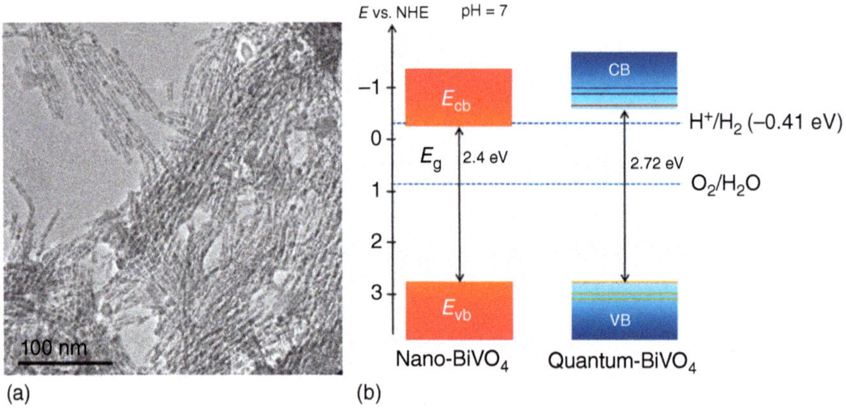

Figure 6.9 (a) TEM image quantum-sized $BiVO_4$. (b) Comparison of the band position of nanoscale $BiVO_4$ and quantum-sized $BiVO_4$. Source: Reproduced with permission. Sun et al. [89]. Copyright 2014, American Chemical Society.

(down to 1 nm). A well-defined visible-light response was observed on the resulting high-quality $BiVO_4$, revealing the evident superiority of quantum-sized NTs to bulk materials [88]. Inspired by this study, $BiVO_4$ quantum tubes were utilized as excellent photocatalysts for solar light water oxidation [88–90]. Interestingly, Sun et al. [89] reported that the quantum-sized $BiVO_4$ could function as the catalysts for photocatalytic water oxidation in the absence of cocatalysts or sacrificial reagents (Figure 6.9a). Band structure analysis indicated that quantum confinement effect results in the negative shift of CB edge (E_{CB}) and reduces water oxidation overpotentials, resulting in a significant photocatalytic activity (Figure 6.9b). These findings provided new insights to enhancing water splitting performance by controlling the sizes of photocatalysts.

As mentioned above, once the sizes of semiconductors are below a certain nanometer scale, the band gap will begin to increase, known as the quantum size effects. To meet this demand, Serpone et al. reported that the exciton radius (r_{ex}) of TiO_2 should be in the range of 0.35–1.9 nm [91]. In a density functional theory (DFT) study by Iacomino et al., r_{ex} was calculated to be at least 1.5 nm [92]. Specifically, compared with the bulk value, the blueshift of NWs grown along the [001] direction can vary from 0.18 to 0.67 eV, depending on the size variations. Therefore, the quantum confinement can be adopted as an effective approach to improve the photocatalytic water oxidation capability, and many studies have been performed on TiO_2 inspired by this.

Although TiO_2 performs well in the ultraviolet (UV) region, it hardly has absorption for visible light that represents the most abundant part of solar light energy, making it a poor choice for practical applications. Some attempts have been made to modify the band gap of TiO_2, including nitrogen incorporation, which can upshift the VB edge of TiO_2 by hybridizing substitutional N 2p with O 2p states, while the CB edge can be almost remained unchanged [93]. For stronger visible-light absorption, semiconductors with intrinsic narrow band gaps (E_g) like hematite (α-Fe_2O_3,

E_g = 2.0–2.2 eV), WO_3 (E_g = 2.5–2.8 eV), or $BiVO_4$ (2.4 eV) are especially attractive in spite of their respective demerits. Additionally, ZnO with an electron mobility rate of 10–100 times higher than that of TiO_2 is another common 1D nanostructured candidate.

6.3.4 Two-Dimensional Semiconductor Materials for Photocatalytic Water Oxidation

2D nanosheets or nanolayered materials have been explored in a wider range of photocatalysis than 0D or 1D materials, attributed to their ultrahigh specific surface area and unique physical, optical, and chemical properties [94–96]. However, as mentioned above, the main drawback of 2D semiconductors is the high recombination rate caused by randomly migrated electrons and larger specific interfacial areas. This can be mitigated by suitable control over the morphology, thickness, surface defects, and the vacancies. In this section, representative 2D materials, including metal oxide nanosheets, layered double hydroxide (LDH) nanosheets, metal-based chalcogenides, and oxyhalides, for photocatalytic water oxidation are summarized to better explore key approaches to enhance the photocatalytic performance of 2D materials.

6.3.4.1 2D Metal Oxide Nanosheets for Photocatalytic Water Oxidation

Discoveries on the optoelectronic properties of 2D metal oxide nanosheets can be traced back to the 1990s [96, 97]. In an earlier example, Sasaki and coworkers conducted the exfoliation of layered titanates and acquired $Ti_{0.91}O_2$ nanosheets at 0.93 nm thickness, which astonishingly showed a semiconducting activity similar to bulk TiO_2 anatase. Specifically, $Ti_{0.91}O_2$ nanosheets exhibited a sharp photoabsorption peak at 265 nm (4.77 eV), revealing a larger band-gap energy than bulk TiO_2 [96]. Since then, the photoelectrochemical properties of various 2D metal oxides have been extensively studied. Here we focus on recent advances on modification and application of these 2D metal oxide nanosheets for photocatalytic water oxidation.

Tailoring exposed crystal facets is a conventional strategy for optimizing the heterogeneous reactivity of 2D materials, which is closely related to the bonding environment and surface atomic configuration. Recent investigations explained that nanosheets exposed with the (001) facets give more excellent catalytic activities than that with the other facet exposures resulting from the highest oxygen atom density [99]. Taking TiO_2 as an example, it was reported that the formation energies were 0.90, 0.53, and 0.44 J m^{-2} for the (001), (100), and (101) facets, respectively [100]. The super-high Ti=O=Ti bond angles and coordinatively unsaturated O and Ti atoms on the (001) endow the facets with the highest surface energy. Using a modified hydrothermal strategy, Li and coworkers reported anatase TiO_2 nanosheets with ultrathin thicknesses (1.6–2.7 nm) and a high percentage of reactive (001) facet exposure (82%) [48]. Attributed to these features, the resulting products showed a high UV–vis light-driven water splitting performance. Another typical 2D semiconductor whose catalytic activity is relevant to exposed facets is

WO_3. Waller et al. investigated WO_3 nanosheets (0.75 nm thick) with the exposed {001} facets obtained by $Bi_2W_2O_9$ exfoliation for wireless photochemical water oxidation [98]. It is worth noting that this work also discovered the extended band gap of nano-WO_3 by quantum confinement, providing the designing opportunities of thickness-dependent 2D semiconductors for adjustable band-edge positions and band-gap energies.

Defect and vacancy engineering also performs a vital part in improving the water oxidation performance of 2D materials [58, 101]. On the one hand, grain boundaries, surface defects, and bulk defects in the 2D materials are generally considered as the electron–hole trapping sites and recombination centers. On the other hand, proper introduction of defects to semiconductors such as by doping with foreign elements or via creating oxygen vacancies can improve the electrical conductivity significantly. It is worth noting that oxygen vacancies have more important implications on 2D nanosheets than on bulk materials, owing to the LSPR. In 2015, Gong and coworkers synthesized WO_3 nanosheets rich of surface oxygen vacancies for solar-driven water oxidation [58]. In this study, layered tungstic acid was first exfoliated to WO_3 nanosheets, followed by a vacuum (WO_{3-x}-VT) or hydrogen treatment (WO_{3-x}–HT) to introduce oxygen vacancies and surface disorder of oxides (Figure 6.10a–c). As obtained from UV–vis–NIR spectra, the LSPR absorption peaks were observed at the NIR region for WO_{3-x}-VT (≈1520 nm) and WO_{3-x}–HT (≈1450 nm). As demonstrated in Figure 6.10d, three different channels were theoretically proposed for solar light adsorption: (i) electrons activated from VB maximum to CB minimum, (ii) electrons generated from VB maximum to oxygen vacancies under CB, and (iii) electrons activated by LSPR. The created oxygen vacancies caused SPR of WO_3, which not only resulted in enhanced photocatalytic activity in UV and visible region but also induced light adsorption in the NIR range through LSPR. Consequently, the as-prepared substoichiometric WO_3 nanosheets (WO_{3-x}-HT) display the most distinguished oxygen evolution activity, followed by WO_{3-x}-VT and then WO_3 under UV and visible-light irradiation. Meanwhile, the effect of LSPR on the photocatalytic activity was further proved when testing the oxygen evolution performance of WO_{3-x}-HT under different monochromatic lights. Attributed to the enhanced electron–hole separation by LSPR, it can be recognized that added NIR light could raise the oxygen evolution rate to some extent, depending on the applied monochromatic light wavelength. Ultrathin nanosheets of α-Fe_2O_3 were reported to be active for photocatalytic water oxidation [102]. These α-Fe_2O_3 nanosheets own attractive properties, e.g. ultrathin thickness (3.5 nm, which is comparable with the hole diffusion length) and largely exposed active (110) facets. The photocatalytic oxygen evolution rate can achieve 70 mmol h^{-1} g^{-1} and maintain for 15 hours.

6.3.4.2 Layered Double Hydroxide (LDH) Nanosheets for Photocatalytic Water Oxidation

LDHs refer to a group of 2D ionic composites composed of stabilized anions and solvation molecules intercalated inside the positively brucite-like layers [103]. Since Garcia et al. first proposed that LDHs can act as band-gap-controllable semiconductors for splitting water into oxygen [104], a series of Zn/Ti, Zn/Ce, and Zn/Cr LDHs

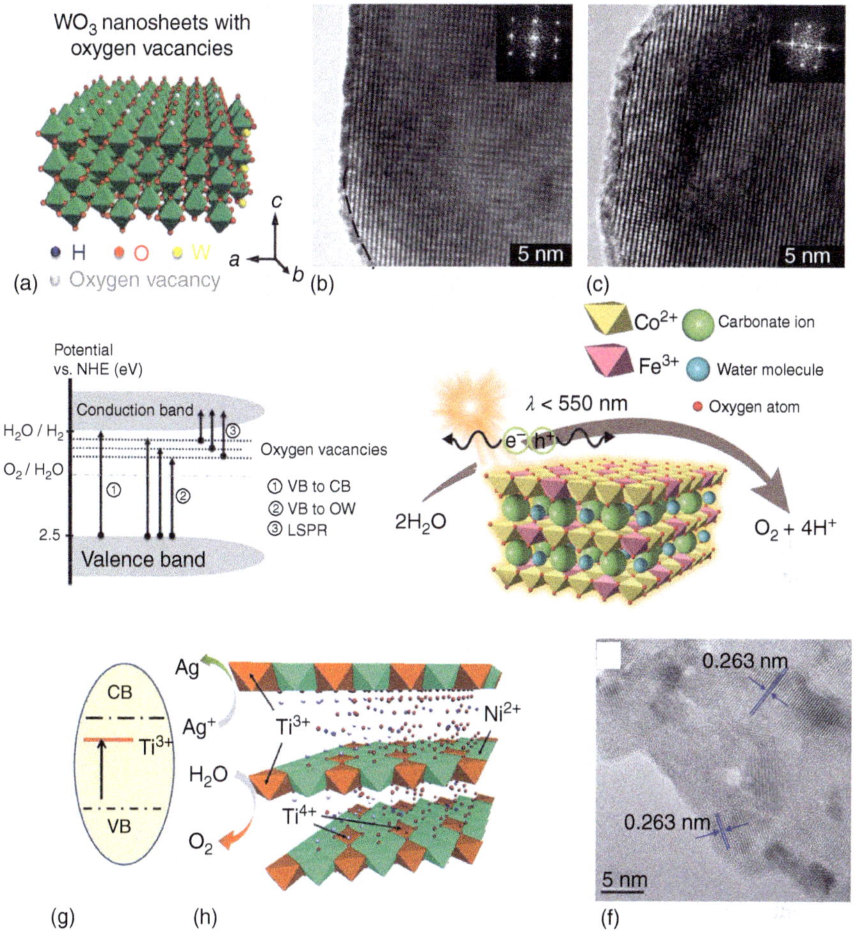

Figure 6.10 (a) Tungsten oxide single-crystal nanosheets. (b,c) HRTEM images of WO_3, WO_{3-x}-VT, and WO_{3-x}-HT nanosheets. (d) Band position regulation of WO_3 nanosheets with oxygen vacancies. Source: From Yan et al. [58]. © 2015 John Wiley & Sons. (e,f) Schematic model of a Co–Fe LDH structure and corresponding water oxidation pathway; HR-TEM image of Co–Fe LDH. Source: From Kim et al. [105]. © 2014 Royal Society of Chemistry. (g,h) Structural illustration of energy level for the Ti^{3+} self-doped NiTi-LDH nanosheet and the corresponding oxygen evolution pathway over the NiTi-LDH. Source: From Zhao et al. [106]. © 2014 Royal Society of Chemistry.

have been synthesized and applied for the visible-light-driven wireless photocatalytic OER. The most active Zn/Cr LDH nanosheets showed an impressive quantum yield up to 60.9% at 410 nm for O_2 generation, which was one of the highest rates ever recorded for the visible-light water oxidation at that time. Kim et al. synthesized Co–Fe LDH photocatalysts with different Co–Fe ratios, and the sample with the highest Fe content exhibited the highest efficiency for visible-light water oxidation (Figure 6.10e,f) [105]. It was indicated that Co–O–Fe bridges existed in the LDH structure can effectively suppress the electron–hole recombination, which was responsible for the superior photocatalytic efficiency.

The catalytic activity of LDHs can be readily tuned by adjusting the interlayer distance through the balanced anions of different ionic diameters. It has been proven that a larger interlayer spacing will lead to a higher OER activity, possibly attributable to the increased active site exposure and the enhanced exchange of the intercalated reactive anions. Exfoliating LDHs to several layers and even to monolayer is an attractive approach to shorten the charge diffusion path and increase the active electrical surface area (ECSA). As such, nanosized LDHs have received much attention. Lee et al. [103] prepared a series of nanosized Ti-embedded LDHs for excellent photocatalytic water oxidation. The abundant surface Ti^{3+}–O defects served as the trapping sites to improve charge separation, while TiO_6 octahedra confined within the 2D matrix can suppress the electron–hole recombination under visible-light irradiations. Distinctly, the high-surface-area Ni/Ti-LDH with light absorption in blue and red wavelengths exhibited a higher O_2 evolution rate than that of the Cu/Ti-LDH by the visible-light catalytic reaction.

During the synthesis process of ultrathin LDHs, a certain amount of surface atoms tends to be easily removed or lost, creating local structural distortions and vacancy defects, which will exert important impacts on the photochemical activity of LDH nanomaterials. To gain a deeper understanding of the types of defects and corresponding effects, extensive works have been conducted [106]. In an earlier example, Zhao et al. [106] synthesized ultrathin Ti^{3+}-containing LDH layers (thickness of about 2.0 nm) incorporated with M^{2+} cations (e.g. Ni^{2+} or Zn^{2+}) through a reverse microemulsion method. The detection of Ti^{3+} species in NiTi-LDH nanosheets indicated the presence of oxygen defects, which affected the electronic structure and facilitated efficient separation of electron–hole pairs (Figure 6.10g,h). Theoretical calculations revealed that the electronic property was modified and the band-gap energy was reduced to boost stronger light absorption. As a result, the NiTi-LDH displayed a super-high photocatalytic oxygen evolution activity (2148 $\mu mol\ g^{-1}\ h^{-1}$) with a quantum yield of 65.0% at 400 nm, which is one of the most effective OER efficiencies to date.

6.3.4.3 Metal-Based Oxyhalide Semiconductors for Photocatalytic Water Oxidation

As a new class of 2D materials, it is facile to obtain single- or few-layer structures in bismuth oxyhalide (BiOX, X = I, Cl, Br). BiOCl nanolayer with an atomically thin thickness and the (001) exposed facets was firstly developed by Liu and coworkers [107]. DFT calculations revealed that the (001) facets showed lower VB and CB positions than the (110) facets in BiOCl. This means that, under light irradiations, photogenerated electrons tend to gather on the (001) facets, while holes are inclined to migrate along the (110) facets, as depicted in Figure 6.11a. It was also discovered that defects created in basal planes of 2D BiOCl lowered CB edge, while VB edge remained unchanged. As shown in Figure 6.11b,c, the defects in BiOCl nanosheet could promote more efficient charge separations while they had little effect on the oxidizing ability of photoinduced holes. Under the synergetic function of quantum size effect and defect-triggered properties, defect-rich BiOCl exhibited dramatically enhanced wireless photocatalytic OER activity with OER rate of 56.85 $mmol\ g^{-1}\ h^{-1}$

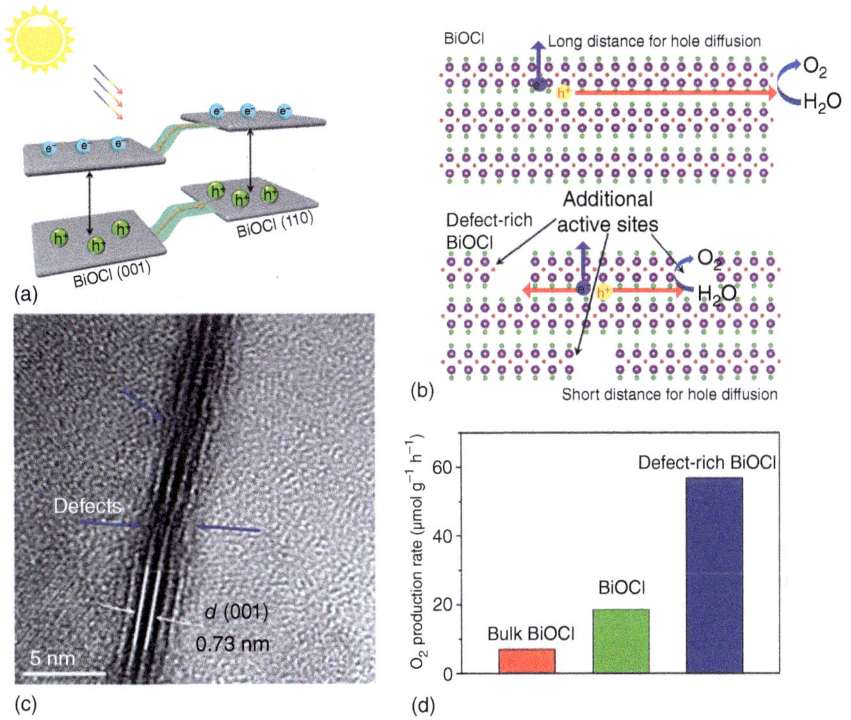

Figure 6.11 (a) Representation of charge transfer between (001) and (110) facets. (b) Detachment and transfer of photoexcited charges in the BiOCl and defect-rich BiOCl. (c) TEM image of the defect-rich BiOCl ultrathin nanosheets. (d) Photocatalytic OER for BiOCl and defect-rich BiOCl. Source: Reproduced with permission. Di et al. [107]. Copyright 2017, the Royal Society of Chemistry.

in $AgNO_3$ solution (Figure 6.11d), which was approximately 3 and eight times higher than that of 2D BiOCl and bulk BiOCl, respectively.

6.3.5 LD Semiconductor-Based Hybrids for Photocatalytic Oxygen Evolution

As mentioned above, due to the fast recombination rate of photogenerated holes, pristine photocatalysts generally show limited photocatalytic activity [108, 109], and we have summarized possible modification methods to tackle this problem. Apart from the structural adjustment, synthesizing LD hybrid materials is another appealing approach, which allows the construction of cascading heterojunctions with atomically sharp interfaces to promote charge dissociation. The sensible screening of LD materials with various dimensionalities for building heterojunctions in the hybrids can further accelerate charge carrier transport and increase the light absorption (Figure 6.12). In this section, we will focus on the synergistic effect caused the LD hybrids for photochemical water oxidation.

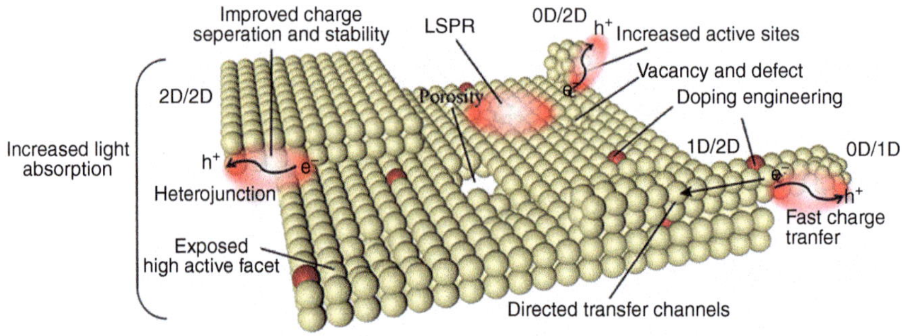

Figure 6.12 The photocatalytic properties of LD materials including light absorption and charge transfer can be improved by doping engineering, active facet exposure, defect or vacancy creation, and LSPR. Hybrid LD materials with heterojunction can also be constructed in the form of 0D/1D, 1D/1D, 0D/2D, and 2D/2D configurations.

6.3.5.1 1D-Based (0D/1D and 1D/1D) Semiconductor Hybrids for Enhanced Photocatalytic Water Oxidation

As mentioned above, 1D nanoarray structure is mostly applied in PEC rather than the photocatalytic system, as it offers a vectorial and confined channel for transmit of photogenerated electrons to conductive substrate and external circuit, thereby separating electron and hole efficiently. By contrast, electron transport in 0D or 2D nanostructures is relatively slow since electrons are scattered or trapped at grain boundaries. However, the variety of 1D semiconductors is limited, and their photocatalytic activity is sometimes low with small number of active sites. As such, a suitable hybridizing of 1D with 0D or 2D nanostructures can solve their respective drawbacks and realize material optimization for photocatalytic water oxidation. 1D channels can serve as excellent supports for 0D and 2D semiconductors. Furthermore, appropriate 0D or 2D nanostructure can act as OER cocatalysts to reduce reaction barriers for 1D photocatalysts. Heterojunctions can be built among 1D/0D or 1D/2D photocatalysts, which further enhance the separation and transport of charge carriers.

Taking 1D TiO_2 as example, several groups have prepared TiO_2-based 0D/1D hybrids that exhibited higher OER efficiencies in the visible-light region [108, 110, 111]. For instance, Maeda et al. [110] discovered that rutile TiO_2 NRs decorated with 0D $Co(OH)_2$ nanoclusters (about 2 nm) were able to oxidize water into O_2 under visible light with wavelengths up to 850 nm (Figure 6.13a–c), which is the longest wavelength ever recorded for heterogeneous photocatalytic water oxidation. It is worth noting that $Co(OH)_2$ nanoclusters could function not only as a catalyst for water oxidation but also as visible-light absorbers, as presented in Figure 6.13d.

Representative structures for 1D/1D hybrids mainly include a combination of nanobundles or branched NRs. For example, Wang et al. fabricated WO_2–WO_3 hybrid NRs to show a profoundly effective visible-light-driven photocatalytic water oxidation performance [53]. The platinum-like attribute of metallic WO_2 caused LSPR, which extended the visible-light absorption to the NIR and facilitated the

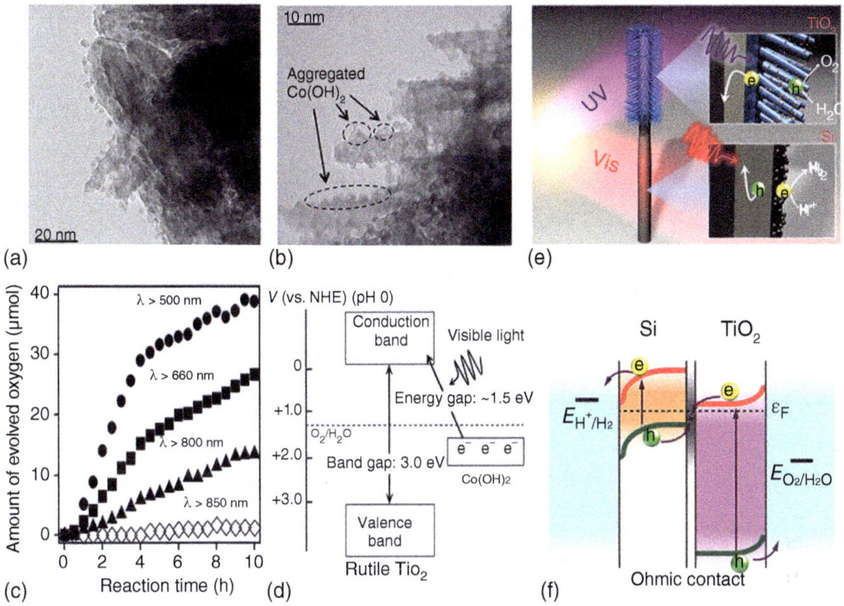

Figure 6.13 (a,b) TEM images. (c) Time dependence of the O_2 evolution. (d) Proposed energy diagram for $Co(OH)_2/TiO_2$. Source: Reproduced with permission. Maeda et al. [110]. Copyright 2016, Wiley-VCH. (e) Representation of the nanotree heterostructure. The TiO_2 nanowires (blue) were spread on the top half of the Si nanowire (gray), and the two materials absorb light with different wavelengths. (f) Representation of the electron–hole separation and transfer in the nanotree heterostructure for photocatalytic water splitting. Source: From Liu et al. [112]. © 2013 American Chemical Society.

charge carrier detachment on WO_3. As a result, the photocatalytic oxygen evolution rates reached approximately 220 ($\lambda = 700$ nm) and 200 ($\lambda = 800$ nm) mmol g^{-1} h^{-1}. Yang's group describes a nanoscale treelike light-harvesting hierarchical structure, in which the Si and TiO_2 1D NW serves as the trunk and the branches, respectively (Figure 6.13e,f) [112]. Once dispersing these nanotree heterostructures in the electrolyte, they can function as a "Z-scheme" to generate H_2 and O_2 simultaneously during illumination. Because of the band bending at the semiconductor–electrolyte interfaces, the photoexcited holes and electrons from Si and TiO_2 would recombine at the ohmic junction. Owing to the cooperation of cocatalysts (yellow and gray dots displayed in the inset picture in Figure 6.13e), the photogenerated electrons from Si NWs would move to the surface and reduce protons to hydrogen; the photogenerated holes from TiO_2 NWs would oxidize water to generate oxygen. Varying the percentage of the TiO_2-branched NWs can optimize the nanotree heterostructures to achieve a high energy conversion efficiency (0.12%), which is comparable with that of the natural photosystem.

6.3.5.2 2D-Based (2D/2D) Semiconductor Hybrids for Enhanced Photocatalytic Water Oxidation

Compared with heterostructures formed in 1D/1D and 0D/1D hybrids, it is generally viewed that 2D/2D heterostructures form relatively better-built hetero-interfaces for charge separation and migration. Recently, Wang's group easily synthesized novel 2D/2D WO_3@$CoWO_4$ bilayer nanosheets as active photocatalytic WOCs [113]. The constructed WO_3@$CoWO_4$ p–n heterostructure and created interfacial oxygen vacancies reduced the energy barriers for OER. DFT calculations indicated that the p–n heterojunction vested the 2D/2D composite with a narrowed band gap for better visible-light absorption and rapid interface charge transfer and separation. Note that the stability of WO_3 that is sensitive to photocorrosion was improved by promoting the migration of holes to $CoWO_4$.

6.3.5.3 Metal-Free-Based Semiconductors for Water Oxidation

Although we mainly illustrated the applications of earth-abundant metal-based LD semiconductors for water oxidation, a series of LD metal-free conductors like 0D carbon QDs [90] and 2D g-C_3N_4 [114] can also be hybridized with metal-based materials like WO_3 [115] and α-Fe_2O_3 [116] and used for water oxidation application. To some extent, some strategies summarized above for modifying photocatalytic properties of metal-based semiconductors such as building suitable heterojunctions, morphology control, doping, and synthesizing thin films or dots to generate quantum effect can also be applied for LD metal-free materials modifications [117–120]. For example, Zhang et al. reported that 2D g-C_3N_4 supported 0D Co_3O_4 QDs can be applied for photocatalytic water oxidation. Co_3O_4 QDs were dispersedly loaded onto porous g-C_3N_4 nanosheets, and a tight heterojunction could be realized at the interface between g-C_3N_4 and Co_3O_4. Due to the effective photogenerated electron–hole separation and enhanced charge carrier transfer, Co_3O_4 QDs decorated g-C_3N_4 nanosheets exhibited enhanced water oxidation activity compared with pristine g-C_3N_4. Carbon dots decorated $BiVO_4$ QD show improved photocatalytic water splitting activity without any cocatalysts or sacrificial reagents. The introduction of carbon dots could enlarge the photoadsorption range and enhance the charge separation efficiency of $BiVO_4$. More importantly, the energy level position of $BiVO_4$ could be hence adjusted to complete the efficient overall water splitting via a two-electron pathway [90]. Through loading 0D α-Fe_2O_3 NPs onto the surface of 2D reduced graphene oxide (rGO) nanosheets, photocatalytic water oxidation activity of α-Fe_2O_3 could be highly enhanced. Transient characterizations exhibited that the photoexcited electrons could be directly transferred from α-Fe_2O_3 to rGO, which stimulates the charge separation and suppresses the charge recombination [121].

6.4 Catalytic Active Site–Catalysis Correlation in LD Semiconductors

The differences in dimensionality define the various catalytic attributes of 0D, 1D, and 2D semiconductor materials, and we tried to point out their most unique properties, respectively. As discussed in this chapter, 0D semiconductors, which are easier to be dispersed in solution, are more favored in photocatalytic water oxidation. They are also commonly applied to fabricate hybrid materials because of the small size, easy dispersion, and deposition on other materials. Without a doubt, quantum size effect exists mostly on 0D materials to trigger changes in the band structure and enable more active OER, due to the low dimensions in all directions. Owing to the explicit growing orientation, 1D NWs or NTs can be easily assembled into ordered vertical arrays that are more suitable for PEC applications rather than photocatalytic, mainly thanks to the directed electron transfer channels. In addition, 1D NW/or NT arrays can serve as ideal supports for other catalysts. 2D materials have the highest photoconversion efficiency due to the short migration distance of the photoexcited carriers in the 2D nanosheet, and with the large section area, enormous photons can be absorbed in transient time under a low photon flux density. 2D materials are also widely used in photocatalytic water oxidation. The more exposed surface atoms bring out more possibilities in adjusting the active facets and porosities and creating surface defects or oxygen vacancies, which can induce unique LSPR and active sites on 2D materials.

6.5 Conclusions and Perspectives

Photocatalytic water oxidation, as the primary half-reaction in artificial photosynthesis, has attracted significant attention. In the past decades, considerable progress has been reported in designing homogeneous and heterogeneous WOCs. For homogeneous WOCs, great activities have been made in molecular complexes or clusters based on transition metals. The definite molecular structure and excellent oxygen evolution performance observed in homogeneous medium stimulate further work in a more profound perception of the water oxidation mechanism. Moreover, in this chapter, we have reviewed that the photocatalytic water oxidation performance of these complexes can be altered by adjusting the metal nuclearity and coordinated ligands. However, the stability and retrievability of homogeneous WOCs are always the most concerned flaw. Therefore, it is very desirable to synthesize more stable molecular complexes that own high photocatalytic water oxidation activity in various reaction conditions. In terms of heterogeneous WOCs, semiconductors with low dimensions have provided us fundamental insights and new possibilities into the burgeoning photocatalytic field, showing well-defined technological promises. By means of regulating the nanostructures and/or electronic properties of nanosized semiconductors, the photocatalytic performances can be readily enhanced, allowing for the well establishment of superior photocatalysts for water oxidation. Specifically, we have outlined some prominent strategies for better LD semiconductors in this chapter including quantum size effect, LSPR, active facet exposure, defect engineering, and heterojunction construction in LD hybrids. As a result, inspiring progress including enhanced efficiencies, improved activities, and prolonged stabilities has been achieved in LD photocatalysts.

For future research in selecting suitable LD semiconductors for water oxidation, several factors such as band-edge position, charge carrier diffusion length, charge separation, and stability need to be considered. Although nanosized materials possess superior photochemical advantages in terms of fast charge transport, excellent light-harvesting ability, improved reaction kinetics, and vigorous photocatalytic activity, they have some inherent restrictions that may result in poor durability and diminished power conversion efficiency. For single-component nanosized materials with sizes lower than certain values, carrier separation is even more difficult than in bulk materials [45, 122]. Taking n-type semiconductor as an example, when the diameter of nanosized materials is lower than twice the value of surface space charge layer width (L_d) of bulk materials, it might be difficult to relax the bands completely to the bulk level. In that case, small nanosized materials tend to possess a lower potential drop and electric field across the space charge layer, which can elevate the recombination rate of photogenerated electrons/holes. Therefore, a suitable size balance needs to be established to maximize the exposed surface active sites while optimizing charge separations at the same time. Another drawback of nanomaterials is the declined thermodynamic stability along with the increased surface energy. Hence, it is vital to identify the optimal operation conditions (e.g. light and optimum pH values in aqueous solution) of specific LD semiconductors to ensure that they are both active and stable. Because of the sustainable energy requirements in society, further efforts are needed to be given in reducing the cost of earth-abundant LD semiconductors, because the projected expenditure of artificial photosynthesis systems is principally determined by light harvesters.

In the end, the photocatalytic water oxidation system has its drawbacks because the charge carrier separation efficiency is much lower than that in the PEC system. Besides, there are difficulties to separate the stoichiometric mixture of O_2 and H_2 effectively and to avoid backward reactions. The water oxidation efficiency in a wireless photocatalytic system remains far below than that in wired PEC system powered by photovoltaic cells. It was reported that PEC system can afford high solar-to-hydrogen efficiencies up to 30% in a three-junction solar cell, approaching the Shockley–Queisser limit [13]. Hence, how to improve the catalytic activity, especially in the heterogeneous system, will always be the most challenging point and the hottest topic in future studies. At last, the mechanism of photocatalytic water oxidation has not been fully unveiled and will still be a challenging subject in the foreseeable future research. It can be predicted that photocatalytic and photoelectrochemical water oxidation will play significant roles together in the future sustainability.

References

1 Akira Fujishima, K.H. (1972). *Nature* 238: 37.
2 Bard, A.J. (1979). *J. Photochem.* 10: 59.
3 Bard, A.J. (1980). *Science* 207: 139.
4 Han, X.B., Li, Y.G., Zhang, Z.M. et al. (2015). *J. Am. Chem. Soc.* 137: 5486.
5 Matheu, R., Ertem, M.Z., Gimbert-surin, C. et al. (2019). *Chem. Rev.* 119: 3453.
6 Cowan, A.J. (2016). *Nat. Chem.* 8: 740.
7 Bergmann, A., Jones, T.E., Moreno, E.M. et al. (2018). *Nat. Catal.* 1: 711.
8 Kärkäs, M.D., Verho, O., Johnston, E.V., and Åkermark, B. (2014). *Chem. Rev.* 114: 11863.

9 Limburg, J., Vrettos, J.S., Chen, H. et al. (2001). *J. Am. Chem. Soc.* 3: 1010.
10 Wang, T. and Gong, J. (2015). *Angew. Chem. Int. Ed.* 54: 10718.
11 Hisatomi, T., Kubota, J., and Domen, K. (2014). *Chem. Soc. Rev.* 43: 7520.
12 Reza Gholipour, M., Dinh, C.-T., Béland, F., and Do, T.-O. (2015). *Nanoscale* 7: 8187.
13 Voiry, D., Shin, H.S., Loh, K.P., and Chhowalla, M. (2018). *Nat. Rev. Chem.* 2: 0105.
14 Du, P. and Eisenberg, R. (2012). *Energy Environ. Sci.* 5: 6012.
15 Ran, J., Zhang, J., Yu, J. et al. (2014). *Chem. Soc. Rev.* 43: 7787.
16 Deng, X. and Tüysüz, H. (2014). *ACS Catal.* 4: 3701.
17 Dismukes, G.C., Brimblecombe, R., Felton, G.A.N. et al. (2009). *Acc. Chem. Res.* 42: 1935.
18 Umena, Y., Kawakami, K., Shen, J., and Kamiya, N. (2011). *Nature* 473: 55.
19 Zhang, C., Chen, C., Dong, H. et al. (2015). *Science* 348: 690.
20 Al-oweini, R., Sartorel, A., Bassil, B.S. et al. (2014). *Angew. Chem. Int. Ed.* 53: 11182.
21 Kanady, J.S., Tsui, E.Y., Day, M.W., and Agapie, T. (2011). *Science* 333: 733.
22 Panda, C., Debgupta, J., Díaz Díaz, D. et al. (2014). *J. Am. Chem. Soc.* 136: 12273.
23 Hong, D., Jung, J., Park, J. et al. (2012). *Energy Environ. Sci.* 5: 7606.
24 Hetterscheid, D.G.H. and Reek, J.N.H. (2012). *Angew. Chem. Int. Ed.* 51: 2.
25 Wang, N., Zheng, H., Zhang, W., and Cao, R. (2018). *Chin. J. Catal.* 39: 228.
26 Yan, Y., Lee, J.S., and Ruddy, D.A. (2015). *Inorg. Chem.* 54: 4550.
27 McCool, N.S., Robinson, D.M., Sheats, J.E., and Dismukes, G.C. (2011). *J. Am. Chem. Soc.* 133: 11446.
28 Smith, P.F., Kaplan, C., Sheats, J.E. et al. (2014). *Inorg. Chem.* 53: 2113.
29 Berardi, S., La Ganga, G., Natali, M. et al. (2012). *J. Am. Chem. Soc.* 134: 11104.
30 Lin, W., Su, Z.-M., Li, Y.-G. et al. (2014). *J. Am. Chem. Soc.* 136: 5359.
31 Huang, Z., Luo, Z., Geletii, Y.V. et al. (2011). *J. Am. Chem. Soc.* 133: 2068.
32 Lin, J., Meng, X., Zheng, M. et al. (2019). *Appl. Catal., B* 241: 351.
33 Song, F., More, R., Schilling, M. et al. (2017). *J. Am. Chem. Soc.* 139: 14198.
34 Wasylenko, D.J., Ganesamoorthy, C., Borau-garcia, J., and Berlinguette, C.P. (2011). *Chem. Commun.* 47: 4249.
35 Yin, Q., Tan, J.M., Besson, C. et al. (2010). *Science* 328: 342.
36 Stracke, J.J. and Finke, R.G. (2011). *J. Am. Chem. Soc.* 4: 14872.
37 Stracke, J.J. and Finke, R.G. (2013). *ACS Catal.* 3: 1209.
38 Vickers, J.W., Lv, H., Sumliner, J.M. et al. (2013). *J. Am. Chem. Soc.* 4: 14110.
39 Yoffe, A.D. (1993). *Adv. Phys.* 42: 173.
40 Zhao, Y.S., Fu, H., Peng, A. et al. (2008). *Adv. Mater.* 20: 2859.
41 Mino, L., Agostini, G., Borfecchia, E. et al. (2013). *J. Phys. D: Appl. Phys.* 46: 423001.
42 Rossetti, R., Ellison, J.L., Gibson, J.M., and Brus, L.E. (1984). *J. Chem. Phys.* 80: 4464.
43 Gerischer, H. and Lübke, M. (1986). *J. Electroanal. Chem.* 204: 225.
44 Dingle, R., Wiegmann, W., and Henry, C.H. (1974). *Phys. Rev. Lett.* 33: 827.

45 Osterloh, F.E. (2013). *Chem. Soc. Rev.* 42: 2294.

46 Robel, I., Kuno, M., and Kamat, P.V. (2007). *J. Am. Chem. Soc.* 129: 4136.

47 Wang, X., Li, Z., Shi, J., and Yu, Y. (2014). *Chem. Rev.* 114: 9346.

48 Yang, X.H., Li, Z., Liu, G. et al. (2011). *CrystEngComm* 13: 1378.

49 Mohapatra, S.K., John, S.E., Banerjee, S., and Misra, M. (2009). *Chem. Mater.* 21: 3048.

50 Liu, X. and Swihart, M.T. (2014). *Chem. Soc. Rev.* 43: 3908.

51 Lee, J., Mubeen, S., Ji, X. et al. (2012). *Nano Lett.* 12: 5014.

52 Hou, W. and Cronin, S.B. (2013). *Adv. Funct. Mater.* 23: 1612.

53 Wang, S.L., Mak, Y.L., Wang, S. et al. (2016). *Langmuir* 32: 13046.

54 Zhou, S., Pi, X., Ni, Z. et al. (2015). *ACS Nano* 9: 378.

55 Manthiram, K. and Alivisatos, A.P. (2012). *J. Am. Chem. Soc.* 134: 3995.

56 Li, W., Zamani, R., Rivera Gil, P. et al. (2013). *J. Am. Chem. Soc.* 135: 7098.

57 Liu, X., Wang, X., Zhou, B. et al. (2013). *Adv. Funct. Mater.* 23: 1256.

58 Yan, J., Wang, T., Wu, G. et al. (2015). *Adv. Mater.* 27: 1580.

59 Wang, Z., Yang, C., Lin, T. et al. (2013). *Adv. Funct. Mater.* 23: 5444.

60 Liu, Y., Xiao, C., Huang, P. et al. (2018). *Chem* 4: 1263.

61 Polman, A. and Atwater, H.A. (2012). *Nat. Mater.* 11: 174.

62 Semonin, O.E., Luther, J.M., Choi, S. et al. (2011). *Science* 334: 1530.

63 Guo, Y., Mei, S., Yuan, K. et al. (2018). *ACS Catal.* 8: 6203.

64 Hocking, R.K., Brimblecombe, R., Chang, L.Y. et al. (2011). *Nat. Chem.* 3: 461.

65 Jiao, F. and Frei, H. (2009). *Angew. Chem. Int. Ed.* 48: 1841.

66 Jiao, F. and Frei, H. (2010). *Energy Environ. Sci.* 3: 1018.

67 Yusuf, S. and Jiao, F. (2012). *ACS Catal.* 2: 2753.

68 Gomez-mingot, M., Roch-marchal, C., Lassalle-kaiser, B. et al. (2018). *J. Am. Chem. Soc.* 140: 3613.

69 Grzelczak, M., Zhang, J., Pfrommer, J. et al. (2013). *ACS Catal.* 3: 383.

70 Wang, J., Zhang, N., Su, J., and Guo, L. (2016). *RSC Adv.* 6: 41060.

71 Xu, R. and Zeng, H.C. (2004). *Langmuir* 20: 9780.

72 Zhang, N., Shi, J., Mao, S.S., and Guo, L. (2014). *Chem. Commun.* 50: 2002.

73 Shi, N., Cheng, W., Zhou, H. et al. (2015). *Chem. Commun.* 51: 1338.

74 Fominykh, K., Feckl, J.M., Sicklinger, J. et al. (2014). *Adv. Funct. Mater.* 24: 3123.

75 Chang, C.M., Orchard, K.L., Martindale, B.C.M., and Reisner, E. (2016). *J. Mater. Chem. A* 4: 2856.

76 Blakemore, J.D., Gray, H.B., Winkler, J.R., and Müller, A.M. (2013). *ACS Catal.* 3: 2497.

77 Liao, L., Zhang, Q., Su, Z. et al. (2014). *Nat. Nanotechnol.* 9: 69.

78 Maron, Z.O., Gardner, G., Greenblatt, M. et al. (2015). *ACS Catal.* 5: 3403.

79 Plaisance, C.P. and Van Santen, R.A. (2015). *J. Am. Chem. Soc.* 137: 14660.

80 Wang, L. and Van Voorhis, T. (2011). *J. Phys. Chem. Lett.* 2: 2200.

81 Hutchings, G.S., Zhang, Y., Li, J. et al. (2015). *J. Am. Chem. Soc.* 137: 4223.

82 Kwak, I.H., Im, H.S., Jang, D.M. et al. (2016). *ACS Appl. Mater. Interfaces* 8: 5327.

83 Townsend, T.K., Sabio, E.M., Browning, N.D., and Osterloh, F.E. (2011). *Energy Environ. Sci.* 4: 4270.
84 Yehezkeli, O., Harguindey, A., Domaille, D.W. et al. (2015). *RSC Adv.* 5: 58755.
85 Lijima, S. (1991). *Nature* 354: 56.
86 Zhang, H., Tian, W., Li, Y. et al. (2018). *J. Mater. Chem. A* 6: 24149.
87 Zhang, H., Tian, W., Guo, X. et al. (2016). *ACS Appl. Mater. Interfaces* 8: 35203.
88 Sun, Y., Xie, Y., Wu, C. et al. (2010). *Nano Res.* 3: 620.
89 Sun, S., Wang, W., Li, D. et al. (2014). *ACS Catal.* 4: 3498.
90 Wu, X., Zhao, J., Guo, S. et al. (2016). *Nanoscale*: 17314.
91 Serpone, N., Lawless, D., and Khairutdinov, R. (1995). *J. Phys. Chem.* 99: 16646.
92 Iacomino, A., Cantele, G., Trani, F., and Ninno, D. (2010). *J. Phys. Chem. C* 114: 12389.
93 Hoang, S., Guo, S., Hahn, N.T. et al. (2012). *Nano Lett.* 12: 26.
94 Liu, Y., Zhou, W., and Umezawa, N. (2017). *J. Phys. Chem. C* 121: 18683.
95 Sasaki, T. and Watanabe, M. (1997). *J. Phys. Chem. B* 5647: 10159.
96 Osada, M. and Sasaki, T. (2009). *J. Mater. Chem.*: 2503.
97 Sakai, N., Ebina, Y., Takada, K., and Sasaki, T. (2005). *J. Phys. Chem. B* 109: 9651.
98 Waller, M.R., Townsend, T.K., Zhao, J. et al. (2012). *Chem. Mater.* 24: 698.
99 Liu, Y., Liang, L., Xiao, C. et al. (2016). *Adv. Energy Mater.* 6: 1600437.
100 Ge, M., Li, Q., Cao, C. et al. (2017). *Adv. Sci.* 4: 1600152.
101 Zhao, Y., Chang, C., Teng, F. et al. (2017). *Adv. Energy Mater.* 7: 1700005.
102 Zhu, J., Yin, Z., Yang, D. et al. (2013). *Energy Environ. Sci.* 6: 987.
103 Lee, Y., Choi, J.H., Jeon, H.J. et al. (2011). *Energy Environ. Sci.* 4: 914.
104 Gomes Silva, C., Bouizi, Y., Fornés, V., and García, H. (2009). *J. Am. Chem. Soc.* 131: 13833.
105 Online, V.A., Kim, S.J., Lee, Y. et al. (2014). *J. Mater. Chem. A* 2: 4136.
106 Zhao, Y., Li, B., Wang, Q. et al. (2014). *Chem. Sci.* 5: 951.
107 Di, J., Chen, C., Yang, S.-Z. et al. (2017). *J. Mater. Chem. A* 5: 14144.
108 Rodenas, P., Song, T., Sudhagar, P. et al. (2013). *Adv. Energy Mater.* 3: 176.
109 Sun, W.-T., Yu, Y., Pan, H.-Y. et al. (2008). *J. Am. Chem. Soc.* 130: 1124.
110 Maeda, K., Ishimaki, K., Tokunaga, Y. et al. (2016). *Angew. Chem. Int. Ed.* 55: 8309.
111 Chaguetmi, S., Mammeri, F., Nowak, S. et al. (2013). *RSC Adv.* 3: 2572.
112 Liu, C., Tang, J., Chen, H.M. et al. (2013). *Nano Lett.* 13: 2989.
113 Zhang, H., Tian, W., Li, Y. et al. (2018). *J. Mater. Chem. A* 6: 6265.
114 Liu, J., Liu, Y., Liu, N. et al. (2015). *Science* 347: 970.
115 Shi, W., Zhang, X., Brillet, J. et al. (2016). *Carbon* 105: 387.
116 Su, T., Shao, Q., Qin, Z. et al. (2018). *ACS Catal.* 8: 2253.
117 Zhang, J., Grzelczak, M., Hou, Y. et al. (2012). *Chem. Sci.* 3: 443.
118 Zhang, G., Zang, S., and Wang, X. (2015). *ACS Catal.* 5: 941.
119 Zhou, L., Zhang, H., Sun, H. et al. (2016). *Catal. Sci. Technol.* 6: 7002.
120 Ma, T.Y., Dai, S., Jaroniec, M., and Qiao, S.Z. (2014). *Angew. Chem. Int. Ed.* 53: 7281.
121 Meng, F., Li, J., Cushing, S.K. et al. (2013). *ACS Catal.* 3: 746.

122 Tian, J., Zhao, Z., Kumar, A. et al. (2014). *Chem. Soc. Rev.* 43: 6920.

123 Bazzan, I., Di Valentin, M., Galloni, P., and Conte, V. (2013). *Chem. Commun.* 49: 9941.

124 Nakazono, T., Parent, A.R., and Sakai, K. (2013). *Chem. Commun.* 49: 6325.

125 Nakazono, T., Parent, A.R., and Sakai, K. (2015). *Chem. Eur. J.* 21: 6723.

126 Younus, H.A., Ahmad, N., Chughtai, A.H. et al. (2017). *ChemSusChem* 10: 862.

127 Das, B., Orthaber, A., Ott, S., and Thapper, A. (2015). *Chem. Commun.* 51: 13074.

128 Zhang, H., Tian, W., Zhou, L. et al. (2018). *Appl. Catal., B* 223: 2.

7

Photoelectrochemical Oxygen Evolution

Fumiaki Amano

The University of Kitakyushu, Department of Chemical and Environmental Engineering, 1-1 Hibikino, Wakamatsu-ku, Kitakyushu 808-0135, Japan

7.1 Introduction

The photoelectrochemical (PEC) oxygen evolution is a half reaction to induce water splitting to evolve H_2 and O_2 using photo-energy. The standard electrode potential of oxygen evolution reaction (OER, $2H_2O \rightarrow O_2 + 4H^+ + 4e^-$) is 1.23 V vs. the standard hydrogen electrode (SHE). The PEC water oxidation is usually performed by using n-type semiconductor oxide electrodes. The semiconductor electrode is called "photoanode" since anodic oxidation of water to evolve O_2 occurs under light irradiations. The PEC water oxidation was first reported by using a rutile TiO_2 single-crystal photoanode in the year 1969 (Figure 7.1) [1]. Since the finding, the photoinduced chemical reaction at the semiconductor surface has attracted much attention in terms of the conversion, storage, and transportation of solar energy [2–8].

When a semiconductor material is irradiated with light of energy larger than the band-gap energy (E_g), the pair of a photoexcited electron (e^-) and a hole (h^+) is generated by the interband transition. The recombination of the e^- and h^+ pairs does not produce an electric current, but the e^- flows from the photoanode to the cathode via an external circuit are observed when the h^+ in the valence band (VB) promote charge transfer reactions at the electrode/electrolyte interface. The photocurrent corresponds to a rate of an anodic oxidative reaction under light irradiations. The PEC water oxidation by photogenerated h^+ frequently competes with undesired oxidation of the photoanode material itself, which is often called photocorrosion.

Developing highly efficient photoanode material for OER is a challenge to establish a solar-to-hydrogen (STH) conversion technology since the PEC water oxidation is usually the bottleneck in energy efficiency and productivity. In this chapter, design strategies of the highly efficient photoanode materials for OER are described after the explanation about the fundamental points and the factors affecting the PEC performances of photoanode [4, 9–13]. The PEC properties of photoanode provide information about the photoinduced process in points of view of thermodynamics and surface kinetics of OER.

Solar-to-Chemical Conversion: Photocatalytic and Photoelectrochemcial Processes, First Edition.
Edited by Hongqi Sun.

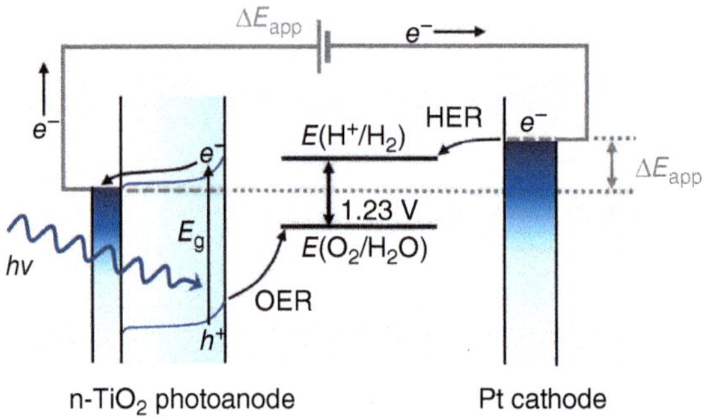

Figure 7.1 PEC water splitting by n-type TiO_2 photoanode for oxygen evolution reaction (OER) and Pt cathode for hydrogen evolution reaction (HER). The TiO_2 with band-gap energy (E_g) of 3.0 eV is photoexcited under ultraviolet light irradiation. External voltage (ΔE_{app}) is applied between the electrodes to induce water photoelectrolysis.

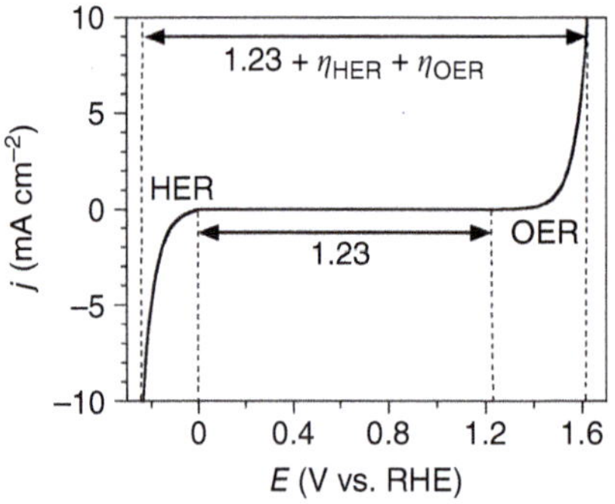

Figure 7.2 Current–potential curves for conventional water electrolysis in the dark. Over potentials (η) are required for HER and OER over each electrocatalyst electrode ($\eta_{HER} + \eta_{OER}$).

7.2 Honda–Fujishima Effect

For H_2 production from water, a potential difference between OER anode and hydrogen evolution reaction (HER) cathode should be more than 1.23 V, which is a thermodynamically theoretical voltage for water electrolysis (Figure 7.2). The electrocatalytic anode for OER requires overpotentials (η) relative to standard electrode potential $E°(O_2/H_2O) = 1.23\,V$ vs. the reversible hydrogen electrode (RHE). The η is more than 0.2 V at least, since the OER is induced by the transfer of multiple electrons and protons ($2H_2O \rightarrow O_2 + 4H^+ + 4e^-$), which requires a high activation energy. The HER cathode also requires η relative to $E°(H^+/H_2) = 0\,V$ vs. RHE.

In the case of the PEC reaction, the applied electric bias can be reduced by the photon energy. When OER on photoanode occurs at a potential more negative than that at which HER occurs, water splitting can be induced without applied electric bias (Figure 7.3a). Fujishima and Honda found that n-type semiconducting electrode of rutile TiO_2 ($E_g = 3.0\,eV$) single crystal shows an anodic current in the aqueous electrolyte under irradiation of light with wavelengths shorter than 415 nm [1]. The photon energy, $h\nu$ (eV), can be calculated from light wavelength, λ (nm):

$$h\upsilon\,(\text{eV}) = \frac{1240}{\lambda\,(\text{nm})} \tag{7.1}$$

The anodic current of the TiO_2 electrode was not observed in dark, proportional to the irradiation light intensity, and accompanied by O_2 gas evolution. These facts

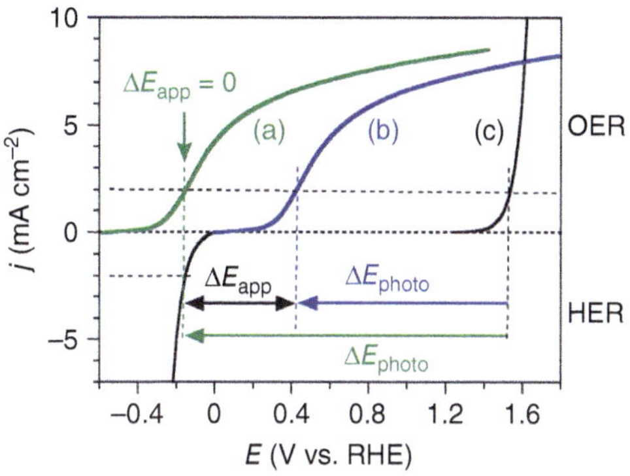

Figure 7.3 Current–potential curves for water photoelectrolysis using photoanodes for OER and a cathode for HER at the current density (j) of 2 mA cm^{-2}. (a) Unbiased PEC water splitting and (b) PEC water splitting with externally applied voltage (ΔE_{app}). The ΔE_{photo} is the shift of the potential by using photoanodes in comparison with (c) anode electrocatalyst for OER. The absolute value of the current is equal between the series-connected electrodes.

indicate that the photoanodic current is attributed to the oxidative decomposition of water by the h^+ photogenerated in the VB:

$$2H_2O + 4h^+ \rightarrow O_2 + 4H^+ \tag{7.2}$$

The water splitting without applied external voltage (ΔE_{app}) can be promoted by connecting the photoanode with H_2-evolving Pt cathode when the anodic photocurrent occurs at more negative potential than 0 V vs. RHE. It is reported that the PEC system of using TiO_2 photoanode and Pt cathode, which is the best electrocatalyst for HER, showed a small photocurrent even in the absence of any external voltage. This "Honda–Fujishima effect" is the first report of the PEC water splitting using an O_2-evolving photoanode. However, many photoanode materials require the ΔE_{app} as shown in Figure 7.3b. Therefore, several tandem cell systems have been proposed to apply the potential difference to the OER photoanode in combination with a photovoltaic (PV) cell or a p-type semiconductor electrode as an HER photocathode. The bias voltage also can be applied by chemically using the difference in the pH solutions for two electrodes: the chemical potential corresponds to 59 mV per one pH unit at 25 °C. Even if a ΔE_{app} is needed, the shift of OER potential by using photoanode (ΔE_{photo}) is recognized as the accruing of photo-energy.

7.3 Factors Affecting the Photoanodic Current

The photocurrent of photoanode material depends on several factors of experimental conditions such as incident light intensity, light wavelength, applied potential, and solution composition. The current–potential (J–E) relationship of photoanode materials is generally investigated in aqueous electrolyte solutions using linear sweep voltammetry (LSV) and cyclic voltammetry (CV). In the PEC measurement, the applied potential is varied linearly with time at sweep rates of about 20–50 mV s^{-1}. In contrast to conventional electrochemistry, the sweep rate does not affect the photocurrent of the semiconductor electrode because the photocurrent is not usually limited by diffusion and mass transfer. Besides, the electrolyte solution may be stirred when the PEC reaction is limited by diffusion and gas evolution. For example, gas bubble formation on the photoanode surface may affect the photocurrent for OER. The PEC cell is purged with an inert gas when side reactions are concerned to be involved by O_2. The photocurrent density, j_{photo} ($mA\,cm^{-2}$), normalized to the geometrical area of photoanode is more suitable than photocurrent, J_{photo} (mA), because J_{photo} is proportionally changed with the photoanode area. The j_{photo} is obtained by subtraction of the j in the dark from the j under photoirradiation.

The j–E behavior of the semiconductor electrode is considerably different from that at conductive electrodes such as metals and carbon, where there is always a high density of charge carriers in the conductor [9]. The j_{photo} of the semiconductor electrode is affected by the density of minority carriers: e^- for p-type semiconductor and h^+ for n-type semiconductor. The PEC water oxidation on an n-type semiconductor electrode is promoted by the charge transfer between the photogenerated h^+ and water at the semiconductor–electrolyte interface. After the generation of charge carriers by light absorption, the photoexcited e^- in the conduction band (CB) transports to the inside of the semiconductor and moves into the external circuit, while the h^+ in the VB transports to the surface to oxidize water to O_2. The charge separation efficiency is decreased by bulk recombination and surface recombination. Based on the reaction mechanism, the rate law of the h^+ concentration, $[h^+]$, can be expressed by simply assuming that the charge recombination and the charge transfer are the first-order reactions with respect to $[h^+]$:

$$\frac{d[h^+]}{dt} = I_0\alpha - k_{rec}[h^+] - k_{ct}[h^+] \tag{7.3}$$

Here I_0, α, k_{rec}, and k_{ct} are the intensity of incident light intensity, photoabsorption efficiency, the rate constant of recombination, and rate constant of charge transfer to the water. When the steady-state approximation is applied to the h^+ concentration ($d[h^+]/dt = 0$), we obtained the rate of the charge transfer, r_{ct}:

$$r_{ct} = k_{ct}[h^+] = \frac{k_{ct}}{k_{rec} + k_{ct}} I_0\alpha \tag{7.4}$$

The corresponding j_{photo} is obtained by multiplying the charge transfer rate by Faraday constant, $F = 9.65 \times 10^{-4}\,C\,mol^{-1}$:

$$j_{photo} = Fr_{ct} = \frac{k_{ct}}{k_{rec} + k_{ct}} FI_0\alpha \tag{7.5}$$

Therefore, the j_{photo} is ideally proportional to the I_0 owing to the increase of the concentration of h^+ by photoexcitation (Figure 7.4). Generally, the j_{photo} increases with I_0. However, proportional relation is not always observed because the charge recombination may change to second-order reaction with respect to the concentration of the photogenerated carriers when I_0 is high [14, 15]. The irradiation wavelength also should be paid attention since the useful wavelength is dependent on the photoabsorption properties of the photoanode materials: E_g of semiconductor materials, highest occupied molecular orbital (HOMO)–lowest unoccupied molecular orbital (LUMO) energy gaps of molecular photosensitizers, and their absorption coefficient.

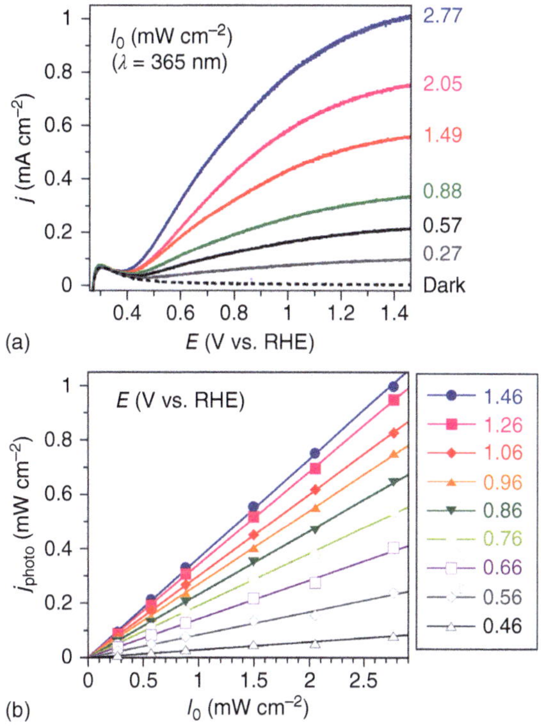

Figure 7.4 (a) Current–potential curves of WO_3 photoanode in 0.1 mol l^{-1} Na_2SO_4 (pH = 7) with 10 vol% methanol under UV irradiation (monochromatic light, λ = 365 nm) with different intensities of incident light. The light intensity (I_0) was controlled by using a neutral density filter. (b) Effect of the I_0 on the photocurrent density (j_{photo}) of the WO_3 photoanode in the linear sweep voltammetry. The j_{photo} was linearly increased with I_0 at each applied potential.

In the PEC water oxidation, the electron transfer at the semiconductor–electrolyte interface competes with the recombination of photoexcited e^- and h^+ in bulk and surface. The recombination rates are strongly dependent on the structural and surface properties of the photoanode materials. In general, high crystallinity has a positive effect on the photocurrent due to the decrease in the density of crystalline defects and surface trap states, which act as recombination centers for e^- and h^+. The recombination occurs when the lifetime of h^+ is shorter than the charge transfer kinetics. The surface recombination can be reduced by the passivation of surface traps and the electrocatalytic effect promoting water oxidation.

The flat band potential (E_{fb}) of n-type semiconductors is very important for PEC water oxidation since j_{photo} started to increase at potentials higher than the E_{fb} of photoanode. The potential where j_{photo} starts increasing from zero is called onset potential (E_{onset}). The flat band means the state where there was no potential gradient in the energy bands of the semiconductor (Figure 7.5a). The E_{fb} is the potential of the flat band state. The recombination of e^- and h^+ easily occurs at the flat band condition since there was no driving force to separate the charge carriers. When the potential applied to the semiconductor electrode is more positive than the E_{fb}, the potential drop ($\Delta\phi$) mainly occurs in the semiconductor side rather than the electrolyte side (Figure 7.5b). This is considerably different from the case of conducting electrodes, where the potential drops occur in the electric double layer in the electrolyte. The electric double layer is composed of the Helmholtz layer and the diffuse

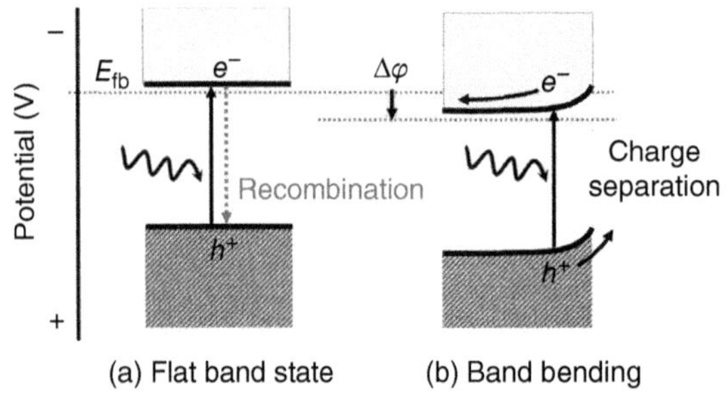

Figure 7.5 (a) Flat band state and (b) band bending state of n-type semiconductor. The recombination of the photoexcited e^- and h^+ pair easily occurs in the flat band state. In contrast, the e^- and h^+ pair is efficiently separated in the space charge layer (SCL). In the absence of surface states, the potential drop in the SCL ($\Delta\phi$) is linear to the anodic shift of the applied potential from the flat band potential (E_{fb}).

layer. Since the capacitance of semiconductor (C_{SC}) is much smaller than the capacitance of the Helmholtz layer (C_H), the total capacitance is close to C_{SC}:

$$\frac{1}{C} = \frac{1}{C_{SC}} + \frac{1}{C_H} \tag{7.6}$$

$$C = \frac{C_{SC}C_H}{C_{SC} + C_H} \approx C_{SC} \tag{7.7}$$

$$\because C_{SC} \ll C_H$$

Therefore, the potential gradient is formed in the region of the semiconductor surface, which is called the space charge layer (SCL). The potential gradient in SCL facilitates the separation of the photoexcited e^- and h^+. The positive shift of the applied potential increases the $\Delta\phi$ in the SCL, which facilitates the charge separation. Since the energy bands of semiconductors are bent upward with respect to the energy level in the bulk semiconductor, this effect is also called band bending.

The electronic properties of photoanode materials are important to affect the photoabsorption, the charge separation, and the charge transport. Apart from the bulk properties of photoanode materials, the j_{photo} (or reaction rate) is also governed by the surface properties of the photoanode. The OER activity is enhanced by the decrease in the activation energy for the charge transfer process on the surface. Therefore, the electrocatalytic materials such as RuO_2, IrO_2, and cobalt-phosphate (CoPi) are frequently used to modify the surface kinetics of the photoanode materials. The surface properties are also related to the recombination at surface trap state and instability during PEC reactions.

7.4 Electrode Potentials at Different pH

The applied potential is strongly related to the thermodynamics and kinetics of water photoelectrolysis. The electrode potential of OER is dependent on pH in the aqueous

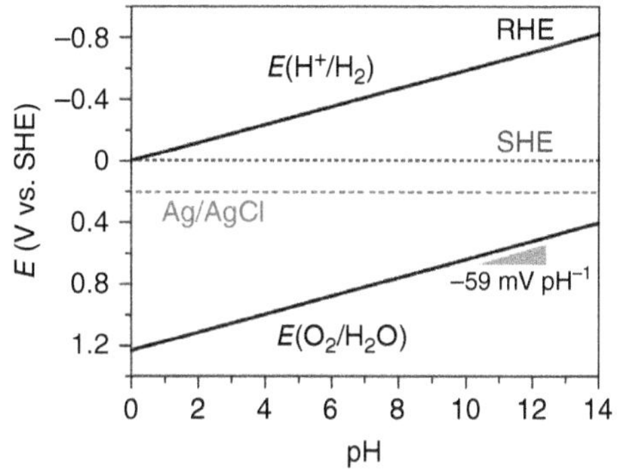

Figure 7.6 The pH dependence of the electrode potentials at 25 °C. The potentials of SHE and Ag/AgCl reference electrode are constant, but the potentials of RHE and the potentials for HER and OER are linearly increased by an increase in pH value with a slope of 59 mV (ln 10 × *RT*/*F*) at 25 °C.

solution (Figure 7.6) as expressed by the Nernst equation at 25 °C:

$$E(O_2/H_2O) = E°(O_2/H_2O) - \frac{RT}{4F}\ln\frac{1}{[H^+]^4} = 1.23 - 0.059\text{pH [V vs.SHE]} \tag{7.8}$$

The potential of HER is also negatively shifted with a slope of 59 mV per pH unit at 25 °C. Thus, the theoretical voltage for water splitting reaction ($\Delta E° = 1.23$ V) is not dependent on pH value. The theoretical voltage is calculated from the reaction Gibbs energy of water splitting ($H_2O \rightarrow H_2 + 1/2O_2$, $\Delta_r G° = 237.1$ kJ mol^{-1} at 25 °C):

$$\Delta E^{°} = -\frac{\Delta rG^{°}}{2F} = -\frac{237.1 \times 10^3\ (\text{J})}{2 \times 9.65 \times 10^4\ (\text{C})} = 1.23\ (\text{V}) \tag{7.9}$$

The energetically uphill reaction ($\Delta_r G° > 0$) induced by the PEC system represents the conversion of the light energy into chemical energy in the form of H_2, which is storable and useful owing to fuel cell technology.

The PEC properties of photoanode are usually investigated in a three-electrode system. The applied potential to the working photoanode is controlled and measured with respect to the reference electrode. Silver–silver chloride (Ag/AgCl) reference electrodes are frequently used in neutral electrolyte solution owing to the easy handling and the relatively constant potential. The $E°$(Ag/AgCl) = ~0.20 V vs. SHE is constant relative to the pH values of the solution (Figure 7.6). On the other hand, the potential of the reactions involving proton and electron transfer is changed with the pH value. Therefore, the electrode potentials may be changed with pH when it is measured relative to the Ag/AgCl electrode. The E_{fb} of semiconductor electrodes is also dependent on the pH values when the protonation of the surface is changed by pH values. Since the hydroxyl group on the metal oxide surface (M–OH) interacts with H^+ and OH^-, the surface is positively charged ($M{-}OH_2^+$) when the pH value is less than the isoelectric point and negatively charged ($M{-}O^-$) in the opposite case.

In contrast, the potential measured relative to RHE is constant regardless of pH, because the potential of RHE itself is linearly increased with pH with a slope of 59 mV pH^{-1} at 25 °C. Therefore, it is useful to convert the applied potential vs. Ag/AgCl to the potential vs. RHE in order to understand the performance of the photoanode for OER:

$$E\text{ [V vs. RHE]} = E\text{ [V vs. Ag–AgCl]} + E°(\text{Ag/AgCl}) + 0.059 \tag{7.10}$$

7.5 Evaluation of PEC Performance

STH conversion efficiency should be measured in the two-electrode system in the short circuit condition, where an external bias voltage is zero [16–18]. When an external voltage is necessary to induce the PEC water splitting system, the efficiency is evaluated by applied bias (AB)-STH conversion efficiency [4]:

$$\mathrm{AB-STH} = \frac{j_{\mathrm{photo}} \times (1.23 - \Delta E_{\mathrm{app}}) \times \eta_{\mathrm{FE}}}{P_{\mathrm{total}}} \tag{7.11}$$

Here, j_{photo}, ΔE_{app}, η_{FE}, and P_{total} are the steady-state j_{photo} (mA $\mathrm{cm^{-2}}$), the applied bias between two electrodes (V), Faraday efficiency (current efficiency) of H_2 evolution, and the incident power of the solar-simulated light (mW $\mathrm{cm^{-2}}$). The light should be close to the shape of the Air Mass 1.5 Global (AM1.5G) in STH measurement. The P_{total} of AM1.5G is 100 mW $\mathrm{cm^{-2}}$. For evaluation of photoanode in the three-electrode system, the ΔE_{app} may be calculated from the applied potential (E) hypothetically assuming that the overpotential for HER is negligible at the counter electrode:

$$\Delta E_{\mathrm{app}} = E(\mathrm{O_2/H_2O}) - E \tag{7.12}$$

The hypothetical AB-STH is called half-cell STH efficiency, which is independent of the properties of the counter cathode. It should be noted that the half-cell STH efficiency is different from AB-STH efficiency measured in a two-electrode system.

The Faraday efficiency (η_{FE}) of O_2 evolution is essential to analyze the selectivity of OER at photoanode. The quantification of the amount of H_2 evolved on the cathode is not important in the three-electrode system. However, the measurement of η_{FE} of H_2 evolution might be useful to check if there is a reaction between the evolved H_2 and O_2 back into the water. The amount of O_2 evolution was quantified by gas chromatography, mass spectroscopy, and O_2 gas sensor. In the gas chromatography, a thermal conductivity detector is used to measure O_2, which is often through a molecular sieve 5A column with argon carrier. The η_{FE} of O_2 is calculated from the O_2 evolution rate, rate (O_2), assuming 4-electron reaction:

$$\eta_{\mathrm{FE}} \text{ of } \mathrm{O_2} = \frac{4F \times \mathrm{rate(O_2)}\ (\mathrm{mmol\ s^{-1}\ cm^{-2}})}{j_{\mathrm{photo}}\ (\mathrm{mA\ cm^{-2}})} \tag{7.13}$$

When $\eta_{\mathrm{FE}} = 100\%$, the rate($O_2$) is ~9.3 μmol $\mathrm{h^{-1}\ cm^{-2}}$ at j_{photo} of 1.0 mA $\mathrm{cm^{-2}}$. The OER by water oxidation may compete with the formation of hydroxyl radical ($^{\bullet}$OH) and H_2O_2. The reactions to form $^{\bullet}$OH and H_2O_2 are more difficult than OER in thermodynamics: $E°(^{\bullet}\mathrm{OH/H_2O}) = 2.38$ V, $E°(\mathrm{H_2O_2/H_2O}) = 1.76$ V, and $E°(\mathrm{O_2/H_2O}) = 1.23$ V vs. RHE [19]. However, the reaction rates are controlled by kinetics rather than thermodynamics. The formation of $^{\bullet}$OH is analyzed by electron spin resonance (ESR) spectroscopy using a spin trapping agent. The formation of H_2O_2 is quantified by redox titration, colorimetric method, and rotating ring-disk electrode (RRDE) method [20].

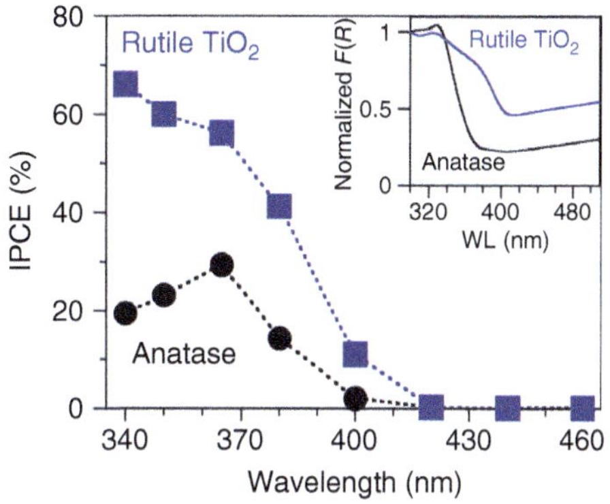

Figure 7.7 IPCE action spectra of rutile and anatase TiO_2 electrodes in 0.2 mol l^{-1} Na_2SO_4 with phosphate buffer (pH = 7) at 0.9 V vs. RHE. The inset shows their diffuse reflectance UV–vis spectra using Kubelka–Munk function, F(*R*). Source: From Amano et al. [21]. © 2018 The Electrochemical Society.

The measurement of the apparent quantum efficiency (AQE), also called the external quantum efficiency (EQE), is important to compare the results obtained by using different light sources with different conditions. The incident photon-to-current conversion efficiency (IPCE) is a simple and easy method to evaluate EQE, but the IPCE is different from the EQE of O_2 evolution because the η_{FE} may be less than unity. Therefore, O_2 evolution should be separately confirmed. The IPCE is the ratio of J_{photo} at a static potential vs. the incident photon of monochromatic light:

$$\mathrm{IPCE} = \frac{j_{\mathrm{photo}}\ (\mathrm{mA\ cm^{-2}})}{I_0\ (\mathrm{mW\ cm^{-2}})} \times \frac{1240}{\lambda\ (\mathrm{nm})} \tag{7.14}$$

The IPCE measurement is a standard method to evaluate photoanode performance. Moreover, a plot of IPCE as a function of the wavelength of the incident light, often called an action spectrum, is useful to obtain information about the properties of photoabsorber. When the action spectrum of IPCE resembles the absorption spectrum of the semiconductor material, the photocurrent is attributed to the reactions induced by the photoexcited e^-–h^+ pairs of the material. Figure 7.7 shows the IPCE action spectra of TiO_2 photoanodes for water oxidation [21]. The IPCE response of the rutile TiO_2 electrode was shifted to a longer wavelength compared with the anatase TiO_2 electrode. This shift corresponded to that of the photoabsorption edge of rutile TiO_2 in diffuse reflectance UV–vis spectra. The IPCE action spectrum can determine the E_g, which is similar to the reported optical E_g of rutile TiO_2 (~3.0 eV) and anatase TiO_2 (~3.2 eV).

The IPCE must be calculated from the carefully evaluated I_0 and j_{photo}. The incident light should be monochromatic, and the optical power meter should be guaranteed the calibration at the wavelength. The j_{photo} is measured at the steady-state condition in chronoamperometry at static potential since the j–E behavior in LSV and CV may be transient. It is necessary to subtract the dark current, if which exists, from the j under photoirradiation. Noted that the IPCE may decrease with an increase in the light intensity when the charge recombination does not obey first-order kinetics at high e^- and h^+ density.

7.6 Flat Band Potential and Photocurrent Onset Potential

In theory, the E_{onset} of photocurrent response is close to the E_{fb} of the semiconductor electrode. Therefore, the estimation of E_{fb} is meaningful to understand the photoanode behavior in the *j*–*E* curve measurement. A Mott–Schottky analysis is a common method to evaluate the E_{fb} of semiconductor electrodes in aqueous solution [10, 17]. The analysis is based on the measurement of the SCL capacitance (C_{sc}) as a function of the E. The Mott–Schottky relationship is expressed by the equation

$$\frac{1}{{C_{\text{sc}}}^2} = \frac{2}{q\varepsilon\varepsilon_0 N_{\text{D}}}\left(E - E_{\text{fb}} - \frac{kT}{q}\right) \tag{7.15}$$

Here, q is an elementary electric charge, ε is the dielectric constant of the material, ε_0 is the permittivity of vacuum, N_{D} is the donor density, k is Boltzmann's constant, and T is the temperature. The E_{fb} and N_{D} can be obtained from the Mott–Schottky plot, which is a linear relationship between the ${C_{\text{sc}}}^{-2}$ and the E (Figure 7.8). The E_{fb} is determined from the *x*-axis intercept at ${C_{\text{sc}}}^{-2} = 0$, which equals $E_{\text{fb}} + kT/q$. The C_{sc} is measured in the dark by electrochemical impedance spectroscopy (EIS) technique and estimated from the imaginary part of the impedance ($Z = Z' + iZ''$) assuming a series capacitor–resistor model:

$$C_{\text{sc}} = -\frac{1}{\omega Z''} = -\frac{1}{2\pi f Z''} \tag{7.16}$$

Here, ω is angular modulation frequency (rad s^{-1}) and f is frequency (Hz) in the EIS measurement. The positive slope of the Mott–Schottky plot indicates the n-type semiconductivity. The N_{D} measured by the slope of the Mott–Schottky plot is related to the electrical conductivity (ρ) of n-type semiconductor:

$$\rho = qN_{\text{D}}\mu_e \tag{7.17}$$

Here, μ_{e} is electron mobility. The N_{D} is also related to the Fermi level and the width of SCL (W_{SCL}):

$$W_{\text{SCL}} = \sqrt{\frac{2\varepsilon_0\varepsilon\Delta\phi}{qN_{\text{D}}}} \tag{7.18}$$

The W_{SCL} will be narrower as the N_{D} is increased by heavily doping. In contrast, the W_{SCL} becomes wide when the $\Delta\phi$ (potential drop in the SCL) is increased. Since the recombination of e^- and h^+ is suppressed in the SCL, the matching W_{SCL} and the light penetration depth results in the enhancement of IPCE value.

The photoexcited e^- and h^+ are separated at the band bending in SCL when the applied potential is more positive than E_{fb}. However, it is generally rare to observe a clear relationship between the E_{fb} and the E_{onset}, particularly for particulate photoanodes because of the fast recombination compared to the slow kinetics of OER (Figure 7.9). The OER requires additional overpotential across the activation barrier of the multi-electron transfer. In contrast, methanol oxidation is more facile than water oxidation since it is one-electron transfer oxidation at the first step [22]:

$$CH_3OH + h^+ \rightarrow {}^{\bullet}CH_2OH + H^+ \tag{7.19}$$

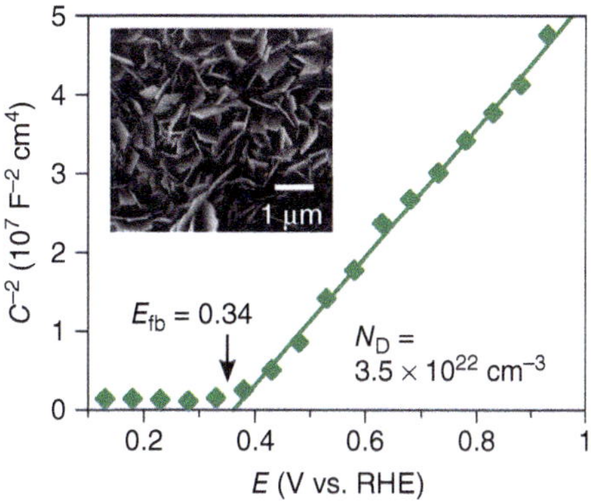

Figure 7.8 Mott–Schottky plots of WO_3 electrode in 0.1 mol l^{-1} H_2SO_4 (pH = 1) in the dark to estimate the flat band potential (E_{fb}) and the donor density (N_D). The capacitance of the space charge layer (C_{sc}) was measured at 1 kHz with a sinusoidal amplitude of 10 mV in the dark. The inset shows the SEM image of the WO_3 electrode surface. Source: Adapted with permission Amano et al. [22]. Copyright 2011, Springer-Verlag.

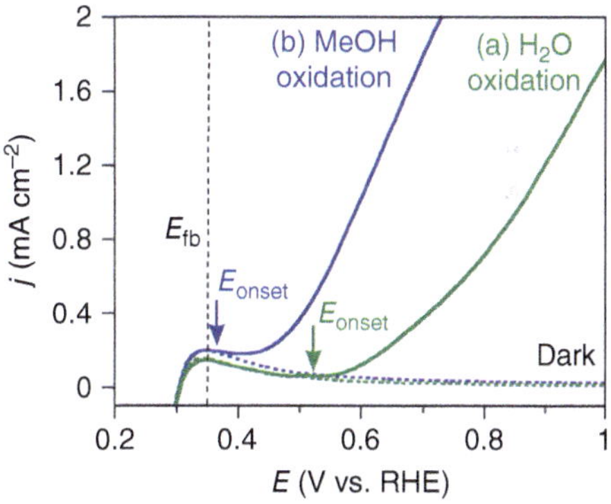

Figure 7.9 Linear sweep voltammograms of the WO_3 electrode for (a) water oxidation in 0.1 mol l^{-1} H_2SO_4 and (b) methanol oxidation in the solution with 10 vol% methanol under photoirradiation (solid *j*–*E* curves) and in the dark (dashed *j*–*E* curves). The applied potential where photocurrent response starts is denoted as E_{onset}. The E_{fb} is the value measured in Figure 7.8. Source: Based on Amano et al. [22].

The E_{onset} of the WO_3 electrode for methanol oxidation was found close to the E_{fb} measured by Mott–Schottky analysis. This implies that the time constant of photogenerated h^+ is sufficient to induce the oxidation of methanol even at weak anodic polarization. The reaction rate of the photogenerated h^+ with methanol (<50 ns) is reported to be much faster than that with water (<2 μs) by time-resolved infrared (IR) spectroscopy [23, 24]. In addition to methanol, the strong reductants such as Na_2SO_3, H_2O_2, I_3^-, and $[Fe(CN)_6]^{4-}$ have also been used as a hole scavenger in the PEC study [25–27]. The charge transfer efficiency (η_{trans}) for water oxidation can be estimated by comparison with the j_{photo} for the oxidation of Na_2SO_3 and H_2O_2, assuming that the η_{trans} of the scavenger oxidation is 100%:

$$\eta_{trans} = \frac{j_{photo}\ \text{for}\ H_2O\ \text{oxidation}}{j_{photo}\ \text{for scavenger oxidation}} \tag{7.20}$$

7.7 Selection of Materials

The E_g, CB minimum, and VB maximum are considerable parameters for the selection of photoanode materials. The electronic structures of the well-studied n-type semiconductors for PEC water oxidation are listed in Figure 7.10. The E_g

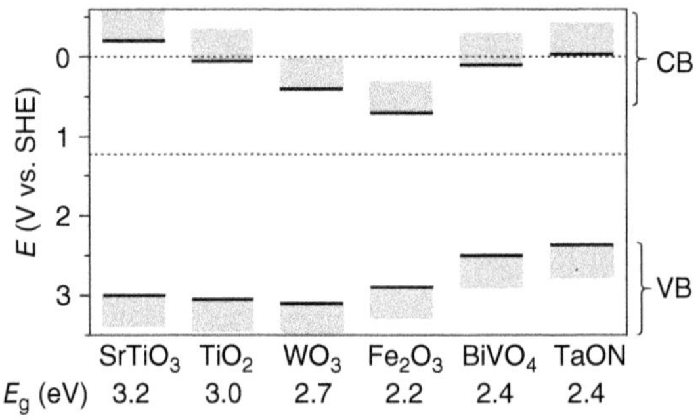

Figure 7.10 Band diagrams of n-type semiconductors for PEC oxygen evolution: $SrTiO_3$, rutile TiO_2, monoclinic WO_3, α-Fe_2O_3 (hematite), monoclinic scheelite-type $BiVO_4$ (clinobisvanite) [28], and TaON [29] with $E°(H^+/H_2)$ and $E°(O_2/H_2O)$. Here, the reported E_{fb} is considered as the conduction band (CB) minimum, although the CB minimum is more negative than E_{fb} by 0.1–0.3 eV. The valence band (VB) maximum is determined from the CB minimum and the optical E_g.

should be narrower than 3.1 eV to absorb visible light with $\lambda > 400$ nm. Since the visible light (400 nm $< \lambda <$ 800 nm) is the main component of the solar spectrum (42% of the total energy), the maximum j_{photo} under AM1.5G solar illumination is gradually increased with a decrease in E_g: 1.9 mA cm^{-2} for rutile TiO_2 (E_g = 3.0 eV) < 4.0 mA cm^{-2} for WO_3 (E_g = 2.7 eV) < 7.5 mA cm^{-2} for scheelite-type $BiVO_4$ (E_g = 2.4 eV) < 10.7 mA cm^{-2} for α-Fe_2O_3 (E_g = 2.2 eV). In addition to the oxide semiconductors, (oxy)nitride and (oxy)sulfide such as Ta_3N_5 (E_g = 2.1 eV), $BaTaO_2N$ (E_g = 1.9 eV), and $Sm_2Ti_2S_2O_5$ (E_g = 2.1 eV) are the candidates for the photoanodes with narrow band gaps [2, 6].

For PEC water oxidation, the VB maximum should be more positive than the $E°(O_2/H_2O)$. The CB minimum is ideally higher than $E°(H^+/H_2)$ because the cathodic shift of the E_{onset} provides the decrease of the ΔE_{app} in PEC water splitting. The VB of the oxide semiconductor is generally composed of O 2p orbital, which is located at around 3 V vs. SHE. Therefore, the oxides with decreased E_g commonly exhibit the E_{fb} with positive potentials. On the other hand, the VB maximums of $BiVO_4$ and TaON are higher than that of simple O 2p orbital. The negative shift of VB of $BiVO_4$ may be caused by the crystal distortion rather than the presence of Bi 6s levels above the O 2p orbital [30]. In the case of (oxy)nitrides such as TaON and Ta_3N_5, the VB composed of N 2p orbital is located at more negative potential than VB of O 2p orbitals owing to the lower electronegativities of nitrogen compared with oxygen [4]. The durability of photoanode materials is also an important factor in practical usage (Figure 7.11). Pourbaix diagram is a guidance of the stability of materials in the water with different pH values and at different potentials [31]. WO_3 is soluble to WO_4^{2-} in alkaline and neutral solutions. Therefore, the WO_3 photoanode should be studied in the acidic solution below pH = 4. The WO_3 is stable in acidic solution at anodic potentials, but it is easily reduced at $E°(H^+/H_2)$. For Fe_2O_3, Fe^{3+} and Fe^{2+} are stable in acidic solution, and Fe_3O_4 is easily formed at cathodic potentials. Therefore, the Fe_2O_3 photoanode is used in alkaline solution at an anodic potential. The pH of the electrolyte should be selected considering

Figure 7.11 Pourbaix diagram, which is a potential–pH equilibrium diagram, for (a) tungsten–water system and (b) iron–water system at 25 °C. The molar concentration of WO_4^{2-}, Fe^{2+}, and Fe^{3+} is 10^{-6} mol l^{-1}. The dashed lines show the potentials of HER and OER relative to pH value in water. Source: From Pourbaix [31]. © 1974, NACE.

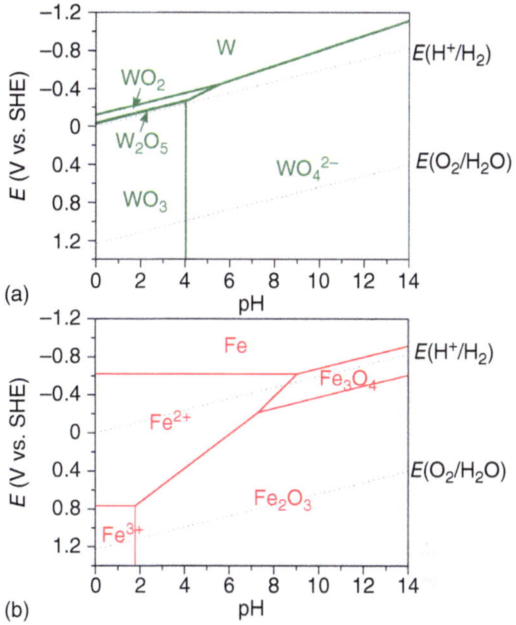

the stability of the photoanode materials. However, the possibility should be remembered that the reactant in OER is changed with the change of pH values. For example, the reactant for OER in alkaline solution is hydroxyl ions rather than water molecules since OH^- is more easily oxidized than H_2O in general [13].

The Pourbaix diagram provides useful information on the chemical instabilities. However, the stability of photoanodes is not predicted only from the considerations of the thermodynamic equilibrium in the dark. The deactivation is also induced by the oxidation of the photoanode material itself during PEC reactions. The undesired oxidation by photogenerated h^+ is known as photocorrosion, which is the reason that Si, CdS, and ZnO are not suitable for photoanode materials for OER. However, it is possible to mitigate the photocorrosion by the coating of protective layers and the OER electrocatalysts on the photoanode materials [8].

The photoanode materials should be composed of earth-abundant elements to achieve global scalability. The costs of production and operation must not be high. The most challenging and critical task is the enhancement of the PEC performances of the inexpensive and earth-abundant photoanode materials.

7.8 Enhancement of PEC Properties

The PEC performances of the photoanode are dependent on many factors such as materials, composition, morphology, conductive substrates, preparation method, preparation conditions (temperature, time, and precursor), surface states, defects, and doping. The purpose of this section is not to provide exhaustive descriptions of the screening results, but to give the guidelines to design photoanode materials rationally.

The j_{photo} for water oxidation can be expressed by the product of the light absorption efficiency (η_{abs}), the charge separation efficiency (η_{sep}), and η_{trans}:

$$j_{photo} = I_0 \times \eta_{abs} \times \eta_{sep} \times \eta_{trans} \tag{7.21}$$

The η_{abs}, which is related to the generation of photoexcited e^-–h^+ pairs, can be estimated from the reflectance (R) and the transmittance (T) of photoanode measured by using an integrating sphere [26]:

$$\eta_{abs} = 1 - R - T \tag{7.22}$$

The η_{sep}, which is the efficiency of the transport of the photogenerated carriers to the current collector and the electrolyte, is affected by the bulk recombination, electrical conductivity, carrier mobility, carrier diffusion length, and W_{SCL}. The η_{trans}, which is the efficiency of h^+ transfer to water at the semiconductor–electrolyte interface, is limited by slow OER kinctics and surface recombination. The improvement of the individual efficiencies provides the increase of j_{photo} and the negative shift of the E_{onset}. The long-term durability of photoanode is also a negligible factor in enhancing PEC properties.

7.8.1 Nanostructuring and Morphology Control

Nanostructured porous films composed of nanoparticles are one of the typical photoanode structures. The nanostructuring increases the electroactive surface area for OER at the electrode/electrolyte interface and decreases the distance that h^+ must move to the surface. It is reported that the silicon-doped Fe_2O_3 photoanode exhibits high efficiency for OER owing to the formation of a smaller dendritic Fe_2O_3 nanostructure [32]. The smaller grain size is corresponding to the larger surface area. The distance for h^+ to reach the electrode/solution interface is shortened to meet the diffusion length of h^+ in Fe_2O_3 (2–4 nm) [32]. The carrier diffusion length (L) is expressed by the diffusion coefficient (D) and the lifetime (τ) of the carries:

$$L = \sqrt{D\tau} \tag{7.23}$$

The nanostructuring enables h^+ to transport to the surface before the recombination with e^-.

Low-dimensional nanostructures also exhibit a large surface area and a short distance for the photogenerated h^+ to transport to the electrode/solution interface. The one-dimensional (1D) and two-dimensional (2D) anisotropic nanostructures are nanowires, nanorods, nanotubes, nanosheets, nanoplates, nanoflakes, and so on. Vertically aligned TiO_2 nanotube arrays have attracted much attention owing to their high efficiency in the PEC reaction [33]. The 1D nanostructure provides excellent electron transport pathways for the photoexcited e^- to the conductive substrate. The low resistance of e^- transport can increase the nanotube length to maximize photoabsorption without significant loss due to recombination.

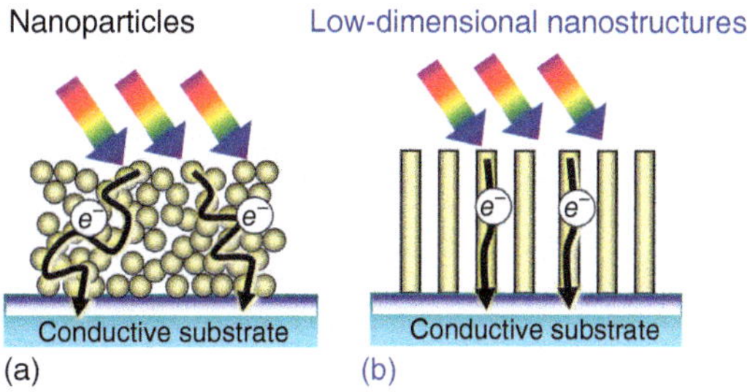

Figure 7.12 Schematic illustrations of photoexcited electron transport in photoanodes composed of (a) nanocrystalline particles with grain boundaries and (b) single-crystalline materials with anisotropic nanostructures. Source: From Amano et al. [35]. © 2011 The Electrochemical Society.

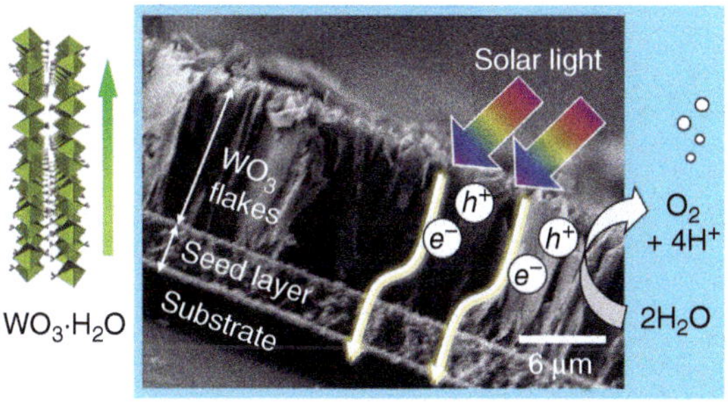

Figure 7.13 Crystal structure of $WO_3{\cdot}H_2O$ and cross-sectional sideview SEM image of vertically aligned WO_3 flakes deposited on transparent conductive oxide (TCO) glass substrate for PEC water oxidation under visible-light irradiation. The seed layer, which is WO_3 nanoparticles, is essential for heterogeneous nucleation of $WO_3{\cdot}H_2O$ flakes. The $WO_3{\cdot}H_2O$ flakes with two-dimensional nanostructure were converted to monoclinic WO_3 by heat treatment. Source: From Amano et al. [36]. © 2010, RCS.

The WO_3 films composed of well-crystallized nanoparticles exhibit a large surface area for semiconductor/liquid junctions [34]. However, the increase in thickness to maximize photoabsorption would increase the grain boundaries, which act as e^- transport resistance and recombination center (Figure 7.12). On the other hand, the vertically aligned WO_3 with a low-dimensional nanostructure is expected to show a higher PEC performance like TiO_2 nanotube arrays. The vertically aligned WO_3 nanoflakes were fabricated on a transparent conductive oxide (TCO) glass by using a seed layer and an anisotropic crystal growth of layered compound, $WO_3{\cdot}H_2O$ (Figure 7.13) [35, 36]. The $WO_3{\cdot}H_2O$ phase was easily converted to monoclinic WO_3 without the change of the 2D flake-like nanostructure. The vertically aligned WO_3 flakes exhibited PEC efficiency for water oxidation higher than that of horizontally stacked WO_3 flakes (Figure 7.14) [35]. The vertically aligned WO_3 flakes exhibited a small difference between j_{photo} under front-side irradiation and back-side irradiation ($j_{front}/j_{back} = 0.7$). Moreover, the j_{photo} of the horizontally stacked WO_3 flakes was significantly dependent on the photoirradiation side ($j_{front}/j_{back} = 0.2$). This indicates that the excellent e^- transport is achieved in the vertically aligned WO_3 flakes owing to the low density of grain boundaries. Fast e^- transport is suggested to increase the lifetimes of h^+ to induce better water oxidation. It is suggested that nanostructuring and low-dimensional morphology control is a useful method to fabricate highly efficient photoanodes.

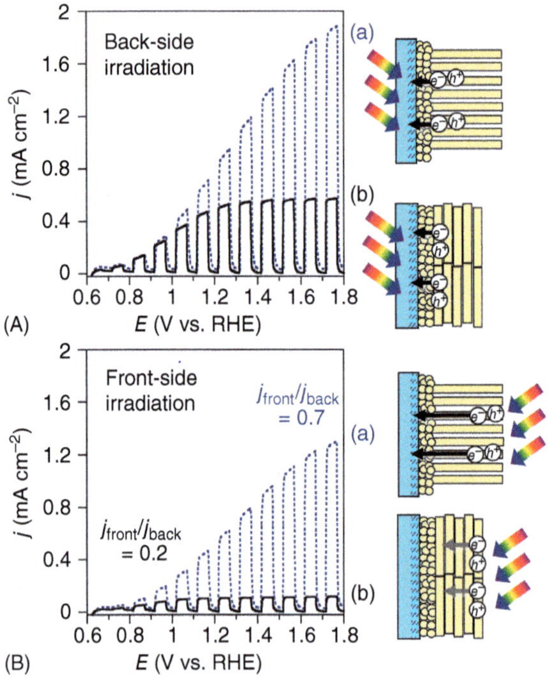

Figure 7.14 Current–potential curves of (a) vertically aligned WO_3 flakes (blue dashed) and (b) horizontally stacked WO_3 flakes (black solid) deposited on TCO glass substrate in $0.1\,mol\,l^{-1}$ Na_2SO_4 (pH = 7) under visible-light irradiation ($\lambda > 400$ nm). (A) Back-side and (B) front-side illumination were performed through the TCO glass substrate and the WO_3 layer, respectively. Chopped illumination was used to show the transient photocurrent response for water oxidation. Source: Based on Amano et al. [35].

7.8.2 Donor Doping

The doping of impurities affects the optical and electrical properties of semiconductors owing to the formation of midgap states and the increase of the carrier density. In the donor doping, the donor atoms (D) doped in the crystal lattice produce e^- to increase N_D:

$$D \rightarrow D^+ + e^- \tag{7.24}$$

The N_D affects Fermi energy, electrical conductivity, and W_{SCL}. In the case of TiO_2, the addition of cations with valence higher than that of the Ti^{4+} lattice (e.g. Nb^{5+}, Ta^{5+}, and W^{6+}) is donor doping [37]:

$$Nb_2O_5 \overset{TiO_2}{\rightarrow} 2Nb^{\cdot}_{Ti} + 2e' + 4O^{\times}_{O} + 1/2O_2 \tag{7.25}$$

The equation of Nb-doped TiO_2 is expressed by Kröger–Vink notation, where $Nb_{Ti^{\cdot}}$ is the Nb^{5+} ion in the Ti^{4+} lattice site, e′ is an electron, and $O^{\times}_{O}$ is an O^{2-} ion in the oxygen lattice site. The increased carrier density is reported to enhance the PEC efficiency of n-type semiconductor electrodes. For example, the PEC performance of the α-Fe_2O_3 (hematite) photoanode is enhanced by the doping of higher-valent cations such as Si^{4+} and Ti^{4+} [25, 32]. The PEC performance of the $BiVO_4$ photoanode is also enhanced by the doping of Mo^{6+} and W^{6+} into the V^{5+} site [30]. The effects of the donor doping are generally attributed to the improvement in the electrical conductivity to facilitate charge carrier diffusion. However, other contributions such as surface charge transfer rate (electrocatalytic

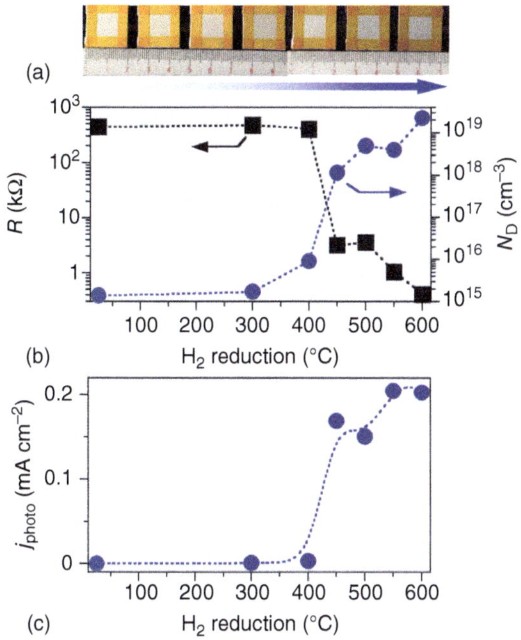

Figure 7.15 (a) Pictures of the thermally oxidized TiO_2 films (Ti plate calcined in air at 900 °C for two hours) and the films treated in H_2 stream at different temperatures (300–800 °C). (b) Effect of H_2 treatment temperature on the sheet resistance (*R*) and the donor density (N_D) of TiO_2 films. The *R* was measured by a four-point probe. The N_D was measured by the Mott–Schottky analysis. (c) Effect of H_2 treatment temperature on the j_{photo} of TiO_2 films in 0.1 mol l^{-1} H_2SO_4 (pH = 1). The j_{photo} for water oxidation was obtained in LSV measurement at 0.87 V vs. RHE under UV irradiation (λ > 300 nm). Source: Adapted with permission Amano et al. [38]. Copyright 2016, American Chemical Society.

activity), suppression of surface recombination, nanostructuring, defect formation, and lattice distortion have been proposed by researchers.

In addition to the cation doping, H_2 reduction treatment of oxide materials is another type of donor doping method:

$$O_O^{\times} + H_2 \rightarrow V_O^{\cdot\cdot} + 2e' + H_2O \tag{7.26}$$

Here, $V_O^{\cdot\cdot}$ is an oxygen vacancy with a double positive charge in Kröger–Vink notation. The effect of H_2 reduction treatment was studied for the thermally oxidized Ti plate (Figure 7.15) [38]. The rutile TiO_2 layers were treated with H_2 to create both oxygen vacancies and electrons. A part of the electrons is trapped in a Ti^{4+} lattice site to form Ti^{3+} ions. The color of the TiO_2 layer was gradually changed at H_2 treatment temperature above 450 °C. At the same time, the surface resistance was greatly reduced, and the N_D was increased by 2–3 orders of magnitude by H_2 treatment. Increasing the density of crystal defects such as oxygen vacancies and Ti^{3+} is generally expected to increase the recombination loss. In contrast, the PEC property of the reduced TiO_2 films was enhanced with an increase in the N_D. This indicates that the improvement in the n-type conductivity resulted in the enhancement of the PEC activity of the TiO_2 photoanode. The bulk electronic property would be an important factor controlling the charge transport in the photoanode.

The doping strategy has also been applied to the wide-band-gap semiconductors to provide a visible-light response. Nonmetal doping such as nitrogen and sulfur has been an interesting topic to activate TiO_2 photomaterials toward effective visible-light utilization in the solar spectrum [39].

7.8.3 Modification of Photoanode Surface

The effect of surface modifications has been extensively studied for photoanode materials. The role of surface modification can be classified into (i) OER electrocatalyst, (ii) heterojunction formation, (iii) sensitizer, (iv) surface protection, and (v) surface passivation to reduce trap states.

Electrocatalysts, such as RuO_2, IrO_2, CoPi, FeOOH, NiOOH, and CoO_x, are loaded on the photoanode to facilitate the insufficient transfer of photogenerated h^+ from the surface to the electrolyte solution (Figure 7.16). The electrocatalysts loaded on photoelectrodes are often called cocatalyst, which is the cooperative component to improve the activities of a photocatalyst and a photoelectrode [3]. The loading of cocatalysts is expected to increase the number of electrocatalytic sites to enhance j_{photo}. Moreover, the reduction of the activation energy of the multi-electron transfer process increases the rate constant of h^+ injection to the water molecules. The decrease of the overpotential for OER results in the cathodic shift of the E_{onset} to near E_{fb}. The cocatalyst loading improves not only the kinetics of water oxidation but also the charge separation at the surface. The time-resolved absorption study of CoO_x-loaded $LaTiO_2N$ revealed that the h^+ are captured rapidly by CoO_x a few picoseconds and the lifetimes of e^- are dramatically prolonged to the millisecond region [40]. This implies that the photogenerated e^- and h^+ are separated effectively at the cocatalyst/photoanode interface. Multiple h^+ should be captured in the cocatalyst to induce four-electron transfer oxidation of water. The cocatalysts also contribute to the stabilization of the non-oxide materials against the photocorrosion during PEC water oxidation. The bare oxynitride photoanodes such as TaON and $BaTaO_2N$ are easily deactivated with time under steady photoirradiation owing to the self-oxidation of the lattice nitrogen by the photogenerated h^+ ($2N_3{}^- + 6h^+ \rightarrow N_2$) [2, 41]. However, the loading of CoO_x efficiently scavenged the h^+ and also suppressed the self-oxidation of the surface. The post-loading of Rh oxide shows synergistic effects on the efficiency and stability of the CoO_x-loaded $BaTaO_2N$ photoanode [41]. The role of the Rh_2O_3 particles is considered to promote h^+ transfer between the photoanode and the CoO_x cocatalyst.

The formation of semiconductor–semiconductor heterojunction is another important strategy to improve charge separation and extend the absorption spectrum [42]. The heterostructure design is based on the combination of semiconductor materials with different band alignments (Figure 7.17). The heterojunction with different energy levels can be categorized into three classes: (type I) straddling gap, (type II) staggered gap, and (type III) broken gap. The most developed system is

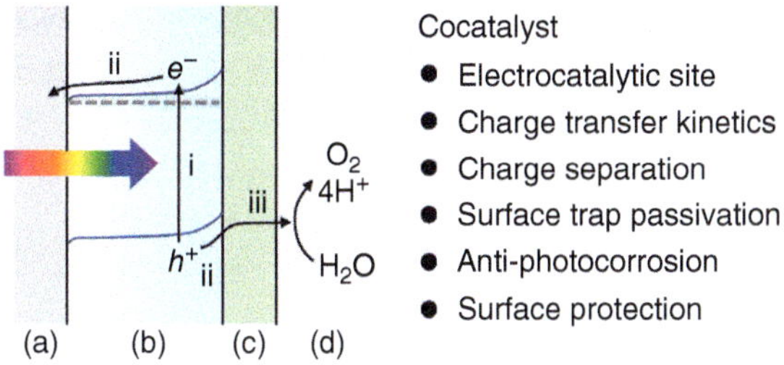

Figure 7.16 Schematic illustration of the PEC water oxidation process at the photoanode with cocatalyst layer: (a) conductive substrate, (b) n-type semiconductor as a photoabsorber, (c) cocatalyst layer for OER, and (d) aqueous electrolyte solution. The number indicates the order of the PEC reaction process: (i) the generation of photoexcited e^-–h^+ pairs; (ii) charge separation, carrier diffusion, and carrier transport; and (iii) h^+ transfer from semiconductor to water.

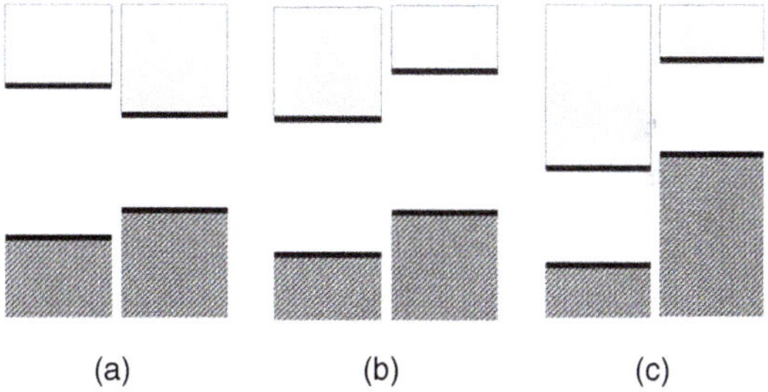

Figure 7.17 Band alignments in three types of semiconductor heterojunctions: (a) straddling gap (type I), (b) staggered gap (type II), and (c) broken gap (type III).

n-WO_3/n-$BiVO_4$ heterojunction with staggered band alignment (type II) [43, 44]. Since the energy levels of CB minimum of the WO_3 underlayer are lower than that of $BiVO_4$ overlayer, the photoexcited e^- is transferred from $BiVO_4$ to the WO_3 inside. Meanwhile, the h^+ generated in WO_3 is diffused toward the $BiVO_4$ surface since the VB maximum of $BiVO_4$ is higher than that of WO_3. The driving force of the charge separation is the band offset at the staggered type II heterojunction. In contrast, the WO_3/Fe_2O_3 system was ineffective because the CB minimum and VB maximum of WO_3 are straddling those of Fe_2O_3 (type I heterojunction).

The cocatalyst loading and the heterojunction formation are useful methods to enhance the OER kinetics, charge separation, and the anti-photocorrosion property. The sensitization of the semiconductor surface is also the topic of academic interest to enhance visible-light response. The sensitizers are quantum dots, organic dyes, metal complexes, and gold colloids with localized surface plasmon resonance.

7.8.4 Electron-Conductive Materials

The photoanode films are prepared on current-collecting conductive substrates such as TCO-coated glasses and metal plates. Fluorine-doped tin oxide (FTO) is often used as a TCO. In addition to the conventional 2D flat substrates, 3D porous substrates such as carbon fiber and Ti fiber felts are used for the conductive substrate [45, 46]. In the case of porous nanoparticle films, the resistance of grain boundaries reduces electrical conductivity. The interparticle electron transfer can be improved by a necking treatment, which is coating a conductive material between the particle gaps. The necking treatment is reported to increase the PEC properties of particulate photoanodes owing to the decrease in the resistance of electron transport [2, 6]. The electrical conductivity of the particulate photoanode films is also enhanced by the use of reduced graphene oxide (rGO) nanosheets and carbon nanotubes (CNTs).

Recently, a technique to fabricate a film composed of the monolayer of semiconductor particles, which is called particle transfer method, is developed to investigate the PEC properties of the semiconductor particle without grain boundaries [4, 6, 47]. The monolayer films of the semiconductor particles are deposited with a thin contact layer and a subsequent thick conductor layer of metals. It is confirmed that the ohmic-like junction between the semiconductor particle and the metal contact is useful in collecting the photoexcited e^- [48]. Larger work functions of metal resulted in the formation of a Schottky barrier, which retards e^- collection at the metal–semiconductor interface. The particle transfer method can be applied to the many semiconductor particles, even if the material is not suitable for the preparation of good particulate photoanode.

7.9 PEC Device for Water Splitting

PEC cell for water splitting has been investigated using multi-light absorbers to increase the photoabsorption of the solar spectrum [49–51]. In the PEC system using two light absorbers, photoanode for OER and photocathode for HER are connected by a conductive metal wire or a transparent conductive substrate (PEC–PEC tandem system). The overlapping of the E_{onset} of the two photoelectrodes should be increased to enhance STH efficiency. The separate arrangement of photoanode and photocathode is called a "parallel cell." A "tandem cell" is the arrangement that the front photoelectrode is transparent for a part of the incident light and the other photoelectrode is set behind it. The tandem cell of a semitransparent $BiVO_4$ photoanode and a copper indium gallium selenide (CIGS)-based photocathode achieved 3.7% of STH efficiency in unbiased water splitting [7]. For further improvement of the PEC–PEC system, the j_{photo} of photoanode should be increased at a more cathodic applied potential.

The combination of a PEC cell with a single or multi-junction PV cell (PV–PEC tandem system) is proposed for the unbiased water splitting, in which PV cell supplies the bias voltage required for the PEC reaction. An STH efficiency of 7.7% was achieved on the PV–PEC tandem system composed of a series-connected two pieces of crystalline silicon PV cell and dual oxide photoanodes (parallel-connected α-Fe_2O_3 and $BiVO_4$) [52]. The dual photoanodes were a Ti-doped Fe_2O_3 and a Mo-doped $BiVO_4$ loaded with Ni- and Fe-based cocatalysts. In general, the PV–PEC system exhibits a higher STH efficiency and more flexibility in the selection of the materials compared with the PEC–PEC system. However, the cost of the PV cells should be considered for large-scale applications.

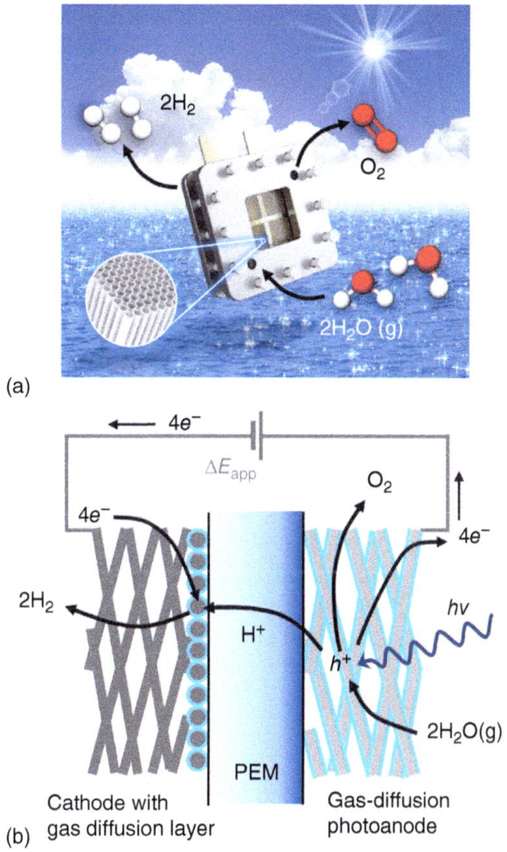

Figure 7.18 Schematic illustrations of (a) PEC cell for solar H_2 production using water vapor from the air over the sea and (b) vapor-fed PEC water splitting system using a gas-diffusion HER cathode, a proton exchange membrane (PEM), and a gas-diffusion photoanode for OER. The photoanode is composed of a TiO_2 nanotube array decorated on porous Ti felt. The surface of TiO_2 nanotubes is coated with Nafion ionomer thin film for the gas-phase operation. Source: (a) Adapted by permission of Wiley-VCH Verlag. Amano et al. [53]; (b) Based on Amano et al. [54].

Apart from the water splitting using liquid water, the water vapor in the air may be a substitute for the H_2 source. The PEC facilities for large-scale H_2 production require long sunshine hours and vast areas, but there is a concern regarding the water supply in the arid area. Seawater is abundant but needs to be purified by removing impurities. In contrast, there is no need for water treatment when gaseous water in the air over the sea is used. Furthermore, no liquid transport system is required because water vapor is supplied by natural convection of the humidified air at the sea surface. Based on the above consideration, vapor-fed PEC cells for water splitting have been investigated using a gas-diffusion OER photoanode, a proton exchange membrane (PEM) as a solid polymer electrolyte, and an HER cathode (Figure 7.18) [53–56]. In the all-solid PEM–PEC system, the photocurrent response was drastically enhanced by the coating of a Nafion perfluorosulfonic acid ionomer on the gas-diffusion photoanodes. The enhancement is because of the increase of proton conductivity and water adsorption in the gaseous environment. It is reported that the vapor-fed photoelectrolysis of water was induced at low ΔE_{app} using the Nafion ionomer-coated $SrTiO_3$ photoanode (Figure 7.19) [54]. The IPCE in the PEM–PEC cell was 3.9% under UV irradiations ($\lambda = 365$ nm) at $\Delta E_{app} = 0.3$ V. The overall water splitting was proved by the evolution of

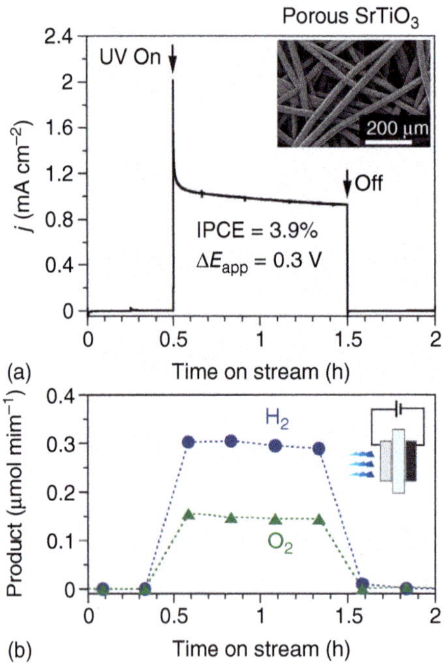

Figure 7.19 Vapor-fed water photoelectrolysis by porous $SrTiO_3$ photoanode |PEM| Pt-carbon black cathode under humidified argon (3 vol% water vapor). The response of (a) photocurrent density and (b) the formation rate of H_2 evolved in the cathode compartment and O_2 evolved in the photoanode compartment at $\Delta E_{app} = 0.3$ V under 365-nm UV irradiation ($I_0 = 42$ mW cm^{-2}, photoanode area 2 cm^2). The porous $SrTiO_3$ photoanode was coated by Nafion ionomer thin films for the gas-phase operation. The evolved gas in each compartment was separated by a PEM and analyzed by on-line gas chromatographs. Source: (a, b) Based on Amano et al. [54].

H_2 and O_2 with a ratio of 2:1. The η_{FE} of H_2 and O_2 was almost 100% at each electrode.

7.10 Conclusions and Outlook

Factors affecting PEC performances of photoanode materials for OER were reviewed in this chapter. The design strategies of highly efficient photoanode have been advanced in recent years: control of morphology and nanostructure, donor doping, cocatalyst loading, and heterojunction layer. The further rational design that combines many factors is required for the development of the photoanode materials with narrow band gap, high quantum efficiency, low E_{onset} for OER, and long-term durability. Another requirement is the use of low-cost earth-abundant materials for global scalability.

One of the advantages of the PEC system is that the photoanode and (photo)cathode can be designed separately and individually. The PEC water splitting device can separate the H_2 evolved on the cathode and the O_2 evolved on the photoanode using a membrane. These are the significant merits of PEC systems compared with powdered photocatalytic systems.

Because the single PEC cell usually requires additional electrical or chemical bias to induce water splitting, a tandem cell device such as PEC–PEC and PV–PEC system is proposed for unbiased water splitting. In parallel with this, the development of water electrolyzer systems powered by conventional photovoltaics (PV-electrolyzer system) is progressing toward practical use for STH conversion [57]. The superiority of the PEC-based system to the PV-electrolyzer system may need to be defined for further research and development.

Considering a solar H_2 production facility on a large scale, the cost of the semiconductor materials and the supply of liquid freshwater could be problems in the practical operation. The PEC-based system for vapor-fed water splitting is a promising method since the water vapor is inexhaustible over the sea. The use of water vapor is limited by low concentration, but it matches the low density of solar light intensity. The key component in the PEC-based system would be the high-performance photoanodes for PEC oxygen evolution. The photoanodes must be inexpensive.

References

1 Fujishima, A. and Honda, K. (1972). *Nature* 238: 37–38.
2 Abe, R. (2010). *J. Photochem. Photobiol., C* 11: 179–209.
3 Yang, J., Wang, D., Han, H., and Li, C. (2013). *Acc. Chem. Res.* 46: 1900–1909.
4 Hisatomi, T., Kubota, J., and Domen, K. (2014). *Chem. Soc. Rev.* 43: 7520–7535.
5 Kuang, Y., Yamada, T., and Domen, K. (2017). *Joule* 1: 290–305.
6 Seo, J., Nishiyama, H., Yamada, T., and Domen, K. (2018). *Angew. Chem. Int. Ed.* 57: 8396–8415.
7 Yamada, T. and Domen, K. (2018). *ChemEngineering* 2: 36.
8 Yao, T., An, X., Han, H. et al. (2018). *Adv. Energy Mater.* 8: 1800210.
9 Sato, N. (1998). *Electrochemistry at Metal and Semiconductor Electrodes*. Amsterdam: Elsevier Science.
10 Rajeshwar, K. (2007). *Semiconductor Electrodes and Photoelectrochemistry* (ed. S. Licht). Weinheim: Wiley-VCH.
11 Walter, M.G., Warren, E.L., McKone, J.R. et al. (2010). *Chem. Rev.* 110: 6446–6473.
12 Peter, L.M. (2016). *Photoelectrochemical Solar Fuel Production: From Basic Principles to Advanced Devices* (eds. S. Giménez and J. Bisquert), 3–40. Cham: Springer International Publishing.
13 Takanabe, K. (2017). *ACS Catal.* 7: 8006–8022.
14 Hisatomi, T., Minegishi, T., and Domen, K. (2012). *Bull. Chem. Soc. Jpn.* 85: 647–655.
15 Rothenberger, G., Moser, J., Gratzel, M. et al. (1985). *J. Am. Chem. Soc.* 107: 8054–8059.
16 Chen, Z., Jaramillo, T.F., Deutsch, T.G. et al. (2010). *J. Mater. Res.* 25: 3–16.
17 Chen, Z., Dinh, H.N., and Miller, E. (2013). *Photoelectrochemical Water Splitting: Standards, Experimental Methods, and Protocols*. New York: Springer.
18 Smith, W.A. (2016). *Photoelectrochemical Solar Fuel Production: From Basic Principles to Advanced Devices* (eds. S. Giménez and J. Bisquert), 163–197. Cham: Springer International Publishing.
19 Bard, A.J., Parsons, R., and Jordan, J. (1985). *Standard Potentials in Aqueous Solution*. New York: CRC Press.

20 Yang, M., He, H., Du, J. et al. (2019). *J. Phys. Chem. Lett.* 10: 6159–6165.

21 Amano, F., Mukohara, H., and Shintani, A. (2018). *J. Electrochem. Soc.* 165: H3164–H3169.

22 Amano, F., Tian, M., Ohtani, B., and Chen, A. (2012). *J. Solid State Electrochem.* 16: 1965–1973.

23 Yamakata, A., Ishibashi, T.A., and Onishi, H. (2002). *J. Phys. Chem. B* 106: 9122–9125.

24 Yamakata, A., Ishibashi, T.A., and Onishi, H. (2003). *J. Mol. Catal. A: Chem.* 199: 85–94.

25 Amano, F., Ohtani, B., and Yoshida, H. (2016). *J. Electroanal. Chem.* 766: 100–106.

26 Zhou, L., Zhao, C., Giri, B. et al. (2016). *Nano Lett.* 16: 3463–3474.

27 Dotan, H., Sivula, K., Grätzel, M. et al. (2011). *Energy Environ. Sci.* 4: 958–964.

28 Kim, T. and Choi, K.S. (2014). *Science* 343: 990–994.

29 Chun, W.J., Ishikawa, A., Fujisawa, H. et al. (2003). *J. Phys. Chem. B* 107: 1798–1803.

30 Abdi, F.F., Berglund, S.P., and van de Krol, R. (2016). *Photoelectrochemical Solar Fuel Production: From Basic Principles to Advanced Devices* (eds. S. Giménez and J. Bisquert), 355–391. Cham: Springer International Publishing.

31 Pourbaix, M. (1974). *Atlas of Electrochemical Equilibria in Aqueous Solutions.* Houston: NACE International.

32 Sivula, K., Le Formal, F., and Grätzel, M. (2011). *ChemSusChem* 4: 432–449.

33 Shankar, K., Basham, J.I., Allam, N.K. et al. (2009). *J. Phys. Chem. C* 113: 6327–6359.

34 Santato, C., Ulmann, M., and Àugustynski, J. (2001). *J. Phys. Chem. B* 105: 936–940.

35 Amano, F., Li, D., and Ohtani, B. (2011). *J. Electrochem. Soc.* 158: K42–K46.

36 Amano, F., Li, D., and Ohtani, B. (2010). *Chem. Commun.* 46: 2769–2771.

37 Amano, F., Tosaki, R., Sato, K., and Higuchi, Y. (2018). *J. Solid State Chem.* 258: 79–85.

38 Amano, F., Nakata, M., Yamamoto, A., and Tanaka, T. (2016). *J. Phys. Chem. C* 120: 6467–6474.

39 Asahi, R., Morikawa, T., Irie, H., and Ohwaki, T. (2014). *Chem. Rev.* 114: 9824–9852.

40 Yamakata, A., Kawaguchi, M., Nishimura, N. et al. (2014). *J. Phys. Chem. C* 118: 23897–23906.

41 Higashi, M., Domen, K., and Abe, R. (2013). *J. Am. Chem. Soc.* 135: 10238–10241.

42 Moniz, S.J.A., Shevlin, S.A., Martin, D.J. et al. (2015). *Energy Environ. Sci.* 8: 731–759.

43 Chatchai, P., Murakami, Y., Kishioka, S.y. et al. (2009). *Electrochim. Acta* 54: 1147–1152.

44 Hong, S.J., Lee, S., Jang, J.S., and Lee, J.S. (2011). *Energy Environ. Sci.* 4: 1781–1787.

45 Homura, H., Ohtani, B., and Abe, R. (2014). *Chem. Lett.* 43: 1195–1197.

46 Amano, F., Shintani, A., Tsurui, K., and Hwang, Y.M. (2017). *Mater. Lett.* 199: 68–71.

47 Minegishi, T., Nishimura, N., Kubota, J., and Domen, K. (2013). *Chem. Sci.* 4: 1120–1124.

48 Ham, Y., Minegishi, T., Hisatomi, T., and Domen, K. (2016). *Chem. Commun.* 52: 5011–5014.

49 Peerakiatkhajohn, P., Yun, J.H., Wang, S., and Wang, L. (2017). *J. Photonics Energy* 7: 012006.

50 Jiang, C., Moniz, S.J.A., Wang, A. et al. (2017). *Chem. Soc. Rev.* 46: 4645–4660.

51 Chen, Q., Fan, G., Fu, H. et al. (2018). *Adv. Phys.* 3: 863–884.

52 Kim, J.H., Jang, J.W., Jo, Y.H. et al. (2016). *Nat. Commun.* 7: 13380.

53 Amano, F., Mukohara, H., Shintani, A., and Tsurui, K. (2019). *ChemSusChem* 12: 1925–1930.

54 Amano, F., Mukohara, H., Sato, H. et al. (2020). *Sustainable Energy Fuels*: 4, 1443–1453.

55 Amano, F., Mukohara, H., Sato, H., and Ohno, T. (2019). *Sustainable Energy Fuels* 3: 2048–2055.

56 Zafeiropoulos, G., Johnson, H., Kinge, S. et al. (2019). *ACS Appl. Mater. Interfaces* 11: 41267–41280.

57 Luo, J., Im, J.H., Mayer, M.T. et al. (2014). *Science* 345: 1593–1596.

8

Photocatalytic and Photoelectrochemical Overall Water Splitting

Nur Aqlili Riana Che Mohamad[1], Filipe Marques Mota[1] and Dong Ha Kim[1,2]

[1]*Ewha Womans University, College of Natural Sciences, Division of Molecular Life and Chemical Sciences, Department of Chemistry and Nano Science, 52, Ewhayeodae-gil, Seodaemun-gu, Seoul 03760, Republic of Korea*
[2]*Ewha Womans University, College of Engineering, Division of Chemical Engineering and Materials Science, 52, Ewhayeodae-gil, Seodaemun-gu, Seoul 03760, Republic of Korea*

8.1 Introduction

Arising environmental awareness has emphasized the urgency to uncover promising solutions to decarbonize the world economy through a shift from dwindling oil supplies in a context of an ever-increasing global energy demand. As a promising non-carbon-emitting source of energy, molecular hydrogen (H_2) possesses an attractive gravimetric heating value (141.9 MJ kg^{-1}) significantly superior to most conventional fossil fuels, e.g. methane (55.5 MJ kg^{-1}), gasoline (47.5 MJ kg^{-1}), diesel (44.8 MJ kg^{-1}), and methanol (20.0 MJ kg^{-1}) [1, 2]. To date, however, H_2 is primarily derived from conventional fossil fuels, e.g. steam reforming of methane (48%), reforming of heavy oils and naphthas (30%), and coal gasification (18%) [3, 4], significantly adding to global anthropogenic CO_2 emissions (~35.5 Gt/yr) and the increasing atmospheric CO_2 concentrations (>400 ppm) [5–7]. To be pinpointed as an environment-friendly solution toward a sustainable future, a large-scale production of high purity H_2 by renewable approaches (e.g. electrolysis, biomass, through the use of bacteria, or utilizing solar, geothermal, wave, and wind powers) has to be realized. Among these, producing H_2 through an HOH bond cleavage driven by electrical (electrolysis) or (solar) light energy (photolysis) has gathered distinct focus in the past decades. With ~100 000 TW reaching our planet at each moment and ~36 000 TW reaching the surface, solar power is today the most promising sustainable clean energy source [8]. On the other hand, among the emerging processes exploiting a solar-to-chemical conversion reviewed in this book, solar-driven water splitting is also, at the same time, the most promising pathway to a groundbreaking shift in the sustainability of our societies. Photo(electro)catalytic overall water splitting (OWS) technologies primarily rely on the utilization of light-responsive materials, capable of harvesting photons

Solar-to-Chemical Conversion: Photocatalytic and Photoelectrochemcial Processes, First Edition.
Edited by Hongqi Sun.

and generating charge carriers, e.g. electrons and holes. These charge carriers subsequently drive simultaneous H^+ reduction and H_2O oxidation reactions, with an ideal stoichiometrical evolution of H_2 (hydrogen evolution reaction [HER]) and O_2 (oxygen evolution reaction [OER]), respectively. Along with the OWS mechanistic details, introduced below are the principles, key performance indicators, benchmark materials, and a review of conventional methodologies or arising strategies in both photocatalytic (PC) and photoelectrochemical (PEC) platforms. Whereas a clear focus has been ascribed in this chapter to systems for OWS (with simultaneous production of H_2 and O_2), this effervescent research field continuously shares to date valuable reports on either HER or OER, which should serve as a complementary and valuable guideline to the reader.

8.2 Photocatalytic Overall Water Splitting

Splitting the water molecule was first demonstrated by Honda and Fujishima under an external bias, using a single-crystalline ultraviolet (UV)-responsive TiO_2 photoanode and an electrically connected Pt as counter electrode [9]. By later demonstrating the decomposition of the water molecule into H_2 and O_2 without the use of external voltage using UV light, the same authors have since stimulated worldwide interests in the development of PC water splitting. The idea was later further exploited by Bard in the construction of PC systems introducing semiconductor particles or powders [10, 11]. The development of appropriate PC OWS systems demands the fabrication of reliable and reproducible photocatalysts that are able to meet imposing requirements toward their commercial feasibility, e.g. suitable band gap and position, efficient charge carrier efficiency, and maximum photostability. An alternative approach is to establish heterojunctions between individual photocatalysts, for which proposed Z-scheme mechanistic pathways have, in particular, shown extended OWS efficiency.

OWS using cost-effective particulate photocatalysts in scalable reactors with simple design is today considered a low-cost technology with the potential to enable large-scale H_2 production and concurrent O_2 evolution [12, 13]. This scalability with a clear promise of commercial implementation is further emphasized upon comparison against PEC OWS technologies, as further discussed in Section 8.3. To break the commercialization barrier, research works primarily focus on the design and development of robust and stable photo-responsive catalytic systems composed of earth-abundant elements to stoichiometrically split the water molecule. In addition, many reviews have hence consistently covered the fundamental aspects of PC water splitting, engineering, and technological groundworks, as well as a meticulous analysis of underlying synthesis strategies in the design of efficient photocatalysts [14–18]. In the sections below, a focus on benchmark photocatalysts and a guide for synthesis strategies with emerging interest has been conveniently provided. To date, however, no PC system has unveiled the necessary potential for practical application.

8.2.1 Principles and Mechanism

To drive the OWS reaction, suitable photocatalysts must meet three primary requirements. With a desirably wide light absorption ability range capable of driving the generation of excitons, semiconductors can be excited by photons having energy greater or equal to its band-gap energy. Electrons will then move from the valence band (VB) to the conduction band (CB), with holes remaining in the VB. A suitable photocatalyst must then possess a matching band structure and alignment with the redox potentials for a simultaneous HER and OER. The photogenerated electrons and holes are capable to drive the reduction of H^+ and H_2O oxidation, respectively, only if the bottom of the CB is more negative than the redox potential of H^+/H_2 (0 V_{NHE}, at pH 0), while the top of the VB is more positive than the redox potential of O_2/H_2O (1.23 V_{NHE}, at pH 0) (see Figure 8.1a) [15]. The thermodynamic minimum requirement of the band-gap energy is therefore 1.23 eV, the equivalent to approximately 1000 nm of light wavelength. However, a semiconductor having a larger band gap in the range of 1.6–2.4 eV will be a practical photocatalyst to drive the OWS since both of the half-reactions require a significant amount of kinetic overpotentials [19]. To trigger the redox reactions involved in OWS, an effective separation and transfer of photoexcited carriers are finally necessary for the rational selection of appropriate photocatalysts. OWS into H_2 and O_2 is an energetically uphill chemical reaction with regard to its Gibbs free energy values ($\Delta G^\circ = 237\ \mathrm{kJ\ mol^{-1}}$, or a potential of 1.23 eV per electron) as shown in Eq. (8.1): [20]

$$H_2O \rightarrow H_2 + \frac{1}{2}O_2 \quad (\Delta G^\circ = 237\ \mathrm{kJ\ mol^{-1}}) \tag{8.1}$$

The photoreduction of protons for HER is generally believed to be a kinetic facile process, with only two electrons being required to produce an H_2 molecule. By comparison, however, with a four-electron transfer process coupled with the removal of four protons from water molecules to form an O—O bond, the sluggish OER is a remarkably challenging step in the PC OWS. In addition, photogenerated electron–hole pairs must have a sufficient lifetime to react with both the dissociative H^+/OH^- species from the water molecule to allow the photocatalysis to happen [21, 22].

The OWS process in one-step photoexcitation can therefore be described according to Eqs. (8.2–8.5):

$$\text{Absorption of light}: \ \mathrm{SC} + h\nu\ (> E_g) \rightarrow e^-_{CB} + h^+_{VB} \tag{8.2}$$

$$\text{Water oxidation (OER)}: \ 2H_2O + 4h^+_{VB} \rightarrow O_2 + 4H^+ \tag{8.3}$$

$$\text{Proton reduction (HER)}: \ 2H^+ + 2e^-_{CB} \rightarrow H_2 \tag{8.4}$$

$$\text{Overall water splitting}: \ 2H_2O \rightarrow 2H_2 + O_2 (\Delta G^\circ = 237.2\ \mathrm{kJ\ mol^{-1}}) \tag{8.5}$$

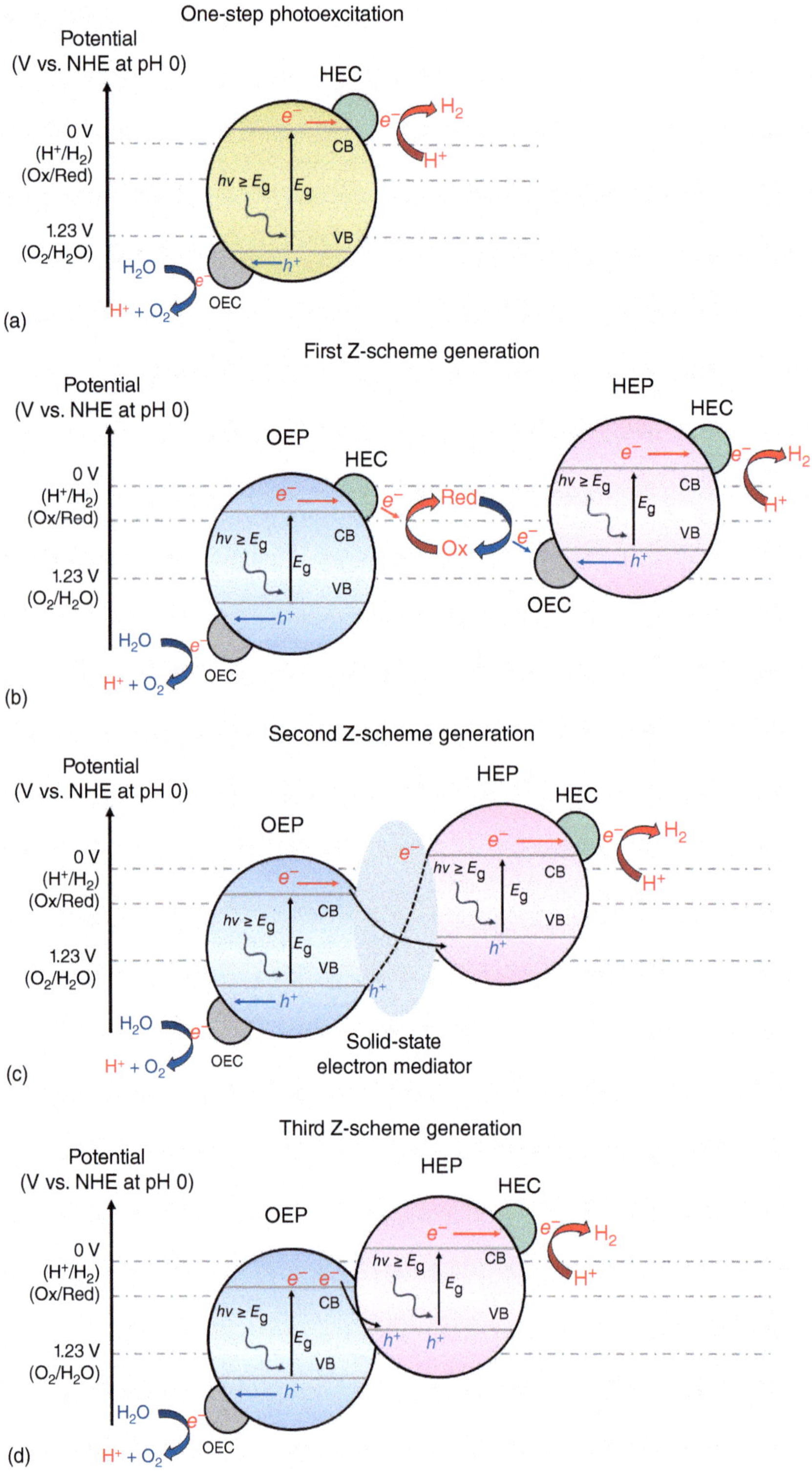

One-step photoexcitation
Potential
(V vs. NHE at pH 0)
0 V
(H+/H2)
(Ox/Red)
1.23 V
(O2/H2O)
HEC
CB
VB
hv ≥ Eg
Eg
H2O
H+ + O2
OEC
(a)
First Z-scheme generation
OEP
HEP
HEC
Red
Ox
OEC
(b)
Second Z-scheme generation
Solid-state
electron mediator
(c)
Third Z-scheme generation
(d)

Figure 8.1 Different types of photocatalytic overall water splitting systems including (a) one-step photoexcitation, (b) first-generation Z-scheme, (c) second-generation Z-scheme, and (d) third-generation Z-scheme. HEP: hydrogen evolution photocatalyst; OEP: oxygen evolution photocatalyst; HEC: hydrogen evolution cocatalyst; OEC: oxygen evolution cocatalyst; CB: conduction band; VB: valence band; E_g: energy band gap; NHE: normal hydrogen electrode; Ox: oxidant; Red: reductant.

8.2.2 Key Performance Indicators

As often underlined on representative photo-driven platforms, factual comparisons of the performance of reported systems are prone to suffer from distinct reaction conditions in the literature, including a broad selection of light sources (e.g. sunlight, solar simulators, visible light) with varying intensity, position, and distance placement. The importance of a comparison of novel photocatalysts with results under dark and against reference reports, along with an assessment of the impact of irradiation on the reaction temperature, and a detailed mass balance and product distribution (stoichiometry evaluation) is thus here emphasized. The performance of PC systems is primarily evaluated based on corresponding conversion efficiency, catalytic activity (i.e. stoichiometry and production rates [$\text{mol h}^{-1}\ \text{g}_{\text{cat}}{}^{-1}$] of evolved H_2 and O_2 gases), and catalytic lifetime. The associated quantum yield is defined as the ratio of photons contributing to the activity and the number of total absorbed photons. Taking into consideration the inexact number of absorbed photons due to light transmission and scattering phenomena, corresponding apparent quantum yields (AQYs) suitably serve as a performance indicator according to Eq. (8.6):

$$\text{AQY} = nR/I \tag{8.6}$$

with n corresponding to the number of photogenerated carriers (electrons or holes) consumed in one-step photoexcitation systems and Z-scheme configurations (between two semiconductors) for the formation of H_2 (2 and 4 electrons, respectively) and O_2 (4 and 8 electrons, respectively) gas molecules. R and I correspond to the amounts of gas molecules evolved in established time intervals and the number of incident photons reaching the photocatalyst during the same time interval, respectively [15].

An estimation of the energy conversion efficiencies can, alternatively, serve as an indicator of the catalytic performance for which all water splitter systems can be reliably compared. In detail, the standard solar-to-hydrogen (ηSTH) is defined as the ratio between the total energy generated and the total energy input from standard sunlight irradiation (AM 1.5G, 100 mW cm^{-2}). Based on witnessed H_2 evolution rates, the estimation can be further translated in Eq. (8.7) [23–29]:

$$\eta\text{STH}(\%) = \frac{(\text{moles H}_2/\text{s}) \times (237\ \text{kJ mol}^{-1})}{P_{\text{sun}}(100\ \text{mW cm}^{-2}) \times \text{Area}\ (\text{cm}^2)} \tag{8.7}$$

Based on reported techno-economic analyses, commercially feasible PC OWS technologies for solar H_2 production should attain ideal ηSTH conversion efficiency levels of 5–10% [12, 13]. When assuming a 10% STH efficiency and a lifetime of

10 years for benchmark-particulate photocatalysts, PC platforms can be notably distinguished cost-wise from PEC technologies introduced in Section 8.3. Most importantly, OWS PC systems show the potential to meet the target hydrogen price of 2.00–4.00 USD kg^{-1} set by the US Department of Energy [12]. However, the current value is far behind, typically around 1% obtained in small-scale trials [30, 31].

8.2.3 Materials for One-Step Photoexcitation Toward Overall Water Splitting

In addition to extended light harvesting, representative one-step photocatalysts for OWS possess suitable CB and VB levels straddling the redox potentials of both H^+/H_2 and H_2O/O_2 redox couples. Effective separation and transfer of photogenerated charge carriers (i.e. electron and hole) to surface HER and OER centers play an additional role in the selection of these materials. Over the years, the promise of oxides (metal and perovskites), nitrides, oxynitrides, chalcogenides, and carbonaceous materials to split the water molecule has been actively surveyed. Among these, transition metal (e.g. Ti^{4+}, Zr^{4+}, Nb^{5+}, Ta^{5+}, and W^{6+}) and post-transition metal cations (e.g. Ga^{3+}, In^{3+}, Sn^{4+}, and Sb^{5+}) with d^0 and d^{10} electronic configurations are photoactive in OWS [14, 18]. As the reports introduced below reflect, however, an overall poor AQY of pristine semiconductors has emphasized the need to consider emerging synthesis strategies for superior catalytic activity, e.g. nano-structuralization and surface/interface engineering [32]. Of additional concerns are examples of light-harvesting semiconductors with promising photoactivity (e.g. Si, GaAs, InP, chalcogenides) possessing, however, poor photostability in the presence of photogenerated holes, prone to corrosion by water, and possessing overall short lifetimes.

8.2.3.1 Semiconductors

Oxides

Since their pioneering examination by Fujishima and Honda [9], TiO_2-based photo-responsive systems have found consistent applications in water splitting and an extended focus in other applications, e.g. environmental remediation, dye bleaching, and the reduction of CO_2 [33–38]. The semiconductor TiO_2 displays high efficiency, cost-effectiveness, nontoxicity, and resistance to photocorrosion. With a 3.2 eV wide band gap, the desirable exploitation of TiO_2 in large-scale applications has spurred substantial efforts to extend its UV-limited optical absorption edge onto the Vis–near infrared (NIR) light ranges, which individually account for ~45% and ~50% of the total solar spectrum, respectively [39]. Several studies surveying the promise of incorporating foreign metal/nonmetal species [40, 41], bulk single-electron-trapped oxygen vacancies [42], Ti^{3+} self-doping [43, 44], the effective suppression of back reactions through NaOH coating [45], and the addition of scavenger species during PC have been reported [46]. Representatively, Mao and coworkers reported the synthesis of a deeply hydrogenated black TiO_2 with an extended Vis light absorption. The disorder-engineered TiO_2 nanocrystals showed impressive solar-driven PC activity and stability in the H_2 production under sunlight in the presence of a sacrificial reagent [47]. Among promising strategies, oxygen vacancies and nitrogen doping are claimed to provide intermediate band levels or additional electron trap states, enhancing Vis light absorption [48, 49]. Generated surface oxygen vacancies can, however, be gradually replenished by oxygen sources in air, hindering the photoactivity of the resulting materials (e.g. N-doped TiO_2 or hydrogenated TiO_2) [22]. Regarding the doping of metal ions, these can often act as recombination centers, with a relatively small doping or simultaneous metal co-doping being preferable to maintain the charge balance and hinder recombination [50, 51]. TiO_2 anatase and rutile polymorphs with variations in their crystal structures, different band-edge positions, and resulting band gaps of 3.2 and 3.0 eV distinctly require minimal photon energies of 389 and 413 nm, respectively. It is well known in defect-free single-crystalline TiO_2, anatase TiO_2 exhibits a lower recombination rate than in rutile TiO_2 [52–54]. Establishing TiO_2 phase junctions has served as a strategy to enhance, in a facile manner, the separation of charge carriers at the polymorph interface. In 1977, Schrauzer and Guth found that the dissolution of Fe_2O_3 in the TiO_2 lattice at a high temperature (1000 °C) could accelerate the conversion between polymorphs, producing strongly phototropic materials with four times superior H_2 and O_2 production rates (9.2 and 4.5 μmol h^{-1} g^{-1}, respectively) against a pristine TiO_2 counterpart (2.3 and 1.2 μmol h^{-1} g^{-1}, respectively) [55].

Since the early stages, leading works on TiO_2-based photo-driven platforms swiftly inspired the investigation of other metal oxides. With analogous drawbacks in the application of representative large band-gap d^0- or d^{10}-type oxide metal cations with analogously deep VB levels, synthesis strategies introduced above for TiO_2 were rapidly pursued [18, 56, 57]. In the wide range of oxide-based photocatalysts surveyed in the past decades, n-type semiconductor gallium oxide (Ga_2O_3) with a d^{10} electron configuration has received unparallel attention notwithstanding its ultrawide band gap of 4.8 eV. Possessing five polymorph phases, some of which undergo facile interconversion at elevated temperatures, finely tunable α–β phase junctions have uncovered induced charge separation and facile transfer of photogenerated carriers [58]. The prepared phase junctions depicted drastically enhanced activity for an H_2 and O_2 stoichiometric evolution, compared with individually assessed α and β phases. The incorporation of divalent cations (e.g. Ca^{2+}, Sr^{2+}, Ba^{2+}, and, in particular, Zn^{2+}) to Ga_2O_3 disclosed a remarkable ~20-fold activity enhancement compared with Ni/Ga_2O_3. This improvement was ascribed to an increase in the mobility and concentration of photogenerated holes induced by an electronic reconstruction due to the incorporation of Zn sites [59]. The subsequent incorporation of $Rh_{2-y}Cr_yO_3$ as a stable HER cocatalyst capable of effectively separating surface intermediates and reaction products resulted in the highest AQY of 71% at 254 nm with a high H_2 rate of 32.0 mmol h^{-1} g^{-1}, reflecting a 2.5-fold enhancement against the pristine material without Zn dopant (11.0 mmol h^{-1} g^{-1}) [60].

Finding parallelism with Vis light-responsive WO_3 and Fe_2O_3, $BiVO_4$ is primarily recognized as a Vis light-driven OER photocatalyst, due to its thermodynamically insufficient CB potential more positive than the HER standard potential [61, 62]. In and Mo co-doped $BiVO_4$, on the contrary, with a partial transformation of the crystal structure unveiled a negative shift in the CB levels, enabling simultaneous OWS reactions with an AQY of 3.2%, the equivalent to a 56.7 μmol H_2 h^{-1} g^{-1} evolution rate under an extended Vis light range (420–800 nm) [63]. $BiYWO_6$, with a contribution of both Bi6s and Y4d to resulting VB and CB, respectively, showed analogous OWS capability under Vis light ($\lambda < 420$ nm) with H_2 and O_2 evolution rates of 13.7 and 6.2 μmol h^{-1} g^{-1}, respectively [64].

Groundbreaking reports utilizing sole pristine semiconductor materials remain scarce to date. In 2014, however, Bao and coworkers disclosed a groundbreaking STH efficiency of 5% with a single-component cobalt(II) oxide possessing a d^7 electronic configuration [65]. The inactive cobalt oxide micropowders were transformed to highly active cobalt(II) oxide nanoparticles through coupled femtosecond laser and ball milling and decomposed pure water under Vis light without any cocatalysts or sacrificial reagents with H_2 evolution rate of 71.4 mmol h^{-1} g^{-1}. The CB edge of CoO nanocrystals rises above the hydrogen evolution potential, and the band edge of the VB lies below the oxygen evolution potential. Limited lifetimes (~1 hour) were nonetheless reflected on collected transmission electron microscopy (TEM) and X-ray photoelectron spectroscopy (XPS) results evidencing clear corrosion or oxidation of the surfaces of these CoO nanoparticles.

Emerging as a clear benchmark OWS photocatalyst active under UV light since 1980, perovskite-type strontium titanate oxide ($SrTiO_3$) possesses a band gap of 3.2 eV [66, 67]. During pivotal works by Domen et al., incorporation of 1.7 wt% NiO on the surface of $SrTiO_3$ showcased a stoichiometric production of H_2 (~0.1 μmol h^{-1} g^{-1}) and O_2 from water vapor for >100 hours [66]. $SrTiO_3$ remains today widely researched, with fundamental and critical studies on the effects of morphology and doping being reviewed below. In 2016, Li and coworkers investigated the effects of charge separation and transfer on cubic phase $SrTiO_3$ with exposed isotropic facets (6 {001} faceted $SrTiO_3$) or anisotropic facets (18 {001} and {110} faceted $SrTiO_3$) [68]. When water reduction (Pt) and oxidation (Co_3O_4) cocatalysts were, respectively, deposited on electron- and hole-transferred sites ({110} and {001}, respectively) of the 18-faceted $SrTiO_3$ (Figure 8.2A), a notably hindered charge recombination resulted in a fivefold enhancement of the AQY with an H_2 evolution rate of 4.1 mmol h^{-1} m^{-2} as compared with 0.66 mmol h^{-1} m^{-2} on isotropic exposed $SrTiO_3$. Doping of foreign metal/nonmetal species on $SrTiO_3$ has also been extensively investigated in the past years. In 2004, successful doping of $SrTiO_3$ with metal cations including Mn, Ru, Rh, and Ir unveiled intense absorption in the Vis light range [72]. Absorption bands were ascribed to excitation from discontinuous levels generated by the metal cations to the CB of the $SrTiO_3$ material. With remarkably superior HER evolution rates, the Rh-doped $SrTiO_3$ system has since inspired subsequent co-doping studies on Rh/Sb and Rh/La on $SrTiO_3$ [51, 73, 74]. From the original works, the presence of Rh^{3+} created an electron donor level (with possible transitions to the CB of $SrTiO_3$ under Vis light), whereas present Rh^{4+} contributed to optical absorption while serving, however, as potential recombination sites for photogenerated carriers. Similarly, the doping

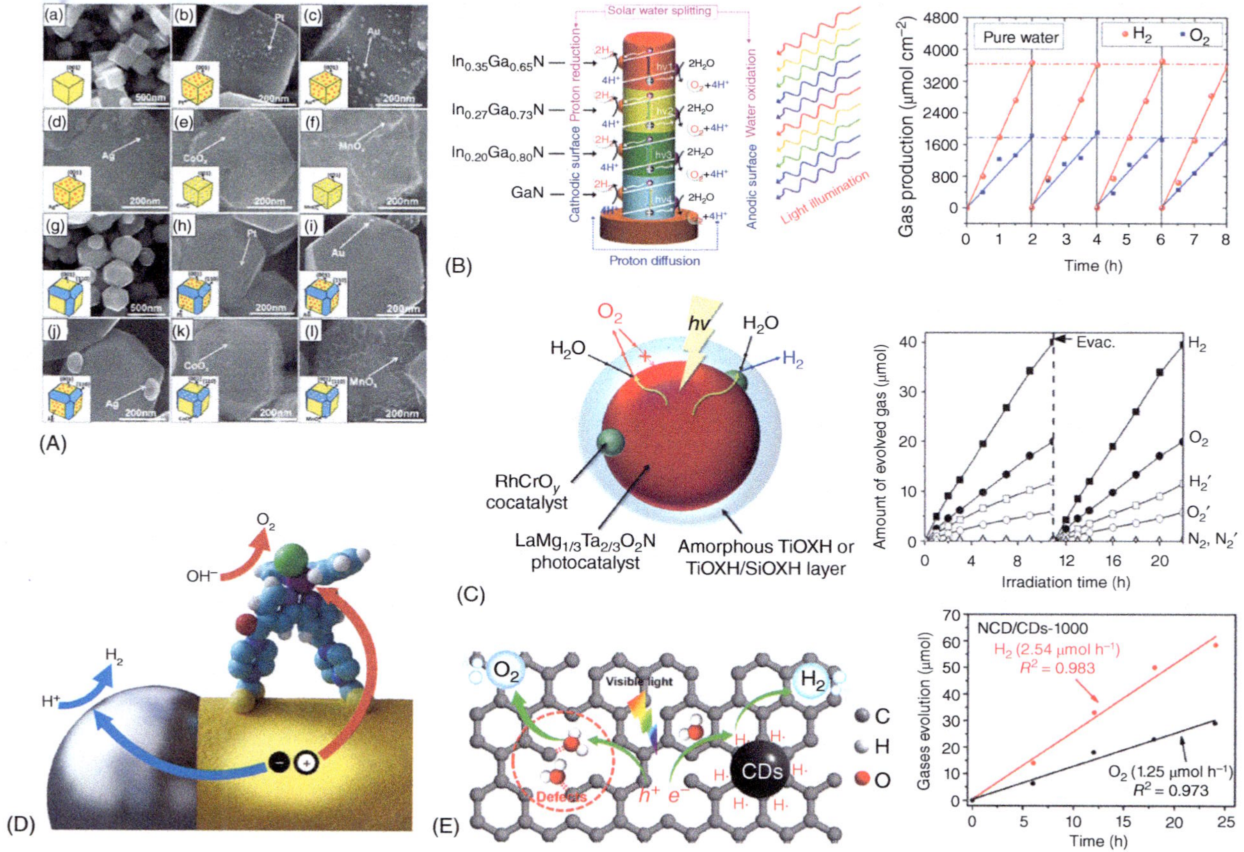

Figure 8.2 Representative material types for one-step photocatalytic overall water splitting systems. (A) $SrTiO_3$ with 18 facets and selective deposition of HER and OER cocatalysts. Source: Mu et al. [68]. Copyright 2016, the Royal Society of Chemistry. (B) Quadruple-band InGaN nanowire arrays and corresponding photocatalytic overall water splitting performance in pure water for eight hours. Source: From Wang et al. [69]. © 2019 RCS. (C) $LaMg_{1/3}Ta_{2/3}O_2N$ with dual coating of amorphous oxyhydroxide layers and photocatalytic performance under Vis light for 22 hours. Source: From Pan et al. [70]. © 2015 John Wiley & Sons. (D) CdS composite with tip-deposited Pt as HER cocatalyst and Ru-based complex as OER cocatalyst. Source: From Wolff et al. [21]. © 2018 Springer Nature. (E) CdS anchored on g-C_3N_4 and stoichiometric evolution of H_2 and O_2 in water under Vis light. Source: Source: From Zhu et al. [71]. © 2018 Elsevier.

of lower-valence cations, e.g. Na^+ into Sr^{2+} and Ga^{3+} into Ti^{4+} on $SrTiO_3$ [75], encouraged later works reporting remarkable AQY (>30% at $\lambda = 360$ nm) and a 5.5 mmol H_2 h^{-1} g^{-1} rate using Al-doped $SrTiO_3$ prepared in alumina crucibles under continuous $SrCl_2$ flux at 1100 °C for 10 hours [76]. Crucible-derived Al ions replaced, homogeneously and efficiently, the Ti^{4+} sites of $SrTiO_3$, thermodynamically occupying most of the stable state of $SrTiO_3$. Incorporation of an $Rh_{2-y}Cr_yO_3$ cocatalyst in the suppression of back reactions was further highlighted. Continuous assessment would later unveil an increase of AQY up to 56% (H_2 evolution rate of 15.6 mmol h^{-1} g^{-1}), by suppressing the grain growth during flux reaction in the presence of Al_2O_3 nanopowder [77], and up to 69%, the highest AQY (H_2 evolution rate of 18.0 mmol h^{-1} g^{-1}) in the selected wavelength region ($\lambda = 365$ nm), by further loading $RhCrO_x$/Al-doped $SrTiO_3$ with MoO_y (0.03 wt% Mo species) [78]. The incorporated Mo species, oxidized to MoO_3 upon calcination at 573 K, would be partially reduced in situ during the PC reaction. Interestingly, the resulting Mo species were tentatively proposed to modify the chemical states of Rh and Cr species in the cocatalyst, indirectly promoting HER.

Among n-type alkali tantalates, $NaTaO_3$ emerges as one of the most promising semiconductors investigated to date [79, 80]. Reported La-doped $NaTaO_3$ loaded with a NiO cocatalyst achieved an AQY of 56% at the shortest wavelength ever reported ($\lambda = 270$ nm) and assured a stable performance for more than 400 hours. Effectively separated reaction sites at the edge and groove of the step structure of NiO ultrafine nanoparticles were identified as highly active for HER (19.8 mmol h^{-1} g^{-1}) and OER (9.7 mmol h^{-1} g^{-1}), respectively [81]. Another case of increasing interest regards layered perovskite materials with superior PC activity under UV light. Reported performances have been tentatively correlated with their interlayer framework capable of favorably inducing charge separation and subsequent HER and OER [82–84]. In 1998, Domen and coworkers first discussed the underlying role of distinct $K_4Nb_6O_{17}$ interlayers as tentative reaction sites and enabling the selective incorporation of Pt nanoparticles as an HER cocatalyst. The strategy allowed a facile separation of oxidation and reduction centers for both O_2 and H_2 evolution, respectively [85]. Oshima et al. later showed the intercalation of Pt metal ions into negatively charged $KCa_2Nb_3O_{10}$ nanosheets through electrostatic attractions, with an estimated AQY for OWS of ~3% at $\lambda = 300$ nm [86]. Several studies have since evaluated the promise of other layered perovskite materials for OWS, including $H_2La_{2/3}TaO_7$ [87], $BaLa_4Ti_4O_{15}$ [83], and $CsCa_2Ta_3O_{10-x}N_y$ [88]. The interlayer nanospace of lamellar solids assures advantageous trapping of reaction species with unusually high catalytic activity [89–91].

Nitrides

The application of non-oxide photocatalysts for OWS was first claimed nearly three decades after the seminal works of Honda and Fujishima. With the top of the VB (i.e. highest occupied molecular orbital) consisting of N2p orbitals possessing higher energy levels than O2p counterparts, transition metal nitrides exhibit narrower band gaps and the promise of Vis light-driven photoactivity [92]. A representative well-studied example is GaN, possessing a band gap of 3.4 eV suitable for unassisted OWS. Solar-driven overall seawater splitting on immobilized p-GaN-based nanowires (NWs) without any sacrificial agents was, for instance, reported to attain a stable STH efficiency of 1.9% [93]. By adjusting the surface Fermi level through optimized p-type Mg doping, the same group later reported a reachable quantum efficiency of ~18% with twofold enhancement of an H_2 production rate (~4.0 mol h^{-1} g^{-1}) against pristine GaN nanowires under UV light [94, 95]. Introduction of indium in group III nitrides serves as a widely investigated side step for limited UV light responsivity. Tuning the energy band of InGaN induces an extension of the light absorption range of this semiconductor within the wide UV–vis–NIR range [31, 96, 97]. In 2013, a pivotal Vis light (400 nm)-responsive photoactivity was reported in the presence of an InGaN/GaN multiband nanowire heterostructure [98]. Its upward band bending reflected, however, a clear drawback against superior catalytic performances. Surface band bending induced by Fermi-level pinning can create an additional energy barrier for charge carrier transport to the photocatalyst–water interface [31, 99, 100]. The inherent upward band bending of the nanowire architecture was reported to repel photogenerated electrons toward the bulk region creating an electron depletion (i.e. hole accumulation) on the nanowire surface [95, 101, 102]. By precisely engineering the surface band bending through controlled Mg doping, p-type GaN/$In_{0.2}Ga_{0.8}N$ nanowire could later attain one of the highest AQY (~12.3%) under Vis light (400–475 nm) with an STH efficiency of 1.8% and ~0.78 mol h^{-1} g^{-1} of HER rate [31]. The precise control of Mg^{2+} doping effectively minimized the potential barrier near the nonpolar catalyst surface, leading to rapid and efficient diffusion of photogenerated excitons. The reported PC system further demonstrated an STH efficiency of 1.9% (HER rate of 6.15 mol h^{-1} g^{-1}) in seawater for a possible large-scale and environmentally friendly utilization of solar energy [93]. Unique double-band InGaN nanosheets with a photochemical diode (PCD) structures [103], capable of inducing a precise control of the carrier flow at the nanoscale, exhibited an STH efficiency of 3.3% (HER rate of ~0.5 mmol h^{-1} cm^{-2} at $\lambda > 420$ nm) [104]. Following an analogous strategy, Mg doping induced a large built-in electric field along the lateral dimension of these nanosheets, where photogenerated electrons and holes could be promptly separated, minimizing surface and bulk recombination and suppressing undesirable back reactions [104]. Following a similar approach, an STH efficiency increment up to 5.2% (HER rate of ~1.84 mmol h^{-1} cm^{-2} under concentrated light illumination) was reported when quadruple-band InGaN nanowire arrays consisting of $In_{0.35}Ga_{0.65}N$, $In_{0.27}Ga_{0.73}N$, $In_{0.20}Ga_{0.80}N$, and GaN segments (band-gap energies of ~2.1 eV, 2.4 eV, 2.6 eV, and 3.4 eV, respectively) were employed (Figure 8.2B) [69].

With a narrow 2.1 eV band gap and suitable VB and CB lying at c. 1.6 and −0.4 V_{NHE} (pH = 0), respectively, Ta_3N_5 is highly regarded as a promising candidate for solar H_2 production [105]. Hindered OWS on particulate photocatalysts without any sacrificial agents has, however, been attributed to strong recombination and trapping of photogenerated charges on the surface defects generated during thermal nitridation of oxide precursors [106]. As a promising side step, the incorporation of a small amount of alkaline metal salts induced better crystallinity and smaller particles with smoother surfaces of the resulting Ta_3N_5, with a sixfold improvement in the OER activity under Vis light [107]. Alternatively, Ta_3N_5 nanorod single crystals, generated on the edges of cubic $KTaO_3$, possessing virtual absence of grain boundaries and surface defects, unlocked superior activity under Vis light when modified with a core–shell-structured Rh/Cr_2O_3 cocatalyst with AQY of 0.22% at 420 nm [106]. Alternatively, conducting polymers, polyaniline- and polythiophene (PANI and PTh)-protected Ta_3N_5, showcased hindered recombination rates, superior activity under Vis light with a high HER rate (60.5 and 45.1 $\mu mol\,h^{-1}\,g^{-1}$, respectively), and superior stability against self-photocorrosion through an accelerated migration of the generated holes to the surface [108].

Upon incorporation of 1 wt% RuO_2 nanoparticles as HER sites, β-Ge_3N_4, with a 2.1 eV band gap estimated by plane wave density functional theory calculations, was also reported to readily split water into H_2 and O_2 with an estimated AQY of 9% at around 300 nm [109]. Comparatively, metallic black tungsten nitride (WN) was recently reported for direct OWS, at $\lambda < 765$ nm. By employing PtO_x/WN as a photocatalyst under Vis light irradiation ($\lambda > 420$ nm), a stoichiometric H_2 and O_2 production could also be detected for 24 hours (0.9 and 0.5 $\mu mol\,g^{-1}$, respectively) [110].

Oxynitrides

With a partial or integral replacement of O with N atoms in metal oxides, the highest occupied molecular orbital of the resulting material is upshifted, without significantly altering the level of the bottom of the CB (i.e. lowest unoccupied molecular orbital) [57]. Oxynitrides exhibit, as a result, narrower band gaps against metal oxides, serving as promising pristine materials for OWS. At the early stages, oxynitride-based water splitter photocatalysts operating under Vis irradiation (500–750 nm) were primarily composed of transition metal cations with d^0 electronic configuration, e.g. Ti^{4+}, Nb^{5+}, or Ta^{5+}, with band-gap energies within the 1.7–2.5 eV range [18, 57, 111]. Along with the above introduced Ta_3N_5, introduced in the previous section, tantalum (oxy)nitride semiconductors, e.g. TaON, $LaTaON_2$, and $ATaO_2N$ (A = Ca, Sr, Ba), possess a narrow band gap and promising band positions for OWS [111, 112]. As the VB is upshifted with a continuous replacement of O atoms, corresponding band-gap narrowness follows the order $Ta_2O_5 < TaON < Ta_3N_5$ [111].

To date, Vis light-responsive TaON (~2.5 eV) remains mainly explored among d^0-based semiconductors, with reported OWS upon surface modification of the pristine material or through cocatalyst incorporation. In 2013, a modified ZrO_2/TaON was the first example of a successful non-oxide photocatalyst with d^0 electronic configuration for OWS under Vis light ($\lambda > 400$ nm). The presence of RuO_2 and Cr_2O_3 dopants were, however, found critical to drive both HER and OER, with an estimated AQY < 0.1% at 420 nm and an HER rate of 15.0 $\mu mol\,h^{-1}\,g^{-1}$ [113]. Manipulation of Ta_2O_5 with d^{10} p-type blocking metal cation In^{3+} ($InTaO_4$) has been demonstrated to extend the absorption wavelength to the Vis region with an eightfold HER rate increment following additional Ni doping [114]. Inspired by these reports, In-Ni-Ta-O-N (general formula) pellets prepared via nitridation of the mixed oxides (In_2O_3, NiO, and Ta_2O_5) were active for OWS in the absence of cocatalyst under Vis light ($\lambda > 400$ nm) with an H_2 evolution rate of 12.5 $\mu mol\,h^{-1}\,g^{-1}$

[115]. Through compositional fine-tuning and surface coating of novel Ta-based oxynitrides, $LaMg_{1/3}Ta_{2/3}O_2N$ enabled OWS for the first time at a wavelength $\lambda < 600$ nm with AQY of ~0.03% at 440 ± 30 nm [70]. Both oxygen reduction reaction (ORR) and N_2 evolution reaction on the incorporated HER cocatalyst ($RhCrO_y$) on $LaMg_{1/3}Ta_{2/3}O_2N$ were efficiently suppressed by sequential incorporation of titania and silica thin amorphous oxyhydroxide layers (see Figure 8.2C). The resulting architecture disclosed superior PC activity with 18.2 μmol H_2 h^{-1} g^{-1} evolution rates at $\lambda \geq 400$ nm for >22 hours.

From the viewpoint of the electronic band structure, d^{10}-based semiconductors are advantageous as the bottom of the CB, composed of hybridized sp orbitals of corresponding metals, possess large dispersion, facilitating the mobility of photogenerated electrons. GaN/ZnO solid solutions with d^{10} metal cations were first reported in 2005, following NH_3 treatment of a mixture of Ga_2O_3 and ZnO [116]. Whereas both GaN and ZnO have wide band gaps (>3.0 eV) and do not absorb Vis light, the band gap of the resulting Vis light-responsive yellow powder, tentatively $(Ga_{1-x}Zn_x)(N_{1-x}O_x)$, could be tuned within a 2.4–2.8 eV range [112]. Since then, an extensive line of works to enhance the intrinsic properties of $(Ga_{1-x}Zn_x)(N_{1-x}O_x)$ has been achieved through the incorporation of suitable cocatalysts [117], optimized synthesis conditions [117–119], and rational nano-structuralization. Introduction of $Rh_{2-y}Cr_yO_3$ nanoparticles as HER active centers showed, representatively, an improved AQY of 2.5% at λ = 420–440 nm [120]. A suppression of Zn- and/or O-related defects, serving as recombination centers for photogenerated electrons and holes, could be further attained upon post-calcination treatments. A twofold enhancement of the AQY up to ~5.9% with an HER rate of 3.1 mmol h^{-1} g^{-1} within the same wavelength region was reported [117]. Promising $(Ga_{1-x}Zn_x)(N_{1-x}O_x)$ nano-structuralization routes for superior Vis light activity were illustrated with grain sizes as small as 6 nm and surface areas boosted up to 100 m^2 g^{-1}, leading to a decrease in the carrier diffusion distances and to an increase in the number of surface reaction centers. The highest quantum efficiency of the $(Ga_{1-x}Zn_x)(N_{1-x}O_x)$ nanostructure for water splitting was 17.3% (HER rate of 1.1 mmol h^{-1} g^{-1}) at 400 nm illumination, reflecting a threefold enhancement against a micro-scaled counterpart [117, 121]. Vis light-responsive d^{10} $ZnGeN_2$–ZnO oxynitrides were further prepared following a similar methodology, with ZnO and GeO_2 as starting materials [122, 123]. Corresponding band gaps, estimated to be ~2.5–2.7 eV based on the absorption edges, are narrower than those of β-Ge_3N_4 (~3.8 eV), $ZnGeN_2$ (~3.3 eV), and ZnO (~3.2 eV). Due to the p–d repulsion between Zn3d and N2p and O2p electrons, the absorption edge of $(Zn_{1+x}Ge)(N_2O_x)$ is shifted toward longer wavelengths with increasing ZnO content. Stable activity for 60 hours under Vis light and an AQY of ~0.20% at 420 nm (HER rate of 2.1 mmol h^{-1} g^{-1} at $\lambda > 300$ nm) were claimed upon the additional doping of $(Zn_{1+x}Ge)(N_2O_x)$ with $Rh_{2-y}Cr_yO_3$ [123].

Metal Chalcogenides

With their VB typically consisting of p orbitals of corresponding chalcogenide ions, Vis light-responsive metal chalcogenides possess a narrow band gap with enhanced solar light utilization [124, 125]. Notwithstanding the interest of this class of materials, further italicized by their effective charge transfer properties and suitable band alignment for OWS, poor photostability and resulting short lifetimes are generally reported. Among attractive candidates for Vis light-driven photocatalysis, cadmium sulfide with a narrow band gap (2.4 eV) has been suitably exploited for application on water splitting, CO_2 reduction, and dye degradation [14, 125]. The CB edge of CdS is relatively more negative than that of benchmark TiO_2, with photogenerated electrons possessing a stronger reduction capability. Nonetheless, in addition to a facile oxidation of S^{2-} by holes near the VB in the absence of sacrificial reagents (self-photocorrosion), interparticle aggregation and limited PC activity of pristine CdS accentuate an hinder commercial implementation of the latter [21, 125]. In this optic, the inherent photocorrosion of CdS has spawned extensive efforts in the development of appropriate protection strategy cocatalyst incorporation draining the accumulation of photogenerated holes. An example of interest is the incorporation of CdS with tip-anchored Pt nanoparticles (HER cocatalyst) and entire decoration of the nanorod side surface with Ru-based complex (ruthenium dithiocarbamate) as a molecular OER cocatalyst (Figure 8.2D). The selective tip-decorated design enhanced the fast electron transfer to Pt, strongly hindering the radiation recombination in the sulfide semiconductor, while the Ru-based complex, firmly anchored on CdS hole trapping sites, further hindered the self-photodegradation of the CdS semiconductor [21].

Against sulfides, oxysulfides inherently unveil superior durability against self-oxidation attributed to the stabilization of S^{2-} surface ions through the hybridization of both S3p and O2p orbitals [126, 127]. Examples of semiconductors showcasing OWS without noticeable degradation include La_3GaS_5O [128], $La_2Ta_2ZrS_2O_8$ [129], and $La_5Ti_2CuS_5O_7$ [130], in aqueous solutions under Vis light. With a band gap of 1.9 eV, $Y_2Ti_2O_5S_2$ was recently reported to stoichiometrically split water under extended light absorption ranges, with optimum loading of IrO_2 and Rh/Cr_2O_3 as OER and HER cocatalysts, respectively [131]. The outstanding light responsivity up to $\lambda = 600$ nm was sustained for 20 hours with an HER rate of ~30 μmol h^{-1} g^{-1} at $\lambda > 420$ nm. Recently, oxyhalides have also gathered attention as potential narrow-band-gap Vis light-responsive photocatalysts. As a promising semiconductor, bismuth oxyhalide ($Bi_lO_mX_n$, with X = Cl, Br) has been further demonstrated to improve the insufficient bulk charge separation and transportation in benchmark TiO_2, $SrTiO_3$, $(Ga_{1-x}Zn_x)(N_{1-x}O_x)$, CdS, and g-$C_3N_4$ through an induced enhancement in the internal electric field (IEF) [18, 132–136]. Upon the replacement of Cl with carbon dopants, a 126-fold increment in IEF leading to highly efficient charge separation and an O_2-evolving rate of 0.9 mmol h^{-1} g^{-1} could be further achieved [137]. By selectively assembling 1T MoS_2 into $Bi_{12}O_{17}C_{12}$ monolayers, HER was later reported up to 33 mmol h^{-1} g^{-1} at 420 nm [138]. Despite a remarkable IEF enhancement in both reports, solar water splitting toward stoichiometric H_2 and O_2 could not be witnessed without the additional incorporation of HER and OER cocatalysts.

Carbon-Based Materials

Lightweight cost-effective carbon-based materials (e.g. graphene [139, 140], graphene oxide (GO) [141, 142], and carbon nanotubes [143]) possess high conductivity, architectural versatility, and tunable surface area. Over the years, the exploitation of carbon materials has conspicuously served a wide range of photo(electro)catalytic applications, extending light absorption ranges, improving light harvesting, and favoring charge separation and transport. In a parallel way, the promise of conjugated polymers based on abundant C, H, and N elements as cost-effective and easily processable candidates for stable PC OWS has been further italicized [144].

In a recent line of works, graphitic carbon nitrides (g-C_3N_4), with two major substructures based on heptazine and poly(triazine imide) units, have gathered particular attention in several PC applications [145, 146]. Abundance of nitrogen lone-pair electrons and grain boundary defects, highly favorable for the construction of metal/carbon nitride heterojunctions for enhanced charge transfer at the interface, has played key roles for the emergence of C_3N_4 as excellent photocatalysts for OWS [147]. C_3N_4 also possess a narrow band gap (2.7 eV) derived from comparatively high VB and CB consisting of N2p_Z and C2p_Z orbitals in which, upon excitation, nitrogen and carbon atoms provide holes and electrons for OER and HER [148–150]. C_3N_4 with higher crystalline poly(triazine imide) frameworks were shown to facilitate the transfer and to hinder the recombination rate of the photogenerated carriers [151].

Enhancing the intrinsic activity of C_3N_4 has been realized upon the integration of oxides [152], sulfides [153], metal nanoparticles [154], graphene [155, 156], and carbon nanotubes [157]. Upon photo-deposition of appropriate cocatalysts, e.g. Pt, PtO_x and CoO_x, NiO_x, and WC, OWS was shown on g-C_3N_5 with a stoichiometric H_2 and O_2 evolution [148, 158–161]. The incorporation of dual cocatalysts (Pt/Co) on highly crystalline carbon nitride was further demonstrated to suppress self-oxidation by the holes on the VB of carbon nitride ascribed to an undesirable N_2 evolution, with a reported AQY of 2.1% (HER rate of 0.65 mmol h^{-1} g^{-1}) for OWS [146]. A novel design to extend the light absorption and simultaneously promote the electron–hole pair separation and transportation of carbon nitride was demonstrated via intimately connecting the in-plane π-conjugated heterostructure with different electron affinity [162]. Carrier separation efficiency was remarkably enhanced when the homogeneous sp^2-hybridized π-conjugated bonding of the carbon ring in the (C_{ring})–C_3N_4 plane heterostructure competently trapped and transfer the photoexcited electrons to a local micro-π-conjugated connection unit. The (C_{ring})–C_3N_4 plane heterostructure possessed a high electron state density around the Fermi level, significantly prolonging the photocarrier diffusion length and lifetime by 10-fold against those of pristine g-C_3N_4, with a resulting AQY of 5% at 420 nm and an H_2 production rate up to 0.37 mmol h^{-1} g^{-1}.

Carbon quantum dots (CQD) (<10 nm in diameter) have gathered increasing attention due to their unique photophysical properties, widen Vis light responsivity, photoluminescence (PL), electron reservoir properties, surface functional groups (e.g. –OH, C=O, –COOH), and excellent photoinduced electron transfer and interfacial charge separation [163, 164]. Various attempts have thus been made to incorporate CQD into a wide range of metal-, metal oxide-, and C_3N_4-based nanostructures [165–167]. Incorporating CQD on a metal-free carbon nitride with induced surface defects showed a prominent catalytic activity, with an H_2 production of 25.4 μmol h^{-1} g^{-1}, an AQY of 2.04% at $\lambda = 420 \pm 20$ nm, and an STH efficiency of 0.1% [71]. The carbon defects, intensified via annealing up to 1100 °C, served as OER active centers, whereas the introduced CQD worked as the HER sites (Figure 8.2E). A similar approach was reported on 1D carbon nitride nanowire bundles, fabricated through vertical stacking of oxygen-rich C_3N_4 quantum dots. The intrinsic properties of these subunits enabled the catalytic activity for H_2O_2 decomposition, thereby driving a stepwise 4-e^- OWS reaction with an AQY up to 5.2% at λ = 420 nm, the correspondent to 35.6 μmol H_2 h^{-1} g^{-1} [168]. The unconventional stepwise 2-e^-/2-e^- two-step pathway (Eqs. (8.8) and (8.9)):

$$2H_2O \rightarrow H_2O_2 + H_2 \tag{8.8}$$

$$2H_2O_2 \rightarrow 2H_2O + O_2 \tag{8.9}$$

Similarly, metal-free carbon nanodot–C_3N_4 nanocomposites showed a stepwise pathway toward OWS, with remarkable quantum efficiencies (4.42%) at λ = 600 ± 10 nm, an STH efficiency of 2%, long-term stability (>200 days), and HER rates of 0.58 mmol h^{-1} g^{-1} at $\lambda > 420$ nm [147]. Herein, C_3N_4 was claimed to be responsible for the first step (photocatalytically driven), whereas CDots were responsible for the second step (chemically driven). The neighboring nature of both C_3N_4 and incorporated carbon nanodots assured the efficiency of the two-step process [147]. Works on N-doped graphene oxide-quantum dots (NGO-QDs), displaying both p- and n-type conductivities and a resulting internal charge transfer at the QD interface, have been further reported. Here, the distinct p- and n-type domains were claimed to serve as HER and OER sites, respectively [169].

Atomically dispersed active centers simultaneously benefiting from the advantages of homogeneous and heterogeneous platforms have received increasing attention based on the successful maximization of the catalytic activity of incorporated (transition) metal atoms [170, 171]. To this end, carbon-based materials have played a crucial role as representative support materials, with N, S, P, and B coordinating ligands offering highly stabilized single-atom centers [172–174]. A wide range of reported synthesis strategies; the versatility offered by the combination of carbon supports, ligands, and metal sites; and the unique catalytic activity and selectivity of the resulting systems have underlined a difficulty in establishing a guideline for a rational design of single-atom nanostructures with predetermined properties [175]. A report on single Co_1–P_4 centers confined within g-C_3N_4 nanosheets showed a remarkable AQY of 2.2% at longer wavelengths $\lambda < 580$ nm, with an equivalent H_2 evolution rate of 0.41 mmol h^{-1} g^{-1}. Introduced Co sites were hypothesized to effectively suppress the recombination of charge carriers, to assure 20-fold prolonged lifetimes against pristine g-C_3N_4, and to boost the overall OWS activity [176].

8.2.3.2 Incorporation of Cocatalysts

Rationally integrated nanocomponents can magnify the separation of photogenerated charge carriers toward superior quantum yields and PC activity while serving as distinct (neighboring) active centers. Incorporation of these cocatalysts has been a widely surveyed approach for semiconductor materials over which, despite a promising light-harvesting capability, simultaneous HER and OER remain low. The semiconductor acts as an antenna for light absorption, whereas the optimized charge separation induced by the introduced cocatalyst enhances the availability of electrons and holes serving both HER and OER, respectively. This simple strategy is alternatively valuable for a wide range of semiconductors with an unfitting band-edge position to drive both oxidation and reduction reactions alone. The activity of the resulting hybrid materials is highly dependent on the interface/interaction between the cocatalyst and the semiconductor support, with fundamental studies at the resulting interface being crucial. It is important to understand that the incorporated cocatalyst itself may, however, act as a light absorber (i.e. compete for light harvesting) or alter the desired product selectivity, unexpectedly impacting the process efficiency. Assessing the prospect of photoexcitation of carriers on the surface of these cocatalysts is, in this sense, of crucial importance.

Upon light irradiation in the presence of an HER cocatalyst with a larger work function at the interface, the electrons are expected to flow from the semiconductor to the cocatalyst, which acts as an electron sink and provides enriched electron density for swift HER. Over the cocatalyst surface, overall HER rates are markedly determined by the hydrogen adsorption free energy. The incorporation of Pt over TiO_2 was demonstrated at the early stages by Sato and White [45]. Suitable Pt and analogous Pd, Ir, and Ru metals possess, however, limited commercial applicability based on the scarcity and cost-ineffectiveness. In electrocatalytic (EC) water splitting platforms, earth-abundant materials with equally promising HER functionalities over a wide range of pH (e.g. transition metal phosphides [NiCoP, CoFeP, and MoWP], MoS_2, and WS_2) have been reported in recent years notwithstanding their often poor stability [142, 177–183]. Conversely, the utilization of cocatalysts with low overpotential in EC platforms is particularly desirable to boost the OER efficiency, the greatest challenging step in the PC water splitting [184–191]. The incorporation of NiO_2 on Ga_2O_3 was, for instance, found to stoichiometrically evolve H_2 and O_2 under light irradiation with production rates linearly dependent on the NiO loading up to 2 wt% (HER and OER rates of ~45 and ~20 μmol h^{-1} g^{-1}, respectively) [192].

A simultaneous incorporation of HER and OER cocatalysts reflected in the research works above summarized, pinpoints the increasing complexity of PC OWS systems, and brings to light the need to control the spatial arrangement of incorporated phases. Spatial separation of oxidation (3.0 wt% Co_3O_4) and reduction sites (1 wt% Pt) at their nanodomains was, for instance, shown to play a key role in the activity of the designed hybrids. The nanoparticles, incorporated into the interior and exterior surface of hollow spheres of a permeable semiconductor with optimized shell thickness, showed stoichiometric OWS with enhanced HER rates (150 μmol h^{-1} g^{-1}) against a randomly distributed nanoparticle-based counterpart (45 μmol h^{-1} g^{-1}) [193]. Neighboring HER and OER active sites can additionally serve as a promising pathway toward cooperative or synergetic effects. The spatial control of generated catalytic interfaces is often disregarded in the design of OWS catalysts [194, 195]. Most importantly, individually optimized synthesis steps for a consecutive integration of distinct nanocomponents easily induce phase transformations, morphological variations, an ambiguity of the spatial arrangement of catalytic sites, and an overall heterogeneity of the final materials [196].

8.2.3.3 Plasmonic Nanostructures

As a strategical platform, incorporating metal nanomaterials possessing broadly tunable optical surface plasmon resonance (SPR) properties derived from the collective coherent oscillation of surface electrons have received increasing research interest [197]. These plasmonic metal nanostructures (e.g. Au, Ag, and Cu) and their resulting light–matter interaction have been explored in areas of catalysis toward a sustainable future. This includes water splitting, CO_2 reduction, and fuel cell reactions, in PEC platforms and as plasmon-enhanced EC systems [198, 199], as well as in optical sensing [200, 201], light emission [202, 203], photodetectors [204, 205], waveguiding [206], and data storage [207]. Directly reflecting their optical properties, corresponding PC quantum efficiencies can be efficiently tuned by the selected light intensity and operating temperature. In agreement, extensive investigations of the effect of the metal nanoparticle geometry, size, and shape on their SPR wavelength/primary plasmon decay have been reported in the past years. The incorporation of these metal nanostructures on a benchmark semiconductor (e.g. TiO_2) can be achieved through a wide range of synthesis methods including chemical, thermal and hydrothermal reduction, photoreduction, deposition–precipitation of metal precursors, and a direct incorporation or sputtering of pre-synthesized nanoparticles [208–211]. The resulting hybrid architectures and the interaction established between the plasmonic structure and the semiconductor offer distinct properties that can be distinctly explored.

To date, the catalytic exploitation of plasmonic metal nanoparticles is typically achieved through the incorporation of these nanostructures on the surface of semiconductors in both PC and PEC systems, as discussed in the following section (section 8.3). Alternatively, the SPR excitation can be transferred to molecular photocatalysts or neighboring metal structures via a plasmon-mediated electron transfer pathway. Works inspecting direct exploitation of the plasmonic metal surface as active catalytic centers (i.e. the reactants/reaction intermediates adsorb on the metal surface) remain, however, scarce. Questions correlating the effect of metal nanoparticle geometry, size, and shape with the resulting catalytic activity, selectivity, and stability in selected catalytic reactions have remained unanswered. With the emergence of more sophisticated plasmonic metal nanoparticles including recent works on the synthesis of chiral nanostructures with a high versatility degree (morphology and optophysical and electrochemical properties), a clear need to understand the extent of the properties of plasmonic materials as active centers is emphasized [212, 213]. It is important to rationalize the properties of exploited metal plasmonic nanoparticles. For the representative Au, besides the well-known SPR absorption in the Vis region (>500 nm), these nanoparticles also have strong absorption below 450 nm corresponding to the interband transitions from 5d to 6sp and are able to direct the OWS without the assistance of cocatalysts [214].

Among non-radiative plasmonic effects including near-field enhancement [215, 216], hot carrier transfer [217, 218], plasmon resonance energy transfer, and photothermal, the former two are typically ascribed to emerging PC activity levels. Unique morphological features and synthesis approaches have been further shown to play a key role in the resulting performances of the prepared hybrid systems, as recently demonstrated over 111 facet-oriented gold nanoplatelets on a multilayer graphene substrate, with strong Au–G interaction [219]. The exploitation of photoinduced hot charge carriers is a function of the (limited) occurrence of single-electron excitations, corresponding (short) lifetimes of these species, and the (relatively long) period between excitation events (in particular at low light intensities). Driving multi-electron reactions through necessary multiple electronic excitations is an issue under controversy especially in reactions requiring a swift reduction of involved intermediate species (e.g. CO_2 requires 8 and 12 electrons for the production of CH_4 or C_2H_4, respectively) [220]. In the same optic, there is an increasing need toward the detection of hot carriers serving PC reactions at the locations where these hot carriers are generated and transferred. Whereas a direct imaging or an investigation of underlying pathways is still challenging to date, primarily due to the associated ultrafast dynamics of these entities, a number of research efforts have been reported and summarized to date [221].

8.2.4 Hybrid Systems for Two-Step Photoexcitation Toward Overall Water Splitting

Notwithstanding the focus of the scientific community to uncover novel semiconducting materials capable of efficiently harvest light, recent reports have increasingly italicized the progress toward complex hybrids with multiple incorporated functionalities. One of the limitations of single semiconductor-based systems is their inherent limitation for light harvesting, which has encouraged a focus on rational synthesis strategies enhancing the intrinsic properties of state-of-the-art semiconductors. Alternatively, the integration of an additional light-responsive material, as plasmonic nanostructures (see above), upconversion nanocrystals [222–225], or black phosphorous [199], can extend the light absorption range of the resulting hybrid architecture across the UV–vis–NIR ranges for efficient utilization of inherently lost photons. In the same optic, the development of platforms incorporating two or more semiconductors driving a two-step photoexcitation process is also of great promise. These systems naturally benefit from the individual optimization of each semiconductor for enhanced light-harvesting efficiency, improved charge carrier separation and migration, enlarged surface area, and promoted diffusion of reactive molecules [16]. On the other hand, a strategy to induce the separation and hinder the recombination of photogenerated charge carriers is also the construction of heterojunctions between semiconductors with distinct and suitable band-edge positions. Charge carriers are separated at the generated interface with electrons in the higher CB of one semiconductor being transferred to the lower CB of the second semiconductor, while holes from the lower VB are oppositely transferred to the higher VB. This approach has been strategically applied over distinct polymorphs of the same semiconductor [55, 226, 227].

8.2.4.1 Z-Schemes

In 1979, Bard first introduced the concept of a Z-scheme PC system composed of two semiconductors with a staggered alignment of band structures [10]. In most Z-scheme cases, the CB and VB of both materials do not fulfill the requirements to drive the OWS and are solely suitable toward individual HER and OER [228]. Selected semiconductor-based architectures serve as hydrogen evolution photocatalyst and oxygen evolution photocatalyst, from hereafter designated as HEP and OEP, respectively. This strategy may additionally suppress undesirable back reactions, enhancing the overall process efficiency. Successful Z-schemes integrate a component with higher CB and VB positions and smaller work function (higher Fermi level). However, as emphasized in section 8.2.3.1, individual fingerprints of semiconductors, e.g. morphology, exposed facets, and surface states, play key roles in the overall efficiency of the rationally designed architecture. The development of Z-schemes was first demonstrated in the presence of redox couples acting as aqueous redox mediators between coupled HEP and OEP individual systems. This line of work was later followed by a tentative incorporation of conductive solid mediators, e.g. Au nanoparticles and graphene (second generation), and most recently in the complete absence of mediating species (third generation). Detailed discussion and representative examples, systematically identifying each HEP and OEP, are provided below.

First-Generation Z-Schemes (Liquid Phase)

Early Z-scheme system, utilizing Pt-loaded anatase TiO_2 (HEP) and pristine rutile TiO_2 (OEP) in the presence of an IO_3^-/I^- aqueous redox mediator and operating under UV light, was reported with simultaneous HER and OER (360 and 180 μmol h^{-1} g^{-1}, respectively) [229]. The redox potential of the selected mediators (e.g. IO_3^-/I^- couple [0.67 V_{NHE} at pH 7]) suitably lied in between those of H^+/H_2 and O_2/H_2O (see Figure 8.1b). The redox mediator was further confirmed to inhibit undesirable back reactions that swiftly occur in one-step photoexcitation systems. Extensive efforts estimating the promise of combining narrow-band-gap HEP (e.g. Rh-doped $SrTiO_3$, TaON, $ATaO_2N$ [A = Ca, Ba], g-C_3N_4, and $Sm_2Ti_2S_2O_5$) and OEP (e.g. WO_3, $BiVO_4$, $AgNbO_3$, Rh/Sb-co-doped TiO_2, TaON, and Ta_3N_5) have since been reported [14, 17, 230, 231]. A comprehensive library of I, Fe, Co, and Mn-based shuttle redox pairs and the corresponding appropriateness of respective redox potentials and applied conditions (e.g. pH of solution reactant) have been also investigated [14]. For representative I-based mediators, I_3^-/I^- and IO_3^-/I^- redox couples are selected in acidic and basic conditions, respectively. Conversely, Fe^{3+}/Fe^{2+} pairs are preferably exploited in strong acidic conditions (pH < 2.5) for the stabilization of Fe^{2+} ions, whereas higher pH conditions (6.0–7.0) are suitable for $Fe(CN)_6^{3-}/Fe(CN)_6^{4-}$ complexes [232, 233].

A Pt/ZrO_2/TaON (HEP) and Pt/WO_3 (OEP) Z-scheme in the presence of IO_3^-/I^- showcased an AQY of 6.3% at λ = 420 nm (H_2 formation rate of 0.35 mmol h^{-1} g^{-1} at a wavelength range of $300 < \lambda < 800$ nm). In opposition to previous reports in which a Pt/WO_3 activity gradual decrease was ascribed to the competitive photooxidation of the present I^- ions [234], herein a swift oxidization of I^- ions is assured by $Pt/ZrO_2/TaON$ [235]. In 2015, Domen and coworkers reported an analogous PC system consisting of an optimized $Pt/MgTa_2O_{6-x}N_y/TaON$ heterostructure (HEP) and PtO_x/WO_3 (OEP) in the presence of IO_3^-/I^- with an AQY of 6.8% at $\lambda = 420$ nm and an HER rate of 0.48 mmol h^{-1} g^{-1}, in which the spatial charge transfer between TaON and $MgTa_2O_{6-x}N_y$ played a key role [236]. As a redox couple, $Fe(CN)_6{}^{3-}/Fe(CN)_6{}^{4-}$ was representatively incorporated on a $ZrO_2/TaON$ (HEP) and $BiVO_4$ (OEP) Z-scheme, with an unprecedented AQY = 10.3% being claimed at $\lambda = 420$ nm, equivalent to an HER rate of 0.83 mmol h^{-1} g^{-1}, as shown in Figure 8.3a [237]. In this report, distinct OER enhancement was further attributed to the highly efficient charge separation driven by dual incorporation of Au and CoO_x cocatalysts on distinct $BiVO_4$ facets.

Simple Co^{3+}/Co^{2+} were found inactive as redox mediators. In comparison however, more complex Co-based shuttle mediators such as $[Co(2,2'\text{-bipyridine})_3]^{2+/3+}$, $[Co(1,10\text{-phenanthroline})_3]^{2+/3+}$, and $[Co(2,2':6',2''\text{-terpyridine})_3]^{2+/3+}$ have been successfully exploited [239, 240]. Representatively, a Ru/Rh-doped $SrTiO_3$ (HEP) and $BiVO_4$ (OEP) in the presence of $[Co(2,2'\text{-bipyridine})_3]^{3+/2+}$ unveiled HER and OER photoactivity under Vis light ($\lambda > 420$ nm) with an AQY of 2.1% at 420 nm. In the absence of a Co-based shuttle mediator, a $(Cu\text{–}Ga)_{0.8}Zn_{0.4}S_2$ (HEP) and $BiVO_4$ (OEP) Z-scheme revealed sole HER, due to the facile photocorrosion of the sulfide photocatalyst and subsequently suppressed OER. Following the incorporation of $[Co(2,2':6',2''\text{-terpyridine})_3]^{2+/3+}$, continuous stoichiometric HER and OER were witnessed for 17 hours, with enhanced photostability of the metal sulfide photocatalyst [239].

Notwithstanding the promise of these pioneering research works and their underlying strategy, the exploitation of first-generation Z-schemes languishes from the poor stability of incorporated redox mediators, a high pH dependence, and a restricted application solely to liquid-phase systems [241–243]. In addition, applied mediators (e.g. Fe based) have been shown to undesirably induce thermodynamically favored back reactions with Fe^{3+} (acceptor species) and Fe^{2+} (donor species) competing with the reactants for oxidation and reduction.

Second-Generation Z-Schemes (All Solid State)

Conductor materials, e.g. noble metal nanoparticle (such as Au, Ag, Cu, Ir, Rh, Pt, and Ni) and carbonaceous materials (such as graphene, carbon nanotubes, and carbon dots) have been proposed as suitable electron mediators [228, 241, 244, 245]. Representatively, noble metal incorporation forms an ohmic contact with low interfacial resistance between semiconductors, in which photogenerated electrons from the CB of one semiconductor can be directly recombined with photogenerated holes from the VB of a second semiconductor (see Figure 8.1c) [202, 241, 246]. Second-generation Z-schemes offer a plethora of advantages, with particular emphasis on an extension of the application of this strategy to gas-phase PC systems and feasible photocatalyst recovery. Noble metal electron mediators absorbing light may

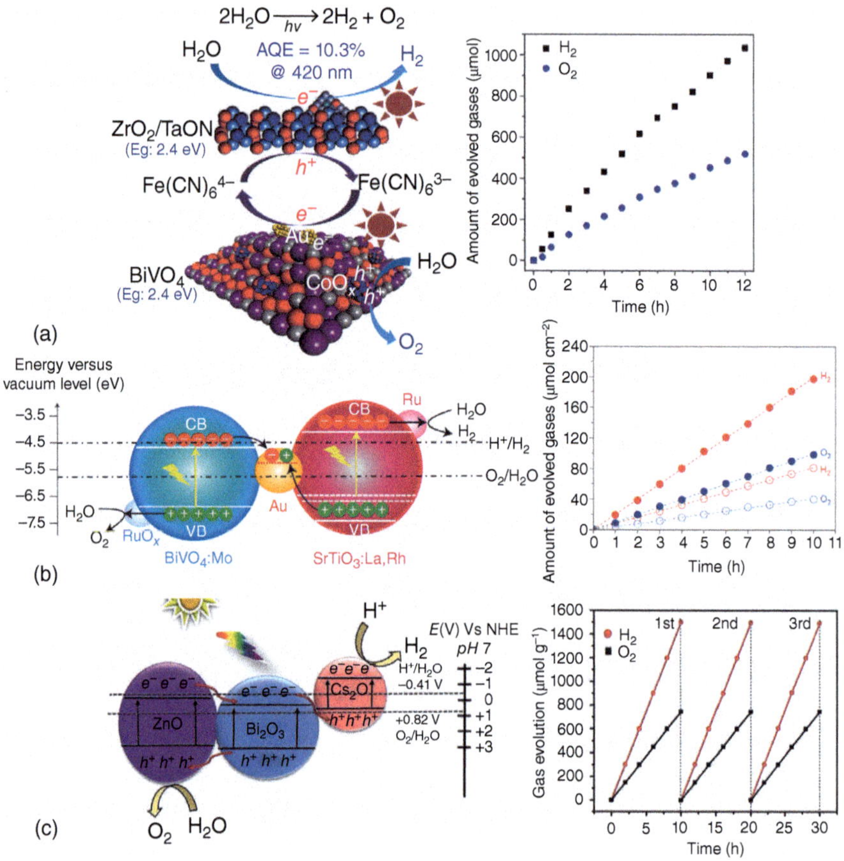

Figure 8.3 Representative Z-scheme types for photocatalytic overall water splitting. (a) First-generation Z-scheme coupling ZrO_2/TaON(HEP) and $BiVO_4$(OEP) with Fe^{3+}/Fe^{2+} redox pairs as aqueous electron mediator and corresponding performance for 12 hours. Source: From Qi et al. [237]. © 2018 Elsevier. (b) Second-generation Z-scheme system coupling La:Rh/$SrTiO_3$(HEP) and Mo:$BiVO_4$ with Au as solid electron mediator. Source: From Wang et al. [30]. © 2016 Springer Nature. (c) Third-generation Z-scheme system within a Cs_2O–Bi_2O_3–ZnO heterostructure and photocatalytic activity for overall water splitting under Vis light. Source: From Hezam et al. [238]. © 2018 RCS.

reduce, however, the light absorption of semiconductor photocatalysts, lessening the overall photoactivity of the system [228, 241]. In addition to light-shielding and potential photocorrosion, operational costs of utilizing rare and expensive noble metals impose clear concerns for scalable operational OWS through a Z-scheme mechanism.

A $ZnRh_2O_4$ (HEP) and defective $Ag_{1-x}SbO_{3-y}$ (OEP) estimated band gap of 2.7 eV; Z-scheme inserting Ag as an electron mediator delivered stoichiometric HER and OER under Vis light ($\lambda > 500$ nm) [247]. The works by Irie and coworkers emphasized the necessity to remove excess Ag through HNO_3 treatment, which served as a sacrificial agent for OER. Undesirably, the treatment induced the formation of the defective $Ag_{1-x}SbO_{3-y}$, with expanded band gap and suppressed photoactivity. Alternatively, the same authors later developed a $ZnRh_2O_4$ and

$AgSbO_3$ (OEP, 2.4 eV) Z-scheme by alternatively removing the excess of Ag through treatment with NH_4OH and H_2O_2, extending in this manner a Vis light utilization up to 660 nm and inducing superior OWS activity of the fabricated heterojunction [248]. Following an analogous strategy, the exploitation of $Bi_4V_2O_{11}$ with a preferably narrower band gap (1.6–2.2 eV) was used as OEP, in place of $AgSbO_3$, under red-light irradiation (λ = 700 nm) [249]. Incorporation of single $Bi_4V_2O_{11}$ crystals, obtained through pulverization of the polycrystalline powder, allowed to achieve the longest wavelength (up to 740 nm) in second-generation OWS Z-schemes [250]. Light harvesting up to 740 nm was further attained with the same $ZnRh_2O_4$ and $Bi_4V_2O_{11}$ Z-scheme utilizing Au as an electron mediator [251].

Intrinsic limitations for the utilization of Ag have rationalized the focus of utilizing Au, with induced increasing catalytic activity, as an electron mediator. In 2016, Domen and coworkers reported a La- and Rh-co-doped $SrTiO_3$ (HEP) and Mo-doped $BiVO_4$ (OEP) photocatalyst sheet-based Z-scheme on a conductive Au layer with controlled thickness prepared by particle transfer method [30]. Nanoparticulate Ru species were then integrated on the surfaces of both HEP and OEP materials. The Au layer (~350 nm) induced the SPR below 520 nm and facilitated the electron transfer from the CB of Mo-doped $BiVO_4$ to the donor levels of La- and Rh-co-doped $SrTiO_3$. As displayed in Figure 8.3b, photogenerated electrons in La- and Rh-co-doped $SrTiO_3$ and incorporated Ru species evolved H_2, whereas photogenerated holes along with RuO_x species oxidized water to O_2. Superior OWS activity was reflected by an AQY of 33% and an STH of 1.1% at pH 6.8 under Vis light (λ = 419 nm). Z-Scheme systems incorporating Au as an electron mediator have been further showcased to prevent the shortcoming self-oxidation of sulfides and selenides [252, 253]. In the presence of metallic Au, optimal $(ZnSe)_{0.5}(CuGa_{2.5}Se_{4.25})_{0.5}$ solid solutions (HEP) and $CoO_x/BiVO_4$ (OEP) showed significantly increased photostability for OWS during 17 hours under Vis light irradiation, with an AQY of 0.54% at λ = 420 nm [252].

Drawbacks associated with the application of noble metals as redox mediators are primarily correlated with a cost-ineffectiveness of this strategy, along with a potential occurrence of reverse reactions typically encountered by noble metals, leading to degradation of overall performance. Carbonaceous materials with large surface area, good conductivity, and tunable band gaps have hindered back ORR activity [254, 255], which occurs on noble metal nanoparticles [16, 256]. La- and Rh-co-doped $SrTiO_3$ (HEP) and Mo-doped $BiVO_4$ (OEP) on a conducting carbon film showed stoichiometric HER and OER for 17 hours, with an STH efficiency (1.0%) comparable against an Au-mediated system counterpart [256]. In 2018, the construction of a metal-free Z-scheme heterostructure through efficient van der Waals interaction between aza-fused microporous polymers and C_2N nanosheets with rGO nanosheets as a solid electron mediator was reported. The properly aligned band structures of the involved nanocomponents facilitated an efficient charge separation and enhanced lifetime and density of photogenerated excitons. The Z-scheme system largely benefited from the introduction of rGO as an electron mediator, with a boost of the STH efficiency from 0.23% up to 0.40% with AQY around 4.3% at λ = 600 nm [257].

Third-Generation Z-Scheme (Direct Z-Scheme)

In 2009, intimately contacted ZnO/CdS heterojunctions unveiled superior HER performance under solar light irradiation through a direct Z-scheme electron transfer mechanism in the absence of electron transfer mediators [258]. In contrast with the above introduced Z-scheme systems and as later pinpointed by Yu et al., direct Z-schemes entail, nonetheless, an interfacial contact between two or more semiconductors (Figure 8.1d) [259]. In late 2009, Vis light-excited electrons in the CB of $BiVO_4$ (OEP) were efficiently transferred to the impurity level of Rh-doped $SrTiO_3$ (HEP) in acidic reactant solution [260]. The interparticle electron transfer largely benefitted from the reversibility of the oxidation states of Rh species, previously discussed by the same group in 2004. Years later, a modified synthesis approach by liquid-state reaction inducing a larger interfacial area unveiled superior AQY (1.6%) at $\lambda = 420$ nm in a neutral reactant solution [261]. In a similar fashion, previously reported La- and Rh-co-doped $SrTiO_3$ (HEP) were integrated to core–shell-based architectures with $Ir/CoO_x/Ta_3N_5$ (OEP) showed an AQY of 1.1% at $\lambda = 420$ nm, with photoresponsivity up to $\lambda = 560$ nm [74].

The promise of exploiting g-C_3N_4 properties has been also denoted in Z-schemes [262, 263]. In 2018, Yang and collaborators reported an AQY of 6.2% at $\lambda = 420$ nm with an H_2/O_2 ratio of ~2 (H_2 and O_2 evolution rate of 482 and 232 μmol g^{-1} h^{-1}, respectively) driven by the suitable alignment of the CB of orthorhombic $WO_3{\cdot}H_2O$ (OEP) positioned above the VB of ultrathin g-C_3N_4 nanosheets (HEP) [262]. Retained photoexcited holes and electrons in the VB of $WO_3{\cdot}H_2O$ and the CB of g-C_3N_4 showed superior propensity to catalyze both water oxidation and reduction reactions, respectively. In the same year, Majima described the synthesis of another 2D-based Z-scheme system combining black phosphorus (BP) nanoflakes with $BiVO_4$ nanosheets [264]. Optimal HER and OER rates (~160 and ~102 μmol g^{-1} h^{-1}, respectively) were reported with 20 wt% BP, whereas an H_2/O_2 ratio of ~2 could be achieved with an increase of BP up to 40 wt%. No noticeable decrement in the PC activity was denoted in the first three cycles for a total of nine hours.

An interesting Z-scheme example regards Janus-like γ-MnS/Cu_7S_4 nanostructures fabricated through meticulous cation exchange [265]. By controlling the initial template and incorporated ions, distinct immiscible domains were integrated in situ at an atomic level, with tetrahedrally coordinated Mn^{2+} ions coexisting with the copper chalcogenide substrate. A nonstoichiometric H_2/O_2 ratio caused by the formation of reactive oxygen species italicizes the possibility of further development. A novel Cs_2O–Bi_2O_3–ZnO heterostructure (Figure 8.3c) was fabricated using a simple solution combustion method and displayed an efficient separation of e^-/h^+ pairs by which small difference in ZnO and Bi_2O_3 VB positions pronouncedly affected the redox potential of both photogenerated charges [238]. Cocatalyst-free Bi_2WO_6 (OEP) and Cu_3P (HEP) Z-scheme systems were prepared by ball milling for 20 minutes with the addition of ethanol in the last 5 minutes. The altered energy level of the composite material assisted the transfer of charge carriers from the CB of Bi_2WO_6 to the VB of Cu_3P, thus achieving feasible electron–hole recombination at the solid–solid interface. The enhancement was attributed to the strong hybridization between Bi_2WO_6 and Cu_3P through mechanical ball milling, providing an effective interfacial contact within nanocomponents of the material [266]. Meanwhile, via first-principles calculations, a series of heterojunctions established through 2D van der Waals was investigated as direct Z-scheme OWS photocatalysts. It was found that all six ($MoSe_2/SnS_2$, $MoSe_2/SnSe_2$, $MoSe_2/CrS_2$, $MoTe_2/SnS_2$, $MoTe_2/SnSe_2$, and $MoTe_2/CrS_2$) 2D van der Waals heterostructures possessed band structures and band-edge positions appropriate for OWS, with $MoTe_2/CrS_2$ being capable of driving OWS under NIR light irradiations [267].

8.3 Photoelectrochemical Overall Water Splitting

In 2019, the society took a remarkable step toward a more sustainable future, when U.S. annual energy consumption from renewables surpassed coal to become the primary source of global electricity capacity [268]. Driven by a steep expansion of novel environment-friendly energy technologies, this achievement is particularly attributed to the photovoltaic (PV) industry with a consistent improvement of solar cell efficiencies [268–272]. Essentially free electricity can ideally power today electrochemical platforms, which operate under mild conditions when compared with conventional thermocatalytic processes. H_2 and O_2 production through OWS simultaneously serve as a promising approach for leveling the electricity output from these daylight- and season-dependent renewable sources (e.g. solar, wind, waves).

The underlying principles of EC water splitting systems and state-of-the-art materials have been extensively reviewed in the literature [273–280]. Notwithstanding the advantages of this strategic pathway, high-energy barriers and associated large overpotentials, in particular for water oxidation, are still key issues in the EC device. To suppress the extensive electricity storage with associated imposing costs, direct coupling with PV platforms has been reported since the early years. In the PV–EC device system, several PV cells produce a high minimum potential required to operate an electrolyzer in an indirect route [281]. The PV cells, electrically wired to both H_2 and O_2 evolution electrodes, are kept out of the electrolyte (to avoid water degradation) or are directly integrated in the electrodes (i.e. as buried PEC junctions) [282–284]. The efficiency of PV–EC is determined by the PV open-circuit voltage and water splitting overpotentials [285, 286]. Among all device configurations, water electrolysis driven by coupled PV cells has also shown the highest STH efficiencies to date [287, 288]. Under concentrated solar illumination (~42 suns), the highest STH conversion efficiencies (>30%) was reported for the first 20 hours utilizing a triple-junction solar cell (InGaP/GaAs/GaInNasSb) producing sufficient free energy for self-sustained water cleavage with HER and OER over two Pt–Ir electrolyzers [289].

PV–EC systems denote a high cost due to the complexity of the device system, markedly impeding its commercial viability against competitive steam reforming technologies. Alternatively, PEC water splitting has gained increasing interest over the years. PEC systems rationally integrate light-responsive materials as

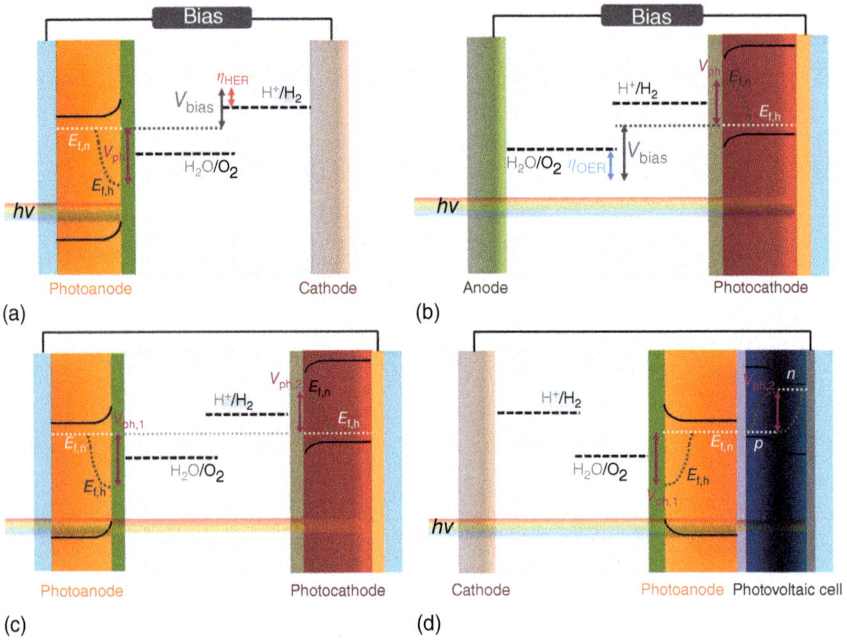

Figure 8.4 Different types of PEC for water splitting comprising (a) photoanode–cathode, (b) photocathode–anode, (c) photoanode–photocathode tandem cell, and (d) photovoltaic–photoanode (PV–PEC). Source: From Kim et al. [19]. © 2019 RCS.

photocathodes, photoanodes, or both to produce in situ photogenerated charge carriers, reducing electricity consumption and boosting the resulting overall process efficiency. The PEC water splitter comprises a lower number of device components, reflecting in this sense superior cost-effectiveness against complex PV–EC devices. Pinpointed estimations by the US Department of Energy for PV–EC and a single-integrated PEC cell (8 and 3 USD kg^{-1}, respectively) underline a difficulty for PEC systems to break the commercialization barrier against the competitive steam reforming alternative [281, 290]. If required developments are achieved, the predicted global annual energy consumption in 2050 (36 TW/yr) could be fulfilled with a 1% land coverage of PEC cells with 10% efficiency [8]. The simplest PEC system, utilizing a single photoelectrode (either a photoanode or a photocathode) with the aid of external bias as illustrated in Figure 8.4a,b, displays a limited theoretical maximum STH efficiency level below 1% [281]. Heterojunctions consisting of two or more semiconductors employed as photoanode and photocathode for OWS have been reported (Figure 8.4c), allowing the generation of high photovoltage with flexibility in the selection of incorporated materials.

Another approach for unbiased OWS is displayed by PV–PEC device configuration that coupled PV device and semiconductor photoelectrode (Figure 8.4d). The PV cell, often composed of single or multiple junctions based on Si [291], group III–V materials [292], and hybrid perovskites [293], supplies the required potential for OWS. As a result, the band-gap positions of the photoelectrodes do not play a direct role in resulting water redox potentials, which makes it a more attractive option for PEC water splitting reaction [19].

8.3.1 Principles and Mechanism

PEC water splitters were demonstrated at the early stages by Honda and Fujishima using TiO_2 as a photoanode material [9]. Based on clear technology parallelism, the know-how of PC devices, principles, and catalyst library has since served as an effective guideline for the development of PEC systems. As in the analogous PC, efficient PEC operation entails the incorporation of semiconductor materials capable of inducing extensive solar light absorption and generating enough free energy to split the water molecule. Narrow-band-gap semiconductors possessing band-edge positions straddling the hydrogen and oxygen redox potentials, and high stability against corrosion in aqueous electrolytes and under light illumination, are required. Representatively, the simple coupling of an n-type semiconductor photoanode with a metal dark cathode is illustrated in Figure 8.4a. Upon light irradiations, the photogenerated charges (electrons and holes) are produced at the photoanode. Photoexcited electrons are then driven to the rear ohmic contact, flowing through external electrical connection to the counter cathode surface, while the photoexcited holes remain in the VB of the photoanode, moving toward the interface between the semiconductor surface and the electrolyte. During this step, nonetheless, the recombination of photogenerated carriers can occur at the bulk semiconductor or at its surface. Highly efficient separation and mobility of photoexcited charges are desirable. Feasible transportation of photogenerated charge carriers to the electrode/electrolyte solution-generated interface is, at last, necessary to minimize HER and OER overpotentials [281]. The above physicochemical steps are analogously involved in PEC water splitting systems coupling photocathodes with dark anodes. The overpotential associated with driving the HER and OER kinetics at the solid–liquid interface and the non-radiative recombination significantly accounted for the losses of the overall PEC efficiency [23]. The PEC OWS device is composed of semiconductor photoelectrodes (photoanode and/or photocathode) connected to charge collectors inside an electrolyte with appropriate membrane separation. Evolved H_2 and O_2 gases are spatially separated on different electrodes. According to Nernst equation, water splitting requires a free energy change of 237.2 kJ mol^{-1} equivalent to 1.23 eV. For a semiconductor to drive the water splitting process toward HER and OER, light-harvesting capability up to ~1000 nm (corresponding to a photon energy of >1.23 eV) is necessary. In this sense, two electron–hole pairs are generated for the evolution of one molecule of H_2 (2×1.23 eV = 2.46 eV), whereas four electron–hole pairs are needed per O_2 molecule (4×1.23 eV = 4.92 eV) [23].

8.3.2 Key Performance Indicators

The energy conversion efficiency is universally acknowledged and used as an indicator to catalyst performance, by which all water systems and devices can be reliably compared. The ηSTH is defined as the ratio between the total energy generated and the total energy input from standard sunlight irradiation (AM 1.5G, 100 mW cm^{-2}), as introduced in the PC section (Eq. (8.6)), and alternatively can be quantified by the generated photocurrent density (J_{SC}) at voltage fixed at 1.23 V as shown in Eq. (8.10): [23, 25]

$$\eta\text{STH}(\%) = \left[\frac{1.23\ \text{V} \times J\ (\text{mA cm}^{-2}) \times \eta_F}{P_{\text{sun}}\ (\text{mW cm}^{-2})}\right] \tag{8.10}$$

where J is the photocurrent density, P_{sun} is power of the incident light, and η_F is faradaic efficiency (FE) of H_2 or O_2 production. FE are defined as the ratio between the charges transferred to a certain product (herein, H_2 and O_2) quantified experimentally and the number of total charges passed through the circuit. A high FE further reflects a high product selectivity of developed PEC platforms and is a key criterion in the selection of suitable photoelectrodes. In PEC systems, the applied bias photon-current efficiency (ABPE) is calculated when an external bias is applied, with subtraction of the electrical energy. ABPE is thus defined according to Eq. (8.11):

$$\text{ABPE} = \frac{J_{\text{ph}}(V_{\text{redox}} - V_{\text{bias}})}{P_{\text{light}}} \tag{8.11}$$

with V_{redox}, V_{bias}, J_{ph}, and P_{light} as the water splitting theoretical redox potential (1.23 V_{NHE}), applied voltage, generated photocurrent density, and light intensity (100 mW cm^{-2}), respectively.

8.3.3 Materials Design

The rational fabrication of photoanodes and photocathodes for PEC devices focuses, in principle, on the selection of semiconductors with appropriate band structure and band gap, capable of inducing efficient charge separation and migration and possessing practical catalytic activity and stability. Often, large band gaps and poor optical properties and conductivity leading to inefficient separation and mobility of photogenerated charge carriers are key bottlenecks directly correlated with modest PEC performances. The construction of appropriate photoelectrodes is typically carried on transparent conductive oxide substrates like glass, e.g. fluorine-doped tin oxide (FTO) or indium tin oxide (ITO). The extensive library of reported systems reflects an analogous complexity emphasized with PC platforms [294, 295]. To date, however, efficient and cost-effective materials and corresponding devices for scalable hydrogen production are still unreported.

8.3.3.1 Photoanode Materials

Several n-type semiconductors capable of driving water oxidation (OER) have been investigated as photoanodes including metal oxides (e.g. TiO_2, $BiVO_4$, CoO, Fe_2O_3, and IrO_2), metal sulfides (e.g. CdS and MoS_2), and group III–IV materials. The underlying appropriateness for the selection of photoanode materials follows the same fundamentals discussed in section 8.2.3. With a large extent of reported photoanodes being metal oxide based, limited light harvesting primarily arises as a critical drawback of several nontoxic materials inherently possessing good photochemical and thermal stability, and remarkable optical, electronic, and catalytic properties. Extending the range of light absorption has received significant literature focus in the past years with relevant works assessing the incorporation of plasmonic structures and upconversion nanocrystals on metal oxides [224, 296–303].

In photoanode-driven systems, particular emphasis has been attributed to the doping of semiconductors with various nonmetal (e.g. C, N, and S) [304] and metal (e.g. Fe, Mo, Mn, W, Ag, and Co) elements. C-doped TiO_2 photoanode thin films [305], S-doped black TiO_2 [306], and N-doped TiO_2 nanowires [307] are representative examples of a remarkable enhancement in light absorption, efficient charge carrier separation and transportation, and excellent PEC activity utilizing a benchmark n-type semiconductor. Fe-doped TiO_2 nanorods [308], Bi-doped WO_3 [309], Ag-doped WO_3 and $CuBi_2O_4$ [310, 311], and Co- and Sn-co-doped Fe_2O_3 [312] are representative examples of metal doping in various semiconductors with induced light harvesting and boosted PEC performance. An unintended formation of surface defects serving as recombination centers upon metal doping has been reported [308]. An alternative to this approach is the co-doping of donor–acceptor pairs. Using density functional theory calculations, modified band gaps and band-edge positions for a range of donor–acceptor pairs were investigated for TiO_2 co-doping, e.g. (W, C) [313, 314], (Mo, C) [313–315], (2Nb, C) and (2Ta, C) [313, 316], (W, 2N) [313, 317], (Ta, N) [313], (Nb, N) [313], (Sb, N) [318], (Cr, N) [319], (Zr, S) [313], and (Nb, P) [313]. Experimentally, W- and C-co-doped TiO_2 nanowires unveiled a twofold photocurrent density enhancement against a mono-doped counterpart due to an improvement in the electrical conductivity of the resulting photoanode [320]. Crystal facet engineering as a key strategy for the enhancement of the efficiency of photoanodes (and photoelectrodes) has allowed to the publication of several reports of interest, which have been comprehensively summarized elsewhere recently [298]. Tuning the crystallographic orientation and optimizing the morphology of materials toward superior surface areas has reflected a significant enhancement of charge carrier transportation and the resulting photocurrent density of n-type semiconductors. Boosting electron transport in 1D TiO_2 nanostructures, e.g. nanorods, nanowires, and nanotubes, was shown to hinder charge recombination at the electrolyte interface, increasing the device performance [321]. Rough TiO_2 film surfaces showed a threefold superior photoactivity against smooth surface films when treated with poly(ethylene glycol), with the number of exposed active centers being a function of the film roughness [322, 323]. Induced surface roughness further plays a role in effective light distribution and abridged hole diffusion length [324]. Regarding other n-type semiconductors, nanoporous Mo-doped $BiVO_4$ on a nanocone substrate with Fe(Ni)OOH coating layer exhibited a photocurrent density of $5.82 \pm 0.36\,mA\,cm^{-2}$ at $1.23\,V_{RHE}$, approaching ~77.8% of its theoretical value. The results were ascribed to the nanocone-templated morphology shortening the carrier diffusion length and boosting charge transport compared with a flat substrate-based counterpart [325]. The unique morphology had shown prior interest as a highly promising candidate for high-efficiency thin film PV [326]. TiO_2/Ti:Fe_2O_3-branched nanorods on FTO were synthesized through atomic layer deposition (ALD) growth of a TiO_2 interlayer acting as a charge carrier density booster and filling the interspace between the FTO substrate and the hematite [327]. The dendrite-like Fe_2O_3 with dramatically enlarged surface area with subsequent FeOOH-layered coating as OER cocatalyst revealed an overall photocurrent density of ~3.1 mA cm^{-2} at 1.23 V_{RHE}.

Incorporation of state-of-the-art OER cocatalysts, primarily consisting of metal oxides (MO_x) (with M = Co, Mn, Ni, Rh, Ir, Ru, and Pd), on representative semiconductors has been actively surveyed, for instance, $Co_3O_4/BiVO_4$ [328], $RhO_2/BiVO_4$ [329], Ni-Bi/$BiVO_4$ [330], Co-Pi/$BiVO_4$ [331], Co-Bi/$BiVO_4$ [332], Co_3O_4/Ta_3N_5 [333], $Co(OH)_x/Ta_3N_5$ [333], and IrO_2/Ta_3N_5 [333]. Photocurrent densities of 5.0 and 2.7 mA cm^{-2} at 1.23 V_{RHE} and 0.6 V_{RHE}, respectively, were reported with dual FeOOH and NiOOH-layered/$BiVO_4$. At 0.6 V_{RHE}, the photocurrent density could be maintained for >48 hours with stoichiometric H_2/O_2 ratios. The performance was ascribed to the simultaneous effects of a hindered charge carrier recombination and a favorable Helmholtz layer potential drop at the generated cocatalysts/$BiVO_4$ interface [334]. Following an analogous approach, dual cocatalyst incorporation (CoO_x and in situ generated hydroxyl-ion permeable NiOOH) on a NiO–$BiVO_4$ p–n heterojunction showcased a photocurrent density at 1.23 V_{RHE} (3.5 mA cm^{-2}) and 0.6 V_{RHE} (2.5 mA cm^{-2}) with stoichiometric H_2/O_2 ratios over 12 hours [335]. State-of-the-art IrO_2 OER cocatalyst introduced on an InGaN photoanode assisted in reducing the hole accumulation on the surface, improving the catalytic activity and photostability, reaching a photocurrent density of 10.9 and 5.2 mA cm^{-2} at 1.23 V_{RHE} and 0.6 V_{RHE}, respectively. In the presence of a hole scavenger (H_2O_2), the photocurrent density at 1.23 V_{RHE} could be remarkably improved up to 21.2 mA cm^{-2}, reaching the theoretical limit for a 1.7 eV band-gap InGaN [336]. Alternatively, incorporation of promising OER cobalt phosphate (Co-Pi) cocatalysts has been widely reported on TiO_2 nanowires [337], $BiVO_4$ [331, 338], Fe_2O_3 [339], and Ta_3N_5 nanorods [340]. In the latter case, interestingly, Co-Pi unveiled superior photostability against a benchmark IrO_2 [340].

As an alternative to conventional n-type metal oxide semiconductors, PV-grade semiconductors (GaAs, InP, Si based, etc.) have disclosed great promises due to their ideal narrow band gaps (e.g. ~1.42 eV for GaAS) with excellent charge carrier mobility. Poor stability against water requires, however, the incorporation of sophisticated protection layers through ALD of coatings on n–p^+-Si photoanode [341], sputtered NiO_x films on Si-heterojunction, NiO_x/n-CdTe and NiO_x/amorphous hydrogenated-Si (a-Si:H)-based photoanodes, or CoO_x thin layer on NiO_x/SiO_x/n-Si photoanodes for PEC water splitting application [342, 343]. Alternatively, layered double hydroxides (LDH) with general formula $[M^{II}_{1-x}M^{III}_x(OH)_2]^{z+}(A^{n-})_{z/n}\cdot\gamma H_2O$ (M^{II} and M^{III} as divalent and trivalent cations, respectively, and A^{n-} as an interlayer non-framework charge compensating inorganic or organic anion) [344–349] have been extensively utilized to co-doped TiO_2, ZnO, and Co_3O_4 semiconductors due to their high surface-to-volume ratio and short diffusion length for charge carrier transportation. Highly stable photocurrent response without noticeable decrement for over 30 cycles on ZnO@CoNi-LDH core–shell NWs was attributed to the effective and high quality of ZnO/LDH interface in which the photogenerated holes from the ZnO core were quickly transferred to the LDH shell, thus suppressed the anodic decomposition/corrosion and increased the stability of the device. Moreover, the LDH also accelerated the water splitting reaction kinetic and recombination rate of photogenerated charges [350]. Another similar LDH OER cocatalyst effect was shown on Ta_3N_5 nanorod arrays co-doped with NiFe LDH as cocatalyst [351].

Finally, the construction of heterojunctions and Z-schemes discussed for PC technologies is of particular importance to efficiently boost the efficiency of photoanodes. Unassisted PEC OWS technologies introduced later in this chapter strikingly emphasize the need to develop complex photo-responsive systems in which heterojunctions established between multiple semiconductors are crucial. Representatively, the intimate contact between a layered $CaFe_2O_4$ and TaON p–n heterojunction induced a ~fivefold photocurrent density enhancement at 1.23 V_{RHE} by reducing the resistance of charge carrier transportation and, as a result, enhancing the electron–hole separation [352]. An extensive library of reported heterojunctions between narrow-band-gap semiconductors (e.g. II–VI chalcogenides) and TiO_2 (e.g. TiO_2/CdSe [321, 353], TiO_2/CdS_xTe_{1-x} [354], and TiO_2/CdS_xSe_{1-x} [355, 356]) or ZnO (e.g. $ZnO/Zn_xCd_{1-x}Se$ [357, 358] and $ZnO/Zn_xCd_{1-x}Te$ [359]) along with $WO_3/BiVO_4$ [360], and core–shell Si-Ta_3N_5 further doped with OER cocatalysts ($CoTiO_x$ and NiO_x) [361] have been reported.

8.3.3.2 Photocathode Materials

Applied photocathodes in PEC platforms representatively consist of p-type semiconductors capable of harvesting light and performing the half-cell HER. The wide material library reported to date includes metal oxide-, oxynitride-, chalcogenide-, perovskite-, and PV-grade material-based photocathodes. Recent research has uncovered the promise of perovskite halides (e.g. $CH_3NH_3PbI_3$) as PEC HER candidates in line with a favorable band alignment for H^+/H_2 [362–364]. The inherent instability of these materials in water requires the integration of protective layers and hydrophobic carbon electrodes. Representatively, encapsulated lead halide $CH_3NH_3PbI_3$ perovskite with fusible InBiSn alloy showed stable performance in aqueous solution during HER for ~1 hour with a photocurrent density of 9.8 mA cm^{-2} at 0 V_{RHE} [365].

Broad attention has been paid to cuprous oxide (Cu_2O), possessing a narrow band gap (2.0 eV), with a CB lying 0.7 V above the HER potential, and a high theoretical photocurrent (14.7 mA cm^{-2}) and STH efficiency (~18%) under AM 1.5G light illumination [366]. A swift electron–hole recombination and cathodic photocorrosion in the electrolyte solution have discouraged. However, a practical utilization of Cu_2O in PEC water splitters attracted several studies focusing on improving the stability of these promising photocathodes. Reports listed below assess suitable protective oxide layers with a staggered type-II band offset inducing facile transfer of photogenerated electrons to the electrolyte; a CB above the H^+/H_2 potential, with no propensity for degradation at potentials within the band gap; and a surface with favorable reaction kinetics for HER. In 2011, the lifetime of electrodeposited Cu_2O could be remarkably enhanced upon sequential incorporation of ultrathin n-type TiO_2 and Al gradually doped ZnO nanolayers by ALD [366]. The resulting nanoarchitecture with electrodeposited Pt as a cocatalyst revealed photocurrents up to 7.6 mA cm^{-2} at 0 V_{RHE} at mild pH, stable up to one hour. A local electrostatic field formed at the resulting p–n junction further assisted in extracting the photogenerated electrons from Cu_2O. When a subsequent RuO_x layer was incorporated on Cu_2O nanowires with Al-doped ZnO and TiO_2-coupled protective layers, a strikingly superior photostability reached up to 55 hour at comparable photocurrent densities (~8.0 mA cm^{-2} at 0 V_{RHE}) [367]. Alternatively, a Cu_2O–Ga_2O_3-buried p–n junction with conformal TiO_2 coating and RuO_x incorporation as a cocatalyst, reaching ~10 mA cm^{-2} at 0 V_{RHE}, demonstrated stable operation exceeding 100 hours [368].

Cu-based chalcogenides, for example, CuIn(S,Se) (CIS), Cu(In_xGa_{1-x})$(S,Se)_2$ (CIGS), and $Cu_2ZnSn(S,Se)_4$ (CZTS) photocathodes with required protection from electrolyte for stable performance, have been extensively reported [369]. Co-evaporated Pt/TiO_2/CdS/CZTS (9 mA cm^{-2} at 0 V_{RHE}) [370], solution-processed Pt/TiO_2/CdS/CZTS (11 mA cm^{-2} at 0 V_{RHE}) [371], Pt/In_2S_3/CdS/CZTS (9.3 mA cm^{-2} at 0 V_{RHE}) [372], and Cd-substituted CZTS with CdS subsequent incorporation and deposition of Ti, Mo, and Pt (Pt/TiMo/CdS/CCZTS) (17 mA cm^{-2} at 0 V_{RHE}) [373] are representative examples with high photocurrent density that could be attained. CIGS-based photocathodes, e.g. S-P-Pt/TiO_2/Al_2O_3/CdS/CIGS and Pt/Mo/Ti/CdS/CIGS (with Mo/Ti conductive layers), showed superior performance (30 and 27 mA cm^{-2} at 0 V_{RHE}, respectively) [374, 375] but suffered from the scarcity and cost-ineffectiveness. In comparison, $CuGaSe_2$ materials are economically attractive and possess an absorption edge near 750 nm. To surpass its limiting shallow CB, Domen demonstrated that partial Ag substitution ($Ag_xCu_{1-x}GaSe_2$) could significantly deepen the VB and induce wider grain size against its pristine counterpart. With additional Pt and CdS co-incorporation, a stable cathodic photocurrent of 8.1 mA cm^{-2} at 0 V_{RHE} was achieved for over 55 hours [376]. Analogously, the same authors reported the introduction of a thin CdS layer on $CuGaSe_2$, inducing the formation of a p–n junction on the surface of the electrode. The optimized thickness of the CdS layer, proposed to tune the balance between the charge separation and the light absorption by CdS, induced a stable HER production for a time period above 10 days [377].

Among other chalcogenide-based materials exploited as p-type photoelectrodes, the utilization of cost-effective antimony triselenide (Sb_2Se_3) as a binary semiconductor with a thermodynamically stable orthorhombic phase has been underlined in the reports below. A designed junction between the Sb_2Se_3 photocathode and the hole-selective layer (HSL), NiO_x based as bottom contact layer, was, for instance, shown to remarkably hinder electron backflow, thus maximizing the utilization of photogenerated charges with reported photocurrent density of ~17.5 mA cm^{-2} at 0 V_{RHE} [134, 378]. Fine-tuning of the Sb_2Se_3 nanostructure using thioglycolic acid and ethanolamine was suitable to synthesize 1D Sb_2Se_3 nanorod arrays with preferred orientation and a resulting fast and efficient charge carriers' transport. The optimized photocathode Pt/TiO_2/Sb_2Se_3 delivered a 12.5 mA cm^{-2} photocurrent density at 0 V_{RHE} [379]. A similar morphology control strategy was recently reported on a $Pt/TiO_2/CdS/Sb_2Se_3$-driven system reaching 13.5 mA cm^{-2} at 0 V_{RHE} [380]. Also in 2019, a simultaneous introduction of optimized Cu-doped NiO_x, TiO_2, and Pt on Sb_2Se_3 was reported to attain a photocurrent density of ~17.5 mA cm^{-2} at 0 V_{RHE} [378]. Cadmium telluride (CdTe) possessing a long absorption edge wavelength of $\lambda = 830$ nm was first assessed as a photocathode in PEC water splitters by Ohashi [381]. A treated $CdCl_2$ CdTe-based photocathode ($Pt/CdS/CdTe(CdCl_2)/Cu/Au$) showed excellent cathodic photocurrent of 22 mA cm^{-2} at 0 V_{RHE} [382].

Being the second most abundant earth element, the wide utilization of single-crystalline silicon (Si)-based photoelectrodes has been rationalized by its narrow band gap (1.1 eV), excellent optical and electrical properties, and a promising application in dual band-gap PV–PEC platforms (see Section 8.3.4.2). A comprehensive comparison of fabrication strategies and performance of Si-based HER photoelectrodes has been reviewed elsewhere [383]. In particular, the utilization of hydrogenated amorphous silicon carbide (a-SiC) photocathodes, with tunable band-gap energy (1.8–2.1 eV) and remarkable stability in a mildly acidic medium under cathodic bias, has seen increasing utilization in multi-junction PV cells with smaller band gap (e.g. 1.1–1.8 eV). Low charge carrier collection and slow reaction kinetics at the established electrode/electrolyte interface have, however, limited reported performances of such triple-junction systems markedly below their theoretical limit [384]. An ALD-deposited n-type amorphous TiO_2 thin layer (25 nm) on p-type/intrinsic (p/i) a-SiC (110 nm), generating a p–i–n heterojunction, was reported to favor an efficient carrier collection for H_2 evolution [385]. With an additionally loaded HER cocatalyst (Ni–Mo), a stable photocurrent density of 8.3 mA cm^{-2} was maintained at 0 V_{RHE} for 12 hours.

8.3.4 Unassisted Photoelectrochemical Overall Water Splitting

As discussed in the introduction of this section, reported unbiased solar-driven overall water splitters are generally grouped into three main categories, i.e. photoanode–photocathode tandem cells, PV–PEC, and PV–EC device systems.

8.3.4.1 Photoanode–Photocathode Tandem Cells

PEC tandem cell systems simultaneously incorporate commonly reported n-type and p-type semiconductor-based systems as suitable photoanode and photocathodes where OER and HER take place, respectively. Ideally, self-sufficient PEC platforms exploiting two photoelectrodes do not require the application of an external bias. Both photoelectrodes are integrated via a conductive metal wire, with a simplified device assembly or wireless connection with a good ohmic contact for superior device compactness. Unassisted overall water splitters fittingly rely on optimized hybrid architectures pinpointed in section 8.3.3, reflecting the increasing complexity of coupled synthesis strategies, while underlining an urge to consider new classes of materials and strategic synthesis pathways. Early reported systems, strongly relying on the use of oxides as photoanodes, showcased low STH levels primarily

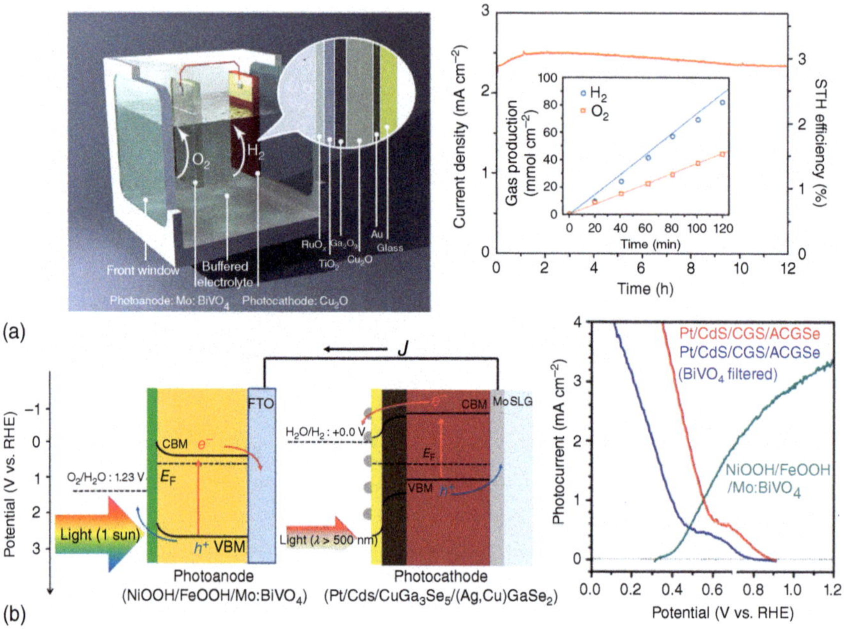

Figure 8.5 Photoanode–photocathode tandem cell for overall water splitting systems using (a) metal oxide-based photoelectrodes (Cu_2O as HEP and Mo:$BiVO_4$ as OEP) and produced photocurrent density with stoichiometric H_2 and O_2 evolution. Source: Reproduced with permission. Pan et al. [368]. Copyright 2018, Macmillan Publishers Limited, part of Springer Nature. (b) PEC tandem cell coupling a NiOOH/FeOOH/Mo-doped $BiVO_4$ photoanode and a Pt/CdS/$CuGaSe_2$/(Ag,Cu)$GaSe_2$ photocathode with plots of anodic and cathodic photocurrent. Source: From Kim et al. [389]. © 2015 John Wiley & Sons.

correlated with the wide band gaps of these semiconductors. Representatively, for photoanode–photocathode tandem cells, a theoretical maximum STH value of 29.7% can be realized with a band-gap combination of 1.60/0.95 eV [386].

Several PEC tandem cell devices incorporating $BiVO_4$-based photoanodes have been reported to date [387]. In 2014, a device coupling Co-Pi/$BiVO_4$ and RuO_x-doped Cu_2O as photoanode and photocathode, respectively, attained an STH conversion efficiency of ~0.5%. A progressive current decay over the course of minutes was denoted with a facile detachment of integrated Co-Pi species [388]. By comparison, an unassisted device with strikingly superior stability would be later reported pairing a $NiFeO_x$/H- and Mo-co-doped $BiVO_4$ photoanode and an all-earth-abundant Cu_2O-based photocathode (Cu/Cu_2O/Ga_2O_3/TiO_2/NiMo) (Figure 8.5a). With a ~2.5 mA cm^{-2} peak current density at pH 9.0, the tandem cell evidenced the highest STH (~3%) and current stability (12 hours) for metal oxide-based photoanode–photocathode PEC system to date [368]. Pairing $BiVO_4$-based photoanodes with chalcogenide-based photocathodes, possessing a long absorption edge wavelength, for unassisted OWS under Vis light, has been of notorious focus. In 2015, Ikeda and coworkers first demonstrated a NiOOH/$BiVO_4$ (photoanode) and Pt/In_2S_3/CdS/Cu_2ZnSnS_4 photocathode-coupled tandem cell, with an STH efficiency of 0.28% at pH 6.5 [372]. The following year, Domen and coworkers reported a parallel system utilizing a NiOOH/FeOOH/Mo-doped $BiVO_4$ photoanode and a Pt/CdS/$CuGaSe_2$/(Ag,Cu)$GaSe_2$ photocathode with a 0.55 mA cm^{-2} photocurrent density and an STH conversion efficiency of 0.67% over two hours at pH 7 (Figure 8.5b) [389]. With an integrated $NiFeO_x/BiVO_4$ (photoanode) and a $(ZnSe)_{0.85}(CuIn_{0.7}Ga_{0.3}Se_2)_{0.15}$ (strip-based photocathode), the resulting PEC tandem cell exhibited a photocurrent density of 1.35 mA cm^{-2} with an STH of 1.0% [390]. In 2018, Domen and coworkers further reported the pairing of Fe/NiO/$BiVO_4$ (photoanode) and highly efficient Pt/CdS/$CuIn_{0.5}Ga_{0.5}Se_2$/Mo-coated soda–lime glass (photocathode) for OWS. In the fabricated device, stoichiometric H_2 and O_2 evolution rate with a photocurrent density of ~3 mA cm^{-2} and an impressive STH conversion efficiency of 3.7% were witnessed at pH 9.5. Since the attained efficiency was hindered by the performance of the semitransparent $BiVO_4$ anode, the authors predicted superior performance in upcoming studies [391].

The promise of a Ga–In alloy semiconductor with high STH was early reported in 1987 [392]. The ZnO/Pt-doped p-type indium phosphide (p-InP) coupled with Mn-oxide thin film coated on n-type gallium arsenide (n-GaAs) could simultaneously evolve H_2, and O_2 under light irradiation with H_2 was the predominant gases in 6.0 M KOH electrolyte. The system gave rise to an STH efficiency of 8.2% with recorded photocurrent density of ~13 mA cm^{-2}, for an overall performance significantly superior to earlier reports using Ga-based photocathodes [393, 394]. More recently, an outstanding STH efficiency of 13.1% was obtained in a series-connected wireless tandem system of both GaAs-based photocathode and photoanode. As illustrated in Figure 8.6A, a thin film stack GaAs/Al_xGa_{1-x}As epitaxial compound was released from the growth wafer and printed on a non-native transparent substrate to form an integrated bifacial PC electrode with decoupled optical and reactive interfaces. The system composed of IrO_x/n-GaAs (photoanode) and Pt/Ti/Pt/Au/p-GaAS (photocathode) exhibited a photocurrent density of 10.6 mA cm^{-2} at pH 0.55 [395]. The optimization of the design of the reported materials and underlying fabrication methods allowed to finely tune the light absorption, carrier transport, charge transfer, and the resulting stability. Most importantly, the delineated strategy was claimed pertinent to a variety of materials (e.g. group IV, III–V, III–N) and extensible to distinct PEC applications.

In line with an extended emphasis on the promise of Si-based photocathodes [397], an optimized nanotree structure with IrO_2/TiO_2 and Pt/Si as photoanode and photocathode, respectively, disclosed an STH efficiency of 0.12% with a photocurrent density intersection of 0.35 mA cm^{-2} in pH of 0.52 [398]. In addition, the first unassisted water splitting using hematite based as photoanode ($NiFeO_x/Fe_2O_3$) and TiO_2/Pt loaded on amorphous Si showed a modest performance with an overall efficiency of 0.91% and stable for 10 hours in pH 11.8 electrolyte [399]. A careful and complex Si photocathode design was prepared by passivating radially doped Si nanowires with SiO_2, which were then spatioselectively removed. The as-prepared Si photocathode (NiMo/SiO_2/n^+/p-Si) coupled with NiOOH/FeOOH/Mo-doped $BiVO_4$ (photoanode) showed an unbiased water splitting efficiency of 2.1% over the

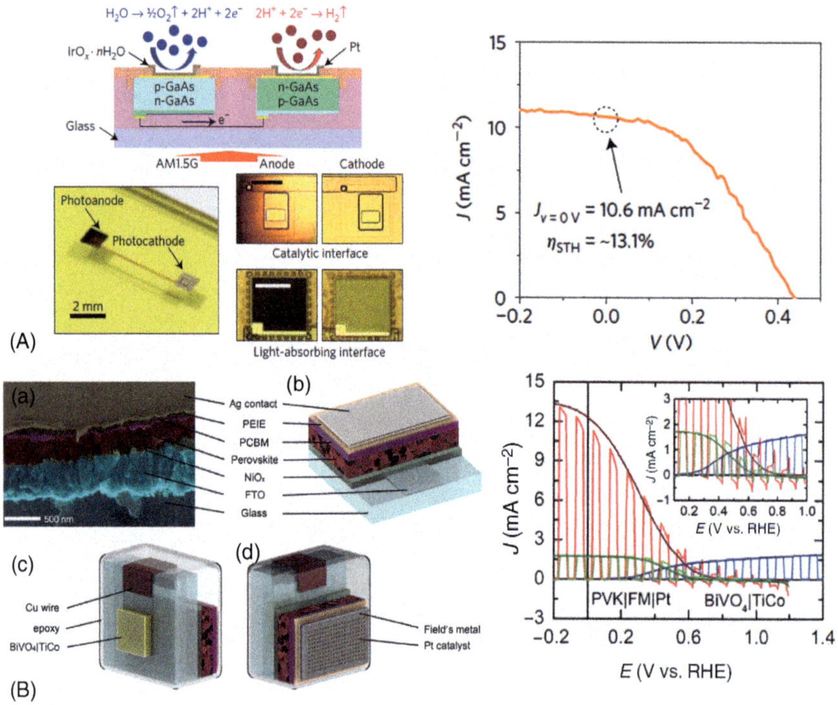

Figure 8.6 Photoanode–photocathode tandem cell for overall water splitting systems using (A) PEC system composed of IrO_x/n-GaAs (photoanode) and Pt/Ti/Pt/Au/p-GaAS (photocathode) and representative *J*–*E* curve. Source: Reproduced with permission. Kang et al. [395]. Copyright 2017, Macmillan Publishers Limited, part of Springer Nature. (B) PEC tandem device with $BiVO_4$/TiCo (photoanode) and encapsulated NiO_x as hole transport layer (HTL) layer on cesium formamidinium methylammonium (CsFAMA) photocathode. Continuous and chopped light linear sweep voltammograms were obtained for the $BiVO_4$|TiCo photoanode (red), perovskite photocathode (blue), and PEC tandem configuration (green). Source: Reproduced with permission. Andrei et al. [396].

course of an hour in the photoelectrode preferred pH solution (photocathode at pH 1 and photoanode at pH 10) [400].

Using highly versatile, cost-effective, and scalable solution-processed single-source precursor (SSP) chemistry technique, TiNi and TiCo protective films were incorporated on a composite photoelectrode for the fabrication of two PEC tandem cells, namely, PEC cell I (p-Si|$TiNi_{HEC}$ with nano WO_3|$TiNi_{OEC}$) and PEC cell II (p-Si|$TiNi_{HEC}$ with nano $BiVO_4$|$TiCo_{OEC}$), operated at pH 9.2. The SSP coating on both photoelectrodes protected and catalytically activated the semiconductors for both half-reactions. Due to an increase in ohmic loss, however, the PEC cell could not be operated without applied bias, whereas low STH efficiency (~0.05%) was reported for the PEC cell II. Applying bias, corresponding STH efficiency levels reached 0.19% (~0.4 mA cm^{-2} at 0.8 V) and 0.59% (~1.0 mA cm^{-2} at 0.6 V), respectively, which were found to be among the highest applied bias STH values reported for noble metal-free dual photoelectrode tandem cells with facile and scalable composite photoelectrode protection [401, 402].

The rapid research growth focusing on lead halide perovskite-based PV has analogously driven the tentative incorporation of these light-harvesting materials into PEC cells. As illustrated in Figure 8.6B, a reported 0.25 cm^2 PEC tandem device with $BiVO_4$/TiCo (photoanode) and encapsulated NiO_x as hole transport layer (HTL) on cesium formamidinium methylammonium (CsFAMA) triple cation-mixed halide perovskite (photocathode) showed a bias-free photocurrent of 0.39 mA cm^{-2}, corresponding to STH efficiency of $0.35 \pm 0.14\%$ for 18 hours. The system also demonstrated remarkable scalability of up to 10 cm^2 devices, for which a bias-free photocurrent of 0.23 ± 0.10 mA cm^{-2} and stability for 14 hours were claimed [396].

8.3.4.2 Photovoltaic–Photoelectrode Devices

PV–PEC technologies encompass the underlying strategies of PEC and PV–EC systems, with two primary pathways being pursued for solar OWS. Device configurations with a spatially separated PV and PEC water splitting devices have achieved to date an STH efficiency up to 16.2% [403]. Alternatively, platforms coupling both devices fully immersed in the selected electrolyte offer superior compactness and deterred design complexity with unrequired wires and external connections, showing greater STH efficiencies >18% [404, 405]. Due to its resemblance to the natural photosynthesis process, this configuration is often referred as "artificial leaf." [406]

Non-artificial PV–PEC Devices

The highest STH conversion efficiency to date (~16%) over non-artificial devices was achieved with an inverted metamorphic multi-junction (IMM) device structure with GaInP/GaInAs (1.8/1.2 eV) tandem PV absorbers and coupled RuO_x NPs (anode) (see Figure 8.7a) [403]. The PV–PEC device, reported early in 2017, was operated in a 3 M H_2SO_4 electrolyte (at pH 1) and remained stable for 1.5 hours under one sun condition. The report further emphasized the notable flexibility of IMMs, capable of integrating individually optimized absorbers for optimal band gap, improved spectral response, enhanced stability, and antireflection properties. In addition, multi-junction designs offer superior free energy for water cleavage, with a balance among device complexity, cost, and efficiency being, nonetheless, required.

Other valuable reports have been shared in recent years. In 2015, a self-biased PEC tandem cell coupling a double-junction GaAs/InGaAsP PV cell and a Co-Pi/$BiVO_4$/WO_3 core–shell heterojunction (photoanode) showed a photocurrent density of 6.56 mA cm^{-2} and an STH efficiency of 8.1% in a neutral electrolyte solution (Figure 8.7b) [408]. A hybrid PV–PEC device consisting of a micromorph Si solar cell (amorphous/nanocrystalline hydrogenated Si, a-Si:H/nc-Si:H) coupled with an optimized Co-Pi/W-doped $BiVO_4$ (photoanode) showed operating photocurrent density of ~4.0 mA cm^{-2}, corresponding to an STH efficiency of 4.9% stable within the course of one hour under AM 1.5 solar spectrum [291]. A similar device system with further optimization of the photoelectrode material would be later reported by some of the previous authors with a slightly

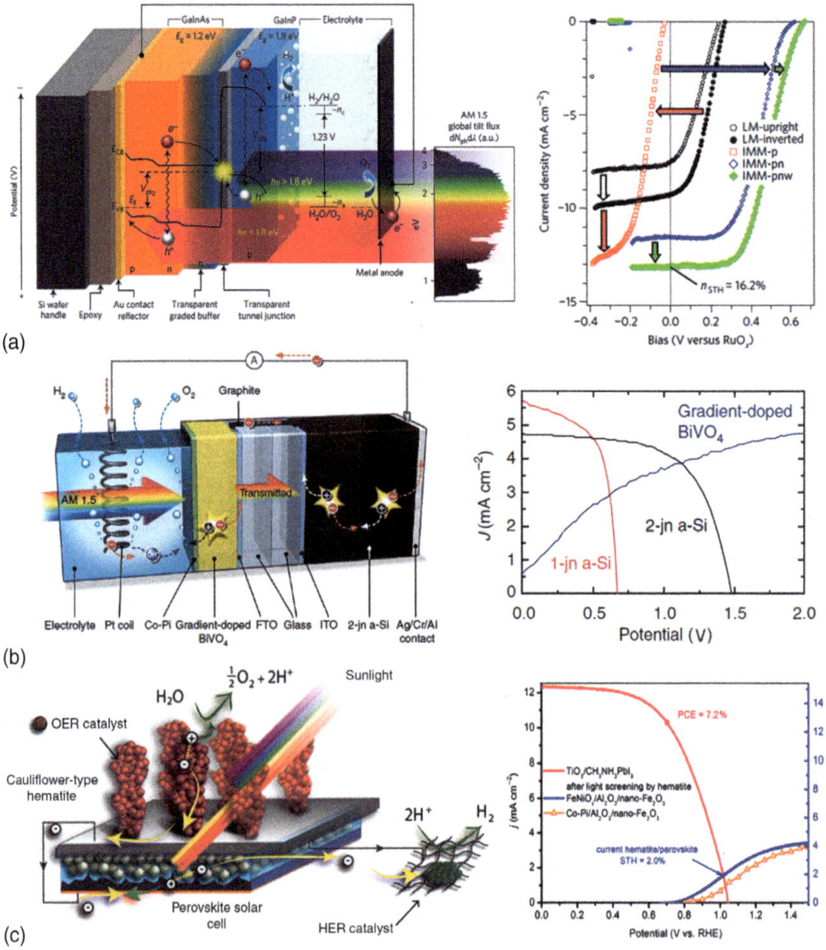

Figure 8.7 Representative PV–PEC devices for overall water splitting. (a) Inverted metamorphic multi-junction (IMM) device structure with GaInP/GaInAs tandem PV cells coupled with RuO_x–Pt and J–V measurements. Source: From Young et al. [403]. © 2017 Springer Nature. (b) PEC tandem cell coupling a double-junction GaAs/InGaAsP PV cell and a Co-Pi/$BiVO_4$/WO_3 core–shell heterojunction (photoanode) and J–V curves of double and single junction of Si solar cells. Source: From Abdi et al. [291]. © 2013 Springer Nature. (c) Hematite-based photoanode coupled with perovskite solar cell and corresponding J–V curve. Source: From Morales-Guio et al. [407]. © 2015 ACS.

higher STH efficiency of 5.2% [409]. Superior STH efficiency was reported in the presence of Pt (dark cathode)-connected Si solar cell coupled with hetero-type FeOOH/NiOOH/Mo-doped $BiVO_4$ and $NiFeO_x$/H-doped TiO_2/Ti-doped Fe_2O_3 (dual photoanodes), achieving a value of 7.7% at pH 9.2 [410]. Highly stable OWS with STH efficiencies of 7.8% was obtained in 1 M KOH for more than 7200 hours by using a triple junction amorphous Si solar cells, which were catalyzed with sputter-deposited $NiFe_yO_x$ (photoanode) and CoMo (photocathode) thin films [411].

Early in 2012, dye-sensitized solar cell coupled with WO_3 and Fe_2O_3 metal oxides (photoanodes) exhibited an STH efficiency of 3.1% (at pH 0) and 1.17% (at pH 13.6), respectively, with stable device operation for more than eight hours [412]. An alternative hematite-based photoanode incorporating an optically transparent amorphous $NiFeO_x$ cocatalyst ($NiFeO_x/Al_2O_3$/Si-doped Fe_2O_3) and coupled with a perovskite solar cell later demonstrated an STH efficiency (1.9%) and a photocurrent density of ~1.54 mA cm^{-2} at the same pH (Figure 8.7c) [407]. By using a similar single-junction perovskite solar cell, a threefold enhancement in STH efficiency was reported in the presence of highly efficient nanocone/Mo-doped $BiVO_4$/Fe(Ni)OOH (photoanode). The inspiring morphology effect (see Section 8.3.3.1) could boost the device performance with STH efficiency of 6.2% and a photocurrent density of 5.01 mA cm^{-2} under 1 sun irradiation for 10 hours at pH 7 (Figure 8.8a) [325]. The above reports further showcase the promise of halide perovskites solar cells in PV–PEC technologies. In 2015, a hybrid $CH_3NH_3PbI_3$-based device wired to an Au-modified Cu_2O (photocathode) and IrO_x (anode) uncovered the role of thin and discontinuous Au layers, capable of significantly increasing the efficiency of the ohmic contact with the conductive substrate while allowing a high degree of transparency (Figure 8.8b) [413]. Al-doped ZnO and TiO_2 overlayers by atom layer deposition enabled heterojunction formation and corrosion protection, respectively, whereas electrodeposited RuO_2 served as an excellent HER cocatalyst. Resulting STH efficiency of 2.5% and a photocurrent density of 2.0 mA cm^{-2} stable were claimed for more than two hours at pH 5. An enhancement above ~2.5-fold with, however, abridged stability was obtained by coupling a $CH_3NH_3PbBr_3$-based solar cell and a Pt-modified novel multilayered $CuIn_xGa_{1-x}Se_2$ (photocathode). Herein, an STH efficiency of 6.3% and a photocurrent density of 5.1 mA cm^{-2} were maintained for one hour at pH 0 [414].

While investigating necessary scalability pathways and corresponding electrochemical engineering enhancement strategies for both PV–EC and PV–PEC, several devices with large photoactive areas have been reported [415]. A designated CoolPEC cell (50 cm^2) with hematite as the photoelectrode and two Si-heterojunction solar cells assessed under non-concentrated sunlight (1000 W m^{-2}) showed stable activity for 1008 hours (42 days) at pH 7, ascribed to the remarkable resistance of the hematite photoelectrode to the accelerated corrosion in a continuous operation mode [416]. A large surface device (50 cm^2) using Co-Pi coated/W-doped $BiVO_4$ photoanodes and (series-connected) silicon heterojunction solar cells showed STH efficiencies of 1.9 and 2.1% for single and dual photoanodes, respectively [417].

Artificial (Wireless) PV–PEC Devices

In 1998, Turner and coworker employed a novel-integrated monolithic tandem device composed of p-type $GaInP_2$ photocathode and a GaAs p–n junction PV [292]. The system exhibited a high STH efficiency of 12.4% operated under 11 concentrated suns in 3 M H_2SO_4 electrolyte, with a severe corrosion being, however, denoted after 20 hours. Whereas this record value would be later corrected

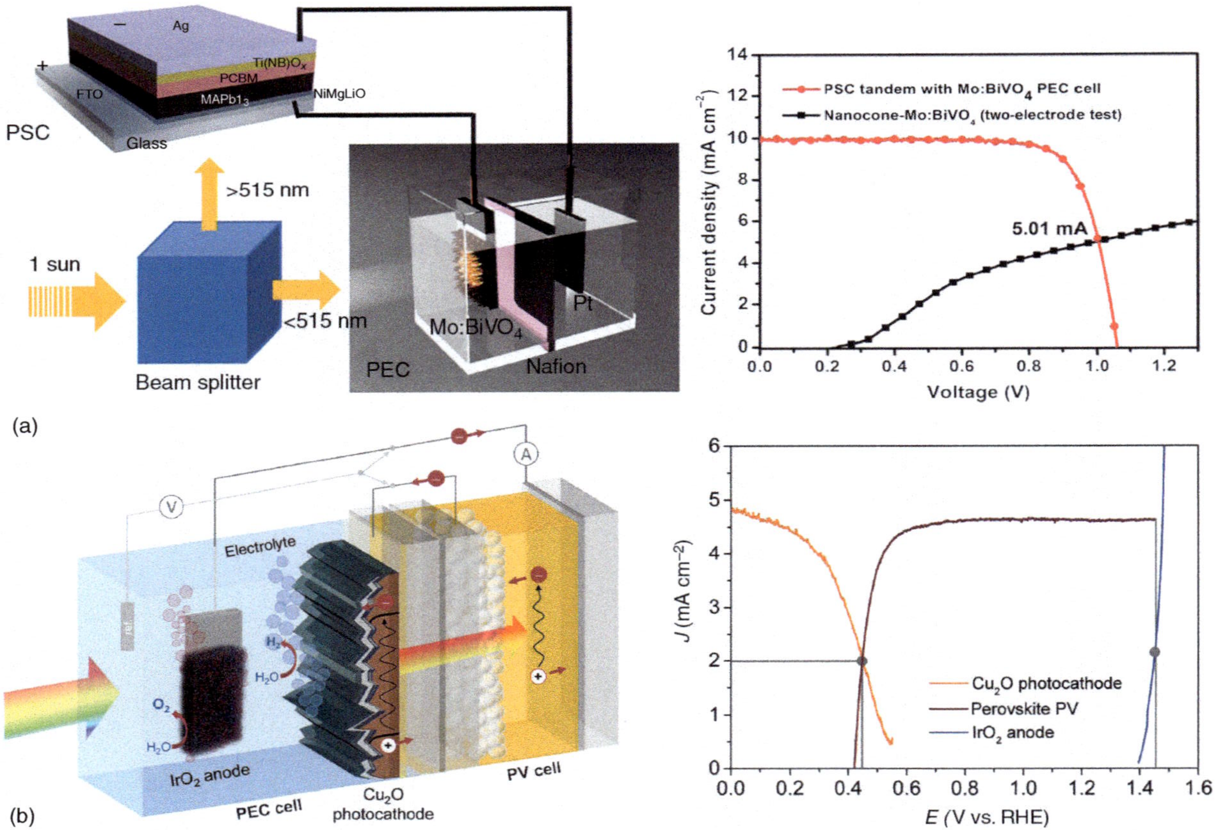

Figure 8.8 Representative PV–PEC devices for overall water splitting. (a) System composed of single-junction perovskite solar cell and a nanocone/Mo-doped $BiVO_4$/Fe(Ni)OOH photoanode, and respective J–V curve for the tandem device. Source: From Qiu et al. [325]. © 2016 Yongcai Qiu. (b) A hybrid $CH_3NH_3PbI_3$-based device wired to an Au-modified Cu_2O (photocathode) and IrO_x (anode) with respective J–V plots. Source: From Dias et al. [413]. © 2015 John Wiley & Sons.

(to ~10%) by using more advanced benchmarking protocols [26], this seminal study has since encouraged extensive efforts on buried junction-type photoelectrodes. Developed wireless devices have simplicity merit and often reflect necessary cost-effectiveness against systems with a solar cell and electrolyzer, even when multiple junctions are involved in the design of the PV–PEC platform [418–420]. In this optic, however, literature reports have remained scarce for the exploitation of single-junction solar cells integrated on wireless PV–PEC devices. An example of interest by Lee and coworkers explored the promise of coupling a tandem $CH_3NH_3PbI_3$ perovskite single-junction solar cell with an H- and Mo-co-doped $BiVO_4$ incorporating cobalt carbonate (Co-Ci) (photoanode). The authors reported a stoichiometric H_2 and O_2 evolution with STH conversion efficiency of 4.3 and 3.0% for the wired and wireless devices, respectively, at pH 7 stable for more than 12 hours (Figure 8.9a) [421].

A wireless device containing illuminated AlGaAs/Si solar cells coupled with large surface areas of RuO_2–Pt_{black} electrodes disclosed an impressive efficiency of 18.3% at pH 0 with a stable operation for more than 12 hours [404]. In another device with far superior device stability, a protective Ga-based cap layer was etched just above the n^+-doped $Al_{0.35}In_{0.65}P$ window layer in a $NH_4OH:H_2O_2:H_2O$ solution providing an appropriate surface condition for the chemical oxidation of AlInP and resulting in the emergence of a smooth mixed In/Al oxide. After that, the subsequent in situ functionalization with a HER cocatalyst (Rh) resulted in a thin Rh covering layer and a partial transformation of the AlIn oxide layer into $InPO_4$ and $In(PO_3)_3$. Thus, the synergistic effects of both In/Al oxide and phosphate/phosphite layers that assisted in reducing the charge carrier combination and provided the surface passivation lead to a high STH value of 14% for the unbiased direct solar water splitting for more than 40 hours in 1 M $HClO_4$ (Figure 8.9b) [422].

A case of interest regards a wireless device with a photoelectrode composed of GaInP and GaInAs subcells on a GaAs substrate, an anatase TiO_2 interphase layer, an Rh-based catalyst layer, and sputtered RuO_2 (anode) (see Figure 8.9c) [405]. The device exhibited the highest STH efficiencies of 19.3 and 18.5% in acidic and neutral electrolytes under AM 1.5G irradiation, respectively, where it reached a value of 0.85 of the theoretical limit for PEC water splitting (ηSTH_{theo} = 22.8%) for the energy gap combination employed in the reported structure. The TiO_2 interphase layer played a key role in the resulting efficiency of the PV–PEC device serving as a corrosion protection layer, an antireflection coating, and a conducting substrate surface for the photoelectrodeposition of Rh and facilitating the electron transport at the cathode–electrolyte interface. A key role was further ascribed to the incorporated Rh nanoparticles with ideal size distribution and spatial arrangement, offering large surface areas, high current exchange, and low light attenuation. Similar to the report above, an oxidized interfacial layer ($AlInPO_x$) resulted from the etching of GaAs cap layer by an $NH_4OH:H_2O_2:H_2O$, efficiently assisted in reducing the charge carrier recombination and increased the optical light coupling into the photoelectrode absorber layer [405].

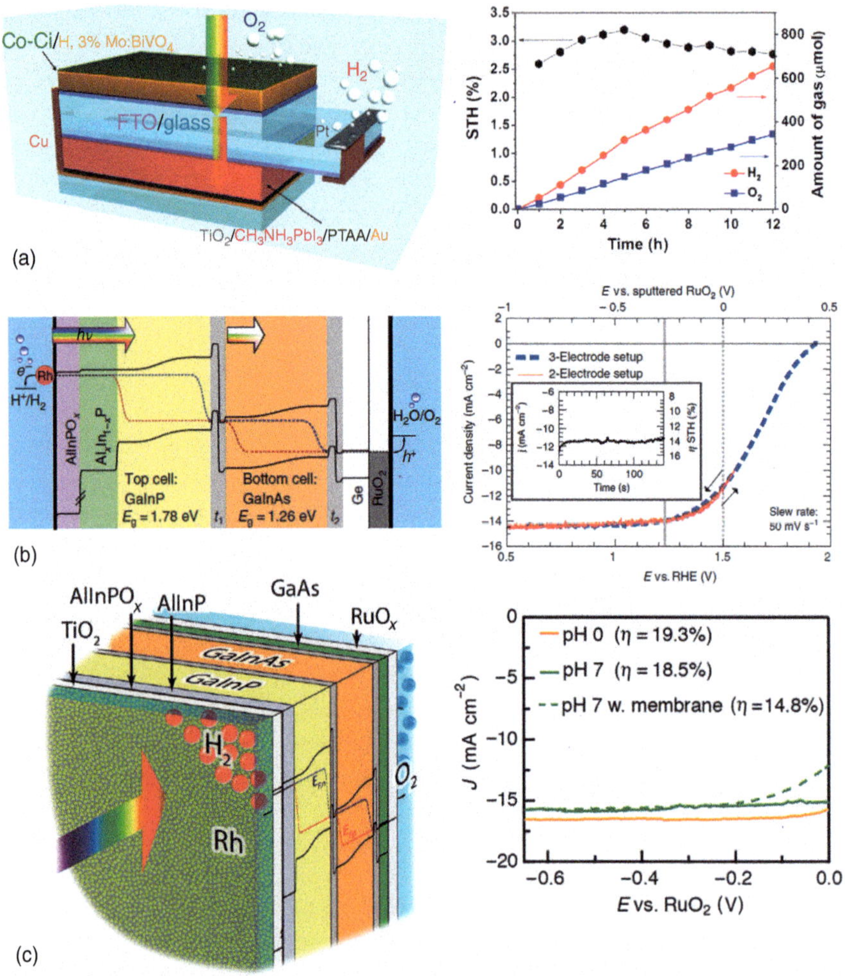

Figure 8.9 Representative artificial (wireless) PV–PEC devices for overall water splitting. (a) Tandem $CH_3NH_3PbI_3$ perovskite single-junction solar cell with Co-Ci incorporated H-doped Mo/$BiVO_4$ photoanode and corresponding gas evolution with calculated STH efficiency. Source: From Kim et al. [421]. © 2015 American Chemical Society. (b) III–V photovoltaic tandem absorber with in situ interface transformation coupled with an Rh HER and a RuO_2 OER electrocatalyst and corresponding output power characteristic plot. Source: May et al. [422]. Licensed under CC BY 4.0. (c) Wireless device coupling a photoelectrode composed of GaInP and GaInAs subcells on a GaAs substrate, an anatase TiO_2 interphase layer, an Rh-based catalyst layer, and a RuO_2-sputtered anode with corresponding STH efficiencies at different pH. Source: From Cheng et al. [405]. © 2018 ACS.

8.4 Concluding Remarks and Outlook

Among emerging processes exploiting a solar-to-chemical conversion, PC and PEC OWS platforms have been pinpointed to serve as auspicious routes toward a sustainable future. Derived H_2 production emerges today as the most promising non-carbon-emitting energy source. Research-wise, OWS has been actively surveyed since Honda and Fujishima first reported their pioneering contribution using a TiO_2 photocatalyst. Exploratory research works of marked magnitude highlighting important mechanistic difficulties, benchmark catalysts, and stability issues have been meticulously reviewed in this chapter for the OWS for both PC and PEC platforms. In the development of photo-driven systems, the extension of absorption wavelengths onto UV–vis–NIR ranges has remained one of the key focuses of the scientific community. Coupled synthesis strategies, e.g. heterojunctions between benchmark semiconductors (e.g. $SrTiO_3$, TaON, CdS) and the integration of OER and HER cocatalysts, have reflected an increasing complexity of the fabricated hybrid materials. Developed systems would, however, benefit from increasing efforts to disentangle mechanistic difficulties, a focus on *operando* characterization techniques, and theoretical efforts to unveil performance descriptors rationalizing the design of novel semiconductor-based architectures. Unifying guidelines for better quantification of H_2 evolution rates and STH efficiencies under standard operating conditions is additionally desirable. Taken together, notwithstanding the steady progress in the development of light-responsive systems for both PC and PEC, to date, the resulting efficiencies remain insufficient to commercially compete with the conventional steam reforming process. Particularly, whereas from an economic point of view, solar-driven OWS PC technologies are ideal, offering a superior potential in scalability against photo(electro)catalytic counterparts; remarkably superior efficiencies are, nonetheless, witnessed with both EC and PEC systems. In addition, PC OWS platforms without using sacrificial agents are still scarce to date, significantly hindering the promise of such approaches. Conversely, a remarkable cost decrease associated with energy storage in both EC and PEC systems has received increasing attention through the coupling of solar cells. For both PC and PEC, efforts are additionally expected for an optimization of the design of cost-effective reactors and corresponding operating conditions.

Acknowledgments

This work was supported by National Research Foundation of Korea Grant funded by the Korean Government (NRF-2020R 1A 2C3003958 and 2017M3 A9D8029942), by Korea Basic Science Institute (National Research Facilities and Equipment Center) grant funded by the Ministry of Education (2020R 1A 6C 101B194) and by Creative Materials Discovery Program through the National Research Foundation of Korea (NRF) funded by Ministry of Science and ICT (2018M 3D 1A 1058536).

References

1 Dincer, I. (2012). *Int. J. Hydrogen Energy* 37: 1954.
2 Nikolaidis, P. and Poullikkas, A. (2017). *Renewable Sustainable Energy Rev.* 67: 597.
3 Levin, D.B. and Chahine, R. (2010). *Int. J. Hydrogen Energy* 35: 4962.
4 Saeidi, S., Fazlollahi, F., Najari, S. et al. (2017). *J. Ind. Eng. Chem.* 49: 1.

5 Chu, S., Cui, Y., and Liu, N. (2016). *Nat. Mater.* 16: 16.
6 BP (2019). BP Statistical Review of World Energy 2019.
7 IPCC (2014). *Climate Change 2014: Synthesis Report. Contribution of Working Groups I, II and III to the Fifth Assessment Report of the Intergovernmental Panel on Climate Change*. Geneva, Switzerland.
8 Barber, J. (2009). *Chem. Soc. Rev.* 38: 185.
9 Fujishima, A. and Honda, K. (1972). *Nature* 238: 37.
10 Bard, A.J. (1979). *J. Photochem.* 10: 59.
11 Bard, A.J. (1980). *Science* 207: 139.
12 Pinaud, B.A., Benck, J.D., Seitz, L.C. et al. (2013). *Energy Environ. Sci.* 6: 1983.
13 Fabian, D.M., Hu, S., Singh, N. et al. (2015). *Energy Environ. Sci.* 8: 2825.
14 Chen, S., Takata, T., and Domen, K. (2017). *Nat. Rev. Mater.* 2: 17050.
15 Wang, Z., Li, C., and Domen, K. (2019). *Chem. Soc. Rev.* 48: 2109.
16 Xia, X., Song, M., Wang, H. et al. (2019). *Nanoscale* 11: 11071.
17 Wang, Y., Suzuki, H., Xie, J. et al. (2018). *Chem. Rev.* 118: 5201.
18 Kudo, A. and Miseki, Y. (2009). *Chem. Soc. Rev.* 38: 253.
19 Kim, J.H., Hansora, D., Sharma, P. et al. (2019). *Chem. Soc. Rev.* 48: 1908.
20 Yang, J., Wang, D., Han, H., and Li, C. (2013). *Acc. Chem. Res.* 46: 1900.
21 Wolff, C.M., Frischmann, P.D., Schulze, M. et al. (2018). *Nat. Energy* 3: 862.
22 Li, Y., Peng, Y.K., Hu, L. et al. (2019). *Nat. Commun.* 10: 4421.
23 Walter, M.G., Warren, E.L., McKone, J.R. et al. (2010). *Chem. Rev.* 110: 6446.
24 Döscher, H., Geisz, J.F., Deutsch, T.G., and Turner, J.A. (2014). *Energy Environ. Sci.* 7: 2951.
25 Chen, Z., Jaramillo, T.F., Deutsch, T.G. et al. (2010). *J. Mater. Res.* 25: 3.
26 Döscher, H., Young, J.L., Geisz, J.F. et al. (2016). *Energy Environ. Sci.* 9: 74.
27 Bak, T., Nowotny, J., Rekas, M., and Sorrell, C.C. (2002). *Int. J. Hydrogen Energy* 27: 991.
28 Varghese, O.K. and Grimes, C.A. (2008). *Sol. Energy Mater. Sol. Cells* 92: 374.
29 Dotan, H., Mathews, N., Hisatomi, T. et al. (2014). *J. Phys. Chem. Lett.* 5: 3330.
30 Wang, Q., Hisatomi, T., Jia, Q. et al. (2016). *Nat. Mater.* 15: 611.
31 Kibria, M.G., Chowdhury, F.A., Zhao, S. et al. (2015). *Nat. Commun.* 6: 6797.
32 Wang, Q. and Domen, K. (2019). *Chem. Rev.* https://doi.org/10.1021/acs.chemrev.9b00201.
33 Marques Mota, F. and Kim, D.H. (2019). *Chem. Soc. Rev.* 48: 205.
34 Guo, Q., Ma, Z., Zhou, C. et al. (2019). *Chem. Rev.* 119: 11020.
35 Zheng, L., Teng, F., Ye, X. et al. (2019). *Adv. Energy Mater.*: 1902355. https://doi.org/10.1002/aenm.201902355.
36 Ma, Y., Wang, X., Jia, Y. et al. (2014). *Chem. Rev.* 114: 9987.
37 Bai, J. and Zhou, B. (2014). *Chem. Rev.* 114: 10131.
38 Dahl, M., Liu, Y., and Yin, Y. (2014). *Chem. Rev.* 114: 9853.
39 Qin, W., Zhang, D., Zhao, D. et al. (2010). *Chem. Commun.* 46: 2304.
40 Asahi, R., Morikawa, T., Ohwaki, T. et al. (2001). *Science* 293: 269.
41 Quan, L.N., Jang, Y.H., Stoerzinger, K.A. et al. (2014). *Phys. Chem. Chem. Phys.* 16: 9023.
42 Hou, L., Guan, Z., Liu, T. et al. (2019). *Int. J. Hydrogen Energy* 44: 8109.

43 Boppella, R., Lee, J.E., Marques Mota, F. et al. (2017). *J. Mater. Chem. A* 5: 7072.

44 Liu, X., Gao, S., Xu, H. et al. (2013). *Nanoscale* 5: 1870.

45 Sato, S. and White, J.M. (1981). *J. Catal.* 69: 128.

46 Guayaquil-Sosa, J.F., Calzada, A., Serrano, B. et al. (2017). *Catalysts* 7: 324.

47 Chen, X., Liu, L., Yu, P.Y., and Mao, S.S. (2011). *Science* 331: 746.

48 Kwon, H., Marques Mota, F., Chung, K. et al. (2018). *ACS Sustainable Chem. Eng.* 6: 1310.

49 Quan, L.N., Jang, Y.H., Jang, Y.J. et al. (2014). *ChemSusChem* 7: 2590.

50 Kato, H. and Kudo, A. (2002). *J. Phys. Chem. B* 106: 5029.

51 Asai, R., Nemoto, H., Jia, Q. et al. (2014). *Chem. Commun.* 50: 2543.

52 Xu, M., Gao, Y., Moreno, E.M. et al. (2011). *Phys. Rev. Lett.* 106: 138302.

53 Luttrell, T., Halpegamage, S., Tao, J. et al. (2015). *Sci. Rep.* 4: 4043.

54 Ozawa, K., Emori, M., Yamamoto, S. et al. (2014). *J. Phys. Chem. Lett.* 5: 1953.

55 Schrauzer, G.N. and Guth, T.D. (1977). *J. Am. Chem. Soc.* 99: 7189.

56 Kudo, A. (2007). *Int. J. Hydrogen Energy* 32: 2673.

57 Maeda, K. and Domen, K. (2007). *J. Phys. Chem. C* 111: 7851.

58 Wang, X., Xu, Q., Li, M. et al. (2012). *Angew. Chem. Int. Ed.* 51: 13089.

59 Sakata, Y., Matsuda, Y., Nakagawa, T. et al. (2011). *ChemSusChem* 4: 181.

60 Sakata, Y., Hayashi, T., Yasunaga, R. et al. (2015). *Chem. Commun.* 51: 12935.

61 Park, Y., McDonald, K.J., and Choi, K.S. (2013). *Chem. Soc. Rev.* 42: 2321.

62 Xie, B., Zhang, H., Cai, P. et al. (2006). *Chemosphere* 63: 956.

63 Jo, W.J., Kang, H.J., Kong, K.J. et al. (2015). *Proc. Natl. Acad. Sci. U.S.A.* 112: 13774.

64 Wang, Q., Liu, H., Jiang, L. et al. (2009). *Catal. Lett.* 131: 16.

65 Liao, L., Zhang, Q., Su, Z. et al. (2014). *Nat. Nanotechnol.* 9: 69.

66 Domen, K., Naito, S., Soma, M. et al. (1980). *J. Chem. Soc., Chem. Commun.*: 543.

67 Domen, K., Kudo, A., Onishi, T. et al. (1986). *J. Phys. Chem.* 90: 292.

68 Mu, L., Zhao, Y., Li, A. et al. (2016). *Energy Environ. Sci.* 9: 2463.

69 Wang, Y., Wu, Y., Sun, K., and Mi, Z. (2019). *Mater. Horiz.* 6: 1454.

70 Pan, C., Takata, T., Nakabayashi, M. et al. (2015). *Angew. Chem. Int. Ed.* 54: 2955.

71 Zhu, C., Zhu, M., Sun, Y. et al. (2018). *Appl. Catal., B* 237: 166.

72 Konta, R., Ishii, T., Kato, H., and Kudo, A. (2004). *J. Phys. Chem. B* 108: 8992.

73 Furuhashi, K., Jia, Q., Kudo, A., and Onishi, H. (2013). *J. Phys. Chem. C* 117: 19101.

74 Wang, Q., Hisatomi, T., Ma, S.S.K. et al. (2014). *Chem. Mater.* 26: 4144.

75 Takata, T. and Domen, K. (2009). *J. Phys. Chem. C* 113: 19386.

76 Ham, Y., Hisatomi, T., Goto, Y. et al. (2016). *J. Mater. Chem. A* 4: 3027.

77 Goto, Y., Hisatomi, T., Wang, Q. et al. (2018). *Joule* 2: 509.

78 Chiang, T.H., Lyu, H., Hisatomi, T. et al. (2018). *ACS Catal.* 8: 2782.

79 Zhang, Q., Li, Z., Wang, S. et al. (2016). *ACS Catal.* 6: 2182.

80 Townsend, T.K., Browning, N.D., and Osterloh, F.E. (2012). *Energy Environ. Sci.* 5: 9543.

81 Kato, H., Asakura, K., and Kudo, A. (2003). *J. Am. Chem. Soc.* 125: 3082.
82 Sun, X., Mi, Y., Jiao, F., and Xu, X. (2018). *ACS Catal.* 8: 3209.
83 Miseki, Y., Kato, H., and Kudo, A. (2009). *Energy Environ. Sci.* 2: 306.
84 Chen, W., Chen, X., Yang, Y. et al. (2014). *Int. J. Hydrogen Energy* 39: 13468.
85 Sayama, O.T., Yase, K., Arakawa, K. et al. (1998). *J. Photochem.* 114: 125.
86 Oshima, T., Lu, D., Ishitani, O., and Maeda, K. (2015). *Angew. Chem. Int. Ed.* 54: 2698.
87 Shimizu, K.I., Itoh, S., Hatamachi, T. et al. (2005). *Chem. Mater.* 17: 5161.
88 Ida, S., Okamoto, Y., Matsuka, M. et al. (2012). *J. Am. Chem. Soc.* 134: 15773.
89 Tagusagawa, C., Takagaki, A., Hayashi, S., and Domen, K. (2008). *J. Am. Chem. Soc.* 130: 7230.
90 Kudo, A., Tanaka, A., Domen, K. et al. (1988). *J. Catal.* 111: 67.
91 Kudo, A., Sayama, K., Tanaka, A. et al. (1989). *J. Catal.* 120: 337.
92 Inoue, Y. (2009). *Energy Environ. Sci.* 2: 364.
93 Guan, X., Chowdhury, F.A., Pant, N. et al. (2018). *J. Phys. Chem. C* 122: 13797.
94 Kibria, M.G., Zhao, S., Chowdhury, F.A. et al. (2014). *Nat. Commun.* 5: 3825.
95 Wang, D., Pierre, A., Kibria, M.G. et al. (2011). *Nano Lett.* 11: 2353.
96 Moses, P.G. and Van De Walle, C.G. (2010). *Appl. Phys. Lett.* 96: 2010.
97 Wu, J. (2009). *J. Appl. Phys.* 106: 011101.
98 Kibria, M.G., Nguyen, H.P.T., Cui, K. et al. (2013). *ACS Nano* 7: 7886.
99 Khanal, D.R., Walukiewicz, W., Grandal, J. et al. (2009). *Appl. Phys. Lett.* 95: 17.
100 Van De Walle, C.G. and Segev, D. (2007). *J. Appl. Phys.* 101: 081704.
101 Zhang, Z. and Yates, J.T. (2012). *Chem. Rev.* 112: 5520.
102 Zhang, S., Connie, A.T., Laleyan, D.A. et al. (2014). *IEEE J. Quantum Electron.* 50: 483.
103 Nozik, A.J. (1977). *Appl. Phys. Lett.* 30: 567.
104 Chowdhury, F.A., Trudeau, M.L., Guo, H., and Mi, Z. (2018). *Nat. Commun.* 9: 1707.
105 Wang, J., Fang, T., Zhang, L. et al. (2014). *J. Catal.* 309: 291.
106 Wang, Z., Inoue, Y., Hisatomi, T. et al. (2018). *Nat. Catal.* 1: 756.
107 Ma, S.S.K., Hisatomi, T., Maeda, K. et al. (2012). *J. Am. Chem. Soc.* 134: 19993.
108 Thanh Truc, N.T., Thi Hanh, N., Nguyen, D.T. et al. (2019). *J. Solid State Chem.* 269: 361.
109 Sato, J., Saito, N., Yamada, Y. et al. (2005). *J. Am. Chem. Soc.* 127: 4150.
110 Wang, Y.L., Nie, T., Li, Y.H. et al. (2017). *Angew. Chem. Int. Ed.* 56: 7430.
111 Chun, W.J., Ishikawa, A., Fujisawa, H. et al. (2003). *J. Phys. Chem. B* 107: 1798.
112 Maeda, K. and Domen, K. (2011). *MRS Bull.* 36: 25.
113 Maeda, K., Lu, D., and Domen, K. (2013). *Chem. Eur. J.* 19: 4986.
114 Zou, Z., Ye, J., and Arakawa, H. (2002). *J. Mater. Res.* 17: 1419.
115 Li, Y., Jiang, S., Xiao, J., and Li, Y. (2014). *Int. J. Hydrogen Energy* 39: 731.
116 Maeda, K., Takata, T., Hara, M. et al. (2005). *J. Am. Chem. Soc.* 127: 8286.
117 Maeda, K., Teramura, K., and Domen, K. (2008). *J. Catal.* 254: 198.
118 Adeli, B. and Taghipour, F. (2016). *Appl. Catal., A* 521: 250.
119 Hou, X., Jiang, S., Li, Y. et al. (2015). *Int. J. Hydrogen Energy* 40: 15448.

120 Maeda, K., Teramura, K., Masuda, H. et al. (2006). *J. Phys. Chem. B* 110: 13107.
121 Li, Y., Zhu, L., Yang, Y. et al. (2015). *Small* 11: 871.
122 Lee, Y., Terashima, H., Shimodaira, Y. et al. (2007). *J. Phys. Chem. C* 111: 1042.
123 Lee, Y., Teramura, K., Hara, M., and Domen, K. (2007). *Chem. Mater.* 19: 2120.
124 Nie, L. and Zhang, Q. (2017). *Inorg. Chem. Front.* 4: 1953.
125 Cheng, L., Xiang, Q., Liao, Y., and Zhang, H. (2018). *Energy Environ. Sci.* 11: 1362.
126 Ishikawa, A., Takata, T., Kondo, J.N. et al. (2002). *J. Am. Chem. Soc.* 124: 13547.
127 Ishikawa, A., Takata, T., Matsumura, T. et al. (2004). *J. Phys. Chem. B* 108: 2637.
128 Ogisu, K., Ishikawa, A., Shimodaira, Y. et al. (2008). *J. Phys. Chem. C* 112: 11978.
129 Goto, Y., Seo, J., Kumamoto, K. et al. (2016). *Inorg. Chem.* 55: 3674.
130 Nandy, S., Hisatomi, T., Sun, S. et al. (2019). *ACS Appl. Mater. Interfaces* 11: 5595.
131 Wang, Q., Nakabayashi, M., Hisatomi, T. et al. (2019). *Nat. Mater.* 18: 827.
132 Schneider, J., Matsuoka, M., Takeuchi, M. et al. (2014). *Chem. Rev.* 114: 9919.
133 Chen, H., Jiang, D., Sun, Z. et al. (2017). *Catal. Sci. Technol.* 7: 1515.
134 Hisatomi, T., Kubota, J., and Domen, K. (2014). *Chem. Soc. Rev.* 43: 7520.
135 Wang, Y., Wang, X., and Antonietti, M. (2012). *Angew. Chem. Int. Ed.* 51: 68.
136 Li, J., Li, H., Zhan, G., and Zhang, L. (2017). *Acc. Chem. Res.* 50: 112.
137 Li, J., Cai, L., Shang, J. et al. (2016). *Adv. Mater.* 28: 4059.
138 Li, J., Zhan, G., Yu, Y., and Zhang, L. (2016). *Nat. Commun.* 7: 11480.
139 Albero, J., Mateo, D., and García, H. (2019). *Molecules* 24: 906.
140 Fadlalla, M.I. and Babu, S.G. (2019). *Graphene-Based Nanotechnologies Energy Environ* (eds. M. Jawaid, A. Ahmad and D. Lokhat), Role of graphene in photocatalytic water splitting for hydrogen production, 81–108. Elsevier Inc.
141 Yeh, T.F., Cihlář, J., Chang, C.Y. et al. (2013). *Mater. Today* 16: 78.
142 Boppella, R., Choi, C.H., Moon, J., and Kim, D.H. (2018). *Appl. Catal., B* 239: 178.
143 Zhang, Y., Liu, Y., Gao, W. et al. (2019). *Front. Chem.* 7: 325.
144 Wang, L., Wan, Y., Ding, Y. et al. (2017). *Adv. Mater.* 29: 1702428.
145 Yoon, M., Oh, Y., Hong, S. et al. (2017). *Appl. Catal., B* 206: 263.
146 Lin, L., Wang, C., Ren, W. et al. (2017). *Chem. Sci.* 8: 5506.
147 Liu, J., Liu, Y., Liu, N. et al. (2015). *Science* 347: 970.
148 Ong, W.J., Tan, L.L., Ng, Y.H. et al. (2016). *Chem. Rev.* 116: 7159.
149 Wang, X., Maeda, K., Thomas, A. et al. (2009). *Nat. Mater.* 8: 76.
150 Martin, D.J., Reardon, P.J.T., Moniz, S.J.A., and Tang, J. (2014). *J. Am. Chem. Soc.* 136: 12568.
151 Zhou, M., Yang, P., Yuan, R. et al. (2017). *ChemSusChem* 10: 4451.
152 Chai, B., Peng, T., Mao, J. et al. (2012). *Phys. Chem. Chem. Phys.* 14: 16745.
153 Hong, J., Wang, Y., Wang, Y. et al. (2013). *ChemSusChem* 6: 2263.
154 Liu, J., Zhang, Y., Lu, L. et al. (2012). *Chem. Commun.* 48: 8826.
155 Xiang, Q., Yu, J., and Jaroniec, M. (2011). *J. Phys. Chem. C* 115: 7355.

156 Rahman, M.Z., Zhang, J., Tang, Y. et al. (2017). *Mater. Chem. Front.* 1: 562.

157 Suryawanshi, A., Dhanasekaran, P., Mhamane, D. et al. (2012). *Int. J. Hydrogen Energy* 37: 9584.

158 Zhang, G., Lan, Z.A., Lin, L. et al. (2016). *Chem. Sci.* 7: 3062.

159 Fu, Y., Liu, C., Zhu, C. et al. (2018). *Inorg. Chem. Front.* 5: 1646.

160 Garcia-Esparza, A.T., Cha, D., Ou, Y. et al. (2013). *ChemSusChem* 6: 168.

161 Zhang, G., Lan, Z.A., and Wang, X. (2017). *Chem. Sci.* 8: 5261.

162 Che, W., Cheng, W., Yao, T. et al. (2017). *J. Am. Chem. Soc.* 139: 3021.

163 Kumari, R. and Kumar Sahu, S. (2018). *ChemistrySelect* 3: 12998.

164 Xia, C., Zhu, S., Feng, T. et al. (2019). *Adv. Sci.*: 1901316.

165 Zhang, Z., Zheng, T., Li, X. et al. (2016). *Part. Part. Syst. Char.* 33: 457.

166 Shi, W., Guo, F., Zhu, C. et al. (2017). *J. Mater. Chem. A* 5: 19800.

167 Liu, J., Zhang, H., Tang, D. et al. (2014). *ChemCatChem* 6: 2634.

168 Zhang, K., Wang, L., Sheng, X. et al. (2016). *Adv. Energy Mater.* 6: 1502352.

169 Yeh, T.F., Teng, C.Y., Chen, S.J., and Teng, H. (2014). *Adv. Mater.* 26: 3297.

170 Li, M., Duanmu, K., Wan, C. et al. (2019). *Nat. Catal.* 2: 495.

171 Yao, Y., Hu, S., Chen, W. et al. (2019). *Nat. Catal.* 2: 304.

172 Cao, L., Luo, Q., Liu, W. et al. (2019). *Nat. Catal.* 2: 134.

173 Cao, Y., Chen, S., Luo, Q. et al. (2017). *Angew. Chem. Int. Ed.* 56: 12191.

174 Li, P., Wang, M., Duan, X. et al. (2019). *Nat. Commun.* 10: 1711.

175 Wang, Y., Mao, J., Meng, X. et al. (2018). *Chem. Rev.* 119: 1806.

176 Liu, W., Cao, L., Cheng, W. et al. (2017). *Angew. Chem. Int. Ed.* 56: 9312.

177 Shi, Y. and Zhang, B. (2016). *Chem. Soc. Rev.* 45: 1529.

178 Pei, Y., Cheng, Y., Chen, J. et al. (2018). *J. Mater. Chem. A* 6: 23220.

179 Zhu, J., Hu, L., Zhao, P. et al. (2019). *Chem. Rev.* https://doi.org/10.1021/acs.chemrev.9b00248.

180 Jaramillo, T.F., Jørgensen, K.P., Bonde, J. et al. (2007). *Science* 317: 100.

181 Voiry, D., Yamaguchi, H., Li, J. et al. (2013). *Nat. Mater.* 12: 850.

182 Li, H., Tsai, C., Koh, A.L. et al. (2016). *Nat. Mater.* 15: 364.

183 Karunadasa, H.I., Montalvo, E., Sun, Y. et al. (2012). *Science* 335: 698.

184 Kim, J.S., Kim, B., Kim, H., and Kang, K. (2018). *Adv. Energy Mater.* 8: 1702774.

185 Schubert, J.S., Popovic, J., Haselmann, G.M. et al. (2019). *J. Mater. Chem. A* 7: 18568.

186 Smith, R.D.L., Prévot, M.S., Fagan, R.D. et al. (2013). *Science* 340: 60.

187 Yoshinaga, T., Saruyama, M., Xiong, A. et al. (2018). *Nanoscale* 10: 10420.

188 Zhang, B., Zheng, X., Voznyy, O. et al. (2016). *Science* 352: 333.

189 Suntivich, J., May, K.J., Gasteiger, H.A. et al. (2011). *Science* 334: 2010.

190 Stoerzinger, K.A., Rao, R.R., Wang, X.R. et al. (2017). *Chem* 2: 668.

191 Hwang, J., Rao, R.R., Giordano, L. et al. (2017). *Science* 358: 751.

192 Yanagida, T., Sakata, Y., and Imamura, H. (2004). *Chem. Lett.* 33: 726.

193 Zheng, D., Cao, X.N., and Wang, X. (2016). *Angew. Chem. Int. Ed.* 55: 11512.

194 Yamada, Y., Tsung, C.K., Huang, W. et al. (2011). *Nat. Chem.* 3: 372.

195 Su, J., Xie, C., Chen, C. et al. (2016). *J. Am. Chem. Soc.* 138: 11568.

196 Marques Mota, F., Choi, C.H., Boppella, R. et al. (2019). *J. Mater. Chem. A* 7: 639.

197 Gellé, A., Jin, T., de la Garza, L. et al. (2019). *Chem. Rev.* https://doi.org/10.1021/acs.chemrev.9b00187.

198 Choi, C.H., Chung, K., Nguyen, T.T.H., and Kim, D.H. (2018). *ACS Energy Lett.* 3: 1415.

199 Zhu, M., Cai, X., Fujitsuka, M. et al. (2017). *Angew. Chem. Int. Ed.* 56: 2064.

200 Mayer, K.M. and Hafner, J.H. (2011). *Chem. Rev.* 111: 3828.

201 Jang, Y.H., Chung, K., Quan, L.N. et al. (2013). *Nanoscale* 5: 12261.

202 Kochuveedu, S.T., Jang, Y.H., and Kim, D.H. (2013). *Chem. Soc. Rev.* 42: 8467.

203 Kochuveedu, S.T. and Kim, D.H. (2014). *Nanoscale* 6: 4966.

204 Wang, H. and Kim, D.H. (2017). *Chem. Soc. Rev.* 46: 5204.

205 Liu, Y., Cheng, R., Liao, L. et al. (2011). *Nat. Commun.* 2: 579.

206 Fang, Y. and Sun, M. (2015). *Light Sci. Appl.* 4: e294.

207 Mansuripur, M., Zakharian, A.R., Lesuffleur, A. et al. (2009). *Opt. Express* 17: 1777.

208 Lu, Y., Yu, H., Chen, S. et al. (2012). *Environ. Sci. Technol.* 46: 1724.

209 Hou, W., Liu, Z., Pavaskar, P. et al. (2011). *J. Catal.* 277: 149.

210 Silva, C.G., Juarez, R., Marino, T. et al. (2011). *J. Am. Chem. Soc.*: 595.

211 Liu, Y., Shu, W., Peng, Z. et al. (2013). *Catal. Today* 208: 28.

212 Ma, W., Xu, L., De Moura, A.F. et al. (2017). *Chem. Rev.* 117: 8041.

213 Hentschel, M., Schäferling, M., Duan, X. et al. (2017). *Sci. Adv.* 3: e1602735.

214 Tian, B., Lei, Q., Zhang, W. et al. (2018). *Chem. Commun.* 54: 1845.

215 Linic, S., Christopher, P., and Ingram, D.B. (2011). *Nat. Mater.* 10: 911.

216 Schuller, J.A., Barnard, E.S., Cai, W. et al. (2010). *Nat. Mater.* 9: 193.

217 Tian, Y. and Tatsuma, T. (2005). *J. Am. Chem. Soc.* 127: 7632.

218 Furube, A., Du, L., Hara, K. et al. (2007). *J. Am. Chem. Soc.* 129: 14852.

219 Mateo, D., Esteve-Adell, I., Albero, J. et al. (2016). *Nat. Commun.* 7: 11819.

220 Aslam, U., Rao, V.G., Chavez, S., and Linic, S. (2018). *Nat. Catal.* 1: 656.

221 Jang, Y.J., Chung, K., Lee, J.S. et al. (2018). *ACS Photonics* 5: 4711.

222 Méndez-Ramos, J., Acosta-Mora, P., Ruiz-Morales, J.C. et al. (2013). *RSC Adv.* 3: 23028.

223 Thi Thuy, T.N., Atabaev, T.S., Thi Vu, H.H. et al. (2017). *J. Nanosci. Nanotechnol.* 17: 7647.

224 Boppella, R., Marques Mota, F., Lim, J.W. et al. (2019). *ACS Appl. Energy Mater.* 2: 3780.

225 Lee, J.E., Marques Mota, F., Choi, C.H. et al. (2019). *Adv. Mater. Interfaces* 6: 1801144.

226 Liu, L., Zhao, H., Andino, J.M., and Li, Y. (2012). *ACS Catal.* 2: 1817.

227 Anpo, M., Yamashita, H., Ichihashi, Y., and Ehara, S. (1995). *J. Electroanal. Chem.* 396: 21.

228 Li, H., Tu, W., Zhou, Y., and Zou, Z. (2016). *Adv. Sci.* 3: 1500389.

229 Abe, R., Sayama, K., and Domen, K. (2001). *H. Arakawa* 344: 339.

230 Abe, R. (2010). *J. Photochem. Photobiol., C* 11: 179.

231 Maeda, K. (2013). *ACS Catal.* 3: 1486.

232 Wang, W., Chen, J., Li, C., and Tian, W. (2014). *Nat. Commun.* 5: 4647.

233 Iwase, Y., Tomita, O., Higashi, M. et al. (2019). *Sustain. Energy Fuels* 3: 1501.

234 Abe, R., Sayama, K., and Sugihara, H. (2005). *J. Phys. Chem. B* 109: 16052.

235 Maeda, K., Higashi, M., Lu, D. et al. (2010). *J. Am. Chem. Soc.* 132: 5858.

236 Chen, S., Qi, Y., Hisatomi, T. et al. (2015). *Angew. Chem. Int. Ed.* 54: 8498.

237 Qi, Y., Zhao, Y., Gao, Y. et al. (2018). *Joule* 2: 2393.

238 Hezam, A., Namratha, K., Drmosh, Q.A. et al. (2018). *J. Mater. Chem. A* 6: 21379.

239 Kato, T., Hakari, Y., Ikeda, S. et al. (2015). *J. Phys. Chem. Lett.* 6: 1042.

240 Sasaki, Y., Kato, H., and Kudo, A. (2013). *J. Am. Chem. Soc.* 135: 5441.

241 Zhou, P., Yu, J., and Jaroniec, M. (2014). *Adv. Mater.* 26: 4920.

242 Low, J., Jiang, C., Cheng, B. et al. (2017). *Small Methods* 1: 1700080.

243 Xu, Q., Zhang, L., Yu, J. et al. (2018). *Mater. Today* 21: 1042.

244 Li, H., Yu, H., Quan, X. et al. (2016). *ACS Appl. Mater. Interfaces* 8: 2111.

245 Li, X., Yan, X., Lu, X. et al. (2018). *J. Catal.* 357: 59.

246 Byun, K.E., Chung, H.J., Lee, J. et al. (2013). *Nano Lett.* 13: 4001.

247 Kobayashi, R., Tanigawa, S., Takashima, T. et al. (2014). *J. Phys. Chem. C* 118: 22450.

248 Hara, Y., Takashima, T., Kobayashi, R. et al. (2017). *Appl. Catal., B* 209: 663.

249 Kobayashi, R., Kurihara, K., Takashima, T. et al. (2016). *J. Mater. Chem. A* 4: 3061.

250 Kobayashi, R., Takashima, T., Tanigawa, S. et al. (2016). *Phys. Chem. Chem. Phys.* 18: 27754.

251 Kamijyo, K., Takashima, T., Yoda, M. et al. (2018). *Chem. Commun.* 54: 7999.

252 Chen, S., Ma, G., Wang, Q. et al. (2019). *J. Mater. Chem. A* 7: 7415.

253 Isimjan, T.T., Maity, P., Llorca, J. et al. (2017). *ACS Omega* 2: 4828.

254 Wu, X., Zhao, J., Wang, L. et al. (2017). *Appl. Catal., B* 206: 501.

255 Iwase, A., Ng, Y.H., Ishiguro, Y. et al. (2011). *J. Am. Chem. Soc.* 133: 11054.

256 Wang, Q., Hisatomi, T., Suzuki, Y. et al. (2017). *J. Am. Chem. Soc.* 139: 1675.

257 Wang, L., Zheng, X., Chen, L. et al. (2018). *Angew. Chem. Int. Ed.* 57: 3454.

258 Wang, X., Liu, G., Chen, Z.-G. et al. (2009). *Chem. Commun.*: 3452.

259 Yu, J., Wang, S., Low, J., and Xiao, W. (2013). *Phys. Chem. Chem. Phys.* 15: 16883.

260 Sasaki, Y., Nemoto, H., Saito, K., and Kudo, A. (2009). *J. Phys. Chem. C* 113: 17536.

261 Jia, Q., Iwase, A., and Kudo, A. (2014). *Chem. Sci.* 5: 1513.

262 Yang, Y., Qiu, M., Li, L. et al. (2018). *Sol. RRL* 2: 1800148.

263 Sepahvand, H. and Sharifnia, S. (2019). *Int. J. Hydrogen Energy* 44: 23658.

264 Zhu, M., Sun, Z., Fujitsuka, M., and Majima, T. (2018). *Angew. Chem. Int. Ed.* 57: 2160.

265 Yuan, Q., Liu, D., Zhang, N. et al. (2017). *Angew. Chem. Int. Ed.* 56: 4206.

266 Rauf, A., Ma, M., Kim, S. et al. (2018). *Nanoscale* 10: 3026.

267 Fu, C.F., Zhang, R., Luo, Q. et al. (2019). *J. Comput. Chem.* 40: 980.

268 U.S. Energy Information Administration (2019). https://www.eia.gov/ (accessed 17 July 2020).

269 Hedley, G.J., Ruseckas, A., and Samuel, I.D.W. (2017). *Chem. Rev.* 117: 796.
270 Jena, A.K., Kulkarni, A., and Miyasaka, T. (2019). *Chem. Rev.* 119: 3036.
271 Gong, J., Li, C., and Wasielewski, M.R. (2019). *Chem. Soc. Rev.* 48: 1862.
272 Jang, Y.H., Jang, Y.J., Kim, S. et al. (2016). *Chem. Rev.* 116: 14982.
273 You, B., Tang, M.T., Tsai, C. et al. (2019). *Adv. Mater.* 31: 1807001.
274 You, B. and Sun, Y. (2018). *Acc. Chem. Res.* 51: 1571.
275 Anantharaj, S., Ede, S.R., Sakthikumar, K. et al. (2016). *ACS Catal.* 6: 8069.
276 Yang, W., Wang, Z., Zhang, W., and Guo, S. (2019). *Trends Chem.* 1: 259.
277 Zhang, J. and Dai, L. (2016). *Angew. Chem. Int. Ed.* 55: 13296.
278 Roger, I., Shipman, M.A., and Symes, M.D. (2017). *Nat. Rev. Chem.* 1: 3.
279 Li, X., Hao, X., Abudula, A., and Guan, G. (2016). *J. Mater. Chem. A* 4: 11973.
280 Hu, C., Zhang, L., and Gong, J. (2019). *Energy Environ. Sci.* 12: 2620.
281 Jiang, C., Moniz, S.J.A., Wang, A. et al. (2017). *Chem. Soc. Rev.* 46: 4645.
282 Ager, J.W., Shaner, M.R., Walczak, K.A. et al. (2015). *Energy Environ. Sci.* 8: 2811.
283 Shukla, P.K., Karn, R.K., Singh, A.K., and Srivastava, O.N. (2002). *Int. J. Hydrogen Energy* 27: 135.
284 Jahagirdar, A.H. and Dhere, N.G. (2007). *Sol. Energy Mater. Sol. Cells* 91: 1488.
285 Carrillo, J., Guerrero, A., Rahimnejad, S. et al. (2016). *Adv. Energy Mater.* 6: 1502246.
286 Winkler, M.T., Cox, C.R., Nocera, D.G., and Buonassisi, T. (2013). *Proc. Natl. Acad. Sci. U.S.A.* 110: E1076.
287 Jacobsson, T.J., Fjällström, V., Edoff, M., and Edvinsson, T. (2014). *Energy Environ. Sci.* 7: 2056.
288 Heremans, G., Trompoukis, C., Daems, N. et al. (2017). *Sustain. Energy Fuels* 1: 2061.
289 Jia, J., Seitz, L.C., Benck, J.D. et al. (2016). *Nat. Commun.* 7: 13237.
290 U.S. Department of Energy (2011). Offices of Fuel Cell Technologies, Hydrogen Threshold Cost Calculation.
291 Abdi, F.F., Han, L., Smets, A.H.M. et al. (2013). *Nat. Commun.* 4: 2195.
292 Khaselev, O. and Turner, J.A. (1998). *Science* 280: 425.
293 Luo, J., Im, J.H., Mayer, M.T. et al. (2014). *Science* 345: 1593.
294 Nellist, M.R., Laskowski, F.A.L., Lin, F. et al. (2016). *Acc. Chem. Res.* 49: 733.
295 Thorne, J.E., Li, S., Du, C. et al. (2015). *J. Phys. Chem. Lett.* 6: 4083.
296 Hou, T.F., Boppella, R., Shanmugasundaram, A. et al. (2017). *Int. J. Hydrogen Energy* 42: 15126.
297 Boppella, R., Kochuveedu, S.T., Kim, H. et al. (2017). *ACS Appl. Mater. Interfaces* 9: 7075.
298 Wang, S., Liu, G., and Wang, L. (2019). *Chem. Rev.* 119: 5192.
299 Miller, E.L. (2015). *Energy Environ. Sci.* 8: 2809.
300 Niu, F., Wang, D., Li, F. et al. (2019). *Adv. Energy Mater.*: 1900399.
301 Park, K., Kim, Y.J., Yoon, T. et al. (2019). *RSC Adv.* 9: 30112.
302 Joy, J., Mathew, J., and George, S.C. (2018). *Int. J. Hydrogen Energy* 43: 4804.
303 Yoon, M., Lee, J.E., Jang, Y.J. et al. (2015). *ACS Appl. Mater. Interfaces* 7: 21073.

304 Kochuveedu, S.T., Jang, Y.H., Jang, Y.J., and Kim, D.H. (2013). *J. Mater. Chem. A* 1: 898.

305 Park, J.H., Kim, S., and Bard, A.J. (2006). *Nano Lett.* 6: 24.

306 Yang, C., Wang, Z., Lin, T. et al. (2013). *J. Am. Chem. Soc.* 135: 17831.

307 Hoang, S., Guo, S., and Mullins, C.B. (2012). *J. Phys. Chem. C* 116: 23283.

308 Wang, C., Chen, Z., Jin, H. et al. (2014). *J. Mater. Chem. A* 2: 17820.

309 Kalanur, S.S., Yoo, I.H., Eom, K., and Seo, H. (2018). *J. Catal.* 357: 127.

310 Liu, Z., Wu, J., and Zhang, J. (2016). *Int. J. Hydrogen Energy* 41: 20529.

311 Kang, D., Hill, J.C., Park, Y., and Choi, K.S. (2016). *Chem. Mater.* 28: 4331.

312 Wang, J., Du, C., Peng, Q. et al. (2017). *Int. J. Hydrogen Energy* 42: 29140.

313 Yin, W.J., Tang, H., Wei, S.H. et al. (2010). *Phys. Rev. B* 82: 045106.

314 Wang, D., Zou, Y., Wen, S., and Fan, D. (2009). *Appl. Phys. Lett.* 95: 10.

315 Gai, Y., Li, J., Li, S.S. et al. (2009). *Phys. Rev. Lett.* 102: 23.

316 Ma, X., Wu, Y., Lu, Y. et al. (2011). *J. Phys. Chem. C* 115: 16963.

317 Long, R. and English, N.J. (2009). *Appl. Phys. Lett.* 94: 132102.

318 Niu, M., Xu, W., Shao, X., and Cheng, D. (2011). *Appl. Phys. Lett.* 99: 203111.

319 Zhu, W., Qiu, X., Iancu, V. et al. (2009). *Phys. Rev. Lett.* 103: 226401.

320 Cho, I.S., Lee, C.H., Feng, Y. et al. (2013). *Nat. Commun.* 4: 1723.

321 Bang, J.H. and Kamat, P.V. (2010). *Adv. Funct. Mater.* 20: 1970.

322 Liu, M., De Leon Snapp, N., and Park, H. (2011). *Chem. Sci.* 2: 80.

323 Zhang, Z. and Wang, P. (2012). *Energy Environ. Sci.* 5: 6506.

324 Ibadurrohman, M. and Hellgardt, K. (2015). *ACS Appl. Mater. Interfaces* 7: 9088.

325 Qiu, Y., Liu, W., Chen, W. et al. (2016). *Sci. Adv.* 2: e1501764.

326 Brongersma, M.L., Cui, Y., and Fan, S. (2014). *Nat. Mater.* 13: 451.

327 Luo, Z., Wang, T., Zhang, J. et al. (2017). *Angew. Chem. Int. Ed.* 56: 12878.

328 Long, M., Cai, W., and Kisch, H. (2008). *J. Phys. Chem. C* 112: 548.

329 Luo, W., Yang, Z., Li, Z. et al. (2011). *Energy Environ. Sci.* 4: 4046.

330 Choi, S.K., Choi, W., and Park, H. (2013). *Phys. Chem. Chem. Phys.* 15: 6499.

331 Zhong, D.K., Choi, S., and Gamelin, D.R. (2011). *J. Am. Chem. Soc.* 133: 18370.

332 Ding, C., Shi, J., Wang, D. et al. (2013). *Phys. Chem. Chem. Phys.* 15: 4589.

333 Liao, M., Feng, J., Luo, W. et al. (2012). *Adv. Funct. Mater.* 22: 3066.

334 Kim, T.W. and Choi, K.-S. (2014). *Science* 343: 990.

335 Zhong, M., Hisatomi, T., Kuang, Y. et al. (2015). *J. Am. Chem. Soc.* 137: 5053.

336 Chu, S., Vanka, S., Wang, Y. et al. (2018). *ACS Energy Lett.* 3: 307.

337 Ai, G., Li, H., Liu, S. et al. (2015). *Adv. Funct. Mater.* 25: 5706.

338 Pilli, S.K., Furtak, T.E., Brown, L.D. et al. (2011). *Energy Environ. Sci.* 4: 5028.

339 Zhong, D.K., Cornuz, M., Sivula, K. et al. (2011). *Energy Environ. Sci.* 4: 1759.

340 Li, Y., Takata, T., Cha, D. et al. (2013). *Adv. Mater.* 25: 125.

341 Hu, S., Shaner, M.R., Beardslee, J.A. et al. (2014). *Science* 344: 1005.

342 Sun, K., Saadi, F.H., Lichterman, M.F. et al. (2015). *Proc. Natl. Acad. Sci. U.S.A.* 112: 3612.

343 Zhou, X., Liu, R., Sun, K. et al. (2015). *Energy Environ. Sci.* 8: 2644.

344 Fogg, A.M., Green, V.M., Harvey, H.G., and O'Hare, D. (1999). *Adv. Mater.* 11: 1466.

345 Gu, Z., Thomas, A.C., Xu, Z.P. et al. (2008). *Chem. Mater.* 20: 3715.
346 Costantino, U., Coletti, N., Nocchetti, M. et al. (1999). *Langmuir* 15: 4454.
347 Wang, Q. and Ohare, D. (2012). *Chem. Rev.* 112: 4124.
348 Shao, M., Ning, F., Zhao, J. et al. (2012). *J. Am. Chem. Soc.* 134: 1071.
349 Gunjakar, J.L., Kim, T.W., Kim, H.N. et al. (2011). *J. Am. Chem. Soc.* 133: 14998.
350 Shao, M., Ning, F., Wei, M. et al. (2014). *Adv. Funct. Mater.* 24: 580.
351 Wang, L., Dionigi, F., Nguyen, N.T. et al. (2015). *Chem. Mater.* 27: 2360.
352 Kim, E.S., Nishimura, N., Magesh, G. et al. (2013). *J. Am. Chem. Soc.* 135: 5375.
353 Liu, Y., Zhao, L., Li, M., and Guo, L. (2014). *Nanoscale* 6: 7397.
354 Ai, G., Mo, R., Xu, H. et al. (2015). *Sources* 280: 5.
355 Ai, G., Mo, R., Xu, H. et al. (2014). *J. Appl. Phys.* 116: 174306.
356 Chen, Z., Peng, W., Zhang, K. et al. (2012). *Nanoscale* 4: 7690.
357 Cheng, K., Han, X., Meng, J. et al. (2015). *RSC Adv.* 5: 11084.
358 Luo, Q., Wu, Z., He, J. et al. (2015). *Nanoscale Res. Lett.* 10: 181.
359 Zhan, X., Wang, Q., Wang, F. et al. (2014). *ACS Appl. Mater. Interfaces* 6: 2878.
360 Hong, S.J., Lee, S., Jang, J.S., and Lee, J.S. (2011). *Energy Environ. Sci.* 4: 1781.
361 Narkeviciute, I., Chakthranont, P., MacKus, A.J.M. et al. (2016). *Nano Lett.* 16: 7565.
362 Chen, J., Dong, C., Idriss, H. et al. (2019). *Adv. Energy Mater.*: 1902433.
363 Poli, I., Hintermair, U., Regue, M. et al. (2019). *Nat. Commun.* 10: 2097.
364 Xiao, M., Hao, M., Lyu, M. et al. (2019). *Adv. Funct. Mater.*: 1905683.
365 Crespo-Quesada, M., Pazos-Outón, L.M., Warnan, J. et al. (2016). *Nat. Commun.* 7: 6.
366 Paracchino, A., Laporte, V., Sivula, K. et al. (2011). *Nat. Mater.* 10: 456.
367 Luo, J., Steier, L., Son, M.K. et al. (2016). *Nano Lett.* 16: 1848.
368 Pan, L., Kim, J.H., Mayer, M.T. et al. (2018). *Nat. Catal.* 1: 412.
369 Bae, D., Seger, B., Vesborg, P.C.K. et al. (2017). *Chem. Soc. Rev.* 46: 1933.
370 Yokoyama, D., Minegishi, T., Jimbo, K. et al. (2010). *Appl. Phys Express* 3: 101202.
371 Yang, W., Oh, Y., Kim, J. et al. (2016). *ACS Energy Lett.* 1: 1127.
372 Jiang, F., Gunawan, T.H., Kuang, Y. et al. (2015). *J. Am. Chem. Soc.* 137: 13691.
373 Tay, Y.F., Kaneko, H., Chiam, S.Y. et al. (2018). *Joule* 2: 537.
374 Chen, M., Liu, Y., Li, C. et al. (2018). *Energy Environ. Sci.* 11: 2025.
375 Kumagai, H., Minegishi, T., Sato, N. et al. (2015). *J. Mater. Chem. A* 3: 8300.
376 Zhang, L., Minegishi, T., Kubota, J., and Domen, K. (2014). *Phys. Chem. Chem. Phys.* 16: 6167.
377 Moriya, M., Minegishi, T., Kumagai, H. et al. (2013). *J. Am. Chem. Soc.* 135: 3733.
378 Lee, H., Yang, W., Tan, J. et al. (2019). *ACS Energy Lett.* 4: 995.
379 Yang, W., Ahn, J., Oh, Y. et al. (2018). *Adv. Energy Mater.* 8: 1702888.
380 Park, J., Yang, W., Oh, Y. et al. (2019). *ACS Energy Lett.* 4: 517.
381 Ohashi, K. (1977). *Nature* 266: 610.
382 Su, J., Minegishi, T., and Domen, K. (2017). *J. Mater. Chem. A* 5: 13154.

383 Sun, K., Shen, S., Liang, Y. et al. (2014). *Chem. Rev.* 114: 8662.
384 Han, L., Digdaya, I.A., Buijs, T.W.F. et al. (2015). *J. Mater. Chem. A* 3: 4155.
385 Digdaya, I.A., Han, L., Buijs, T.W.F. et al. (2015). *Energy Environ. Sci.* 8: 1585.
386 Hu, S., Xiang, C., Haussener, S. et al. (2013). *Energy Environ. Sci.* 6: 2984.
387 Kornienko, N., Gibson, N.A., Zhang, H. et al. (2016). *ACS Nano* 10: 5525.
388 Bornoz, P., Abdi, F.F., Tilley, S.D. et al. (2014). *J. Phys. Chem. C* 118: 16959.
389 Kim, J.H., Kaneko, H., Minegishi, T. et al. (2016). *ChemSusChem* 9: 61.
390 Higashi, T., Kaneko, H., Minegishi, T. et al. (2017). *Chem. Commun.* 53: 11674.
391 Kobayashi, H., Sato, N., Orita, M. et al. (2018). *Energy Environ. Sci.* 11: 3003.
392 Kainthla, R.C., Zelenay, B., and Bockris, J.O.M. (1987). *J. Electrochem. Soc.* 134: 841.
393 Yoneyama, H., Sakamoto, H., and Tamura, H. (1975). *Electrochim. Acta* 20: 341.
394 Nozik, A.J. (1976). *Appl. Phys. Lett.* 29: 150.
395 Kang, D., Young, J.L., Lim, H. et al. (2017). *Nat. Energy* 2: 17043.
396 Andrei, V., Hoye, R.L.Z., Crespo-Quesada, M. et al. (2018). *Adv. Energy Mater.* 8: 1801403.
397 Lin, Y., Battaglia, C., Boccard, M. et al. (2013). *Nano Lett.* 13: 5615.
398 Liu, C., Tang, J., Chen, H.M. et al. (2013). *Nano Lett.* 13: 2989.
399 Jang, J.W., Du, C., Ye, Y. et al. (2015). *Nat. Commun.* 6: 7447.
400 Vijselaar, W., Westerik, P., Veerbeek, J. et al. (2018). *Nat. Energy* 3: 185.
401 Lai, Y.H., Palm, D.W., and Reisner, E. (2015). *Adv. Energy Mater.* 5: 1501668.
402 Lin, C.Y., Lai, Y.H., Mersch, D., and Reisner, E. (2012). *Chem. Sci.* 3: 3482.
403 Young, J.L., Steiner, M.A., Döscher, H. et al. (2017). *Nat. Energy* 2: 17028.
404 Licht, S., Wang, B., Mukerji, S. et al. (2000). *J. Phys. Chem. B* 104: 8920.
405 Cheng, W.H., Richter, M.H., May, M.M. et al. (2018). *ACS Energy Lett.* 3: 1795.
406 Reece, S.Y., Hamel, J.A., Sung, K. et al. (2011). *Science* 334: 645.
407 Morales-Guio, C.G., Mayer, M.T., Yella, A. et al. (2015). *J. Am. Chem. Soc.* 137: 9927.
408 Pihosh, Y., Turkevych, I., Mawatari, K. et al. (2015). *Sci. Rep.* 5: 11141.
409 Han, L., Abdi, F.F., Van De Krol, R. et al. (2014). *ChemSusChem* 7: 2832.
410 Kim, J.H., Jang, J.W., Jo, Y.H. et al. (2016). *Nat. Commun.* 7: 13380.
411 Rocheleau, R.E., Miller, E.L., and Misra, A. (1998). *Energy Fuels* 12: 3.
412 Brillet, J., Yum, J.H., Cornuz, M. et al. (2012). *Nat. Photonics* 6: 824.
413 Dias, P., Schreier, M., Tilley, S.D. et al. (2015). *Adv. Energy Mater.* 5: 1501537.
414 Luo, J., Li, Z., Nishiwaki, S. et al. (2015). *Adv. Energy Mater.* 5: 1501520.
415 Turan, B., Becker, J.P., Urbain, F. et al. (2016). *Nat. Commun.* 7: 12681.
416 Vilanova, A., Lopes, T., Spenke, C. et al. (2018). *Energy Storage Mater.* 13: 175.
417 Ahmet, I.Y., Ma, Y., Jang, J.W. et al. (2019). *Sustain. Energy Fuels* 3: 2366.
418 Okamoto, S., Deguchi, M., and Yotsuhashi, S. (2017). *J. Phys. Chem. C* 121: 1393.
419 Verlage, E., Hu, S., Liu, R. et al. (2015). *Energy Environ. Sci.* 8: 3166.
420 Shi, X., Zhang, K., Shin, K. et al. (2015). *Nano Energy* 13: 182.
421 Kim, J.H., Jo, Y., Kim, J.H. et al. (2015). *ACS Nano* 9: 11820.
422 May, M.M., Lewerenz, H.J., Lackner, D. et al. (2015). *Nat. Commun.* 6: 8286.

9

Photocatalytic CO_2 Reduction

Maochang Liu[1,2], Guijun Chen[1], Boya Min[1], Jinwen Shi[1], Yubin Chen[1] and Qibin Liu[3]

[1] *International Research Center for Renewable Energy, State Key Laboratory of Multiphase Flow in Power Engineering, Xi'an Jiaotong University, No. 28, West Road, Xi'an, Xianning, Shaanxi 710049, P.R. China*
[2] *Suzhou Academy of Xi'an Jiaotong University, No. 99 Renai Road, Suzhou, Jiangsu 215123, P.R. China*
[3] *Institute of Engineering Thermophysics, Chinese Academy of Sciences, Beijing 100190, P.R. China*

9.1 Introduction

The rapid development of industry has intensified energy utilization, resulting in a sharp depletion of fossil fuels, and this large-scale use of fossil fuels (usually in the form of direct burning) has caused significant CO_2 emission and thus aggravated greenhouse effect. For sustainable energy development and environmental protection, a large number of research activities are devoted to the capture, storage, and utilization of carbon dioxide, among which photocatalytic reduction of carbon dioxide to form hydrocarbon fuel is considered to be a promising method.

Generally, photocatalytic reduction of CO_2 by solar energy is a green technology, which converts carbon dioxide into value-added and renewable fuels, such as CO, CH_4, CH_3OH, HCOOH, etc., under the irradiation of sunlight and in the presence of photocatalysts. This technology has received intensive focus and has been predicted with economic efficiency, reproducibility, safety, etc. Compared with other methods for CO_2 capture, this chemical transformation process can be carried out under relatively mild conditions, such as room temperature and atmospheric pressure, and employs copious amounts of CO_2 as a carbon source, driven by an endless supply of clean solar energy. The photoreduction of CO_2 can directly produce hydrocarbon fuels. In other words, the technology makes it possible to cycle the utilization of fossil fuels in a way of hydrogen-carbon cycling. The energy and environmental benefits of photocatalytic reduction of CO_2 into hydrocarbon fuels are therefore obvious.

The initial research of photocatalytic reduction of CO_2 can be traced back to 1972. At that time, two scientists, Fujishima and Honda, from the University of Tokyo, Japan, discovered a kind of n-type semiconductor TiO_2 electrode, which could photochemically decompose water to produce H_2 and O_2 under ultraviolet light irradiation [1]. This discovery enables the blossoming and popularization of the research

Solar-to-Chemical Conversion: Photocatalytic and Photoelectrochemcial Processes, First Edition. Edited by Hongqi Sun.

field of semiconductor-based photocatalysis. In 1978, Halmann et al. first reduced CO_2 in aqueous solution to generate formic acid, methanol, and formaldehyde by using p-type semiconductor GaP photocathode [2]. In 1979, Inoue et al. [3] employed a suspension system containing WO_3, TiO_2, ZnO, CdS, GaP, or SiC semiconductor photocatalysts. In this system, CO_2 was converted to formaldehyde, formic acid, and methanol in the presence of light irradiation, and the possible reaction mechanism was also discussed. Later in 1987, Soichiro et al. used SiC and ZnSe as the photocatalytic materials to reduce CO_2 to produce ethanol, and dicarbonic acids were also synthesized for the first time [4]. This work significantly promoted the research and development of this field. Since then, many semiconductors have been identified as potential photocatalytic materials for CO_2 reduction under ultraviolet or visible-light irradiation. Notable examples include $SrTiO_3$ [5], Bi_2S_3 [6], ZnS [7], ZrO_2 [8], SnO_2 [9], Fe_2O_3 [10], Cu_2O [11], etc.

However, the latest advances in photocatalytic reduction of carbon dioxide indicate that the development of the technology is far less than anticipated. At the most basic level, it is the relatively poor performance of the photocatalytic materials that restricts the application. In order to develop photocatalysts with high activity, good stability, and high selectivity for artificial photoreduction of CO_2, researchers have made great efforts in enhancing the photo-trapping ability, accelerating the separation of photoinduced charge carriers, enhancing the adsorption and activation of CO_2, and exploring the reaction mechanism of CO_2 reduction. Meantime, it is also necessary to optimize the reaction system for photocatalytic CO_2 reduction, which yet has been less emphasized in the previous research. In principle, control over both energy transfer and mass transfer should be essential to achieve good photocatalytic performance. Briefly, energy transfer means energy flow of solar energy from the sun to the catalyst and further to the as-synthesized hydrocarbons, while mass transfer involves the flow of molecules/ions/radicals from reactants to the outcomes. It is worth pointing out that both energy and mass transfers have different scales with spatiotemporal characteristics. Specifically, for a given photocatalytic CO_2 reduction reaction, the whole process contains spatial scales from nanometers to meters and temporal scales from femtoseconds to seconds. Since mass conversion and energy transfer are usually in a manner of mutual causality, the coupling and matching of the energy flow and the mass flow in the whole process are of considerable significance in reducing unnecessary energy loss and reaction obstacles and thus increasing the conversion efficiency.

This chapter reviews the photocatalytic CO_2 reduction from the perspective of energy and mass transfers. It aims to clarify the whole process of photocatalytic reduction of CO_2 by emphatically pointing out the losses and transformation rules during the energy flow and the influence of the mass flow on the photocatalytic reaction, thus giving the researchers a clear understanding on the overall energy flow and mass flow. In addition, this chapter also provides some guidance for optimizing the coordination of each part involved in the photocatalytic system to achieve a better conversion behavior.

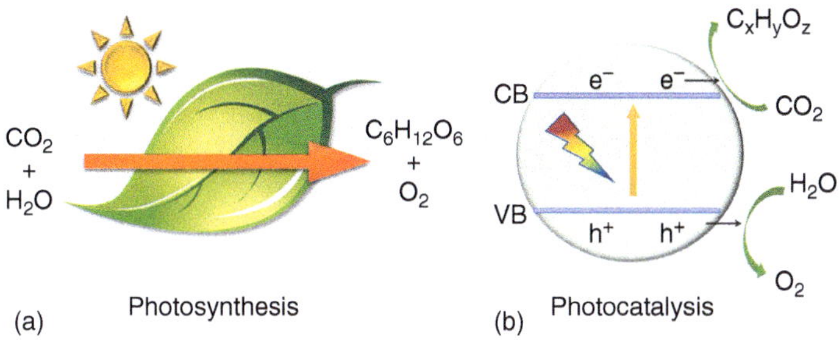

Figure 9.1 Schematic illustration of the natural photosynthesis (a) and the artificial photosynthesis (b) using semiconductor as the photocatalyst.

9.2 Principle of Photocatalytic Reduction of CO_2

CO_2 is an essential substance in the natural carbon cycle. Plants convert the mixture of CO_2 and H_2O into carbohydrates and oxygen through photosynthesis (see Figure 9.1a). This process enables storage of solar energy into carbohydrates. Artificial photocatalytic reduction of CO_2 is a way that imitates the natural process yet by replacing the plants with a certain photocatalyst to achieve CO_2 reduction. Figure 9.1b schematically illustrates the principle of photocatalytic CO_2 reduction. Generally, the photocatalysts used for CO_2 reduction can be excited to generate charge carriers under illumination. These photocatalytic materials usually include semiconductors, metal complexes, and some light-sensitive organic substances. Upon light absorption, the generated free electrons and holes will react with the oxidants and reductants in the reaction environment, respectively. The catalyst thus works in this repeated photoexcitation–oxidation–reduction process. Semiconductor materials are a class of the most widely studied photocatalysts used in photocatalytic reactions. Basically, a valence band (VB) filled with ground-state electrons and an empty conduction band (CB) construct the band structure of a semiconductor. Band gap is defined as the energy gap between the VB maximum and CB minimum, where no electron states are allowed. When light irradiates the semiconductor catalyst, photons with energy larger than the band gap of semiconductors are absorbed. Electrons in valence band are simultaneously excited from the valence band to the conduction band, generating holes in the valence band, and electron–hole pairs are thus formed. These photogenerated electron–hole pairs may combine again via a so-called recombination process. Energy is accordingly dissipated in the form of light and heat. Otherwise, the electron–hole pairs should be utilized in redox reactions. For instance, in a photoelectrochemical reaction, at the photoanode, the holes could reach the interface between the catalyst and electrolyte solution and subsequently have the reductants in the solution (usually in the form of H_2O, organic matters, or reductive inorganic matters such as sulfur ions and hydrogen) oxidized. On the contrary, electrons at the photocathode will reduce CO_2 and H^+ at the interface between the catalyst and electrolyte solution to form intermediate products, leading to the formation of stable reduction products, such as CO, CH_4, etc. Practically, photocatalytic reduction of CO_2 proceeds in the case of two primary conditions: (i) the energy of the incident light must be larger than the band gap of the semiconductor, so as to induce photoexcitation and give rise to the formation of electron–hole pairs. (ii) The potential at the top of the valence band of the semiconductor must be more positive than the oxidation potential of H_2O or sacrificial agents, so that the oxidation reaction can occur. At the same time, the potential of the conduction band minimum of the semiconductor must be more negative than the potential of CO_2 reduction to make the reaction occur. In this regard, the redox potentials of common reactions are extremely important for the design of the photocatalytic reaction. Table 9.1 presents the common reactions

Table 9.1 Reduction potentials of CO_2, H_2CO_3, and CO_3^{2-}.

Reaction	E^0 (V) vs. NHE at pH = 7
Reduction potentials of CO_2	
$2H^+ + 2e^- \rightarrow H_2$	−0.41
$CO_2 + e^- \rightarrow CO_2^-$	−1.9
$CO_2 + 2H^+ + 2e^- \rightarrow HCOOH$	−0.61
$CO_2 + 2H^+ + 2e^- \rightarrow CO + H_2O$	−0.53
$CO_2 + 4H^+ + 4e^- \rightarrow C + 2H_2O$	−0.2
$CO_2 + 4H^+ + 4e^- \rightarrow HCHO + H_2O$	−0.48
$CO_2 + 6H^+ + 6e^- \rightarrow CH_3OH + H_2O$	−0.38
$CO_2 + 8H^+ + 8e^- \rightarrow CH_4 + 2H_2O$	−0.24
$2CO_2 + 8H_2O + 12e^- \rightarrow C_2H_4 + 12OH^-$	−0.34
$2CO_2 + 9H_2O + 12e^- \rightarrow C_2H_5OH + 12OH^-$	−0.33
$3CO_2 + 13H_2O + 18e^- \rightarrow C_3H_7OH + 18OH^-$	−0.32
Reduction potentials of H_2CO_3	
$2H^+ + 2e^- \rightarrow H_2$	−0.41
$2H_2CO_3 + 2H^+ + 2e^- \rightarrow H_2C_2O_4 + 2H_2O$	−0.8
$H_2CO_3 + 2H^+ + 2e^- \rightarrow HCOOH + H_2O$	−0.576
$H_2CO_3 + 4H^+ + 4e^- \rightarrow HCHO + 2H_2O$	−0.46
$H_2CO_3 + 6H^+ + 6e^- \rightarrow CH_3OH + 2H_2O$	−0.366
$H_2CO_3 + 4H^+ + 4e^- \rightarrow C + 3H_2O$	−0.182
Reduction potentials of CO_3^{2-}	
$2H^+ + 2e^- \rightarrow H_2$	−0.41
$CO_3^{2-} + 4H^+ + 2e^- \rightarrow C_2O4^{2-} + 2H_2O$	0.07
$CO_3^{2-} + 3H^+ + 2e^- \rightarrow HCOO^- + H_2O$	−0.099
$CO_3^{2-} + 6H^+ + 4e^- \rightarrow HCHO + 2H_2O$	−0.213
$CO_3^{2-} + 8H^+ + 6e^- \rightarrow CH_3OH + 2H_2O$	−0.201
$CO_3^{2-} + 6H^+ + 4e^- \rightarrow C + 3H_2O$	0.065

Source: Li et al. [12]. © 2014, Springer Nature.

in the photocatalytic reduction of CO_2 and their corresponding redox potentials obtained from Ref. [12]. Obviously, the potential of direct CO_2 reduction can be up to −1.9 V (vs. NHE at pH = 7), which is definitely unfavorable for the practical application. However, Table 9.1 also indicates that the introduction of H_2O in the reaction can effectively reduce the potential requirement.

Generally, in order to achieve efficient, stable, low-cost, and large-scale industrial application of CO_2 photocatalytic reduction to produce fuel, the catalyst materials must have an appropriate band gap and meet the thermodynamic requirements. They also need to possess good conductivity, small carrier recombination rate, and stable physical and chemical properties. Naturally, they should be abundant in Earth reservation, cost effective, safe, nontoxicity contained, etc. The band gap and band-edge positions of common photocatalytic materials used for photocatalytic reduction of carbon dioxide are shown in Figure 9.2. Among these well-studied photocatalytic materials, TiO_2 has been regarded as a promising photocatalyst and thus the widest focused because of its good stability, low cost, and nontoxicity. However, the band gap is too large to use visible light with wavelength larger than 400 nm (e.g. band gap of anatase TiO_2 is 3.2 eV, which can only utilize ultraviolet light, accounting for only about 5% of the energy of sunlight reaching the ground) [13]. To achieve a high energy conversion efficiency, it is necessary to modify pure TiO_2 photocatalyst with different synthetic methods.

9.3 Energy and Mass Transfers in Photocatalytic Reduction of CO_2

Up to now, the main research direction is still to find new photocatalytic materials or to improve property of existing photocatalytic materials. In this case, only lab-scale devices and simulated sunlight sources have been utilized, which are significantly different from a real-space solar irradiation. In fact, energy loss before light entering the reaction is also very large while less attention has been paid. More specifically, although catalytic material plays a crucial role in photocatalytic CO_2 reduction, the other parts involved in the whole photocatalytic reaction system should not be neglected, and they could raise a notable impact contributing to the entire energy conversion efficiency.

In a typical photocatalytic reaction system, sunlight is first collected by a concentrator to improve the light intensity and then enters the reactor cavity through the reactor wall, and the catalyst is distributed in the reactor cavity. Reactants such as CO_2 and H_2O are fed into the reactor from the raw material container. For a photocatalytic reaction based on a solid semiconductor photocatalyst, it generally proceeds in a heterogeneous way where chemical transformations occur on either a gas–solid interface or a solid–liquid interface upon the arrival of photogenerated charge carriers and reactants at surface of the catalyst. This kind of chemical transformation makes effective conversion from solar energy to chemical energy. However, the conversion efficiency is determined by the effectiveness of the reaction. During the reaction, the products and unreacted reactants could be separated by a separation

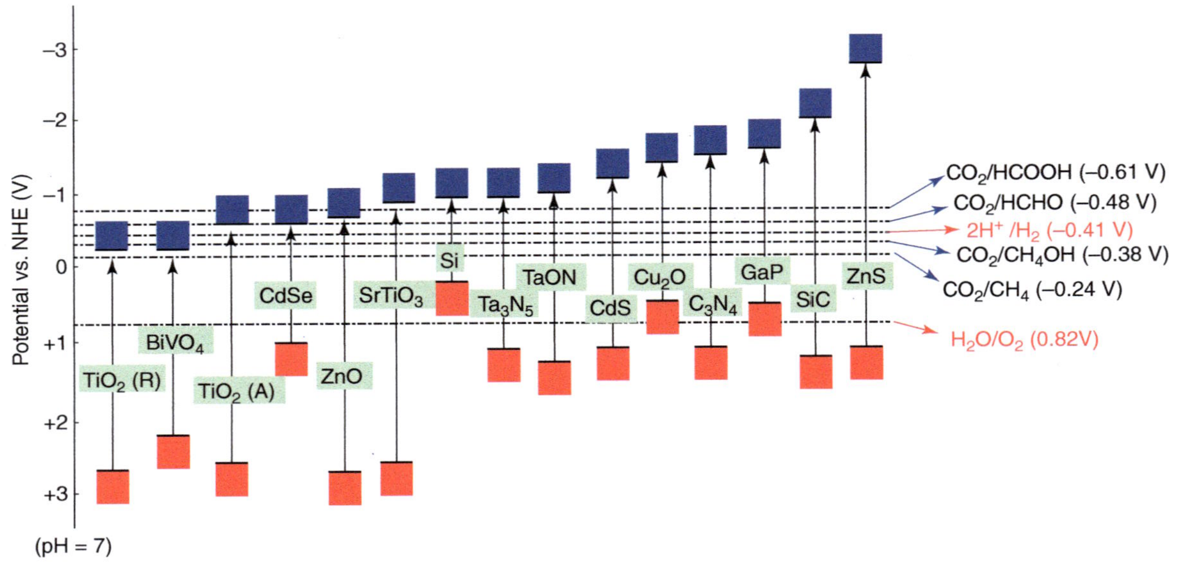

Figure 9.2 Band positions of some semiconductor photocatalysts and the redox potentials of CO_2 reduction at pH 7 in aqueous solution. Source: Li et al. [12]. © 2014, Springer Nature.

device with the products entering a product container and the remained reactants circulating in the reactor.

From the viewpoint of system engineering, each step included in the mass and energy transfer process should be emphasized. The overall photocatalytic reaction performance can also be improved by adequately matching each link of the reaction system. To this end, all parts of the CO_2 photocatalytic reduction system need to be carefully considered and optimized. This notion also explains the fundamentals to achieve the industrial application of a highly efficient, stable, low-cost, and large-scale system of photocatalytic CO_2 reduction. The energy flow from solar energy to chemical energy and mass flow from CO_2 to various small molecule chemicals/fuels are two basic processes that run through the whole system. They are also the direct measurement of the performance of a given system. Achieving smooth and well-matched energy flow and mass flow can reduce unnecessary energy loss, avoid reaction obstacles, and thus increase utilization efficiency of solar energy. The key is to examine the details of energy flow and mass flow to recover the transformation rules in the whole photocatalytic process. The understanding of energy and mass transfers and their role on photocatalytic reaction can give a clear picture of the overall energy and mass flow situations, thus enabling maximized energy utilization and optimized chemical/fuel output by adjusting and coupling various parts involved in the whole process. Obviously, it is a complicated process that requires the collaboration of chemists, physicists, material scientists, and engineers.

9.3.1 Energy Flow from the Concentrator to Reactor

The intensity of the incident light is an important factor affecting the photocatalytic reaction rate. Generally, the solar radiation received by the upper interface of the atmosphere of the Earth perpendicular to the solar rays has an intensity of 1357 $W\,m^{-2}$ [14]. After the scattering, reflection, and absorption of the atmosphere, the solar radiation reaching the ground is lower than the upper boundary of the atmosphere [15]. Basically, at a constant power of the incident light, photons will become more effective upon concentration. This improvement is attributed to the strong interaction effect between these photons. In this case, more effective photogenerated electron–hole pairs could be obtained, leading to increased photocatalytic activity. For instance, Han et al. reported the CH_4 generation from photocatalytic CO_2 reduction over TiO_2 and Pt/TiO_2 photocatalyst [16]. In their research, by simply using a lens as the concentrator, they found that when the concentration ratio is 12.9, the formation rate of CH_4 is seven times than that of non-concentrating conditions. This attempt clearly demonstrated the significance of solar concentration on the performance of photocatalytic reaction. Moreover, the smaller the reaction area, the less photocatalyst material is required. This reduction in material usage is meaningful for the reduction of synthetic cost.

To achieve efficient light concentration, a proper concentrator is very important. Concentrators have been widely studied and applied in photothermal power generation systems. Notable examples include trough, tower, and disc solar power generation systems. The photocatalytic reactors are generally in the form of tubes that require a linear concentrator. There are many types of linear concentrators, including trough parabolic concentrator, linear Fresnel concentrator, composite parabolic concentrator, etc. Among them, the composite parabolic concentrator has many advantages. In practical application of the concentrator, the challenge is how to precisely track the light source. In order to keep the bulky equipment move continuously toward the incident radiation of the sunlight, complicated instruments are required, which in turn further increases the cost and energy consumption. The nature of automatic light concentration makes composite paraboloid widely studied. The composite parabolic concentrator is based upon reflective principle, and the incident light is reflected to the focal area at the bottom. In this case, the shell of the reactor is designed as a cylindrical structure with high mechanical strength. It is worth pointing out that the sunlight coming from the concentrator inlet is not all concentrated on the bottom of the reactor. Some of the sunlight is absorbed by the concentrator paraboloid, and some is reflected from the concentrator inlet. Even though the light has successfully arrived to the reactor, energy loss at the reactor wall and the reaction solution during the subsequent transportation will be still unavoidable. Specifically, the incident photons would be reflected or absorbed by the reactor wall, and the molecules involved in the reaction flow before they could reach the surface of the photocatalyst. For most of the time, the energy loss is in the form of thermal energy. Moreover, H_2O in the reaction also restricts the utilization of long-wavelength photons, especially the infrared part of the sunlight by the photocatalysts.

The reactor can be designed in a variety of ways according to the type of a specific reaction. Generally, photoelectrochemical and photocatalytic reactors are two types of basic reactions. The photoelectrochemical reactor is mainly composed of a photoanode, a photocathode, and an auxiliary power source. Taking a semiconductor as the example, an n-type semiconductor is used as a photoanode, and H_2O or sacrificial agent is oxidized on the surface of the electrode to form an oxidation product such as O_2. A p-type semiconductor is used as a photocathode, and CO_2 is reduced on the surface of the electrode to form a carbon-based compound. The advantage of this type of reactor is that the oxidation product and the reduction product can be readily separated. Moreover, the electrode reactor allows the introduction of an external bias voltage to promote CO_2 reduction. In fact, an external bias is required for the most time since mismatch between the band positions of semiconductors and the redox potentials usually exits. Despite these benefits, the synthesis and fabrication processes of the thin film photoelectrodes are complicated, difficult, and costly. Large-scale utilization of these photoelectrodes becomes a grand challenge. On the contrary, the photocatalytic reactor does not use a bulk photoelectrode. Specifically, the photocatalyst is prepared in the form of fine powders. Each photocatalyst particle can be treated as a microelectrode. In principle, the semiconductor photocatalysts are excited under light illumination to stimulate electrons to the conduction band, and the holes left in the valence band. The conduction band and the valence band can be regarded as the cathode and the anode, respectively. As a result, according to whether the powder photocatalyst is fixed or not, one can

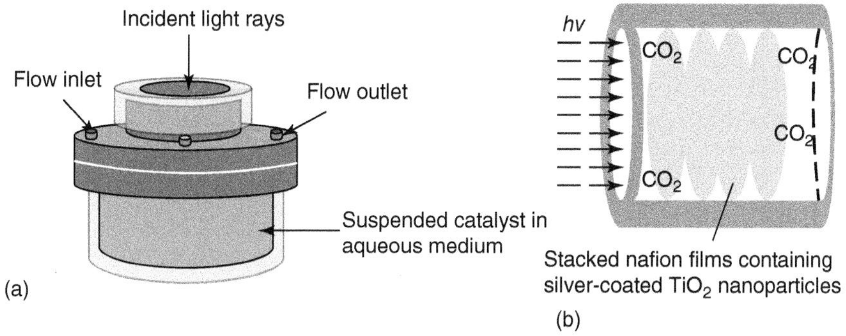

Figure 9.3 Schematic diagram of the slurry reactor (a) and the fixed bed reactor (b). Source: (a) Ola and Maroto-Valer [17]. Licensed under CC BY 4.0. (b) Bakherad et al. [19]. © 2020, Royal Society of Chemistry.

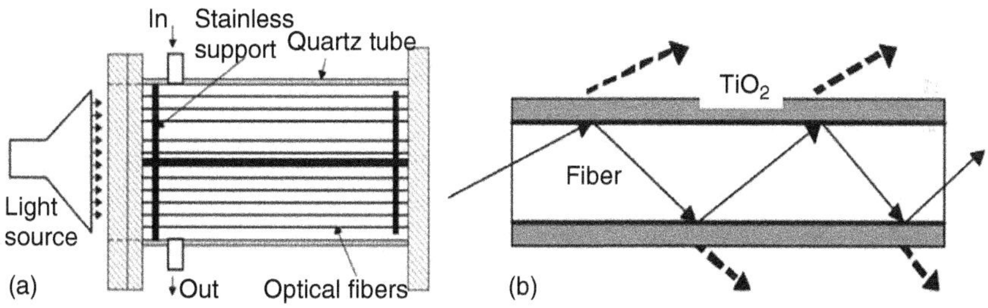

Figure 9.4 (a) Schematic illustration of optical-fiber photo reactor. (b) Light propagation in a TiO_2-coated fiber reactor. Source: Wu et al. [20]. © 2008, Springer Nature.

divide the photo-reactor into a slurry type and a fixed bed type [17, 18], as shown in Figure 9.3. In the slurry reactor, the photocatalyst particles are suspended in the solution and flow with the solution. For the reaction taking place in the fixed bed reactor, the powder photocatalyst is first fixed on the surface of a film or a porous substrate. Redox reactions occur in the event of the reactants flowing through the surface of the photocatalyst.

If a powder photocatalyst is used for the reaction, in the slurry reactor, the inefficient absorption of light in the solution leads to a large loss of light energy. Consequently, light energy reaching the surface of the catalyst is limited. In order to solve this problem, researchers have also invented fiber-optic reactors, which are a class of fixed bed reactors (Figure 9.4a). Structurally, the bundled fibers are placed in a reactor, and the catalyst is coated on the outer surface of the fibers. In this case, light propagates inside the fibers in a predicted direction parallel to the fibers (Figure 9.4b). Notably, the matching of a proper semiconductor and the light with suitable wavelengths is of particular importance to ensure complete absorption of the light. This type of reactor thus reduces the absorption loss of light propagating at the beginning. For example, Wu et al. [20] designed and assembled a fiber-optic reactor consisting of 216 quartz fibers of 112 μm diameter coated with metal-loaded titanium dioxide nanoparticles. When using 1.0 wt%-Ag/TiO_2 photocatalyst, the optimum methanol yield of 4.12 μmol g_{cat}^{-1} h^{-1} was obtained at a given reaction condition containing 1.13 bar of CO_2 and 0.03 bar of H_2O. The mixture was irradiated under UV light with a light density of 10 W m^{-2} and an average residence time of 5000 s. In another experiment, Cu and Fe co-doped TiO_2 photocatalyst was prepared by a sol–gel method and coated on the optical fiber. In this case, the methane yield reached 0.914 mol g_{cat}^{-1} h^{-1}. Under the same reaction conditions, the methane yield was only 0.060 mol g_{cat}^{-1} h^{-1} when the catalyst was simply deposited on a glass plate [21]. The results clearly demonstrated the significance of sufficient light absorption in determining the performance.

9.3.2 Energy Flow on the Surface of the Photocatalyst

The incident light reaching the surface of the catalyst cannot be fully utilized to generate electrons and holes. Only those photons with energy larger than the band-gap energy of the semiconductor can excite electrons from the valence band to the conduction band. A portion of the light is scattered by the surface of the catalyst, and this portion of the light continues to propagate within the reactor, may be absorbed, or may pass through the walls of the reactor. Part of the absorbed light is directly converted into heat dissipation. Photons with energy greater than the band gap of the semiconductor will have the electrons excited from the valence band to the conduction band. It is worth noting that only a certain amount of energy can be stored in the charge carriers. More specifically, for a given photon, only the energy equaling to the band-gap energy can be transferred; the rest will be lost in the form of thermal energy. Consequently, the efficiency of light absorption should be extremely important. One effective way to enhance light absorption is through multiple absorption, which can be achieved by the formation of multiple reflective structures in the catalyst particle. A notable example can be found in the synthesis of the petal-like TiO_2 material, which was fabricated by the self-assembly of the nanosheet [22]. Due to the gap between the nanosheets, the surface of the material is distributed with mesopores of about 10 nm. The incident light is scattered and absorbed multiple times on the surface of the catalyst, which improves the absorption efficiency.

In order to fundamentally improve the light-absorption efficiency, it is necessary to broaden the light response of the material. The conduction band of the transition metal oxide semiconductor is generally composed of 3d orbital from the metal cations, and the valence band is generally composed of 2p orbital from oxygen ions. The doped metal or nonmetal element provides a doping level. If the doping level is located in the forbidden band of the raw material, this intermediate level can lead to enhanced light absorption. Specifically, a slightly positive doping level than the conduction band level can lower the position of conduction band minimum, and a slightly negative doping level than the valence band level can raise the position of valence band maximum. These two conditions can improve the light-absorbing properties of the catalyst, but if the doping level is located at a position near the center of the forbidden band, it becomes a recombination center for photogenerated carriers and is detrimental to the photocatalytic reaction. Therefore, the dopant and the primary photocatalytic material should be matched. There are two doping states

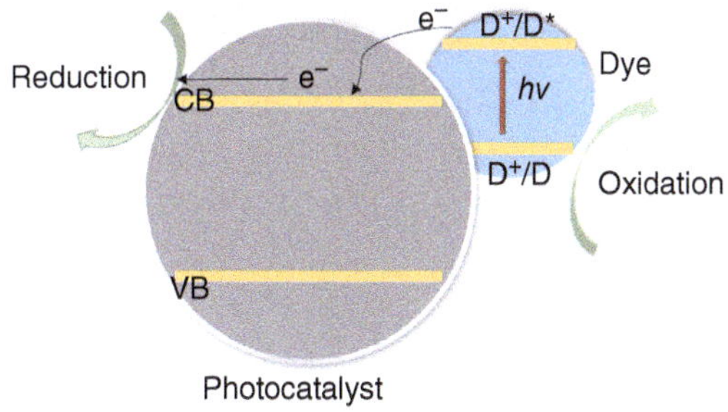

Figure 9.5 Dye-sensitized photocatalytic reaction.

of N ions in TiO_2, e.g. substitution doping and vacancy doping. The energy level of substitution doping is near the valence band, while that of vacancy doping is located near the middle position of the forbidden band. Nitrogen doping leads to a narrow band gap by hybridization of N 2p and O 2p orbits, thus inducing photocatalytic activity under visible-light irradiation. At the same time, oxygen vacancy will be generated, which is conducive to the absorption of CO_2 [23]. In addition, a quantum dot sensitizer or dye sensitizer with a narrow band gap loaded on the main photocatalyst can assist the absorption of more sunlight. Specifically, as shown in Figure 9.5, some organic dye compounds also have strong light-absorption ability. Once they are photoexcited, the electrons can be transferred from the organic matter to the conduction band of the semiconductor if the energy potential of the generated electrons is higher than the conduction band level of the semiconductor catalyst. The compounds as a result would occupy a strong oxidation state with excessive positive charges. The transferred electrons move to the surface of the catalysts to reduce CO_2, and the organic substances in the oxidized state can oxidize H_2O to form O_2 and finally return to their ground state. Consequently, if the organic light-sensitizing materials with appropriate band levels are selected, the light-absorption ability of the wide-band-gap semiconductor can be greatly improved.

Anatase phase TiO_2 has been widely studied because of its many advantages, but its application has been restricted by the limited light utilization ability. Sharma et al. used potassium ferrocyanide as a dye sensitizer to coat TiO_2 powder for photocatalytic reaction in aqueous solution [24]. Formic acid and formaldehyde were generated from the reaction. The study on the particle size of catalyst shows that the smaller the particle size, the higher the yield of reduction products. In another attempt, Wang et al. synthesized and characterized a series of TiO_2 photocatalysts sensitized by CdSe quantum dots [25]. The band structures of CdSe and TiO_2 and relevant redox potentials of CO_2 and H_2O are shown in Figure 9.6. Upon visible-light irradiation ($\lambda > 420$ nm), they obtained CH_4 (48 ppm g^{-1} h^{-1}), CH_3OH (3.3 ppm g^{-1} h^{-1}), and trace amount of CO and H_2 as the final products.

Upon photoexcitation, the light energy is successfully transferred to the photogenerated carriers. However, the photogenerated carriers are not all involved in the redox reaction. Specifically, some of the energy cannot transfer and be stored into

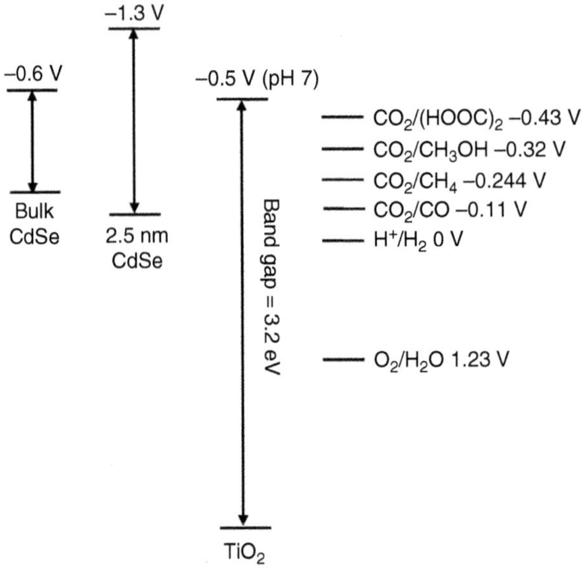

Figure 9.6 Band alignment of bulk CdSe, CdSe QDs with a diameter of 2.5 nm, and TiO_2, as well as relevant redox potentials of CO_2 and H_2O. Source: Wang et al. [25]. © 2010, American Chemical Society.

the fuel because of the recombination of electron–hole pairs. Generally, recombination can occur inside or on the surface of the catalyst to generate fluoresce or heat energy. The time scales of electron excitation, electron transfer, and electron reactions are very different. Among these procedures, electrochemical reactions require the longest time. This time mismatching causes significant accumulation of photogenerated charges on the surface of the photocatalyst and thus increased probability of recombination. Therefore, the primary method for reducing the probability of recombination is the rapid separation and utilization of electron–hole pairs. Efficient utilization of photogenerated charges represents a way of improving heterogeneous reaction kinetics by coupling the energy flow and mass flow. For the rapid separation of electron–hole pairs, firstly, there must be an electron transfer channel with very high conductivity, connected to a suitable electron acceptor, as well as a barrier to prevent the recombination of electron–hole pairs. If a metal with Fermi energy level lower than that of a semiconductor catalyst is deposited on the catalyst, the photogenerated electrons will flow from the semiconductor to the metal, forming a depletion layer and a Schottky barrier at the interface, which prevents the recombination of electrons and holes. For example, Slamet et al. prepared a CuO-coupled TiO_2 catalyst by an improved impregnation method [26]. They found that Cu_2O has the most positive redox potential compared with Cu and CuO, leading to the strongest ability to accept photogenerated electrons. Despite the superiority, its photocatalytic activity is not as good as that of CuO. The reason is that Cu_2O has a strong ability to capture electrons, however, without further transporting these electrons to CO_2 molecules adsorbed on the surface of the photocatalyst. As a result, the highest activity toward photocatalytic reduction was achieved over 3 wt%-CuO/TiO_2 photocatalyst with a quantum efficiency of 19.23%. This value is much higher than that of pure TiO_2 (quantum efficiency: 5.86%).

As a special class of materials, one-dimensional nanocrystals provide a channel for rapid electron transfer. To this end, we can have the anatase phase TiO_2 uniformly deposited on the carbon nanotubes. This unique structure would reduce the recombination of photogenerated carriers and is thus beneficial to the photocatalytic reaction. Compared with TiO_2 supported by graphite, TiO_2 supported by carbon nanotubes has higher catalytic efficiency, because electrons and holes can propagate on carbon nanotubes, inhibiting the recombination of electron–hole pairs [27]. Using nitrogen-doped titanium dioxide nanotube array can reduce CO_2 into hydrocarbon fuels under visible-light irradiation. The hole diffusion length in TiO_2 is about 10 nm, while the electron diffusion length is about 10 μm. Considering that the wall thickness of the nanotube is about 10 nm, photogenerated carriers can easily propagate to the catalyst surface [28]. Figure 9.7 shows the reaction of CO_2 and H_2O molecules in the nanotube arrays. However, it is worth pointing out that excessive carbon nanotubes would also attenuate the light absorption of the primary photocatalyst. In this case, photocatalytic efficiency is sometimes rather reduced.

Heterojunctions are widely used in photocatalytic reactions. There are three types of semiconductor heterojunctions: straddling type (type I), staggered type (type II), and broken type (type III) as shown in Figure 9.8. Type I heterojunction consists of two semiconductors A and B. The conduction band of semiconductor A is higher than that of semiconductor B, and the valence band is lower than that of semiconductor B. Therefore, holes and electrons are transferred and aggregated to semiconductor B. Such heterojunctions cannot function to separate charges. In a type II heterojunction, photogenerated electrons are transferred from semiconductor A to semiconductor B. This is because the conduction band position of semiconductor B is more positive. And holes can be transferred from the valence band of semiconductor B to the valence band of semiconductor A, because the valence band of semiconductor A is more positive than that of semiconductor B. This type of heterojunction achieves a charge separation process in which electrons and holes are separately transferred to two semiconductors, thereby increasing photocatalytic activity. Asi et al. [29] designed $AgBr/TiO_2$ composite materials. As shown in Figure 9.9, under the irradiation of visible light, photogenerated electrons on AgBr can be efficiently

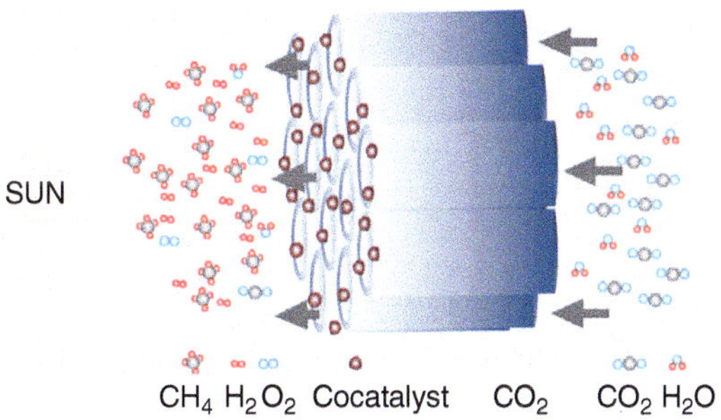

Figure 9.7 Schematic diagram of the reduction of CO_2 and water vapor into hydrocarbon fuels through a nanotube–catalyst array. Source: Varghese et al. [28]. © 2009, American Chemical Society..

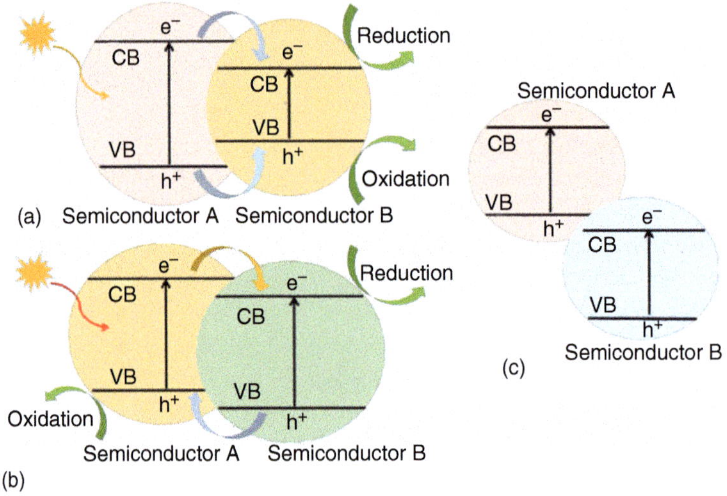

Figure 9.8 Semiconductor heterojunctions in the form of type I (a), type II (b), and type III (c).

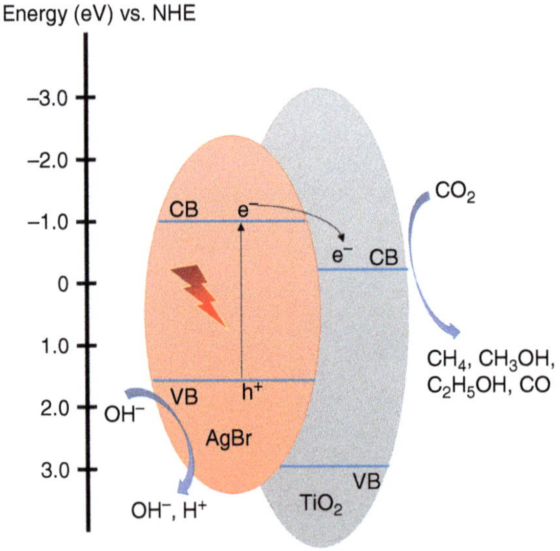

Figure 9.9 Schematic diagram of the photoexcitation process of $AgBr/TiO_2$ composite under visible-light irradiation.

transferred to TiO_2. Type III band alignments are generally unacceptable for photocatalysis since slight impact of charge transfer behaviors can be expected in this kind of heterojunctions.

The band position is not only related to the material but also related to the crystal form and the crystal face. The rutile phase and the anatase phase of TiO_2 are conducted with a difference of 0.2 eV. Time-resolved photoluminescence is used to study the mixed-phase TiO_2 sample with anatase phase and rutile phase, and the photocatalytic activity is enhanced. This enhanced photocatalytic activity is attributed to efficient charge separation and reduced charge recombination at the anatase/rutile junction [30].

Yu et al. [31] proposed a new concept of "surface heterojunction" based on the density functional theory (DFT) calculation. Since the {101} face is more thermodynamically stable than the {001} face, the less reactive {101} face accounts for about 90% of the entire exposed surface of the natural anatase TiO_2. The {001} surface shows a larger Ti–O–Ti bond angle, while the {101} surface shows a smaller Ti–O–Ti bond angle. When Ti–O–Ti bond angle is large, the 2p state of O atoms on the {001} surface is unstable and very active, so it has higher photocatalytic activity. The Fermi level of {001} face is located at the lower end of the valence band top, while the Fermi level of {101} face is still located at the upper end of the valence band top. Therefore, when the {001} face and the {101} face make a contact, an energy band bending is generated, and a heterogeneous junction will be formed (Figure 9.10), which is beneficial to the separation of photogenerated charges.

The addition of HF controls the growth of TiO_2 crystals. Yu et al. reported that larger proportion of {001} crystal faces could be obtained by adding larger amount of HF (Figure 9.11). The results showed that the highest methane yield is 1.35 $\mu mol\,h^{-1}\,g^{-1}$ at a moderate amount of HF. The photocatalytic activity is three times and nine times higher than that of commercially available P25 and TiO_2 without adding HF, respectively [31].

The heterojunction inhibits the recombination of electrons and holes to some extent, but the photocatalytic activity is weakened due to the decrease in energy when electrons and holes are transferred in the heterojunction. Based on this idea, a Z-scheme photocatalytic system was designed. In a Z-scheme photocatalytic system,

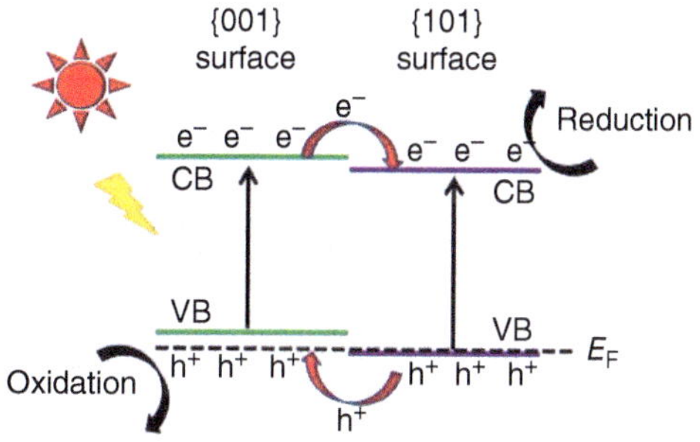

Figure 9.10 Schematic illustration of TiO_2 with {001} and {101} surface heterojunction. Source: Yu et al. [31]. © 2014, American Chemical Society.

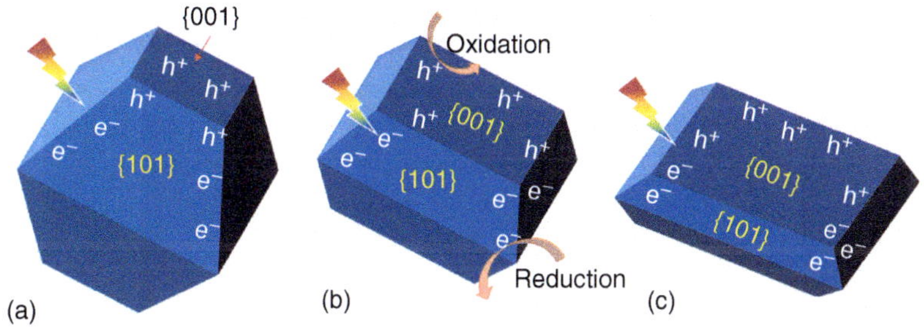

Figure 9.11 Schematic illustration of the spatial separation of redox sites on the TiO_2 photocatalysts prepared without HF (a), by adding a moderate amount of HF (b), and by adding a high amount of HF (c).

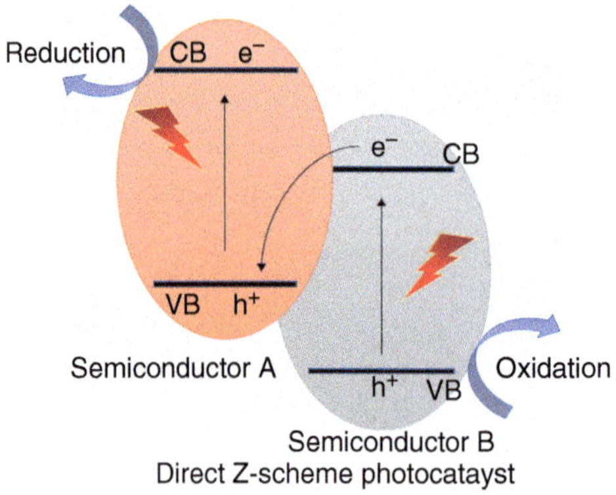

Figure 9.12 Schematic diagram of charge transfer in a Z-scheme semiconductor–semiconductor composite.

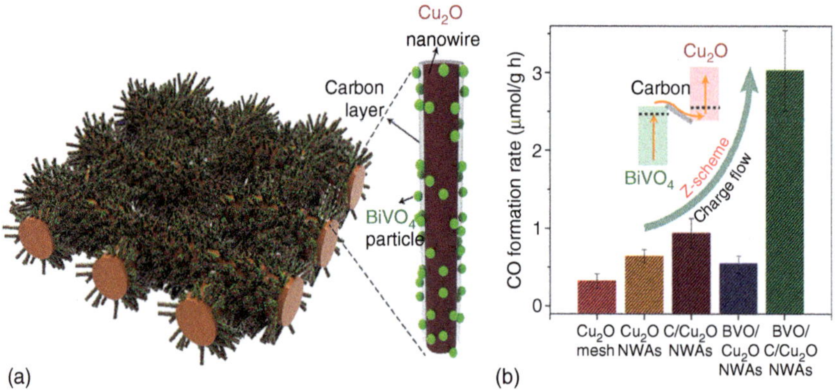

Figure 9.13 Schematic diagram of Z-scheme $BiVO_4/C/Cu_2O$ nanowire arrays. Source: Reproduced with permission. Kim et al. [32]. © 2018, American Chemical Society.

photogenerated electrons from the lower conduction band of semiconductor B will recombine with the photogenerated holes from semiconductor A with higher valence band, so that electrons/holes have a stronger reduction/oxidation ability at different active sites of semiconductor A/semiconductor B (Figure 9.12). Kim et al. [32] prepared the 3D carbon-coated Cu_2O nanowires, with $BiVO_4$ deposited on the nanowires (Figure 9.13a). This structure uses the advantages of Cu_2O that it has high position of the conduction band and narrow band gap. Due to the difference in the energy band positions of $BiVO_4$ and Cu_2O, the Z-scheme structure was formed, and with the favorable carrier transport characteristics of the carbon layer, the problem of easy recombination of electron–hole pair and poor stability of the catalyst was solved. Attributed to the above advantages, the photocatalytic CO_2 conversion performance of $BiVO_4/C/Cu_2O$ nanowires featured a CO formation rate of 3.01 $\mu mol\,h^{-1}\,g^{-1}$, which is 9.4 and 4.7 times those on Cu_2O mesh and Cu_2O nanowires, respectively (Figure 9.13b).

The binary $g\text{-}C_3N_4/ZnO$ photocatalytic system has been constructed by a simple calcination method, whose band positions are shown in Figure 9.14. Compared with pure $g\text{-}C_3N_4$, the prepared photocatalytic system of $g\text{-}C_3N_4/ZnO$ showed a 2.3-fold increase in photocatalytic activity for CO_2 reduction [33].

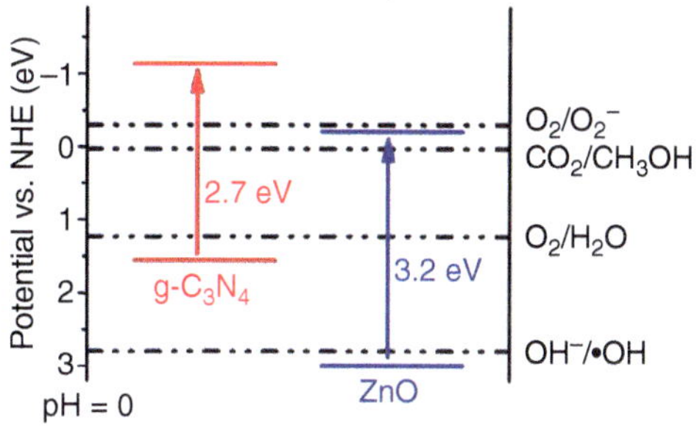

Figure 9.14 Comparison of the energy band positions and the related redox reaction potentials of g-C_3N_4 and ZnO. Source: Yu et al. [33]. © 2015, Royal Society of Chemistry.

When photogenerated electrons are transferred to the surface of the catalyst without recombination, they will be transferred to the electron acceptor. Both CO_2 and a series of intermediate reduction products can accept electrons. H^+ will be reduced into hydrogen radicals, which plays an essential role in the reduction of CO_2 into hydrocarbon fuels, and the details will be discussed during the mass flow. Photogenerated holes accept electrons from the electron donors. In the absence of sacrificial agents, water acts as an electron donor and is oxidized to oxygen or peroxide ions. This process has a large overpotential. More often, it is necessary to add a sacrificial agent to remove holes to reduce the recombination of photogenerated electrons and holes and the self-corrosion of the catalyst. For the photoelectrocatalytic reaction using a thin film electrode, an n-type semiconductor is used as a photoanode, a p-type semiconductor is used as a photocathode, and photogenerated electrons pass through an external circuit to reach the counter electrode. In the process of electron migration, it is necessary to overcome the contact potential of the different materials that will otherwise cause a part of the energy loss.

9.3.3 Mass Flow in CO_2 Photocatalytic Reduction

No matter if it is a gas–solid interface reaction or a liquid–solid interface reaction, photogenerated electrons and holes are transferred to the catalyst surface, and CO_2 and H_2O are also present on the catalyst surface. However, CO_2 is an extremely stable molecule, and the reaction is difficult to occur. The time scale of light absorption and the separation of electron and hole are much smaller than the time scale of redox reaction of CO_2 and H_2O. This timing mismatch makes the photogenerated electron–hole pairs accumulate on the surface of the catalyst, and then a large amount of them combine with each other. Electronic energy is converted into light and heat. The more serious problem is that some photocatalysts have photocorrosion problems, such as CdS materials. If the oxidation reaction on the electrode process is too slow, the holes will oxidize the catalyst causing self-oxidation, resulting in catalyst out of work. Therefore, to study the principle of surface catalytic reaction for CO_2 reduction, designing a suitable catalyst to accelerate the CO_2 reaction is a key step to improve the photocatalytic efficiency.

Considering the reaction of CO_2 in the solution, the CO_2 gas is introduced into the reaction liquid through a syringe pump, and the place where the reaction takes place is the interface where the catalyst and the solution contact. In order to better

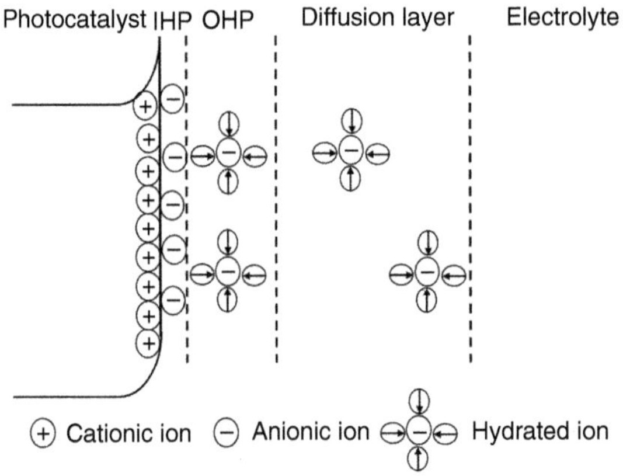

Figure 9.15 Schematic diagram of the solid–liquid interface structure.

understand and improve the mass flow process, it is necessary to have a thorough understanding of the interface between catalyst and solution. When the n-type semiconductor catalysts enter the solution, the Fermi level of semiconductor is higher than the Fermi level of solution, and the electrons are transferred from the semiconductor to the electrolyte adsorbed on the surface of the catalyst. Finally, the semiconductor band is upward bending, and the bent portion called the electric depletion layer is positively charged, while the electrolyte molecules adsorbed on the surface of the catalyst are negatively charged, naturally forming an electric double layer.

In history, scientists have proposed different electric double-layer models. In 1947, Grahame summarized the conclusions of the predecessors and improved the electric double-layer model. As shown in Figure 9.15, the innermost layer close to the solid surface has the ions without hydration, which is called the inner Helmholtz layer (IHP). The outside layer has the partially hydrated ions, which is called the outer Helmholtz layer (OHP). The two are collectively called the Helmholtz layer, and the outside region is the diffusion layer [34].

Reactants have to migrate from the inside of the liquid phase to the Helmholtz layer. In the liquid phase, there is no concentration difference, and the electric field is weak, so the main transfer effect is convection. In the diffusion layer, there is concentration difference, and the main transfer effect is diffusion. In the Helmholtz layer, although the voltage is small, the field strength is very large due to the extremely short distance, and the ions move mainly by electromigration [35].

The complete reaction process consists of five parts, e.g. liquid-phase mass transfer process before reaction, pre-conversion process, electrochemical process, post-reaction conversion process, and post-reaction liquid-phase mass transfer process. Photocatalytic reduction of CO_2 in solution includes liquid-phase mass transfer process, electrochemical process, and liquid-phase mass transfer process after reaction. In the process of mass transfer in liquid phase before the reaction, CO_2 reaches the IHP from the inside of the liquid phase. In the IHP, linear CO_2 molecules interact with the catalyst, and electrons are injected into the CO_2 molecules from the conduction band of the catalyst. If the linear structure of CO_2 is destroyed, the C=O bond becomes longer, and the symmetric stretching vibration frequency becomes weakened. Such CO_2 molecules are prone to various reactions,

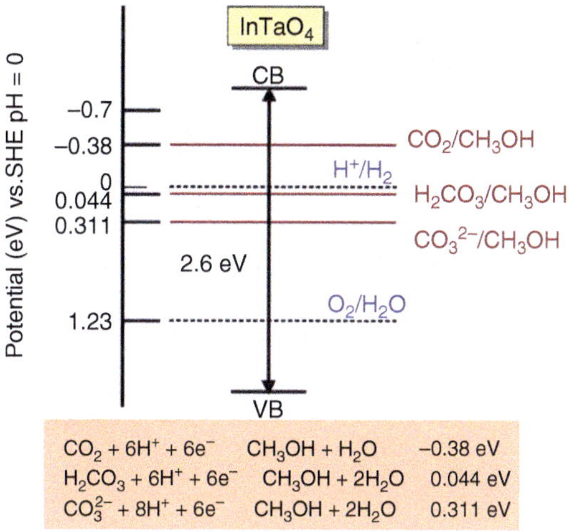

Figure 9.16 Potential comparison about the reduction of CO_2, H_2CO_3, and CO_3^{2-} ions into methanol. Source: Pan and Chen [38]. © 2007, Elsevier.

which are called activated CO_2 molecules [36]. H^+ ions accept electrons in the IHP and become H atoms to be adsorbed on the surface of the catalyst, and activated CO_2 molecules continue to obtain electrons and H atoms to generate various intermediate products and final products. After the reaction, the final product enters the liquid phase or forms bubbles to separate from the solid–liquid interface.

The solubility of CO_2 in pure water is small, and the low concentration of CO_2 is not conducive to photocatalytic reduction reaction. Adding alkaline sacrificial agent can effectively enhance the reaction activity. OH^- has two functions. Firstly, it acts as a hole scavenger to quickly remove holes in valence band and reduce charge recombination. Secondly, in alkaline environment, the solubility of CO_2 is greatly enhanced. After saturation, the solution mainly contains HCO_3^-, with a small amount of CO_2, H_2CO_3, and CO_3^{2-} [37]. When CO_2 becomes CO_3^{2-}, the potential decreases by nearly 0.7 eV (Figure 9.16). For the same catalyst, the potential difference between the redox potential and the bottom of the conduction band becomes larger, and the reduction ability is enhanced [38].

Polar groups and lone pairs on the catalyst surface can adsorb CO_2 molecules efficiently. On the surface of the $BiVO_4$ catalyst, CO_3^{2-} are fixed through weak Bi—O bond, which is favorable for CO_3^{2-} to obtain photoelectrons from V3d orbit. However, Bi ions in monoclinic crystal are more asymmetric than that in tetragonal crystal phase, so lone pair is stronger and Bi–O bonds can be formed, and $BiVO_4$ in monoclinic crystal phase has higher reactivity [39]. Huang et al. [40] used in situ diffuse reflectance infrared Fourier transform spectroscopy (DRIFTS) to study the photocatalytic reduction of CO_2 on TiO_2 and Cu/TiO_2 photocatalysts and proposed a possible adsorption mechanism of CO_2 on the surface of TiO_2. Hydroxyl groups on the catalyst surface play an essential role in the adsorption of CO_2. Oxygen vacancies on the catalyst surface also play a role in the adsorption of CO_2. CO_2 adsorption on the catalyst surface forms HCO_3^- bicarbonate and CO_3^{2-}, respectively. As shown in Figure 9.17, there are three types of adsorption substances on the surface of TiO_2. One way is that CO_2 reacts with free OH groups on the surface and is converted

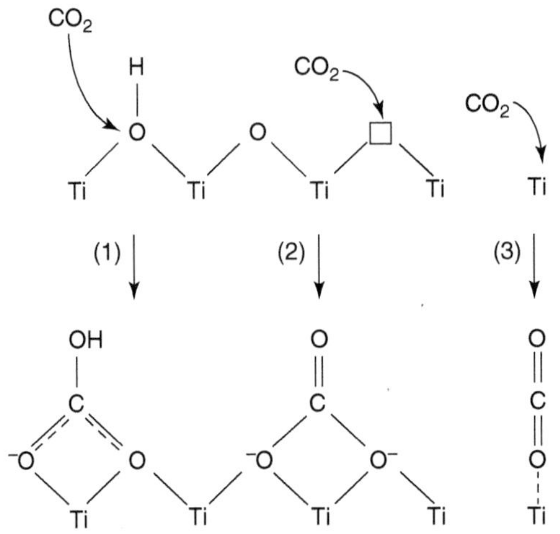

Figure 9.17 Possible adsorption mechanism of CO_2 on TiO_2 surface. Source: Wu and Huang [40]. © 2010, Springer Nature.

into HCO_3^-. The second way is that CO_2 attaches to oxygen vacancies and then be converted into CO_3^{2-}. The third way is that CO_2 is chemisorbed on the surface of TiO_2 and reaches equilibrium quickly. In this adsorption state, CO_2 can easily obtain electrons from the conduction band of catalyst, and the reduced hydrogen ions are also adsorbed on the TiO_2, so the hydrogenation process then occurs to generate methanol.

The cocatalyst can provide a large number of reactive sites, which have strong adsorption for CO_2. The optimized reduced graphene oxide (RGO)–CdS nanorod composite photocatalyst, illustrated in Figure 9.18, showed a high CH_4 production rate of 2.51 μmol h $^{-1}$ g^{-1} when the RGO content was 0.5 wt%. This high photocatalytic activity is attributed to the deposition of CdS nanorods on RGO plates that act as electron acceptors and transporters, thus effectively separating photogenerated charge carriers. On the other hand, the p–p conjugate interaction between CO_2 and graphene leads to enhanced adsorption of CO_2 on RGO–CdS nanorod composites and also leads to instability and activation of CO_2 molecules [41].

9.3.4 Product Selectivity in CO_2 Photocatalytic Reaction

Besides the low energy conversion efficiency, the photocatalytic reduction of CO_2 also has the problem of low product selectivity. According to reported studies, the products of CO_2 reduction mainly include the following categories: H_2, monocarbon compounds (CO, CH_4, HCOOH, HCOH, CH_3OH), and bicarbon compounds (C_2H_6O, C_2H_4O), with the most reported being monocarbon compounds. Among these products, H_2, CO, and CH_4 are water-insoluble gases. During the reaction in aqueous solution environment, the gases undergo nucleation, growth, detachment, and liquid-phase transport and finally get out from the reaction solution. However, other products are all oxygen-containing derivatives of hydrocarbons and can be dissolved in water, which makes product separation difficult; especially in systems employing slurry reactors, catalyst particles are also difficult to separate. Since

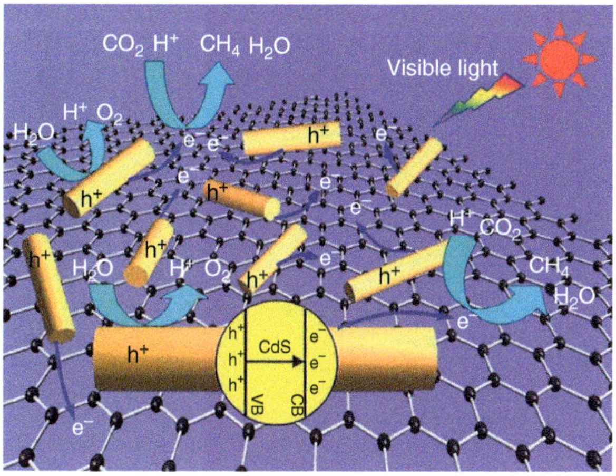

Figure 9.18 Schematic illustration of the charge transfer and separation in RGO–CdS nanorod system under visible-light irradiation. Source: Yu et al. [41]. © 2014, Royal Society of Chemistry.

the research on photocatalytic reduction of CO_2 was carried out in 1970s, many researchers have studied the production mechanism and reaction selectivity. The process of CO_2 reduction has not been fully understood, but some conclusions have been reached.

During the photocatalytic reduction–oxidation reaction of CO_2 and H_2O, two intermediate products are the most important: CO_2^- ion radical and H atom radical. Researchers have proposed two possible pathways for reducing CO_2 to produce methane, formaldehyde pathway, and carbene pathway (Figure 9.19). What the two paths have in common is that CO_2 obtains electrons and hydrogen ions step-by-step, eventually producing methane. In the formaldehyde pathway, CO_2 first obtains two electrons and two protons to generate formic acid. Formic acid obtains two electrons and two protons and removes one water molecule to obtain formaldehyde. Formaldehyde obtains two electrons and two protons to obtain methanol. Methanol obtains two electrons and two protons and removes one water molecule to obtain methane. In the carbene pathway, CO_2 gets two electrons and one proton and removes one hydroxide ion to get CO. The obtained CO gets two electrons and one proton and removes one hydroxide to generate C atom. Subsequently, C atom may undergo a four-hydrogenation process to produce methane or complete a three-hydrogenation process and further add one hydroxyl radical to get methanol [42].

The product type and relative yield are related to the catalyst material, the size of catalyst particles, and the surface morphology of the catalyst. Yamamura et al. found that 100 mesh SiC powder as the catalyst did not produce ethanol, while 1000 mesh SiC powder produced ethanol [4]. Lkeue et al. used OH^- and F^- as structure-directing anions to synthesize Ti-β zeolite by hydrothermal method in different states [43]. Ti-β zeolite prepared by OH^- has hydrophilicity, while that prepared by F^- has hydrophobicity. The photoreduction reaction shows that Ti-β(OH)

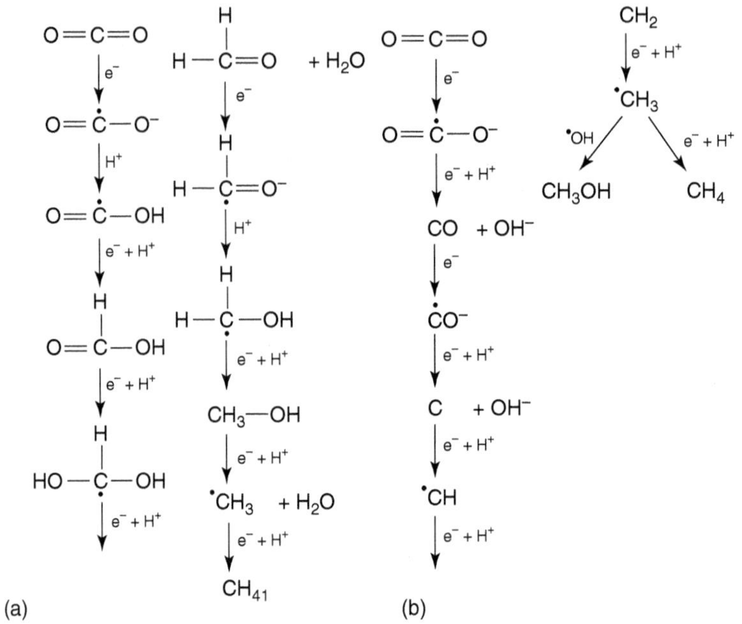

Figure 9.19 Two proposed mechanisms for the photoreduction of CO_2 to methane: formaldehyde (a) and carbine (b) pathways. Source: Izumi [42]. © 2015, American Chemical Society.

has the highest methane and methanol production rate, mainly methane, and also contains trace carbon monoxide. In methanol selectivity, Ti-β(OH) is 11%, and Ti-β(F) is 41%. The results of fluorination show that the stronger the fluorination is, the higher hydrophobicity of the catalyst and selectivity for methanol are achieved. This is because the larger ratio of H_2O to CO_2 leads to higher selectivity for CH_4. The greater the polarity of the catalyst surface is, the more H_2O molecules will be adsorbed, thus showing higher CH_4 selectivity on the catalyst surface with greater polarity.

Metal loading has a significant effect on the selectivity of the product. As shown in Figure 9.20, TiO_2 catalyst mainly produces CO when no metal is loaded, meanwhile CH_4 when Pt is loaded [44].

Methane production requires a large number of electrons and protons. As shown in Figure 9.21, carbon quantum dots (CQDs) decoration and oxygen atom doping are applied to modify g-C_3N_4 (CN). The incorporated CQD not only facilitates charge transfer and separation but also provides CO_2 adsorption and activation sites. In addition, oxygen atom-doped CN (OCN), in which oxygen doping is accompanied by the formation of nitrogen defects, has proved to be a sustainable H^+ provider by promoting dissociation and oxidation half-reaction of water. The continuous supply of H^+ enhances the selective generation of CH_4 [45]. If we aim to produce fuel, then CH_4 and CH_3OH are undoubtedly the best products. It is necessary to prepare catalytic materials with high selectivity for CH_4 and CH_3OH products.

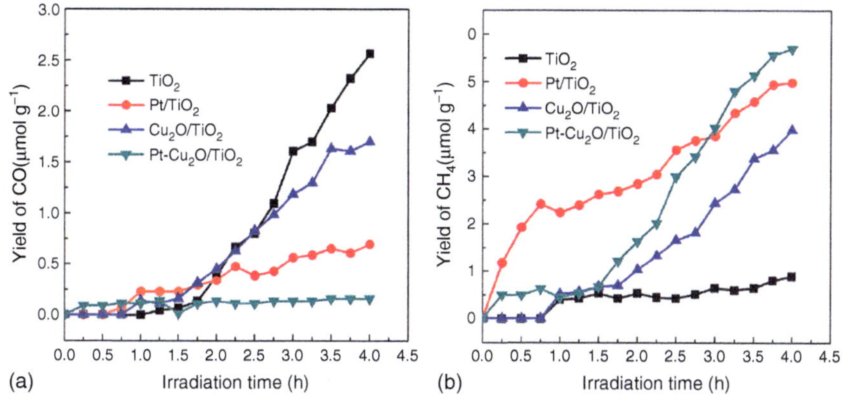

Figure 9.20 Photocatalytic reduction of CO_2 (86.65 kPa) by TiO_2 (a) and Pd(1%)-TiO_2 (b). Source: Xiong et al. [44]. © 2017, Elsevier.

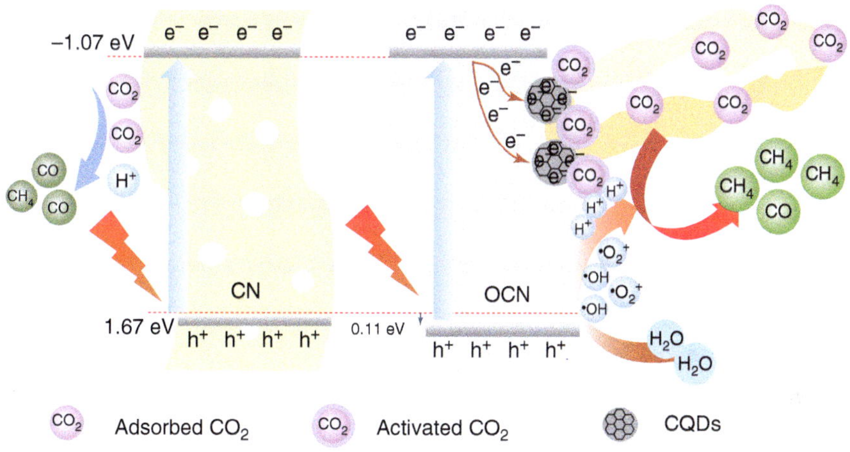

Figure 9.21 Schematic mechanism of photocatalytic CO_2 reduction over CQDs/OCN-x under visible-light irradiation. Source: Li et al. [45]. © 2019, Springer Nature.

9.4 Conclusions

In conclusion, the photocatalytic CO_2 reduction reaction includes many parts with the catalyst as the core. Good coordination of all the parts is the only way to achieve efficient photocatalytic CO_2 reduction. Energy flow and mass flow are two basic processes for the reaction. In this chapter, we start from the point of view of energy flow and mass flow. We analyze the various obstacles and energy loss in the entire process from solar energy to chemical energy stored in the fuel and the influence of the mass-change process from CO_2 to various small molecular carbon-based fuels on the energy transfer. The results show that most of the solar energy will not be absorbed by the catalyst but be dissipated in the form of light or heat, due to the transmission loss and the mismatch between the band gap of semiconductor and the solar spectrum. Moreover, some of the solar energy absorbed by the catalyst is lost due to the recombination of photogenerated charges, leading to a very low efficiency of photocatalytic CO_2 reduction. CO_2 is firstly adsorbed and activated on the surface of the catalyst and then undergoes a process of obtaining electrons and hydrogen step-by-step to be reduced into monocarbon compounds or bicarbon compounds. Good adsorption and activation capability will accelerate the CO_2 reaction rate and reduce the recombination caused by charge accumulation. Due to the difference in the intermediate reaction itself and the difference in the surface state of the catalyst, the reduction products from CO_2 show diversified characteristics. Optimized design of concentrators and reactors can reduce the loss of sunlight during propagation. Targeted modification of catalysts, through structural design, surface loading, doping, sensitization, formation of heterojunctions, etc., can enhance light-absorption rate and reduce carrier recombination rate. Further research on the principle of CO_2 reduction reaction is needed to guide the research of catalytic materials with high reaction activity and high product selectivity.

Acknowledgments

This work is supported by the National Natural Science Foundation of China (No. 51876173 and No. 51906197), the Shaanxi Technical Innovation Guidance Project (No. 2018HJCG-14), the Natural Science Foundation of Jiangsu Province (No. BK20190054), and China Fundamental Research Funds for the Central Universities.

References

1 Fujishima, A. and Honda, K. (1972). *Nature* 238: 37.
2 Halmann, M. (1978). *Nature* 275: 115.
3 Inoue, T., Fujishima, A., Konishi, S., and Honda, K. (1979). *Nature* 277: 637.
4 Yamamura, S., Kojima, H., Iyoda, J., and Kawai, W. (1987). *J. Electroanal. Chem.* 225: 287.
5 Aurian-Blajeni, B., Halmann, M., and Manassen, J. (1980). *Sol. Energy* 25: 165.
6 Aliwi, S. and Al-Jubori, K. (1989). *Sol. Energy Mater.* 18: 223.
7 Inoue, H., Torimoto, T., Sakata, T. et al. (1990). *Chem. Lett.* 19: 1483.
8 Kohno, Y., Tanaka, T., Funabiki, T., and Yoshida, S. (1997). *Chem. Commun.* 33: 841.
9 Wu, S., Cao, H., Yin, S. et al. (2009). *J. Phys. Chem. C* 113: 17893.
10 Jiang, Z., Wan, W., Li, H. et al. (2018). *Adv. Mater.* 30: 1706108.
11 Huang, W., Lyu, L., Yang, Y. et al. (2012). *J. Am. Chem. Soc.* 134: 1261.
12 Li, X., Wen, J., Low, J. et al. (2014). *Sci. China Mater.* 57: 70.
13 Wang, L., Chen, Y., Zhao, H. et al. (2015). *Mater. Rev.* 29: 147.
14 Crouch, A., Charbonneau, P., Beaubien, G., and Paquin-Ricard, D. (2008). *Astrophys. J.* 677: 723.

15 Bo, L. (1978). *Sol. Energy* 20: 143.

16 Han, S., Chen, Y., Abanades, S., and Zhang, Z. (2017). *J. Energy Chem.* 26: 743.

17 Ola, O. and Maroto-Valer, M. (2015). *J. Photochem. Photobiol., C* 24: 16.

18 Pathak, P., Meziani, M., Castillo, L., and Sun, Y.P. (2005). *Green Chem.* 7: 667.

19 Bakherad, M., Moosavi-Tekyeh, Z., Rezaeifard, A. et al. (2020). Metal-free green synthesis of aryl amines in magnetized distilled water: experimental aspects and molecular dynamics simulation. *Green Chem.*, Advance Article, DOI: https://doi.org/10.1039/D0GC01329C.

20 Wu, J., Wu, T., Chu, T. et al. (2008). *Top. Catal.* 47: 131.

21 Nguyen, T. and Wu, J. (2008). *Sol. Energy Mater. Sol. Cells* 92: 864.

22 Parlett, C., Karen, W., and Lee, A. (2013). *Chem. Soc. Rev.* 42: 3876.

23 Zhao, Z., Fan, J., Wang, J., and Li, R. (2012). *Catal. Commun.* 21: 32.

24 Sharma, B., Ameta, R., Kaur, J., and Ameta, S.C. (1997). *Int. J. Energy Res.* 21: 923.

25 Wang, C., Thompson, R., Baltrus, J., and Matranga, C. (2010). *J. Phys. Chem. Lett.* 1: 48.

26 Slamet, N.H., Purnama, E., Kosela, S., and Gunlazuardi, J. (2005). *Catal. Commun.* 6: 313.

27 Xia, X., Jia, Z., Yu, Y. et al. (2007). *Carbon* 45: 717.

28 Varghese, O., Paulose, M., Latempa, T., and Grimes, C.A. (2009). *Nano Lett.* 9: 731.

29 Abou Asi, M., He, C., Su, M. et al. (2011). *Catal. Today* 175: 256.

30 Wang, X., Shen, S., Feng, Z., and Li, C. (2016). *Chin. J. Catal.* 37: 2059.

31 Yu, J., Low, J., Xiao, W. et al. (2014). *J. Am. Chem. Soc.* 136: 8839.

32 Kim, C., Cho, K., Al-Saggaf, A. et al. (2018). *ACS Catal.* 8: 4170.

33 Yu, W., Xu, D., and Peng, T. (2015). *J. Mater. Chem. A* 3: 19936.

34 Wu, X., Jia, Z., Ma, H. et al. (2013). *Energy Storage Sci. Technol.* 2: 152. (in Chinese).

35 Sang, L., Zhang, Y., Wang, J. et al. (2016). *Phys. Chem. Chem. Phys.* 18: 15427.

36 Freund, H. and Roberts, M. (1996). *Surf. Sci. Rep.* 25: 225.

37 Kočí, K., Matějů, K., Obalová, L. et al. (2010). *Appl. Catal., B* 96: 239.

38 Pan, P. and Chen, Y. (2007). *Catal. Commun.* 8: 1546.

39 Liu, Y., Huang, B., Dai, Y. et al. (2010). *Catal. Commun.* 11: 210.

40 Wu, J. and Huang, C. (2010). *Front. Chem. Eng. Chin.* 4: 120.

41 Yu, J., Jin, J., Cheng, B., and Jaroniec, M. (2014). *J. Mater. Chem. A* 2: 3407.

42 Izumi, Y. (2015). *ACS Symp. Ser.* 1194: 1.

43 Ikeue, K., Yamashita, H., Anpo, M., and Takewaki, T. (2001). *J. Phys. Chem. B* 105: 8350.

44 Xiong, Z., Lei, Z., Kuang, C. et al. (2017). *Appl. Catal., B* 202: 695.

45 Li, Q., Wang, S., Sun, Z. et al. (2019). *Nano Res.* 12: 2749.

10

Photoelectrochemical CO_2 Reduction

Zhongxue Yang[1], Hui Ning[1], Qingshan Zhao[1], Hongqi Sun[2] and Mingbo Wu[1]

[1]*China University of Petroleum (East China), College of New Energy, College of Chemical Engineering, State Key Laboratory of Heavy Oil Processing, No. 66, Changjiang West Road, Qingdao 266580, P.R. China*
[2]*Edith Cowan University, School of Engineering, 270 Joondalup Drive, Joondalup, WA 6027, Australia*

10.1 Introduction

The real charm of nature lies in the transformation of CO_2 to life-based organic matter through the work of a "leaf". Therefore, many researchers have mimicked the photosynthesis of nature in order to achieve "artificial photosynthesis", as described in chapters 8 and 9. The driving force in photosynthesis is the light energy, being used for releasing hydrogen from H_2O and carbon from CO_2 for the synthesis of organic matter. The reduction of CO_2 by photocatalysis (PC) [1, 2] and electrocatalysis (EC) [3] also takes H_2O as the source of hydrogen and converts CO_2 into carbonaceous organic matter. The difference is that photocatalysis is driven by light, while electrocatalysis is driven by electricity. In addition to photocatalytic and electrocatalytic reduction of CO_2, in recent years, photoelectrocatalysis technology is more and more favoring by the majority of researchers [4–6].

10.1.1 Introduction of Photoelectrocatalytic Reduction of CO_2

The galloping consumption of fossil resources calls for urgent development of clean and renewable energy. Due to the inexhaustible, solar energy exhibits splendid merits as the most attractive alternative energy source. Remarkable efforts have been focused on photoelectrochemical (PEC) [7, 8] water splitting [9, 10] for maximizing the utilization efficiency of solar energy, which generally employs semiconductors (SC) as light-harvesting antennas [11]. In 1978 [12], Halmann first reported the PEC reduction of CO_2 to HCOOH, HCHO, and CH_3OH on the semiconductor photocathode p-Gap with an external electrical bias in nature. Since then, there have been many studies on the use of p-type semiconductors (e.g. CdTe, GaP, GaAs, InP, Si, and FeS_2) for driving the PEC reduction of CO_2 in aqueous media [13]. Nowadays, a mass of heterogeneous and homogeneous photoelectrocatalysts have been enormously developed and optimized to improve the efficiency of CO_2 reduction [6, 14]. However some achievement has been made; reducing CO_2 directly by photoelectron

Solar-to-Chemical Conversion: Photocatalytic and Photoelectrochemcial Processes, First Edition.
Edited by Hongqi Sun.

reduction is still believed to be a huge challenge because that (i) CO_2 molecule has a stable thermodynamic configuration and a high energy barrier and (ii) the reaction processes are multi-electron transfer process and very complicated, leading to the diversity and poor selectivity of products [15]. For this reason, the PEC reduction of CO_2 conversion efficiency and selectivity is largely limited.

10.1.2 Principles of Photoelectrocatalytic Reduction of CO_2

The principles of generation and utilization of photoinduced carriers for PEC CO_2 reduction are similar to that for water splitting [2]. Under light illuminations, the incident photon energy can be absorbed if it is equal to or higher than the band-gap energy width of a semiconductor (Eg). Then the electrons from valence band (VB) will be excited to conduction band (CB); simultaneously the holes will be left in the VB. In order to make full use of the photogenerated electrons to reduce CO_2, it is necessary to avoid or slow down the combination of electrons and holes to promote the transfer of charge to the interface of semiconductor and electrolyte, so as to satisfy the multiple electron process required by CO_2 reduction. However, in the absence of the external electrical bias, the electrons and holes would recombine rapidly at the nanosecond level. With the assist of external electrical bias, the Fermi Energy level can be controlled to affect band bending and promote electrons and holes directed migration and accumulation at the surface of cathode and anode, respectively. Finally, the photogenerated electrons reduce CO_2 into hydrocarbon, while the holes oxidize H_2O into O_2 or react with sacrificial agents.

10.1.3 System Configurations of Photoelectrocatalytic Reduction of CO_2

Compared with the photocatalytic process, the photoelectrocatalysis, which combines photocatalysis with electrocatalysis, can utilize external electrical bias to overcome the energy barrier and inhibit the recombination of photogenerated electron–holes, achieving a higher efficiency for solar conversion [10]. Compared with electrocatalytic process, photogenerated electron for the reduction of CO_2, which reduces the input of external energy, reduces energy consumption and is also conducive to the removal of passivates on the surface of the electrocatalyst, maximally displaying the electrocatalytic activity [16].

In the PEC system, semiconductors are commonly processed into photoelectrodes to harvest solar energy and catalyze the reactions [17]. Several significant advantages embodied in the PEC system: (i) the external electrical bias drives the photogenerated electrons and holes to migrate and arrive at the surface of cathode and anode, respectively, which can efficiently suppress or slow down their combinations; (ii) the PEC cells are usually divided into a cathode chamber and an anode chamber by a proton membrane in order to avoid the mixing of oxidation and reduction products and the secondary oxidation of cathodic reduction products by the anode; and (iii) compared with electrocatalysis technology, photoelectrocatalysis utilizing solar energy can provide additional charges and significantly

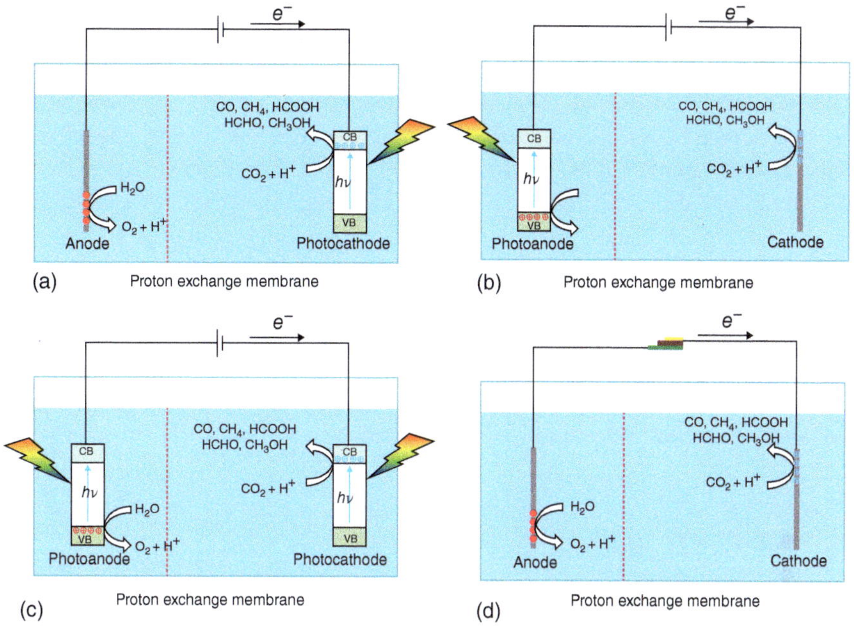

Figure 10.1 Schematic diagrams for PEC CO_2 reduction in water using a semiconductor as (a) photocathode, (b) photoanode, and (c) both photoanode and photocathode. (d) Schematic diagram for the device combining a photovoltaic cell with an efficient electrochemical catalyst for CO_2 reduction. Source: Zhang et al. [18]. © 2018, Springer Nature.

reduce electricity consumption, while powerful electric power is the guarantee of high efficiency. Given the merits, photoelectrocatalysis has been regarded to be an efficacious strategy for CO_2 reduction.

As a reduction reaction, CO_2 conversion must take place at the cathode surface of the electrochemical cell. In fact, the common designs for PEC CO_2 reduction are to construct the photocathode using p-type semiconductors (Figure 10.1a). A number of p-type semiconductors have been developed and investigated, including silicon [19], mental oxides [20, 21], chalcogenide [22], tellurides [23], phosphides [24, 25], and others [26, 27]. Another effective approach for PEC CO_2 reduction is to employ an efficient electrochemical CO_2 reduction catalyst as the cathode and use a photoanode catalyst to capture solar energy and provide photogenerated electrons (Figure 10.1b) [28, 29]. Both methods are to maximally utilize photoelectrodes to harvest solar energy and transmit photogenerated electrons. To maximize the utilization of solar energy and fully capture sunlight to generate photogenic electrons, a PEC CO_2 reduction cell combining photoanode with photocathode was developed for CO_2 reduction (Figure 10.1c) [30]. Recently, a novel tandem device for CO_2 reduction has been developed and assembled by coupling a photovoltaic cell for photoelectrical conversion with high-efficiency electrocatalyst (Figure 10.1d), also showing excellent catalytic effect and promising application prospect [31, 32].

In this chapter, we will systematically outline the recent developments for PEC CO_2 reduction, including the basic principles of photoelectrodes and the application of solar-to-chemical energy conversion in PEC CO_2 reduction. Firstly, the basic principles for PEC CO_2 reduction will be introduced to provide the readers with the fundamentals for CO_2 reduction, including the principles of PEC reduction of CO_2 and four common PEC reactors developed for CO_2 reduction. Secondly, we will summarize the diversiform approaches to evaluate the efficiency and selectivity of CO_2 reduction, such as external influence reaction conditions and evaluation parameters for PEC CO_2 reduction, thermodynamics and kinetics, matters of external reaction conditions, and evaluation parameters. Thirdly, we will examine recent research advances on PEC CO_2 reduction, including some semiconductors, cocatalysts semiconductors, and hybrid semiconductors. Fourthly, we will narrate other novel setups for PEC CO_2 reduction. At last, critical comments will be made, and perspectives for future studies will be presented.

10.2 PEC CO_2 Reduction Principles

10.2.1 Thermodynamics and Kinetics of CO_2 Reduction

The CO_2 molecule containing two C=O bonds with a linear configuration is relative stability with a net negative ΔG_f^o ($-394.4\,kJ\,mol^{-1}$), indicating that the CO_2 activation and C=O bond cleavage must overcome a high activation barrier. To drive the transformation to reduced products, the energy must be introduced into the reaction mixture, and the energy will be furnished by a renewable source such as wind, solar, or hydropower. Moreover, due to the highest chemical state (+4) of C atom in CO_2, the reduction of CO_2 can generate diversity products. The possible products include gases (CO, CH_4, C_2H_4, and other hydrocarbons) and liquids (HCOOH, CH_3OH, and C_2H_5OH) [15], which make the selectivity of the products very low and the types of products detected very complex.

The thermodynamic redox potentials for various CO_2 reduction products and water splitting are listed in Table 10.1 [15]. The first step and rate-limiting step of CO_2 reduction undergoes the one-electron reduction to $CO_2{\cdot}^-$ intermediate [33], because the reaction is an extremely thermodynamically adverse reaction, due to the relatively negative reduction potential of $-1.9\,V$ to the normal hydrogen electrode (NHE) at pH = 7. However, the CB edges of the lots of semiconductors are lower than the redox potential of $CO_2{\cdot}^-$ intermediate, suggesting that this one-electron reduction process is thermodynamically unfavorable and those semiconductors do not have an enough driving force to initiate $CO_2{\cdot}^-$ formation without an external force, partially leading to a limited CO_2 conversion efficiency.

In kinetics, the hydrocarbon products usually require more electron transfer (six and eight electron transfer for CH_3OH and CH_4, respectively) than hydrogen evolution reaction (HER) process (two-electron transfer). The second kinetic challenge of CO_2 reduction is the low solubility of CO_2 in aqueous solution ($0.033\,mol\,L^{-1}$ at 25 °C under 1 atm) [34], which leads to the limitation of mass

Table 10.1 The thermodynamic potentials *vs.* the normal hydrogen electrode at pH = 7 for various CO_2 reduction products and water splitting.

Product	Reaction	E^0 (V *vs.* NHE)
Hydrogen	$2H_2O + 2e^- \rightarrow 2OH^- + H_2$	−0.41
$CO_2{\cdot}^-$ intermediate	$CO_2 + e^- \rightarrow CO_2{\cdot}^-$	−1.90
Methane	$CO_2 + 8H^+ + 8e^- \rightarrow CH_4 + 2H_2O$	−0.24
Carbon monoxide	$CO_2 + 2H^+ + 2e^- \rightarrow CO + H_2O$	−0.51
Methanol	$CO_2 + 6H^+ + 6e^- \rightarrow CH_3OH + H_2O$	−0.39
Formic acid	$CO_2 + 2H^+ + 2e^- \rightarrow HCOOH$	−0.58
Ethane	$2CO_2 + 14H^+ + 14e^- \rightarrow C_2H_6 + 4H_2O$	−0.27
Ethanol	$2CO_2 + 12H^+ + 12e^- \rightarrow C_2H_5OH + 3H_2O$	−0.33
Oxalate	$2CO_2 + 2H^+ + 2e^- \rightarrow H_2C_2O_4$	−0.87

Source: Chang et al. [15]. © 2016, Royal Society of Chemistry.

transfer for hydrocarbon generation. In fact, the photocathode or photoanode in the PEC system is fixed, and there always exists a large reactant concentration gradient between the electrode surface and the electrolyte due to the diffusion layer and dissipation layer of electrolyte.

In the case of PEC CO_2 reduction, the top challenge for optimizing the selectivity of CO_2 reduction heavily relies on the photocatalysts and external environments during the processes of reduction. To largely satisfy the thermodynamic matters, an external electrical bias is usually applied. To improve the overall efficiency by maneuvering reaction kinetics, it is necessary to efficaciously improve the electron utilization, activate CO_2 molecules, and maneuver the products.

10.2.2 Reaction Conditions

The pathways of CO_2 reduction reaction are complex, and the reaction process is susceptibly affected by the external reaction conditions such as reaction temperature and pressure, electrolyte pH value, solvent composition, and external electrical bias [2, 35]. By changing the external conditions to adjust the state of CO_2 molecules and the adsorption of various products, the efficiency and products types of CO_2 reduction are improved immeasurably.

10.2.2.1 Reaction Temperature and Pressure

The reaction temperature significantly influences the reduction of CO_2, as is known to all that a high temperature will reduce the solubility of CO_2 in solution, which will lead to the decrease in reduction production [2]. For this reason, some researchers add a special cooling water system to PEC device to avoid temperature rise during the reaction [36]. However, due to the different adsorption ability of reaction intermediate and products, the reaction temperature may change the selectivity of products [37]. Cook et al. [38] studied the effects of different temperatures on the products of the PEC reduction CO_2 using Cu-loaded p-SiC and proved that methanation occurred only in a temperature range of 30~60 °C.

In addition to the temperature, the pressure of CO_2 affects the solubility too. At a high pressure, the CO_2 concentration in solution can increase and will then lead to the improvement of the current density and CO_2 conversion of PEC [19]. Moreover, Hirota et al. reported that high CO_2 pressure can promote CO_2 reduction compared with proton reduction and enhance the stability of photocathode [39].

10.2.2.2 pH Value

About the pH of electrolyte, it plays a key role in changing the reaction pathways and the final products. In the aqueous solution, the dissolved CO_2 molecules can form carbonic acid (H_2CO_3), bicarbonate (HCO_3^-), or carbonate (CO_3^{2-}). The ratios of H_2CO_3, HCO_3^-, and CO_3^{2-} are determined by the pH of solution. Once pH is lower than 4.5, CO_2 and H_2CO_3 are the main products, and HCO_3^- becomes the major component at pH between 4.5 and 8.5, and as pH is higher than 8.5, CO_3^{2-} is the only existing component. The different chemicals have different adsorption types on the catalyst surface, which can determine the optimal reaction pathways and final products [40].

10.2.2.3 Solvent

Water is the most commonly used solvent for PEC CO_2 reduction and is also employed as the proton source. However, the solubility of CO_2 in water is quite low (0.033 mol L^{-1} at 25 °C under 1 atm) [34], which is not beneficial to the absorption of CO_2 on the catalyst surface and largely hinders the reduction process. It has been reported that CO_2 molecules are more soluble in nonaqueous organic solvents. Miller et al. studied the CO_2 solubility of 15 different low molar compounds and found that the four best solvents, acetone, methyl acetate, 1,4-dioxane, and 2-methoxyethyl acetate, can interact favorably with CO_2 via Lewis acids/Lewis base interactions. The CO_2 solubility in methanol is almost five times higher than that in water. From then, many organic solvents covering methanol [41], acetonitrile [21], and *N,N*-dimethylformamide [42] serve as the electrolyte for PEC CO_2 reduction. Due to the absence of protons, the major product of CO_2 reduction is CO.

Except for organic solvents, ionic liquids have excellent capture capacities. The ionic liquids have been used in electrocatalysis [43] and photocatalysis [44]. Moreover, it has been proven that the unique configuration between CO_2 and ionic liquid can lower the energy barrier of $CO_2 \cdot^-$ intermediate and promote the reduction of CO_2. Lu et al. [45] used the ionic liquid (1-aminopropyl-3-methylimidazolium bromide) aqueous solution as assistant in PEC CO_2 reduction system. Within this PEC reduction strategy, the ionic liquid plays a key role in promoting the conversion of CO_2 to formic acid and suppressing the reduction of H_2O to H_2. The faradaic efficiency (FE) for formic acid production is as high as 94.1%, and the electro-to-chemical efficiency is 86.2%.

10.2.2.4 External Electrical Bias

In order to enhance CO_2 conversion rate, the external electrical bias is often used to provide enough electrons to improve CO_2 conversion efficiency. Similar to electrocatalysis, the external electrical bias has obvious advantages in the selectivity of CO_2 reduction products, especially competition with water reduction. Different external electrical bias can change the electron transfer and energy barrier of the reaction process, providing different energy for reduction.

10.2.3 Performance Evaluation of PEC CO_2 Reduction

The evaluation parameters of PEC CO_2 reduction are usually combined with the evaluation parameters of photocatalysis and electrocatalysis.

10.2.3.1 Product Evolution Rate and Catalytic Current Density

The mass of catalyst or the effective area of photoelectrode ($\mu mol\ h^{-1}\ g_{cat}{}^{-1}$ or $\mu mol\ h^{-1}\ cm^{-2}$) usually normalizes the product evolution rate. Analogously, the area of photoelectrode usually normalizes the catalytic current density at the desired potential comparing with a reference electrode ($mA\ cm^{-2}$ *vs.* RHE). The values for product evolution rate and current density commonly characterize the performance of PEC CO_2 reduction.

10.2.3.2 Turnover Number and Turnover Frequency

Turnover number (TON) and turnover frequency (TOF) can well evaluate the activity of catalytic active centers and have been widely applied to fine catalyst system, such as the metal nanoparticles (NPs) and homogeneous metal complex catalyst. The definitions for PEC CO_2 reduction are as follows (Eqs. (10.1, 10.2)) [21]:

$$TON = \frac{n_{product}}{n_{catalyst}} \tag{10.1}$$

$$TOF = \frac{n_{product}}{n_{catalyst} \times t} \tag{10.2}$$

where $n_{product}$ and $n_{catalyst}$ are the molar numbers of desired products and catalysts, respectively, and t is the reaction time.

10.2.3.3 Overpotential

The overpotential (η) of a catalytic system is a measurement parameter, which is the energy required to drive a reaction. The energy barrier is generated by the inherent limitations of the kinetics. The overpotential is an important measurement parameter in the CO_2 reduction literature, and its definition is given in Eq. (10.3) [46]:

$$\eta = E – E^0 \tag{10.3}$$

where E and E^0 are the formal redox potentials and formal redox potentials, respectively.

A low overpotential means that the catalyst system requires a relatively low electrochemical drive to convert a reactant to the desired product. The ongoing goal in the CO_2 reduction field is to develop catalysts with an overpotential of as low as possible while simultaneously producing the desired products at an acceptable rate.

In some photoelectrode systems, the reaction of CO_2 reduction can be carried out by combining the light energy with the applied electric potentials more positive than the formal redox potential.

10.2.3.4 Faradaic Efficiency

When it comes to the various products, the selectivity is a key evaluation parameter for the evaluation of CO_2 reduction rate. In a photoelectrocatalysis system, the FE is usually used as an evaluation product for selectivity efficiency and calculated by Eq. (10.4) [35]:

$$\mathrm{FE} = \frac{\alpha \times n_{product} \times F}{Q} \tag{10.4}$$

where α is the number of needed electrons for product evolution, n_{product} is the molar number of desired products, F is the Faraday's constant, and Q is the total passed charge.

A high FE means that a reaction pathway of the target product occurs mainly, but other possible side reactions, such as solvent decomposition, catalyst degradation, and even CO_2 reduction to by-products, are significantly inhibited.

Besides the above evaluation indexes, there are lots of assessment criteria, such as polarization curves (linear sweep voltammetry [LSV]), cyclic voltammetry (CV) curve, chronoamperometric responses (stability), photocurrent density, Tafel curve, operando L-edge X-ray absorption near edge structure (XANES) spectra, and so on. All the evaluation indexes are to more accurately explain the experimental phenomenon and mechanism. There is no the best evaluation index, just only the most accurate evaluation method.

10.3 Application of Solar-to-Chemical Energy Conversion in PEC CO_2 Reduction

The direct way to realize "artificial photosynthesis" is to achieve heterogeneous CO_2 chemical reduction on the interface between a semiconductor material with light-absorbing capacity and liquid electrolyte, which has been widely studied since the late 1970s [47]. Finding a stable, highly efficient, and scalable semiconductor-based system that produces organic fuels driven by solar energy and a low overpotential is the primary goal of PEC CO_2 reduction.

10.3.1 PEC CO_2 Reduction on Semiconductors

The most common configuration for PEC CO_2 reduction includes a photocathode semiconductor and a counter anode (Pt or carbon material). The photocathode harvests light and converts solar energy into chemical energy, providing electrons and holes for the catalyst. Theoretically, the utilization rate of light can be essentially improved by enhancing the capture of light and controlling charge kinetics, which ultimately improves the catalytic activity of CO_2 reduction. Therefore, great efforts have been made in the structure design of photocathode.

10.3.1.1 Oxide Semiconductors

Copper-based photoelectrodes are by far the most popular oxide semiconductors, because they produce a wealth of carbon-dioxide-reducing chemicals, including higher-value reduction products such methanol and methane, as well as multicarbon products such as ethylene, ethanol, and *n*-propanol [48]. A lot of verification experiments have been carried out on the importance of Cu-based electrode oxide state and surface structure to CO_2 reduction [49, 50]. Flake and coworkers [51] studied the yield behavior of an electrodeposited cuprous oxide thin film and revealed the relationships of surface chemistry with reaction behavior about air-oxidized and anodized Cu electrode. Compared with air-oxided or anodized Cu electrode, the Cu_2O electrode showed a higher methanol yield and FE, which were 43 μmol cm^{-2} h^{-1} and 38%. This indicated that the Cu(I) played a key role in the selective reduction of CH_3OH. With the PEC and hydrogenation reactions, the results show that CH_3OH yields are dynamic and that the copper oxides are reduced to metallic Cu in a simultaneous process. The authors suggested that the formation of HCO may hydrogenate with carbon atom of CO of adsorbates via proton transfer at −0.74 V (*vs.* reversible hydrogen electrode [RHE]). Once the HCO is formed, the carbon atom continues proton and electron transfer reactions to form H_3CO adsorbates and finally to form CH_3OH. Because of the intermediate stability and ability of H^+ species coordinated with surface bound, oxygen was improved on the surface of the cuprous oxide electrodes. This surface allows hydrogen addition to the oxygen atom of the H_3CO adsorbate rather than the carbon atom. Further studies by Li and coworkers [52] showed that the same oxide-derived Cu electrode can produce multicarbon oxygenates (ethanol, acetate, and *n*-propanol) with up to 57% Faraday efficiency at −0.25 to −0.5 V (*vs.* RHE) in CO-saturated alkaline H_2O. By comparison, when prepared by traditional vapor condensation, Cu nanoparticles, which are similar with an average crystallite size of oxide-derived copper, produce just exclusive H_2 (96% Faraday efficiency) in identical conditions. The results show that it is possible to change the intrinsic catalytic properties of copper by growing interlinked nanocrystals around the confined lattice. However, Jonas and coworkers studied the oxide-derived Cu nanoparticles and found that the Cu_2O layer thickness and local pH change played important roles, not the initial Cu_2O crystal phase. Handoko and Tang [53] showed that the yields from CO_2 to CO as high as 400 ppm g^{-1} h^{-1} were achieved by Cu_2O powders with the RuO_x as a cocatalyst and Na_2SO_3 as a sacrificial hole scavenger in the first 30 minutes of the reduction under illumination from visible light (λ > 420 nm). However, the yields quickly decayed within the first hour. To suppress the photocorrosion, Gong and coworkers [20] employed Cu_2O as a dark cathode, combined with TiO_2 photoanode for harvesting light. This strategy largely improved the stability of Cu_2O and acquired the main products of CH_4, CO, and CH_3OH with the total FE of 87.4% and the selectivity of 92.6% for carbonaceous products.

In addition to Cu_2O, other Cu-based oxide semiconductors also show remarkable performance. Park and coworkers [27] successfully assembled the $CuFeO_2$ and CuO mixed p-type catalysts by annealing of earth-abundant cupric and ferric ions under atmospheric air. The composite electrodes converted CO_2 to formate with over

90% selectivity under simulated solar light (air mass 1.5, 100 mW cm^{-2}) and produced formate for one week at a solar-to-formate energy conversion efficiency of ~1% (selectivity > 90%) without any external electrical bias. To identify the source of carbon in formate, the $CuFeO_2$/CuO was immersed in ^{12}C-bicarbonate solution through which $^{13}CO_2$ gas was purged. And finally, it was verified that the purged CO_2 was a reduction to formate.

Co_3O_4, a spinel semiconductor with the conduction band at −1.53 V and the band gap 2.07 eV, exhibits the ability for visible-light-driven PEC CO_2 reduction. Zhao and coworkers [54] successful synthesized a high surface Co^{3+}-containing hierarchical Co_3O_4 nanoelectrode and used the PEC CO_2 reduction. Results showed that owing to the synergistic PEC reduction process, the formate yield is up to 384.8 ± 7.4 μmol within eight hours (Figure 10.2f). Research further elucidated that the photoelectrocatalytic activity of Co_3O_4 is closely related to the multinanostructure, in which the single-crystalline microflower interconnects with the thin nanopetals on the one-dimensional rhombus nanorod (NR) (Figure 10.2a~d). Figure 10.2e shows distinctive lattice fringe with *d*-spacing of 0.287 nm, which matches with the (202) and (220) crystal planes. Because of the angle of those two planes is c. 75°, the exposed crystal on the nanopetals is $\{12\bar{1}\}$. The strongest $\{12\bar{1}\}$ facet on the Co_3O_4 surface exposes high Co^{3+}, which remarkably improves the transient and steady state during the PEC reduction response of CO_2 and HCO_3^- (Figure 10.2f). In order to vividly describe the $\{12\bar{1}\}$ facet, the authors established the model facets of Co_3O_4 (Figure 10.2g,h,i,j). About $\{12\bar{1}\}$ facet, there have been constituted one building block by a six-coordinated Co^{3+} with three dangling bonds (denoted as $Co_{6c}{}^3$) and two six-coordinated Co^{3+} with one dangling bond (denoted as $Co_{6c}{}^5$). But, on $Co_3O_4\{100\}$ and $Co_3O_4\{110\}$ facet, there only one $Co_{6c}{}^3$ and one $Co_{6c}{}^5$ is observed (Figure 10.2h,i), and there is no Co^{3+} on the $Co_3O_4\{111\}$ facet (Figure 10.2j). It indicates that the amount of Co^{3+} on $\{12\bar{1}\}$ is higher than that on those common low-index facets. With the increment of Co^{3+} on $\{12\bar{1}\}$, the more active sites are exposed on the surface of Co_3O_4 electrocatalysts, which is more beneficial for the PEC reduction of CO_2.

Fe_2O_3 is another oxide semiconductor widely used in PEC reduction of CO_2. Li et al. [55] successfully assembled CuO FCs/Fe_2O_3 NT catalyst by a pulse electrochemical deposition method. After CuO FCs loaded on Fe_2O_3 NTs, the band gap was narrowed from 2.03 to 1.78 eV, and the absorption of visible light was noticeable enhanced. Methanol and ethanol are the major products in the PEC reduction CO_2 process, reached 1.00 mmol l^{-1} cm^{-2} and 107.38 μmol L^{-1} cm^{-2} after six hours, respectively. In order to further improve the PEC reduction performance of CO_2, Li et al. [4] assembled the Cu_2O/Fe_2O_3 NTs by a potentiostatic electrodeposited method. The Cu_2O/Fe_2O_3 NTs with double-layer Cu_2O spheres show excellent PEC properties owing to the suitable energy gap (1.96 eV) and the smaller overpotential (0.18 V). On PEC reduction of CO_2 process, the Cu_2O/Fe_2O_3 NTs show two types of synergisms: (i) combined photocatalysis with electrocatalysis and (ii) combined Cu_2O with Fe_2O_3 NTs. The methanol yield and FE reach 4.94 mmol L^{-1} cm^{-2} and 93% after six hours, respectively. Combining to the experimental data and phenomenon, the mechanism of the PEC reduction of CO_2 on Cu_2O/Fe_2O_3NTs-30

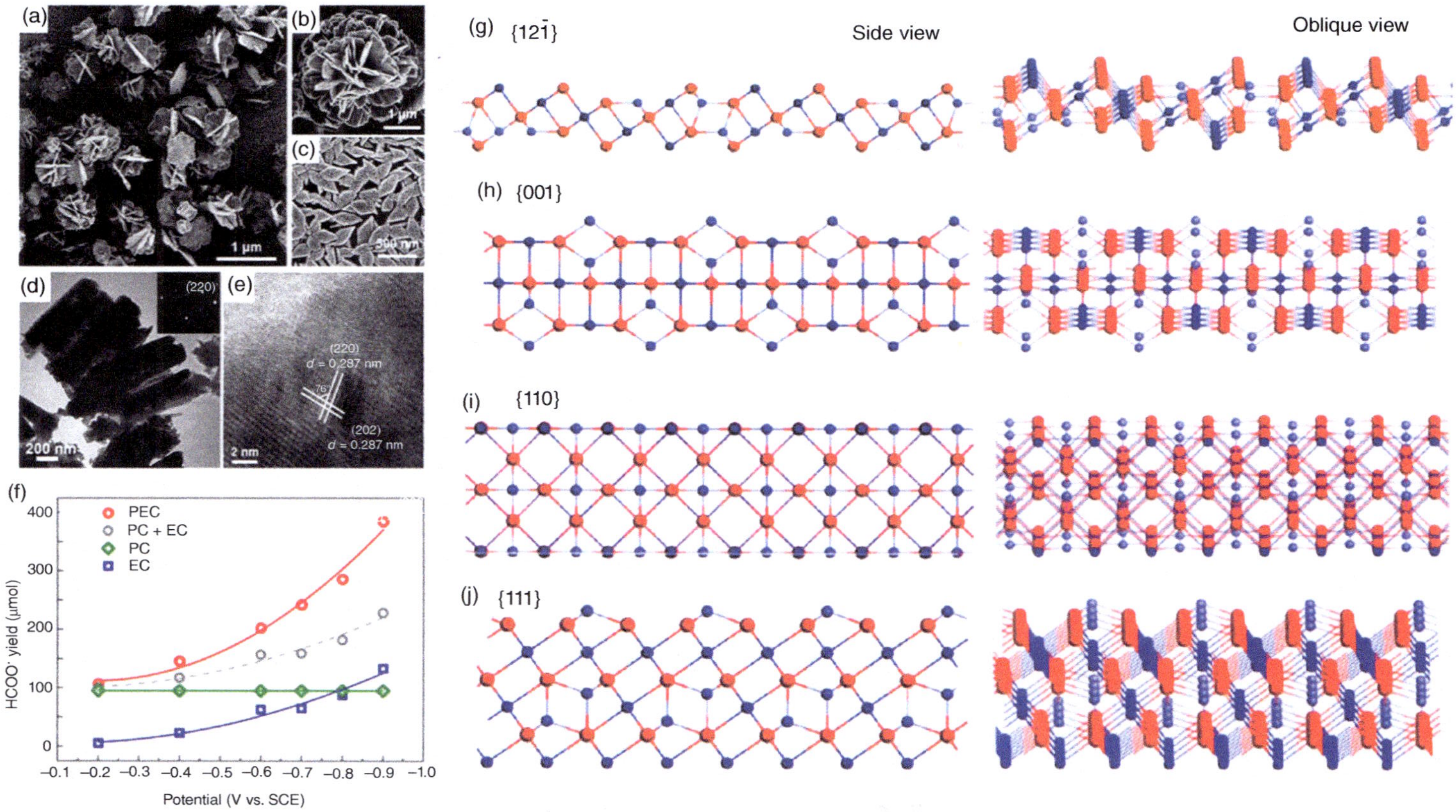

Figure 10.2 (a) Scanning electron microscope (SEM) image of HA-Co_3O_4. (b) Upper layer microflower morphology. (c) Lower layer rhombus nanorod morphology. (d) Transmission electron microscopy (TEM) of petals on the flower; Inset: selected area electron diffraction (SAED) pattern of petal region. (e) High-resolution transmission electron microscopy (HRTEM) of petal region. (f) Formate yields of HA-Co_3O_4 under PEC and EC under different negative potentials. Crystallographic modeling of Co_3O_4 (g) $\{12\bar{1}\}$, (h) {001}, (i) {110}, and (j) {111} facets with side view (left) and oblique view (right). Blue, light blue, and red spheres represent Co^{3+}, Co^{2+}, and O^{2-}, respectively. Source: Huang et al. [54]. © 2013, American Chemical Society.

was deduced. There were two reactions occurred on the electrode surface under irradiations: (i) the photogenerated holes oxidized water to protons and oxygen; (ii) CO_2 was reduced to CH_3OH. Because the conduction band of Cu_2O/Fe_2O_3NTs-30 is −0.87 eV, which is more negative than the reduction potential of CO_2/CH_3OH (−0.38 V *vs.* NHE), it is indicated that this system has enough PC reductive ability to reduce CO_2 to CH_3OH. And the external potential not only enhances the separation of electrons and holes but also offers sufficient electrons to improve the PEC reduction of CO_2. The authors [56] loaded SnO_2 NRs onto the surface of Fe_2O_3 nanotubes. Compared with Fe_2O_3 nanotubes, the SnO_2 NRs/Fe_2O_3 NTs exhibited 7.48 times of CO_2 photoelectric reduction current at −1.1 V (*vs.* RHE), and the FE obtained from CO_2 to methanol was 87.04%. When the SnO_2 NR/Fe_2O_3 NT catalyst is used for photocatalytic reduction of CO_2, the conduction band is negative enough to drive CO_2 reduction, and the VB is positive enough to oxidize H_2O for H^+. On the electrocatalytic reduction point, with the introduction of SnO_2 NRs, the current density of the composite is 7.48 times larger than that of Fe_2O_3 NTs at −1.1 V (*vs.* SCE). There are two synergistic effects on the reduction mechanism. One is the synergistic effect of the photocatalytic reduction and the electrocatalytic reduction; the other is the synergistic effect between SnO_2 NRs and Fe_2O_3 NTs.

10.3.1.2 Non-oxide Semiconductors

$CuInS_2$, a chalcopyrite p-type semiconductor, matches well the solar spectrum with the conduction band at −1.62 V and band gap of 1.3 eV. Yuan and Hao [22] used a chalcopyrite p-$CuInS_2$ thin film as a photocathode for solar-driven PEC reduction reaction of CO_2 by electrodeposition of Cu-In alloy layer followed by sulfurization. Under the pyridinium ion as a cocatalyst, the PEC reduction of CO_2 to methanol can carry out at the overpotential of 20 mV with the FE of 97%. For further improving the reduction efficiency, Yuan et al. [57] fabricated the $CuInS_2$/graphene hybrid thin film as the photocathode by one-step electrodeposition at 30 °C for PEC reduction of CO_2. Comparing with the $CuInS_2$ thin film electrode, the photocurrent densities of $CuInS_2$/graphene hybrid thin films are almost twice, and the rate of methanol formation of $CuInS_2$/graphene hybrid thin film is 1.4 times. In this study, methanol is the only detected reduction product by GC-MS in the PEC reduction of CO_2. During the experiment, an interesting phenomenon was recorded: without pyridine, no methanol was produced in acetic buffer; with just pyridine in solution, no methanol is produced, either. This is because the system formed the intermediate $C_5H_5N{-}H^+\cdots O{=}C{=}O$ during CO_2 reduction in the acetic buffer with pyridine.

There is another semiconductor of graphitic carbon nitride (g-C_3N_4) that has been paid much attention, because the metal-free semiconductor is of sufficiently negative conduction band edge and outstanding PEC performances in CO_2 reduction under solar energy [26]. Shi et al. [58] successfully fabricated g-C_3N_4/$NaNbO_3$ nanowire photocatalysts by introducing g-C_3N_4 on $NaNbO_3$ nanowires. Comparing with g-C_3N_4 or $NaNbO_3$ nanowires, the activity of g-C_3N_4/$NaNbO_3$ nanowires is high for photoreduction of CO_2. Owing to the improved separation and transfer of photogenerated electron–hole pairs at the intimate interface of g-C_3N_4/$NaNbO_3$ heterojunctions, the photocatalytic activity is remarkable enhancement. Ohno and

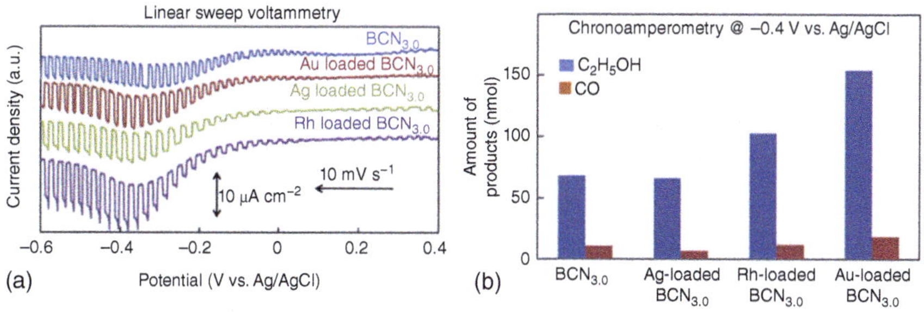

Figure 10.3 (a) Linear sweep voltammetry of the $BCN_{3.0}$ electrodes loaded with or not loaded with cocatalysts. (b) Products analyses of photoelectrochemical reduction of CO_2 over cocatalyst-loaded $BCN_{3.0}$ electrodes. Source: Sagara et al. [26]. © 2016, Elsevier.

coworkers [26] prepared the g-C_3N_4 and boron-doped g-C_3N_4 (B-doped g-C_3N_4) (BCN_x) by heating melamine and a mixture of dicyanodiamide and BH_3NH_3, respectively. Au, Ag, or Rh as a cocatalyst was coated on the surface of g-C_3N_4 and B-doped g-C_3N_4 by using the magnetron sputtering method. Comparing with the pure g-C_3N_4, the photocurrent response of B-doped g-C_3N_4 and Rh/B-doped g-C_3N_4 Rh/B-doped g-C_3N_4 is about 5 times and 10 times, respectively (Figure 10.3a). Under PEC conditions, C_2H_5OH is the main product and a small amount of CO, and no H_2 was detected in gas phase (Figure 10.3b). Among those cocatalysts, Au-loaded $BCN_{3.0}$ shows the highest activity. In the mechanism of CO_2 reduction, the CO_2 adsorption of cocatalyst loaded on $BCN_{3.0}$ is one of the key points for improving property of CO_2 reduction. This is because the cocatalyst-loaded $BCN_{3.0}$ could prevent the recombination and enhance the selectivity for C_2H_5OH generation as a result of multi-electron reduction by electron capturing on metal particles.

10.3.1.3 Chalcogenide Semiconductors

CdTe, a p-type semiconductor with a band gap of 1.5 eV, has been most extensively studied for PEC reduction CO_2. Senftle et al. [59] report a density functional theory (DFT) investigation on the stability of GaP(111) and CdTe(111) surface reconstructions and found that water could readily dissociate on the GaP(111) surface to OH^- and H^+ groups, while water does not dissociate on the GaTe(111) surface to form a stable hexamer structure. About the GaP(111) surface, they found that $1e^-$ reduction of PyH^+ (aq) (aqueous pyridinium) forms the adsorbed Py* and H*, further being reduction to form newly intermediate 2-PyH·*. While on the CdTe(111) surface, they found that the possible reduction steps are $2e^-$ reductions leading to the formation of dihydropyridine. Lessio et al. [60] verified that pyridine plays an important role in CO_2 reduction on surface of p-GaP photoelectrode. They reported that it is likely heterogeneous for the mechanism of CO_2 reduction operating in the PEC reduction system. Because of the different surface displaying a different activity toward CO_2 reduction, the observed difference activity of the different surface may be explained by the differences in adsorption strength for relevant intermediates in the CO_2 reduction mechanism.

In aqueous solutions, the choice of electrolyte is significantly more important in PEC reduction of CO_2. Kaneco et al. [41] investigated metal-modified p-InP photoelectrodes for PEC reduction of CO_2 in the LiOH/methanol-based electrolyte. The deposition metals are led, silver, gold, palladium, copper, and nickel. After metal-modified p-InP photocathodes, the main products from CO_2 are CO and formic acid. The silver-modified p-InP photoelectrode obtained the maximum current efficiency of CO (r_f = 80.4%). In the PEC reduction of CO_2, the distribution of reduction products at metal-modified p-InP photocathodes may be roughly associated with catalytic property of metallic electrode. Kaneco et al. [61] studied a p-InP photoelectrode for PEC reduction CO_2 in copper particle-suspended methanol. During the PEC reduction CO_2, the formic and CO are formed in the absence of copper particles, while hydrocarbons is formed on the addition of copper particles into the catholyte. The maximum FE of methane and ethylene are r_f = 0.56% and 0.8%, respectively. By adding the metal particles into the catholyte during the PEC reduction of CO_2, it is possible to roughly change and control the reduction product distribution.

ZnTe, assembled on the Zn/ZnO nanowire substrate, demonstrates stable photocatalytic activity of PEC reduction of CO_2 to CO [62]. The formation of CO is observed at −0.7 V (*vs.* RHE) in aqueous $KHCO_3$ with high incident photon-to-current conversion efficiency values reaching 85%. Woo and coworkers [63] investigated a polypyrrole-coated p-ZnTe electrode as a new photocathode material for PEC reduction of CO_2. Under the irradiation of visible light, the PPy/ZnTe showed an outstanding performance with 51% of FE at −0.2 V (*vs.* RHE). The PPy/ZnTe electrode improved the photogenerated electron–hole pairs, and the ZnTe and PPy had different functions to introduce an electron pathway from ZnTe to PPy and then to the reactions. Jang et al. [23] investigated a gold-coupled ZnTe/ZnO nanowire (ZOZT) array for selective CO_2 reduction to CO. The photocurrent and incident photon-to-current conversion efficiency of Au-coupled ZnTe/ZnO nanowire array is −16.0 mA cm^{-2} and 97%. In PEC reduction of CO_2, the biggest effect of Au nanoparticles is converting mainly-hydrogen production bare ZnTe/ZnO nanowires into mainly-CO production photocathodes. The main reason is forming the Schottky junction of Au–ZnTe/ZnO to improve charge separation and provide reaction centers for CO_2 reduction.

10.3.2 PEC CO_2 Reduction on Cocatalyst Systems

For PEC CO_2 reduction, semiconductor photocathodes are employed as light-harvesting antennas. For highly efficient solar-driven CO_2 reduction, except for light utilization, the major challenges and limits are the unfeasible CO_2 activation, uncontrollable reduction products, and complicated reaction pathways [15]. However, owing to the deficiency of active sites for CO_2 activation, semiconductor photocathodes often show frustrating activity in CO_2 reduction. It is a promising strategy for effectively surmounting this predicament that highly efficient catalysts such as metal cathode [64] and homogeneous metal complexes [65] as cocatalysts are anchored on the surface of photocathodes. The anchored cocatalysts with high activity at a semiconductor surface can enhance CO_2 activation, lower the overpotential, and reduce the energy barrier, which have been extensively applied to PEC CO_2 reduction.

10.3.2.1 Metal Nanoparticles

It is well believed that anchoring cocatalysts with metal nanoparticles at semiconductors is an effective strategy for enhancing the performance of photoelectrocatalysis reduction CO_2 [66]. In the photocatalytic process, the metal cocatalysts play an important role. First, they can promote charge separation by trapping the photogenerated charges, and second, they can act as catalytic active sites to lower the overpotential and carry out the reaction.

In a PEC system, the p-type semiconductor is usually employed as a photocathode. The electrons will accumulate on the surface of photocathode with the assistance of external electrical bias upon light illumination. At the same time, the photogenerated holes will be separated and migrate toward the counter electrode. When metal nanoparticles are anchored on the surface of photocathode, the electrons will flow to the metal and carry out the reduction reactions.

All metals can act as cocatalysts for photocatalysis, but not all metals are suitable for CO_2 reduction [2]. As shown in Figure 10.4, most metals have a marked effect on hydrogen evolution, such as Pt [67], Fe [68], Ni [69], and so on [70]. Generally speaking, in electrochemical reduction of CO_2 reaction, the metals that can reduce CO_2 to CO are Au [71], Ag [72], and Pd [73] in CO_2-saturated aqueous electrolyte, while Sn [74], In [75], and Pb [76] are beneficial to the formate evolution. Copper electrodes have yielded very rich CO_2 reduction products, including C_1 products such as CO, CH_4, and CH_3OH, as well as multicarbon products such as ethylene, ethanol,

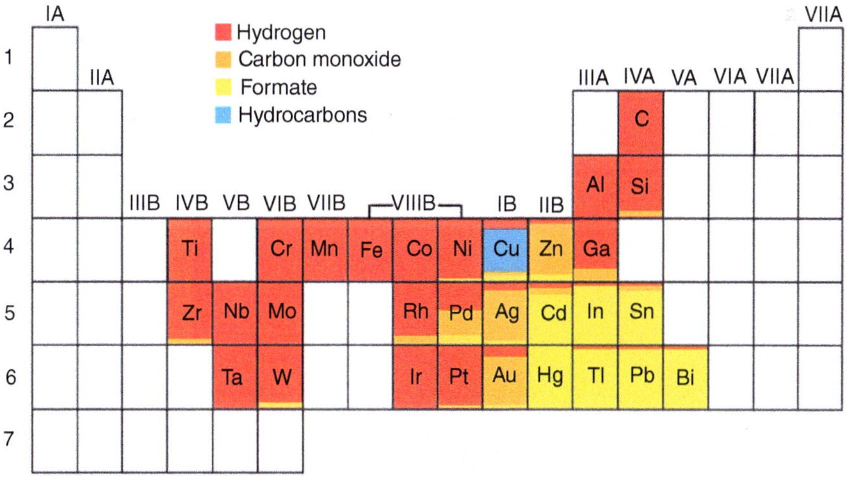

Figure 10.4 Periodic table depicting the primary reduction products in CO_2-saturated aqueous electrolytes on metal and carbon electrodes. Source: White et al. [2]. © 2015, American Chemical Society.

and *n*-propanol [77]. In the PEC reduction of CO_2 system, the metal employed as cocatalysts should consider how to improve the FE and suppress the competitive reactions.

Anchoring metal nanoparticles could lower the overpotential of CO_2 reduction and change the types of products. Li and Kanan [49] prepared the modified Cu electrodes by annealing Cu foil in air and electrochemically reducing the resulting Cu_2O layers. The larger roughness of Cu_2O layers required 0.5 V less overpotential than polycrystalline Cu to reduction of CO_2, while the Cu/Cu_2O electrode resulted in CO_2 reduction current densities >1 mA cm^{-2} at overpotentials <0.4 V. The modified electrode is stable over the course of several hours.

Kaneco et al. [41] investigated the metals, such as Pb, Ag, Au, Pd, Cu, and Ni, modified p-InP photoelectrodes as cocatalysts for PEC reduction of CO_2 in the LiOH/methanol-based electrolyte. After Pb, Au, Ag, and Cu modified p-InP photocathodes, the main products from CO_2 are CO and formic acid, while CO is the only product for Pd. The silver-modified p-InP photoelectrode obtained the maximum current efficiency of CO ($r_f = 80.4\%$). In the PEC reduction of CO_2, the distribution of reduction products at metal-modified p-InP photocathodes may be roughly associated with catalytic property of metallic electrode and different enthalpy and Gibbs energy of dissociative adsorption of CO on metal surface.

It is an effective strategy for efficient CO_2 reduction that rational design semiconductor and cocatalyst is employed. Jang et al. [23] investigated a gold-coupled ZnTe/ZnO nanowire (ZOZT) array for selective CO_2 reduction to CO. Although ZnTe has been verified H_2 evolution during the PEC process, ZnTe is an excellent candidate for CO_2 reduction due to its narrow bang gap and negative CB position. It is found that Au as cocatalyst could effectively enhance the activity and selectivity of CO_2 conversion to CO.

10.3.2.2 Metal Complexes

Apart from metal NPs, metal complexes are widely studied for CO_2 reduction [7, 21, 78]. The multiple and accessible redox states of metal complexes can enhance and altered activity and selectivity of the multi-electron reduction process of CO_2 [79]. In addition, the charge separation can be enhanced by metal complex cocatalysts through trapping the photogenerated electrons from photoelectrodes, which could induce reaction of metal to serve as active sites for activation of CO_2 molecules and adsorption of the intermediates.

Among the various complexes, Re- and Ru-based complexes have been widely employed as cocatalysts for PEC CO_2 reduction. Kubiak and coworkers [80] employed Re(bipy-Bu^t)$(CO)_3$Cl as the cocatalyst to enhance the activity of p-Si photocathode for the selective photoreduction of CO_2 to CO. In the presence of Re(bipy-Bu^t)$(CO)_3$Cl as cocatalyst, the p-Si electrode achieved 60 mV lower than Pt electrode for reducing CO_2 to CO, and a high FE was achieved of 97% for CO production. The similar phenomenon is acquired on polychromatic illumination with an open-circuit voltage of 650 mV and a short-circuit current density of 23 mA cm^{-2}.

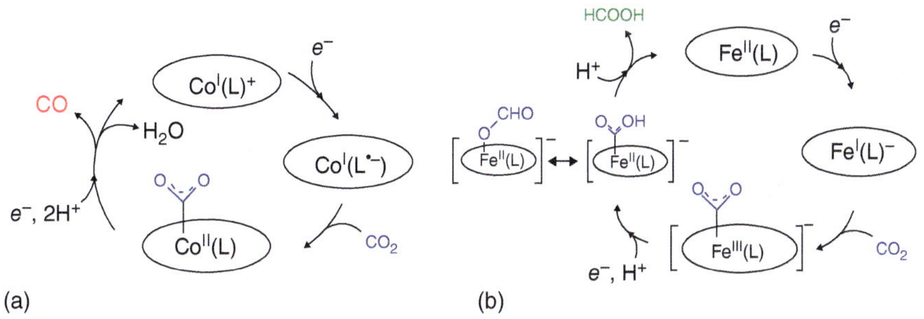

Figure 10.5 Schematic representations for (a) Co^{II} complex and (b) Fe^{II} complex for PEC reduction of CO_2. Source: Chen et al. [83]. © 2015, American Chemical Society.

Sato et al. [81] took Ru complex as the cocatalyst on p-InP photocathode for PEC reduction of CO_2 to $HCOO^-$ and combined Pt with TiO_2 as photoanode for water oxidation. In this reaction system, the selectivity of $HCOO^-$ evolution is over 70%, and H_2O was used as both an electron donor and a proton source. There is no external bias in the so-called Z-scheme (two-step photoexcitation) system, and the conversion efficiency of solar energy to chemical energy is 0.03~0.04%. To confirm the carbon and proton sources of formate, the isotope tracer analysis is identified. The results show that the carbons originate from CO_2 and the electrons and protons are extracted from H_2O during the oxidation process to O_2.

Despite the typical noble metal complexes (e.g. Re and Ru) show the remarkable performance in CO_2 conversion, the high costs impede their practical preparation and application. There is an urgent need to develop low-cost transition metal complex catalysts [82]. Chen et al. [83] employed Co^{II} complex and Fe^{II} complex for PEC reduction of CO_2. Results show that the cobalt complex forms carbon monoxide, while iron complex obtains the formic acid. Selective CO production occurs in a PEC condition with photosensitizer and Co^{II} complex, while selective HCOOH production occurs at a low overpotential with Fe^{II} complex (Figure 10.5). The coordination between CO_2 and metal plays an important role in the reactivity. In the case of Co^{II} complex, CO_2 is bonding to the metal with the carbon atom (η_1-CO_2 coordination), and then the second electron takes part in the cleavage of C—O bond during the protonation and reduction process to finally form an OH^- and a CO molecule. Because of the Co-CO_2H intermediate formed a Co^{II} center to π^* orbitals of CO_2, it could weaken the C—O bond and facilitate the cleavage of C—O bond to get CO product. In the case of Fe^{II} complex, the η_1-CO_2 coordination is preferred, but during the protonation and reduction process, the cleavage of C—O bond does not occur; the Fe-CO_2H intermediate rather evolves to the η_1-OCOH-coordinated complex, finally leading to HCOOH formation.

10.3.3 PEC CO_2 Reduction on Hybrid Semiconductors

To trap electrons and change the product distribution, the semiconductors usually modified with a cocatalyst, which will change the energy level and act as active sites.

The cocatalyst should have some certain extent nucleophilicity to bind CO_2 at low overpotentials to break the C=O bond effectively.

10.3.3.1 Conductive Polymers

The polymers with ammonium groups are the preference to improve the adsorption of CO_2 on the semiconductor surface. The polyaniline [84], polypyrrole [85], and polydopamine [86] have been widely employed for CO_2 reduction with a low overpotential. Through many mechanism derivations and experimental verifications, NR_4^+ has the following advantages for CO_2 reduction. Firstly, when the NR_4^+ ions adsorbed on the photocathode, the surface wettability of photocathode will be changed to impede the adsorption of H_2O and enhance the selectivity of CO_2 reduction. Secondly, the NR_4^+ ions of photocathode surface receive electrons coming from photocathode to form $NR_4\cdot$ radical intermediates, which could improve the activation of CO_2 to $CO_2\cdot^-$. Jeong and coworkers [84] used Cu_2O nanoparticles decorated polyaniline for electrochemical reduction of CO_2. The formic and acetic acid are the main productors with FE of 30.4% and 63.0% at −0.3 V (*vs.* SCE), respectively. The presence of Cu_2O not only enhances the CO_2 adsorption on the polymer surface but also has a synergistic with polyaniline to enable CO_2 reduction. During the electroreduction process, the preliminary step is the adsorption of CO_2 on polyaniline via hydrogen bonding. The –NH groups of polyaniline interact with CO_2 to form hydrogen bonds. At −0.3 V, the H adatoms (H_{ad}) are firstly reduced on the electrode surface, and the H_{ad} atoms transfer to the CO_2 molecules via hydrogen bonding to form $HCOO_{ads}$. With further addition of H_{ad}, the reactions form formic acid, which is attacked by CH_3OH to form acetic acid. Aydin et al. [85] investigated a polypyrrole-coated copper catalyst for electroreduction of CO_2 for the first time. The HCOOH and CH_3COOH are the only products at −0.8 V and −1.0 V, respectively. While at other negative values, CH_4, C_2H_4, C_2H_6, and CO as gaseous products are detected. The coated conducting polymer could cause a shifting of product distribution to hydrocarbons. According to protic and aprotic medium, the authors classified two different mechanism pathways of CO_2 reduction. In case of protic medium on metal electrocatalysts, the HCOOH is preferred as ionic form and the main organic product. While in the case of aprotic medium, the product distribution depends on the employed electrode material, especially copper, and the product main is hydrocarbons, such as CH_4, C_2H_4, and C_2H_6.

The conductive polymers have been taken as cocatalysts for PEC reduction of CO_2. Woo and coworkers [63] investigated a polypyrrole-coated p-ZnTe electrode as a new photocathode material for PEC reduction of CO_2. Under the irradiation of visible light, the PPy/ZnTe shows an outstanding performance with 51% of FE at −0.2 V (*vs.* RHE). Compared with the ZnTe electrode, the PPy/ZnTe electrode markedly enhances the FE, increases the formic acid production from CO_2, and decreases the HER. It indicates that the PPy acts as the active site for the selective conversion of CO_2 to formic acid and hinders the recombination of surface-adsorbed protons. On the other hand, because PPy has a high work function value (reported from 4.6 to 5.2 eV, which is higher than ZnTe, 4.45 eV), the deposition of PPy on the ZnTe can facilitate the electron transfer from ZnTe to PPy.

10.3.3.2 Enzymatic Biocatalysts

Some electroactive enzymes have been employed for reduction of CO_2 to CO and formate [87]. In 1984, Parkinson and Weaver [88] first used CO_2-active enzyme as electron transfer mediator along with semiconductor electrodes for CO_2 reduction. By coupling p-InP photoelectrode with formate dehydrogenase (FDH) enzyme, the selective reduction of CO_2 to formate can be achieved. Photogenerated electrons in the semiconductor can be produced with the light wavelengths (<900 nm), as that much of the solar spectrum energy could be captured by this system. Armstrong and coworkers [89] studied TiO_2 with ruthenium tri(bipyridine) photosensitizer and carbon monoxide dehydrogenase (CODH) for the CO_2 reduction to CO (Figure 10.6a). This system generated 250 μmol g^{-1} h^{-1} of CO under visible light at six hours and 20 °C. Compared with analogous H_2 production system, the rates and stability are low. There are several factors that are influential. One factor is the smaller driving force (~0.1 V) for CO_2 reduction compared with H^+ reduction. The CB of TiO_2 is −0.52 V (vs. standard hydrogen electrode [SHE], at pH = 6), which is only just enough for CO_2 reduction at CODH I. Another factor is the weak bond between CODH I molecules and TiO_2. The key step depends on the efficient electron transfer from TiO_2 to the iron–sulfur clusters leading to the active site. Soon afterward, Armstrong and coworkers [20] again took CODH and NiO for the PEC selective reduction by converting CO_2 to CO at a potential below the thermodynamic equilibrium (Figure 10.6b). The enzymatic electrocatalyst on NiO shows that the electrocatalytic behavior is rectified toward CO oxidation in the dark, with the carrier availability of the reaction governing at the semiconductor–catalyst interface. During the reduction process, the P1 (dye sensitized) acts as a visible-light-responsive sensitizer, which generates excited state electrons, transfers electrons to CODH for catalysis, and finally regenerated by hole injection into NiO. Since the P1 is capable of absorbing visible light, the photocurrent of the P1-NiO electrode markedly increases than NiO. But the P1 does not reduce CO_2 itself, and the lowest unoccupied molecular orbital (LUMO) of P1 is lower than the energy of $CO_2/CO_2\cdot^-$ redox couple (c. 0.7 V). Importantly, P1, as a single-electron-transfer dye that is not capable of accumulating a second electron, cannot undergo the proton-coupled electron transfer reactions.

10.3.3.3 Organic Molecules

Some organic molecules have been employed for PEC CO_2 reduction [2, 90]. The organic molecules play as mediators to transfer electrons and protons to CO_2 in the reaction process. Bocarsly and coworkers [91] employed protonated pyridinium serving as a cocatalyst with p-GaP photocathode for selective conversion of CO_2 to methanol. Solar energy is the driving force of the reaction, and FE is near 100% at potentials well below the standard potential. With different pH, finding that the pyridine molecule plays as active catalyst with p-GaP while without p-GaP could not.

Yuan and Hao [22] used a p-CuInS_2 thin film photocathode with pyridinium ion as the cocatalyst for solar-driven PEC reduction reaction of CO_2 by electrodeposition of Cu-In alloy layer followed by sulfurization. Under the pyridinium ion as a

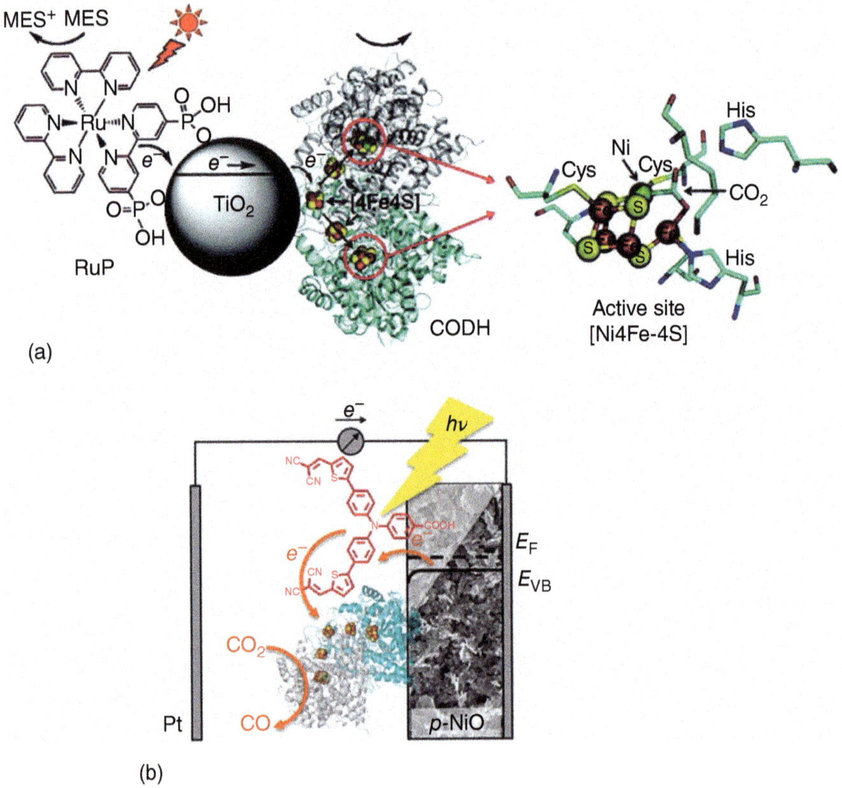

Figure 10.6 Schematic of (a) the ruthenium dye-sensitized TiO_2 nanoparticles with adsorbed CODH enzyme, which catalyzes the reduction of CO_2 to CO. Source: Woolerton et al. [89]. © 2010, American Chemical Society; (b) a photoelectrochemical cell for selective reduction of CO_2 to CO at p-type NiO. Source: Bachmeier et al. [20]. © 2014, American Chemical Society.

cocatalyst, the PEC reduction of CO_2 to methanol can carry out at the overpotential of 20 mV with the FE of 97%. Lessio and Carter [25] studied what the role of pyridinium in pyridine-catalyzed CO_2 reduction on p-GaP photocathodes is. For a CO_2 reduction system, pyridinium reduction to pyridinyl is usually supposed to be an essential mechanistic step. However, theoretical calculations suggest that this step is not feasible in solution. To determine whether such a reduction step could occur by the transfer of photoexcited electrons from the p-GaP photocathode and whether this transfer could be catalyzed by pyridinium adsorption on the p-GaP surface, the cluster and models and accurate periodic boundary condition calculations were used. It was found that both the transfer of photoexcited electrons to pyridinium and pyridinium are not energetically favored. So, a new pathway is proposed, which leads to forming adsorbed species on p-GaP photoelectrode surface and further forming the adsorbed dihydropyridine. Costentin and coworkers [92] further proved that the protonated pyridine (PyH^+) catalyzed reduction CO_2 is not methanol. It appears that CO_2 is further with hydrogen evolution to form carbonic acid instead of being reduced. The study confirms that there is no methanol formation on the bulk electrolysis of PyH^+ solution at Pt surface and suggests that in the presence of acid such as PyH^+ and AcOH, the CO_2 reduction reaction is strongly inhibited. With the cyclic voltammetry and in situ infrared spectroscopy investigations, it finally indicates that the reduction of CO_2 to CO caused the inhibition due to the high affinity in forming Pt–CO film on platinum.

10.4 Other Configurations for PEC CO_2 Reduction

Besides the commonly used photocathodes with dark anode setup system, there are some other setup systems, such as combining photoanodes with dark cathode system (Figure 10.1b), combining photoanodes with photocathode setup system (Figure 10.1c), and photovoltaic-based electrocatalytic devices system (Figure 10.1d).

In photoanodes with a dark cathode system, the photoanode undertakes light harvesting, charge separation and migration, and CO_2 reduction process. How to employ a photoanode to harvest light and an effective cathode is an alternative approach for the reduction of CO_2. Jiao and coworkers [28] studied the PEC cell composing of Ni-coated Si photoanode and nanoporous Ag cathode for CO_2 conversion to CO. This cell system exhibited a markedly photoresponse to visible-light irradiation, and the photovoltage of the photoanode can simultaneously drive the catalytic oxidation reaction at anode and reduction reaction at cathode with the assistance of an external electrical bias. At the external electrical bias 2.0 V (*vs.* RHE), the FE of CO and current density reached 70% and 10 mA cm^{-2}, respectively. Comparing with the pure electrochemical cell, the 0.4 V was determined by the photovoltage of the PEC cell system, and the energy saves 17%. But the stable performance of the PEC cell is just up to three hours, because of the interfacial damage between the nickel catalyst layer and the silicon wafer caused the performance degradation. Lee and coworkers [29] studied the PEC system consisting of WO_3 photoanode under an electrical bias potential and Cu or Sn/SnO_x as a cathode for reduction of CO_2. The FE of CH_4 and all carbon-containing products are 67% and 71.6% by using Cu as the cathode electrocatalyst, while the FE of CO and HCOOH is 44.3% by using Sn/SnO_x at 0.8 V. The results show that the intrinsically stable photoanode system has a higher efficiency for reduction of CO_2 comparing with the conventional photocathode system. In terms of photoanodes, other metal oxides were employed as the component of photoanodes for light harvesting. Cheng et al. [93] studied Pt-RGO||Pt-TNT PEC cell to increase the carbon atom conversion. With the applied voltage increase, the carbon atom conversion shows a sustainable growth, but when the applied voltage is over 2 V, the current efficiency of CO_2 reduction products decreased owing to the enhanced proton interference. Because there have been the synergistic effects between the Pt-RGO catalyst and the nickel foam catalyst support, which could provide plenty adsorption sites for the intermediates by the large specific surface of reduced graphene oxide (RGO) and nickel foam. After then, Cheng et al. [94] studied Pt-RGO/Cu again as the photocathode and TiO_2 nanotube as the photoanode for PEC reduction of CO_2. The dual functional Cu foam, which has well-defined porosity, large specific surface area, and in particular great affinity for CO_2 reduction to hydrocarbons, can markedly improve the carbon atom conversion rate of Pt-RGO photocathode. Through the current density, product generation rate, and carbon atom conversion, the synergistic effect of PC and EC reduction CO_2 in the PEC cell is proven. Under the light illumination, the TiO_2 photoanode generates the photovoltage, which compensates the cathode potential of the required electromotive force for CO_2 reduction, and the photogenerated electrons are transferred to the cathode for CO_2 reduction. The use between photovoltage and photogenerated electrons fully resulted in the synergistic effect between PC and EC in the PEC cell.

In the photoanode and photocathode setup system, the photoanode usually employs an n-type semiconductor with a sufficiently positive VB position for water oxidation, and the photocathode employs a p-type semiconductor with a sufficiently negative CB position for CO_2 reduction, which could minimally supply external electrical bias. Ishitani and coworkers [30] took NiO–RuRe as the photocathode and CoO_x/TaON as the photoanode for PEC CO_2 reduction activity of hybrid photocathode in an aqueous electrolyte solution (Figure 10.7a). The Ru(II) center can sensitize photocathode to enhance light absorption, and the electrons transferred to the Re-based complex that serves as active sites for CO_2 reduction. Combined with the CoO_x/TaON photoanode, the visible-light-driven PEC CO_2 reduction cell can be carried out to generate CO and O_2 using water as a reductant with an external electrical bias of 0.3 V. Sato et al. [81] took p-InP/Ru complex photocathode for PEC reduction of CO_2 to $HCOO^-$ and combined Pt with TiO_2 as the photoanode for water oxidation. In this reaction system (Figure 10.7b), the effective functionalization of the metal complex electrocatalyst (MCE) with the Ru complex anchor ligand markedly improves the rate of two-electron reduction of CO_2 on semiconductors (SC)/MCE. The InP/[MCE2-A + MCE4] and TiO_2/Pt were used as CO_2 reduction photocatalyst and H_2O oxidation photocatalyst, respectively. Except for the photovoltage, no external electrical bias was applied between the two photocatalysts. And the electrons of TiO_2/Pt extracted from H_2O transfer to InP/[MCE2-A + MCE4] through the external circuit. The conversion efficiency from solar energy to chemical energy is 0.03%. By the isotope tracer analysis, the electrons and protons, which extracted from H_2O during the oxidation process to O_2, reduce the CO_2 to formate.

Another system is the photovoltaic-based tandem electrocatalytic device system, which connects a photovoltaic cell with an efficient electrolyzer for CO_2 reduction. Grätzel and coworkers [31] employed series-connected perovskite photovoltaics and high-performance catalyst electrodes for efficient reduction of CO_2 to CO. With the IrO_2 as the anode and Au as the cathode, the solar-to-CO efficiency exceeds 65% with excellent stability over 18 hours. In the PEC system, the solution of anode and cathode must be the same to equilibrate the chemical electrical bias from concentration gradients. In this study, the 0.5 M $KHCO_3$ is employed as the electrolyte, and three connected perovskite cells are employed to capture the sunlight and produce voltage for driving this reaction, which produces an open-circuit voltage of 3.1 V and

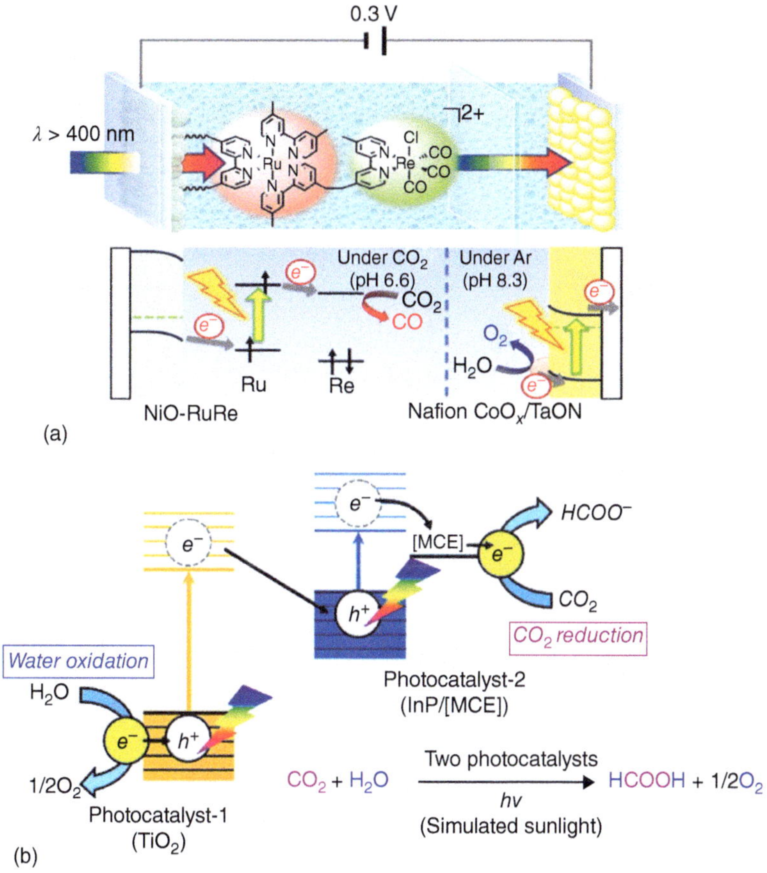

Figure 10.7 Schematic image of (a) NiO–RuRe as photocathode and CoO_x/TaON as photoanode. Source: Sahara et al. [30]. © 2016, American Chemical Society. (b) p-InP/Ru complex photocathode and Pt–TiO_2 as photoanode of hybrid PEC cell with Z-scheme configuration for CO_2 reduction. Source: Sato et al. [81]. © 2011, American Chemical Society.

a short-circuit current density of 6.15 mA cm^{-2}. Jing and coworkers [95] employed Co-Pi/W:$BiVO_4$ as the photoanode and Pd/NR_x@TiO_2 as the photocathode for PEC reduction of CO_2. The main products are methanol and O_2 in the liquid and gas phase, respectively. The CH_3OH yield is 106 μmol L^{-1} h^{-1} cm^{-2}, and quantum efficiency is up to 95% at an external voltage of 0.6 V by a tandem Si solar cell. Without the dye molecules of NR_x, the TiO_2/fluorine-doped tin oxide (FTO) photocathode could not generate methanol. It implies that the dye molecules of NR_x on the surfaces of TiO_2 are not only a good sensitizer to harvest the sunlight but also good ligands to catch CO_2 molecules. Firstly, the CO_2 molecules and the protons are absorbed by the nitrogen atoms of organic moieties and the Pt nanoparticles, respectively. When the dye molecules harvest the photons, the electrons of the dyes highest occupied molecular orbital (HOMO) instantly jump to the LUMO to generate the photoelectrons, which could transfer to the protons of the Pt nanoparticle surface to make abundant active hydrogen atoms for reduction of CO_2 to methanol. At the same time, the generated hydroxyl radicals are transferred to the $BiVO_4$ electrode by the OH^- group and finally releases oxygen gas via H_2O_2.

Apart from the above three setup systems, there exist other photoelectrocatalysis systems, such as flow cell [96], Li-CO_2 cell [97], and Zn-CO_2 cell [98]. Although those cell systems just now are used as electrocatalyst for reduction of CO_2, they should be further developed for PEC CO_2 reduction system in the future.

10.5 Conclusion and Outlook

With the increasing pressure from energy crisis and global warming, the researchers are impelled to exploit the inexhaustible solar energy. Photoelectrocatalysis could combine the merits of photocatalysis with electrocatalysis to exhibit more unexpected performances. The past decades have witnessed the outstanding developments of PEC CO_2 reduction, which are summarized and discussed in this chapter. With respect to the critical factors for PEC CO_2 reduction, there exist confusing questions. Among the various influence factors, the key factors are the reaction temperature and pressure, pH value, solvent species, and external electrical bias. There exist other influence factors, such as protons and electrons transmission rate of proton membrane, density gradient of CO_2 in the solvent, potential difference between two electrolytic cells, and electrolyte caused by the reaction. How to lower the influence of PEC CO_2 reduction during the reaction is an imperative question and a potential research direction.

Some recent advances in the development of semiconductors and setups for PEC CO_2 reduction are summarized in Table 10.2. Due to the fleetly development in recent decades, the studies are still limited in the PEC CO_2 reduction, which is caused by the lower activity and stability of photoelectrode materials and the low effectivity of the setup cell systems. Due to the sluggish thermodynamic and kinetics of the surface multi-electron transfer reactions, the activation energy of rate limiting and overpotential is usually so much high for PEC CO_2 reduction, and the solar-to-chemical energy conversion efficiencies are far lower than the requirement of commercialization in the most PEC CO_2 reduction systems. While with the development of catalyst species, the application of surface cocatalysts could effectively lower the activation energy and overpotential. For the PEC CO_2 reduction, the oxide semiconductors are usually used as the preferred catalysts, such as Cu-based materials, Fe_2O_3 semiconductor and Co_3O_4 semiconductor, some non-oxide semiconductors, such as g-C_3N_4 semiconductor and RGO, and so on. Those catalysts are of high abundance, outstanding selectivity, and unique properties for production of hydrocarbons. To lower the activation energy and overpotential, the metal nanoparticles such as Au NPs or Au-based alloys are the most widely employed cocatalysts owing to their excellent activity and selectivity for CO evolution. Metal complexes such as Ru and Re complexes are also extensively studied as cocatalysts for PEC CO_2 reduction. Apart from metals, metal-free cocatalysts, such as ammonium and pyridinium ions; biological cocatalysts, such as

Table 10.2 Some recent advances reported in the development of semiconductors and setups for PEC CO_2 reduction.

Cathode/photocathode	Anode/photoanode	Electrolyte	Potential	Main product	Faradaic Efficiency (%)	References
Au–ZnTe/ZnO	Pt mesh	0.5 M $KHCO_3$	−0.7 V *vs.* RHE	CO	97	[23]
GaP/TiO_2	Pt mesh	0.5 M NaCl	−0.5 V *vs.* NHE	CH_3OH	55	[24]
$BCN_{3.0}$ (B-g-C_3N_4)	A glassy carbon	0.5 M $KHCO_3$	−0.4 V *vs.* Ag/AgCl	C_2H_5OH	78	[26]
Nanoporous Ag	Ni–Si(110) (photoanode)	0.5 M $KHCO_3$	2.0 V *vs.* RHE	CO	70	[28]
Cu_2O/CuO	WO_3 (photoanode)	0.5 M $KHCO_3$	0.65 V *vs.* RHE	CH_4	67	[29]
CoO_x/TaON	NiO–RuRe	50 mM $NaHCO_3$	−0.7 V *vs.* Ag/AgCl	CO	64	[30]
IrO_2	Au	0.5 M $NaHCO_3$	−0.4 V *vs.* RHE	CO	90	[31]
Ag plate	$WO_3/BiVO_4$ (photoanode)	$KHCO_3$ + [NH_2C_3MIm][Br]	1.7 V *vs.* RHE	HCOOH	94.1	[45]
Cu_2O	TiO_2 (photoanode)	0.1 M $KHCO_3$	0.75 V *vs.* RHE	CO, CH_4, CH_3OH	87.4	[20]
Co_3O_4	Pt plate	0.1 M Na_2SO_4	−0.9 V *vs.* SCE	HCOOH	94.7	[54]
CuO FCs/Fe_2O_3 NTs	Pt plate	0.1 M $KHCO_3$	−1.1 V *vs.* SCE	CH_3OH	91.2	[55]
SnO_2 NRs/Fe_2O_3 NTs	Pt plate	0.1 M $KHCO_3$	−1.1 V *vs.* SCE	CH_3OH	87.04	[56]
CuInS2/graphene	Graphite sheet	0.1 M acetate buff solution	−0.59 V *vs.* SCE	CH_3OH	95	[57]
Zn/ZnO/ZnTe	Pt plate	0.1 M $KHCO_3$	−0.7 V *vs.* RHE	CO	22.9	[62]
PPy/ZnTe	Carbon rod	0.1 M $KHCO_3$	−0.2 V *vs.* RHE	HCOOH	37.2	[63]
Pt-RGO	Pt-TNT	1 M NaCl + 1 M $NaHCO_3$	−0.7 V *vs.* RHE	HCOOH, CH_3OH, C_2H_5OH, CH_3COOH	80.5	[93]

natural enzymes; and organic molecule cocatalysts, such as protonated pyridinium, have also been developed and utilized in PEC CO_2 reduction.

In spite of the above efforts and achievements in PEC CO_2 reduction, there are still many challenges for future confronting:

(i) How to improve the efficiency of PEC CO_2 reduction? The effective pathways are extending light absorption range and improving the efficiency of charge separation and migration to supply sufficient electrons for PEC CO_2 reduction reaction. There are lots of matters that can be used from the photocatalysis and photovoltaic system. The other effective pathway is to lower the overpotential of PEC CO_2 reduction reaction, because the overpotential plays a crucial role in the reaction. An appealing strategy is constructing high active sites to lower the overpotential. So many heterogeneous and homogeneous as cocatalysts are anchored on the photoelectrodes. With the exploitation and design of diversiform cocatalysts, the new challenge is to overcome how to modulate the efficient charge transfer between electrode and cocatalysts and how to avoid the deactivation for high stability.

(ii) How to enhance the selectivity of high-valuable products? The employed cocatalyst seems to be a panacea approach to enhance the performance of PEC CO_2 reduction. But the competitive water reduction is more favorable than CO_2 reduction not only on the thermodynamically but kinetically with the assistance of cocatalysts. It is unavoidable that the PEC CO_2 reduction is companied and competed by hydrogen evolution in aqueous solutions no matter for photocatalysis or electrocatalysis except for CO. However, the hydrogen evolution is not good for nothing during the PEC CO_2 reduction process. If you want to exploit the high-valuable products, H protons are an essential raw material. There are two candidates that have been widely employed for the energy-rich chemicals. One is Cu-based catalysts. Many advanced works have reported that the Cu-based catalysts could generate the high-valuable products, especially the hydrocarbon. However, about the reaction pathway and intermediates for PEC CO_2 reduction, it is still ambiguous so the Cu based is still promising for future development. The other is enzymatic biocatalysts. The enzymatic biocatalyst has unique selectivity for PEC CO_2 reduction, and no side products. But itself has weakness, such as high cost and poor stability, hinder it future application.

(iii) How to make clear the reaction intermediates and pathways during the PEC CO_2 reduction? Through delicately coordinating the surface electronic state, the photoelectrodes could have selective adsorption and desorption for the different intermediates and product molecules to control the selectivity of PEC CO_2 reduction. The reaction barrier of potential and rate-limiting steps could also be lowered by tuning the adsorption of intermediate product in the PEC CO_2 reduction to increase the total conversion efficiency of solar energy. Harvesting light of the photoelectrode and transporting charge are the other hot potato when turning to the surface engineering of photoelectrode.

(iv) How to improve the mass transfer of the PEC CO_2 reduction? In the PEC reaction system, the reactions just occur at the surface of the photoelectrode, and the surface of harvesting light is with back to the counter electrode. Thus, the mass transfer should be carefully considered and overcome. The mass transfer of reactants and products could be partly improved by stirring during the proceeding surface reaction. But there still existed a large concentration gradient between the surface photoelectrode and the electrolyte in the PEC system. So developing new PEC cell setup and optimizing the electrode structure could be taken into consideration for a novel PEC CO_2 reduction system to enhance the mass transfer.

In summary, the PEC CO_2 reduction system can offer a promising platform to utilize solar-to-chemical energy and convert CO_2 into fuels. Up to now, the PEC CO_2 reduction is still in its early stage, and there is still a long way to practical industrial applications.

Acknowledgments

The authors acknowledge the National Natural Science Foundation of China (Nos. 51572296, U1662113), the Fundamental Research Funds for the Central Universities (15CX08005A), the financial support from Taishan Scholar Project, Shandong Provincial Natural Science Foundation (ZR2018ZC1458), and the Scientific Research and the Technology Development Project of PetroChina Co., LTD (2016B-2004(GF).

Conflict of Interest

The authors declare no conflict of interest.

References

1 (a) Hu, B., Hu, M., Guo, Q. et al. (2019). *Appl. Catal., B* 253: 77. (b) Jiang, H., Katsumata, K.-i., Hong, J. et al. (2018). *Appl. Catal., B* 224: 783. (c) Low, J., Cheng, B., and Yu, J. (2017). *Appl. Surf. Sci.* 392: 658.

2 White, J.L., Baruch, M.F., Pander, J.E. III, et al. (2015). *Chem. Rev.* 115: 12888.

3 (a) Bi, J., Hou, P., Liu, F.W., and Kang, P. (2019). *ChemSusChem* 12: 2195. (b) Chatterjee, S., Griego, C., Hart, J.L. et al. (2019). *ACS Catal.* 9: 5290. (c) Nichols, A.W., Chatterjee, S., Sabat, M., and Machan, C.W. (2018). *Inorg. Chem.* 57: 2111. (d) Buckley, A.K., Lee, M., Cheng, T. et al. (2019). *J. Am. Chem. Soc.* 141: 7355.

4 Li, P., Jing, H., Xu, J. et al. (2014). *Nanoscale* 6: 11380.

5 (a) Shen, Q., Chen, Z., Huang, X. et al. (2015). *Environ. Sci. Technol.* 49: 5828. (b) Kuk, S.K., Singh, R.K., Nam, D.H. et al. (2017). *Angew. Chem. Int. Ed.* 56: 3827. (c) Szaniawska, E., Bienkowski, K., Rutkowska, I.A. et al. (2018). *Catal. Today* 300: 145. (d) Li, C., Wang, T., Liu, B. et al. (2019). *Energy Environ. Sci.* 12: 923. (e) Kalamaras, E., Maroto-Valer, M.M., Andresen, J.M. et al. (2019).

Innovative Solutions for Energy Transitions, vol. 158 (eds. J. Yan, H.X. Yang, H. Li and X. Chen), 767. (f) Pawar, A.U., Kim, C.W., Nguyen-Le, M.-T., and Kang, Y.S. (2019). *ACS Sustainable Chem. Eng.* 7: 7431.
6 Xie, S., Zhang, Q., Liu, G., and Wang, Y. (2016). *Chem. Commun.* 52: 35.
7 Kumar, B., Llorente, M., Froehlich, J. et al. (2012). *Annu. Rev. Phys. Chem.* 63: 541.
8 Sivula, K. and Van De Krol, R. (2016). *Nat. Rev. Mater.* 1: 15010.
9 (a) Chen, Z., Dinh, H.N., and Miller, E. (2013). *Photoelectrochemical Water Splitting*. Springer. (b) Jiang, C., Moniz, S.J., Wang, A. et al. (2017). *Chem. Soc. Rev.* 46: 4645.
10 Roger, I., Shipman, M.A., and Symes, M.D. (2017). *Nat. Rev. Chem.* 1: 0003.
11 (a) Chen, B., Zhang, Z., Baek, M. et al. (2018). *Appl. Catal., B* 237: 763. (b) Xiao, F.X. and Liu, B. (2018). *Adv. Mater. Interfaces* 5: 1701098. (c) Passalacqua, R., Perathoner, S., and Centi, G. (2017). *J. Energy Chem.* 26: 219.
12 Halmann, M. (1978). *Nature* 275: 115.
13 Pang, H., Masuda, T., and Ye, J. (2018). *Chem. Asian J.* 13: 127.
14 (a) S. Xu, E. A. Carter. (2019). *Chemical reviews* 11: 6631; (b) Windle, C.D. and Reisner, E. (2015). *CHIMIA Int. J. Chem.* 69: 435. (c) Wang, Y., He, D., Chen, H., and Wang, D. (2019). *J. Photochem. Photobiol., C.*
15 Chang, X., Wang, T., and Gong, J. (2016). *Energy Environ. Sci.* 9: 2177.
16 Yang, Z.X., Xu, J.F., Wu, C.X. et al. (2014). *Appl. Catal., B* 156: 249.
17 Hisatomi, T., Kubota, J., and Domen, K. (2014). *Chem. Soc. Rev.* 43: 7520.
18 Zhang, N., Long, R., Gao, C., and Xiong, Y. (2018). *Sci. China Mater.* 61: 771.
19 Hirota, K., Tryk, D.A., Yamamoto, T. et al. (1998). *J. Phys. Chem. B* 102: 9834.
20 (a) Chang, X., Wang, T., Zhang, P. et al. (2016). *Angew. Chem. Int. Ed.* 55: 8840. (b) Bachmeier, A., Hall, S., Ragsdale, S.W., and Armstrong, F.A. (2014). *J. Am. Chem. Soc.* 136: 13518.
21 Schreier, M., Luo, J., Gao, P. et al. (2016). *J. Am. Chem. Soc.* 138: 1938.
22 Yuan, J. and Hao, C. (2013). *Sol. Energy Mater. Sol. Cells* 108: 170.
23 Jang, Y.J., Jang, J.-W., Lee, J. et al. (2015). *Energy Environ. Sci.* 8: 3597.
24 Zeng, G., Qiu, J., Li, Z. et al. (2014). *ACS Catal.* 4: 3512.
25 Lessio, M. and Carter, E.A. (2015). *J. Am. Chem. Soc.* 137: 13248.
26 Sagara, N., Kamimura, S., Tsubota, T., and Ohno, T. (2016). *Appl. Catal., B* 192: 193.
27 Kang, U., Choi, S.K., Ham, D.J. et al. (2015). *Energy Environ. Sci.* 8: 2638.
28 Zhang, Y., Luc, W., Hutchings, G.S., and Jiao, F. (2016). *ACS Appl. Mater. Interfaces* 8: 24652.
29 Magesh, G., Kim, E.S., Kang, H.J. et al. (2014). *J. Mater. Chem. A* 2: 2044.
30 Sahara, G., Kumagai, H., Maeda, K. et al. (2016). *J. Am. Chem. Soc.* 138: 14152.
31 Schreier, M., Curvat, L., Giordano, F. et al. (2015). *Nat. Commun.* 6: 7326.
32 Schreier, M., Héroguel, F., Steier, L. et al. (2017). *Nat. Energy* 2: 17087.
33 (a) Surdhar, P.S., Mezyk, S.P., and Armstrong, D.A. (1989). *J. Phys. Chem.* 93: 3360. (b) Habisreutinger, S.N., Schmidt-Mende, L., and Stolarczyk, J.K. (2013). *Angew. Chem. Int. Ed.* 52: 7372.

34 Hara, K., Kudo, A., Sakata, T., and Watanabe, M. (1995). *J. Electrochem. Soc.* 142: L57.

35 Yang, C., Yu, X., Heißler, S. et al. (2017). *Angew. Chem. Int. Ed.* 56: 375.

36 (a) Shown, I., Hsu, H.-C., Chang, Y.-C. et al. (2014). *Nano Lett.* 14: 6097. (b) Gao, C., Meng, Q., Zhao, K. et al. (2016). *Adv. Mater.* 28: 6485.

37 Hori, Y., Kikuchi, K., Murata, A., and Suzuki, S. (1986). *Chem. Lett.* 15: 897.

38 Cook, R.L., MacDuff, R.C., and Sammells, A.F. (1988). *J. Electrochem. Soc.* 135: 3069.

39 Hirota, K., Tryk, D.A., Hashimoto, K. et al. (1998). *J. Electrochem. Soc.* 145: L82.

40 de Brito, J.F., Araujo, A.R., Rajeshwar, K., and Zanoni, M.V.B. (2015). *Chem. Eng. J.* 264: 302.

41 Kaneco, S., Katsumata, H., Suzuki, T., and Ohta, K. (2006). *Appl. Catal., B* 64: 139.

42 Kou, Y., Nakatani, S., Sunagawa, G. et al. (2014). *J. Catal.* 310: 57.

43 Medina-Ramos, J., Pupillo, R.C., Keane, T.P. et al. (2015). *J. Am. Chem. Soc.* 137: 5021.

44 Lin, J., Ding, Z., Hou, Y., and Wang, X. (2013). *Sci. Rep.* 3: 1056.

45 Lu, W., Jia, B., Cui, B. et al. (2017). *Angew. Chem. Int. Ed.* 56: 11851.

46 Bard, A.J., Faulkner, L.R., Leddy, J., and Zoski, C.G. (1980). *Electrochemical Methods: Fundamentals and Applications*, vol. 2. New York: Wiley.

47 (a) Li, J. and Wu, N. (2015). *Catal. Sci. Technol.* 5: 1360. (b) Ikeda, S. and Ito, K. (1994). *Int. J. Sol. Energy* 14: 181. (c) Rongé, J., Bosserez, T., Martel, D. et al. (2014). *Chem. Soc. Rev.* 43: 7963.

48 (a) Cook, R.L., MacDuff, R.C., and Sammells, A.F. (1987). *J. Electrochem. Soc.* 134: 2375. (b) Frese, K.W. (1991). *J. Electrochem. Soc.* 138: 3338.

49 Li, C.W. and Kanan, M.W. (2012). *J. Am. Chem. Soc.* 134: 7231.

50 Kas, R., Kortlever, R., Milbrat, A. et al. (2014). *Phys. Chem. Chem. Phys.* 16: 12194.

51 Le, M., Ren, M., Zhang, Z. et al. (2011). *J. Electrochem. Soc.* 158: E45.

52 Li, C.W., Ciston, J., and Kanan, M.W. (2014). *Nature* 508: 504.

53 Handoko, A.D. and Tang, J. (2013). *Int. J. Hydrogen Energy* 38: 13017.

54 Huang, X., Cao, T., Liu, M., and Zhao, G. (2013). *J. Phys. Chem. C* 117: 26432.

55 Li, P., Wang, H., Xu, J. et al. (2013). *Nanoscale* 5: 11748.

56 Yang, Z., Wang, H., Song, W. et al. (2017). *J. Colloid Interface Sci.* 486: 232.

57 Yuan, J., Wang, P., Hao, C., and Yu, G. (2016). *Electrochim. Acta* 193: 1.

58 Shi, H.F., Chen, G.Q., Zhang, C.L., and Zou, Z.G. (2014). *ACS Catal.* 4: 3637.

59 Senftle, T.P., Lessio, M., and Carter, E.A. (2016). *Chem. Mater.* 28: 5799.

60 Lessio, M., Senftle, T.P., and Carter, E.A. (2016). *ACS Energy Lett.* 1: 464.

61 Kaneco, S., Ueno, Y., Katsumata, H. et al. (2009). *Chem. Eng. J.* 148: 57.

62 Jang, J.W., Cho, S., Magesh, G. et al. (2014). *Angew. Chem. Int. Ed.* 53: 5852.

63 Won, D.H., Chung, J., Park, S.H. et al. (2015). *J. Mater. Chem. A* 3: 1089.

64 (a) Hori, Y., Wakebe, H., Tsukamoto, T., and Koga, O. (1994). *Electrochim. Acta* 39: 1833. (b) Muradov, N.Z. and Veziroğlu, T.N. (2011). *Carbon-Neutral Fuels and Energy Carriers*. CRC Press. (c) Li, W. (2012). *ChemInform* 43: 243.

65 (a) Baruch, M.F., Pander, J.E. III,, White, J.L., and Bocarsly, A.B. (2015). *ACS Catal.* 5: 3148. (b) Van de Krol, R. and Grätzel, M. (2012). *Photoelectrochemical Hydrogen Production*, vol. 90. Springer.

66 Bai, S., Yin, W., Wang, L. et al. (2016). *RSC Adv.* 6: 57446.

67 (a) Kuhl, K.P., Hatsukade, T., Cave, E.R. et al. (2014). *J. Am. Chem. Soc.* 136: 14107. (b) Li, H., Wu, Y., Li, C. et al. (2019). *Appl. Catal., B* 251: 305.

68 Sun, T., Fan, J., Liu, E. et al. (2012). *Powder Technol.* 228: 210.

69 (a) Yamada, Y., Miyahigashi, T., Kotani, H. et al. (2012). *Energy Environ. Sci.* 5: 6111. (b) Sun, T., Liu, E., Liang, X. et al. (2015). *Appl. Surf. Sci.* 347: 696.

70 Huerta-Flores, A., Torres-Martínez, L.M., Moctezuma, E., and Ceballos-Sanchez, O. (2016). *Fuel* 181: 670.

71 Zhu, W., Michalsky, R., Metin, Ö. et al. (2013). *J. Am. Chem. Soc.* 135: 16833.

72 (a) Lu, Q., Rosen, J., Zhou, Y. et al. (2014). *Nat. Commun.* 5: 3242. (b) Lee, J.H., Kattel, S., Jiang, Z. et al. (2019). *Nat. Commun.* 10: 3724.

73 Gao, D., Zhou, H., Wang, J. et al. (2015). *J. Am. Chem. Soc.* 137: 4288.

74 Lei, F., Liu, W., Sun, Y. et al. (2016). *Nat. Commun.* 7: 12697.

75 Hegner, R., Rosa, L.F., and Harnisch, F. (2018). *Appl. Catal., B* 238: 546.

76 Alvarez-Guerra, M., Quintanilla, S., and Irabien, A. (2012). *Chem. Eng. J.* 207: 278.

77 (a) Hou, J., Cheng, H., Takeda, O., and Zhu, H. (2015). *Angew. Chem. Int. Ed.* 54: 8480. (b) Long, R., Li, Y., Liu, Y. et al. (2017). *J. Am. Chem. Soc.* 139: 4486. (c) Won, D.H., Choi, C.H., Chung, J., and Woo, S.I. (2014). *Appl. Catal., B* 158–159: 217.

78 (a) Gao, C., Wang, J., Xu, H., and Xiong, Y. (2017). *Chem. Soc. Rev.* 46: 2799. (b) Zhao, J., Wang, X., Xu, Z., and Loo, J.S. (2014). *J. Mater. Chem. A* 2: 15228. (c) Schreier, M., Gao, P., Mayer, M.T. et al. (2015). *Energy Environ. Sci.* 8: 855.

79 Morris, A.J., Meyer, G.J., and Fujita, E. (2009). *Acc. Chem. Res.* 42: 1983.

80 Kumar, B., Smieja, J.M., and Kubiak, C.P. (2010). *J. Phys. Chem. C* 114: 14220.

81 Sato, S., Arai, T., Morikawa, T. et al. (2011). *J. Am. Chem. Soc.* 133: 15240.

82 Liu, X., Inagaki, S., and Gong, J. (2016). *Angew. Chem. Int. Ed.* 55: 14924.

83 Chen, L., Guo, Z., Wei, X.G. et al. (2015). *J. Am. Chem. Soc.* 137: 10918.

84 Grace, A.N., Choi, S.Y., Vinoba, M. et al. (2014). *Appl. Energy* 120: 85.

85 Aydın, R., Doğan, H.Ö., and Köleli, F. (2013). *Appl. Catal., B* 140–141: 478.

86 Coskun, H., Aljabour, A., De Luna, P. et al. (2017). *Sci. Adv.* 3: e1700686.

87 (a) Yan, Y., Chandrasekaran, P., Mague, J.T. et al. (2011). *Inorg. Chem.* 51: 346. (b) Yan, Y., Keating, C., Chandrasekaran, P. et al. (2013). *Inorg. Chem.* 52: 6743. (c) Dobbek, H. (2011). *Coord. Chem. Rev.* 255: 1104.

88 Parkinson, B.A. and Weaver, P.F. (1984). *Nature* 309: 148.

89 Woolerton, T.W., Sheard, S., Reisner, E. et al. (2010). *J. Am. Chem. Soc.* 132: 2132.

90 Oh, Y. and Hu, X. (2013). *Chem. Soc. Rev.* 42: 2253.

91 Barton, E.E., Rampulla, D.M., and Bocarsly, A.B. (2008). *J. Am. Chem. Soc.* 130: 6342.

92 Dridi, H., Comminges, C., Morais, C. et al. (2017). *J. Am. Chem. Soc.* 139: 13922.

93 Cheng, J., Zhang, M., Wu, G. et al. (2015). *Sol. Energy Mater. Sol. Cells* 132: 606.

94 Cheng, J., Zhang, M., Liu, J. et al. (2015). *J. Mater. Chem. A* 3: 12947.
95 Jia, Y., Xu, Y., Nie, R. et al. (2017). *J. Mater. Chem. A* 5: 5495.
96 Gong, Q., Ding, P., Xu, M. et al. (2019). *Nat. Commun.* 10: 2807.
97 Huang, Y., Wang, Y., Tang, C. et al. (2019). *Adv. Mater.* 31: e1803800.
98 (a) Xie, J., Wang, X., Lv, J. et al. (2018). *Angew. Chem. Int. Ed.* 57: 16996.
(b) Wang, X., Xie, J., Ghausi, M.A. et al. (2019). *Adv. Mater.* 31: e1807807.

11

Photocatalytic and Photoelectrochemical Nitrogen Fixation

Lei Shi and Hongqi Sun

Edith Cowan University, School of Engineering, Joondalup, WA 6027, Australia

11.1 Introduction

As one of the most indispensable chemicals in the agriculture society, ammonia (NH_3) is extensively used to produce nitrogen-containing fertilizers [1]. In addition to its crucial role in agriculture, NH_3 is also considered as a renewable and green energy, ascribed to its high energy density, safe storage, and easy transportation, especially comparing with hydrogen (H_2) [2]. At present, massive NH_3 (~200 million tons) is produced industrially each year.

The N source in NH_3 is generally originated from the molecular N_2, which is abundant in the atmosphere (~78%) [3]. However, it cannot be easily and directly utilized for the NH_3 synthesis in ambient conditions, since nonpolar N_2 possesses a high N≡N bond energy (941 kJ mol^{-1}), low proton affinity (−1.9 eV), and strong ionization energy (15.8 eV), as well as the large energy gap (~10.82 eV) between the highest occupied molecular orbital (HOMO) and lowest unoccupied molecular orbital (LUMO), making its oxidation or reduction extremely difficult [4, 5].

In nature, biological N_2 reduction occurs with the assistance of nitrogenases. Nitrogenase contains two key catalytic centers, for example, the Fe- and MoFe-based proteins, which are responsible for the N_2 activation and fixation to generate NH_3 (Eq. (11.1)), consuming the energy supplied by the hydrolysis process of adenosine triphosphate (ATP) to form adenosine diphosphate (ADP) [6]. Despite these attractive merits in the natural processes, the low yield rates could not meet the actual needs and severely restrict their large-scale applications:

$$N_2 + 8H^+ + 8e^- + 16ATP \rightarrow 2NH_3 + H_2 + 16ADP + 16Pi \quad (11.1)$$

To overcome these challenges and meet the great demands of NH_3 in practice, the industrial NH_3 production is realized with the Haber–Bosch process (Eq. (11.2)), which utilizes versatile heterogeneous catalysts (e.g. Fe- and Ru-based catalysts) to efficiently dissociate the strong N≡N bond, at high temperatures (300–500 °C) and pressures (150–300 atm) [7]. Accordingly, around 2–3% of the global energy is consumed by this energy-intensive NH_3 synthesis process. Besides, given the production of required H_2 primarily stemming from the steam reforming (Eq. (11.3))

Solar-to-Chemical Conversion: Photocatalytic and Photoelectrochemcial Processes, First Edition. Edited by Hongqi Sun.

and coal gasification (Eq. (11.4)), tens of thousands of tons of CO_2 is emitted into the atmosphere, thus resulting in serious environmental issues [8]. In view of these highly-concerned problems and challenges, it is imperative to explore promising processes to realize the N_2 reduction in ambient conditions:

$$N_2 + 3H_2 \rightarrow 2NH_3 \tag{11.2}$$

$$CH_4 + 2H_2O \rightarrow 4H_2 + CO_2 \tag{11.3}$$

$$C + 2H_2O \rightarrow 2H_2 + CO_2 \tag{11.4}$$

As the first attempt in 1977, Schrauzer and Guth demonstrated the photocatalytic activity of TiO_2 in N_2 fixation to synthesize NH_3, as well as the hydrazine (N_2H_4) as the by-product, in the presence of gaseous N_2 and water (H_2O) [9]. Since then, photocatalytic NH_3 synthesis has attracted widespread interests in the past few decades. On the one hand, as a clean and sustainable resource, solar energy is abundant on the Earth. Serving as the driving force for sustainable photocatalytic N_2 reduction, the acquisition and utilization of solar energy have opened new horizon for the green NH_3 production under ambient conditions. On the other hand, utilizing the H_2O as the proton supplier (Eq. (11.5)), it could significantly avoid the dependence on H_2 in the Haber–Bosch process, thus effectively addressing these environment issues mentioned previously:

$$2N_2 + 6H_2O \rightarrow 4NH_3 + 3O_2 \tag{11.5}$$

In addition to the photocatalysis, the emerging photoelectrochemical systems have gained great interests in N_2 fixation recently [10]. In essence, the photoelectrochemical process is based on a photocatalytic process with the assistance of an extra bias, which is introduced to promote the reaction forward and enhance the catalytic ability of the photoelectrocatalysts. Taking advantages from both photo- and electro-catalysis, the activities of photo-induced electron–hole (e–h) pairs would be substantially boosted to achieve high conversion efficiencies in NH_3 formation.

Although a rapid development has been observed in this critical research field recently, the limited NH_3 yield rates, as well as low apparent quantum efficiency (AQE)/solar-to-chemical conversion (SCC)/Faradaic efficiency (FE) reported up to now, severely hamper the efficient NH_3 synthesis in real applications. Playing critical roles in the photocatalytic and photoelectrochemical N_2 reduction, photo- and photoelectro-catalysts with desirable electronic structures and surface functionalization have attracted much attention recently [11]. Through the rational design and controllable preparation of promising catalysts, it is highly anticipated to dramatically improve the activity and selectivity in NH_3 production [12, 127].

In this chapter, we present recent advances in the photocatalytic and photoelectrochemical NH_3 synthesis. The fundamental principles and challenges associated with N_2 reduction are firstly introduced. Subsequently, typical strategies for the development of promising catalysts are summarized, such as the defect engineering, structural regulation, interface control, heterojunction construction, co-catalyst engineering, and biomimetic engineering (Scheme 11.1). Critical discussions on the catalyst properties, as well as the structure–activity relationships, are provided. Finally, in the conclusion section, present challenges as well as future developments in the research field are discussed.

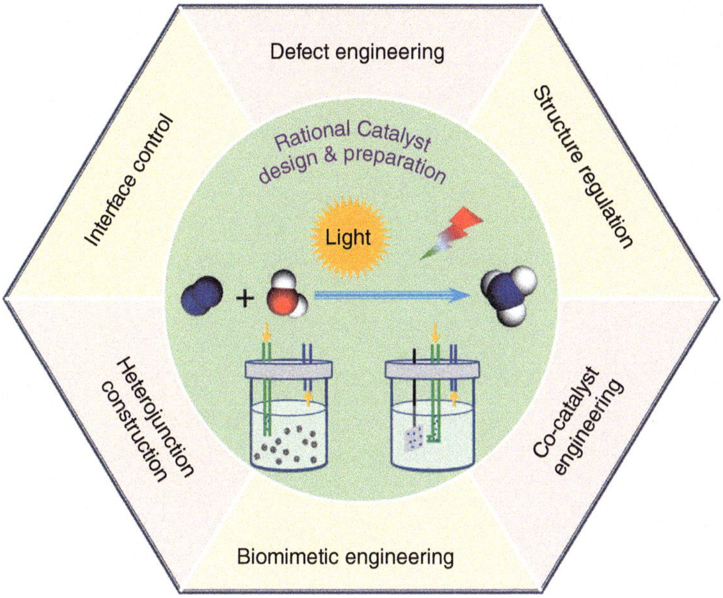

Scheme 11.1 Various strategies for the design and preparation of promising catalysts for the photocatalytic and photoelectrochemical N_2 reduction.

11.2 Fundamental Principles and Present Challenges

11.2.1 Principles in N_2 Reduction for NH_3 Production

In order to design highly efficient catalysts for N_2 fixation, it is urgent to figure out the basic reaction mechanisms involved in the NH_3 formation. Up to now, several classical mechanisms, including the dissociative and associative pathways, have been proposed in N_2 reduction processes (Figure 11.1a) [13]. For the dissociative pathway, a direct N≡N cleavage occurs prior to the hydrogenation process. Thus, a huge energy input is required to meet the dissociation energy of the strong N≡N bond. That is why the Haber–Bosch reaction following this process must be performed under the harsh conditions. As a comparison, the partial hydrogenation on the adsorbed N_2 molecules, following by the breakage of N/N bond is observed in the associative pathway. Based on the different hydrogenation procedures, the associative route can be further classified into the alternating and distal mechanisms. As for the alternating mechanism, the hydrogenation reaction with protons occurs simultaneously on the two adsorbed N atoms, resulting in a route of $HN{=}N \rightarrow HN{=}NH \rightarrow H_2N{-}NH_2 \rightarrow NH_3$. As a common by-product, N_2H_4 will

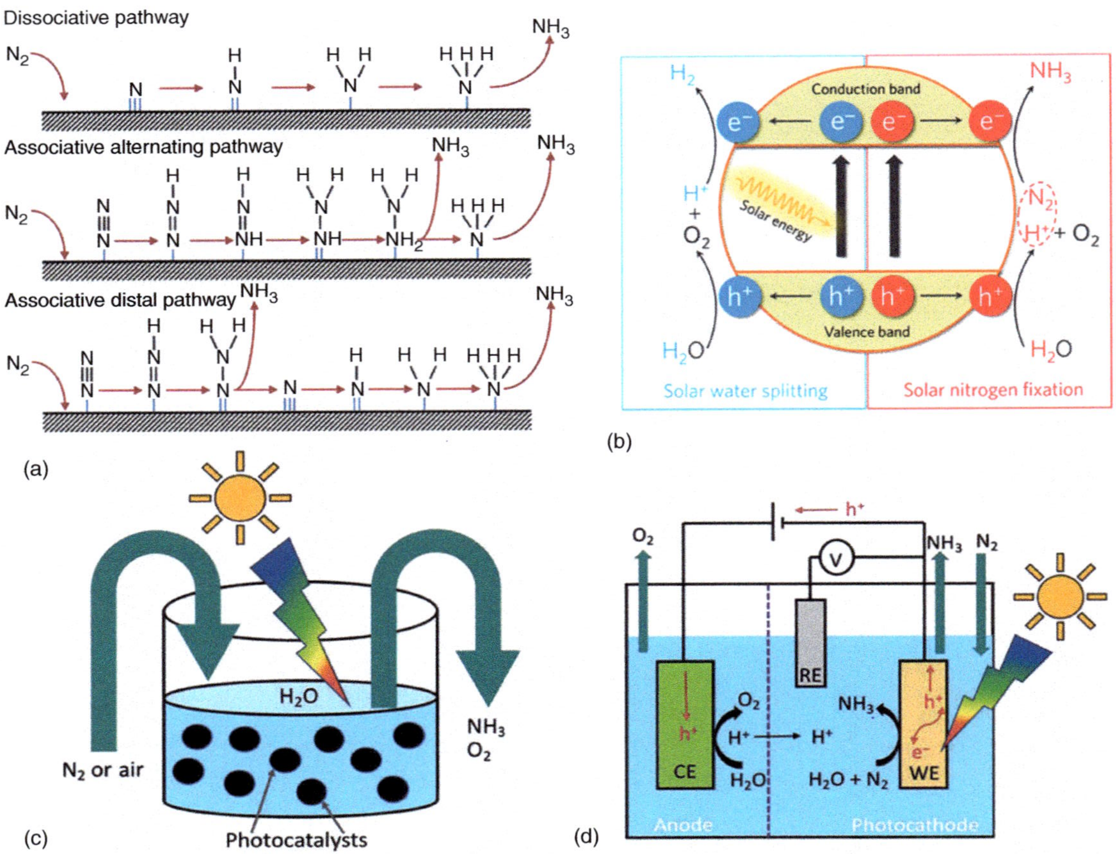

Figure 11.1 (a) Schematic illustrations of different pathways for N_2 reduction, including the dissociative pathway, associative alternating pathway, and associative distal pathway. Source: (a) Shipman et al. [13]. © 2017, Elsevier. (b) Schematic representation of the photocatalytic NH_3 generation and H_2O splitting. Source: (b) Li et al. [14]. © 2017, American Chemical Society. (c) and (d) Setups of photocatalytic and photoelectrochemical reaction systems for NH_3 synthesis. Source: (c–d) Ithisuphalap et al. [15]. © 2019, John Wiley & Sons.

be obtained following this mechanism. In the distal mechanism, the proton will preferably interact with the distal N atom to generate an intermediate of H_2N–N, which further accepts additional electrons and protons, eventually resulting in the generation of NH_3. With the first NH_3 molecule desorbed from the surface, the hydrogenation on the adjacent N atom is triggered on the surface to produce second NH_3 molecule.

Within the photocatalysis-related NH_3 synthesis, it is extensively accepted that these processes generally follow the distal and alternating manners [16], in which the adsorption of N_2 and the electron migration from the catalysts to N_2 molecules provide favorable routes for N≡N bond cleavage (Figure 11.1b) [14]. Under the irradiation of the light that possesses a higher energy than that of the band gap of the photocatalyst, electrons on the valence band (VB) will be activated and injected into its conduction band (CB), forming the photo-generated e–h pairs. Subsequently, these photo-induced carriers are migrated to the catalyst surface, attending the reduction and oxidation reactions, respectively. In the VB, the holes react with H_2O to produce O_2 and proton. Meanwhile, adsorbed N_2 molecules are efficiently reduced on the CB utilizing the active electrons and trapped protons, realizing the production of NH_3.

A typical photocatalytic reaction system for NH_3 production is depicted in Figure 11.1c [15]. Photocatalysts are homogeneously dispersed in the aqueous solutions, and the N_2 or air is applied as the feed gas. Under the light irradiation, the NH_3 will be continuously generated within the system. Generally, instead of pure water, organic scavengers (e.g. ethanol and methanol) are introduced into the system to timely consume the photo-induced holes, eventually improving the conversion efficiency. As for the photoelectrochemical system, an external bias is implemented to accelerate the reaction rate, via improving the separation efficiency of e–h pairs and thus the reduction ability of photoelectrocatalysts. The photoelectrochemical reaction normally involves a three-electrode-based reaction system, containing a working electrode (photocathode), a counter electrode (anode), and a reference electrode (Figure 11.1d). Within the system, the N_2 is reduced on the photocathode modified with the catalysts, while the H_2O splitting occurs at the opposite anode. In view of that the generated NH_3 may be oxidized on the anode within the same cell, two-cell configuration separated with an ion-exchange membrane is recommended to perform the feasible N_2 fixation [17].

11.2.2 Challenges for N_2 Reduction to NH_3

According to the above discussions, to realize such a photocatalytic or photoelectrochemical NH_3 production route using N_2 and H_2O, the synthesized catalysts should possess suitable band positions to realize the N_2 activiation and reduction, as well as the O_2 evolution. However, as the bottleneck among the kinetic steps, the N_2 activation is considered as the most hardest step in photocatalytic NH_3 formation (Table 11.1) [4][18]. Huge energy barriers in the first electron transfer (−4.20 V vs. NHE) and one proton-coupled electron transfer (PCET) (−3.20 V vs. NHE) would definitely postpone the overall reaction rates.

Table 11.1 Applied reduction potentials related to the N_2 reduction [4, 18].

Reaction	E^0 (V)
$N_2 + e^- \rightarrow N_2^-$	−4.20 vs. NHE
$N_2 + e^- + H^+ \rightarrow N_2H$	−3.20 vs. NHE
$N_2 + 2e^- + 2H^+ \rightarrow N_2H_2$	−1.10 vs. RHE
$N_2 + 4e^- + 4H^+ \rightarrow N_2H_4$	−0.36 vs. NHE
$N_2 + 4e^- + 5H^+ \rightarrow N_2H_5^+$	−0.23 vs. NHE
$N_2 + 6e^- + 6H^+ \rightarrow 2NH_3$	0.55 vs. NHE
$N_2 + 8e^- + 8H^+ \rightarrow 2NH_4^+$	0.27 vs. NHE

NHE: normal hydrogen electrode; RHE: reversible hydrogen electrode.
Sources: Bazhenova et al. [4]; Chen et al.[18] .

Therefore, to achieve an efficient photocatalytic or photoelectrochemical N_2 fixation with high NH_3 production rates and conversion efficiencies, several critical challenges are urgently needed to be addressed. The first challenge is the N_2 activation. For most of conventional photocatalysts, it is extremely difficult for them to realize direct N_2 reduction, on account of the unfavorable CB potentials for the generation of intermediate reactions. As one of the few successful examples for direct N_2 activation, H-terminated B-doped diamond with the CB edge located above the vacuum level was demonstrated with direct transfer of electrons to the N_2 molecules [19]. Under an ultraviolet irradiation, solvated electrons in the water were produced after receiving the electrons from the catalyst. These vigorous solvated electrons reacted with protons and dissolved N_2 molecules to generate NH_3. The second consideration is the recombination issue of e–h pairs. In the case of the formation of e–h pairs with light illuminations, a rapid recombination of the carriers is usually observed on the catalyst [20]. A prompt separation of these carriers is highly desirable, thus significantly increasing their availability in the interfacial reactions and further realizing a high conversion efficiency. Meanwhile, a moderate water splitting is another crucial aspect that should be taken into consideration. To achieve a high yield rate in NH_3 production while suppressing the excessive H_2 generation, designed catalysts should possess special adsorption capacity toward N_2, while provide necessary sites for producing protons to accomplish N_2 reduction. Moreover, the controllable bandgaps of the catalysts are of critical importance to realize a broad spectra adsorption, especially in the visible-light region. The last but not least, a robust photocatalyst with high structural stability is extremely important for the feasible applications.

To effectively meet these challenges in photocatalytic or photoelectrochemical N_2 reduction, the design and fabrication of promising catalysts has attracted tremendous interests in the literature [10, 21]. Various strategies, such as defect engineering, interface regulation, and heterojunction engineering, have been widely applied in the design of desirable catalysts. In the following section, comprehensive introduction and in-depth discussions toward these fascinating strategies are presented.

11.3 Strategies for Catalyst Design and Fabrication

11.3.1 Defect Engineering

Various defects, such as vacancy, doping, and non-crystallinity at different scales, have been widely introduced into the photocatalysts to achieve high-performance activities [22, 23]. Rationally engineering these defect structures into the catalysts could substantially regulate their behaviors in N_2 fixation, such as the improvement in the carrier separation and charge transfer, as well as the reduce in the energy barriers.

11.3.1.1 Vacancies

At the atomic scale, a small quantity of vacancies is intrinsically present in the catalysts, which also could be intentionally introduced in the preparation processes or post-synthesis treatments. As the most frequently reported defects, versatile vacancies, such as oxygen, nitrogen, and sulfur vacancies, have been verified as the active centers for achieving the N_2 adsorption and activation [24]. Moreover, the metal cations with low valence adjacent to these vacancies could improve light absorption and enhance the electron transport to adsorbed N_2 molecules.

Oxygen Vacancies

Oxygen vacancies (OVs), one of the mostly studied vacancies, are ubiquitously existed on the photocatalysts, such as typical transition metal oxides (TMOs), (Bi-enriched) bismuth oxyhalides, and layered-double hydroxides (LDHs) [25–27]. With sufficient localized electrons, the presence of OVs usually results in distinct local electronic structures in the catalysts. These active centers have been confirmed as electron traps, which can efficiently capture and activate various gas molecules, e.g. O_2, CO_2, and N_2 [28].

For example, the photocatalytic conversion of N_2 in pure H_2O to produce NH_3 was recently realized on the commercially available TiO_2 with plentiful surface OVs [29]. The observed Ti^{3+} species on the OVs were confirmed as the active centers for N_2 fixation, playing an important role as the trapping sites for the photo-induced electrons and for N_2 adsorption. Attributed to these promising features, a promoted production of NH_3 with a SCC efficiency of 0.02% was achieved. In view of this easily available noble-metal free photocatalyst coupling with H_2O as the proton supplier, this research shows a great potential for the green photosynthesis of NH_3.

Enlightened by this work, Wang et al. reported a "working-in-tandem" pathway for photocatalytic N_2 reduction at ambient conditions, applying the OV-enriched ultrathin TiO_2 nanosheets (NSs) decorated with Au nanocrystals (Au/TiO_2–OV) [30]. On the one hand, the introduction of plasmonic metal is widely applied to boost light absorption and electron transfer [31, 32]. The plasmon excitation of Au resulted in the formation of hot electrons, which were subsequently transferred into the CB of TiO_2 NSs. On the other hand, OVs on the TiO_2 NSs facilitated the N_2 adsorption and activation, and the NH_3 production was successfully realized with the participation of injected hot electrons (Figure 11.2a). The homogeneously dispersed Au nanocrystals on the transparent TiO_2 NSs were confirmed to facilitate the absorption and transformation of solar energy (Figure 11.2b). Meanwhile, the ultrathin morphology of TiO_2 NSs with a high specific surface area (~288 $m^2\ g^{-1}$) could significantly improve availability and accessibility of OVs on the surface. As confirmed in the low-temperature electron paramagnetic resonance (EPR), a typical OV response was observed with a g factor of 1.998 (Figure 11.2c). Benefiting from this synergetic effect, the Au/TiO_2–OV demonstrated a markedly enhanced photocatalytic N_2 reduction activity, with an enhanced NH_3 yield rate of 78.6 $\mu mol\ h^{-1}\ g^{-1}$ (~98 times of Au/TiO_2 and 35 times of TiO_2–OV, Figure 11.2d).

In spite of the attractive properties of OVs observed in the N_2 fixation, the controllable modulation of the OV concentration still remains great challenges in the development of promising photocatalysts. Accordingly, facile and reliable synthetic approaches are highly imperative to achieve this purpose. As an attempt, a precise regulation of the OV concentration in TiO_2 NSs was realized via doping with Cu^{2+} [35]. It was demonstrated that the substitution of Ti^{4+} with Cu^{2+} successfully introduced abundant OVs, accompanying with the generation of compressive strain within nanosheets. Moreover, the presence of Cu^{2+} substantially enhanced the separation efficiency of e–h pairs and dramatically extended the responsive range of light wavelengths (up to 700 nm). Consequently, the designed Cu-modified ultrathin TiO_2 NSs exhibited an impressive photocatalytic activity for NH_3 formation in H_2O, achieving a yield rate of 78.9 $\mu mol\ g^{-1}\ h^{-1}$ with full spectrum illumination. Even treating with the monochromatic excitations at 600 and 700 nm, efficient photoactivity toward NH_3 generation was observed with the yield rates of 1.54 and 0.72 $\mu mol\ g^{-1}\ h^{-1}$, respectively.

Besides the traditional metal oxide-based materials, emerging photocatalysts, e.g. bismuth oxyhalides and LDHs, have received enormous interests in photocatalytic N_2 fixation [33, 34]. On account of their superior optoelectronic properties, such as the narrow bandgap and wide responsive spectrum, bismuth oxyhalides have been extensively used in various photocatalytic applications, e.g. H_2O splitting, pollutant degradation, and CO_2 reduction [36]. For instance, Zhang and coworkers proposed a promising photocatalyst of BiOBr NSs, with plentiful OVs on the (001) facets (BOB–001 – OV), which showed an excellent photocatalytic performance for N_2 fixation (Figure 11.2e) [33]. As confirmed by the theoretical simulations, the active sites of OVs could effectively realize the adsorption and activation of N_2 molecules, prolonging the N≡N bond from initial 1.078 Å to 1.133 Å. Furthermore, sufficient surface OVs with typical defect states could capture excited CB electrons of BiOBr NSs. Therefore, the recombination of these charge carriers was obviously hindered, further promoting the electron transfer to accomplish the N_2 activation. In the absence of the co-catalysts or organic scavengers, the designed catalyst showed a

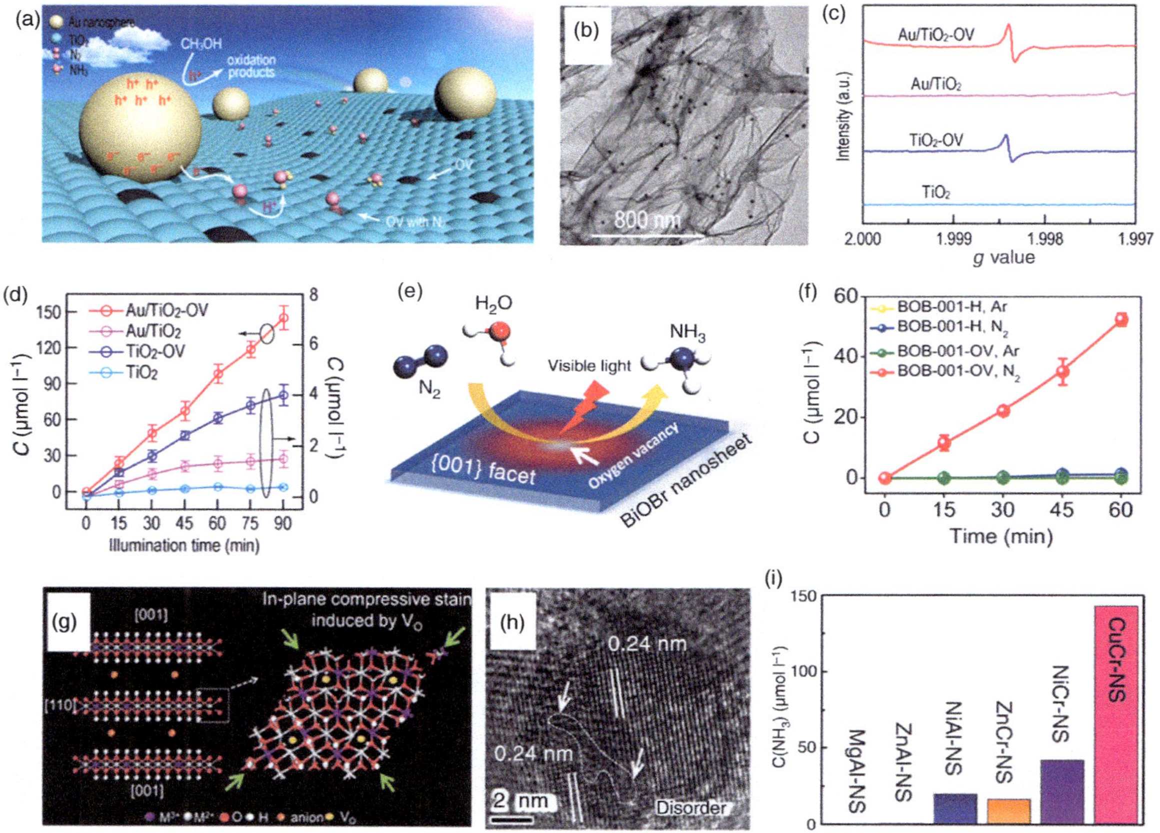

Figure 11.2 (a) Schematic of the plasmon excitation-based photocatalytic N_2 reduction. (b) TEM image of the Au/TiO_2−OV NSs. (c) EPR spectra of four catalysts of TiO_2, TiO_2−OV, Au/TiO_2, and Au/TiO_2−OV. (d) Detected NH_3 concentrations of four catalysts. Source: (a–d) Adapted from Yang et al. [30]. © 2018, American Chemical Society. (e) Schematic of the photocatalytic N_2 fixation on BOB-001−OV. (f) Generated NH_3 concentrations under visible light irradiation ($\lambda > 420$ nm). Source: (e–f) Adapted from Li et al. [33]. © 2015, American Chemical Society. (g) Schematic illustration of the in-plane biaxial compressive strain on the LDH NSs. (h) High-resolution transmission electron microscopy (HRTEM) image of CuCr−NS. (i) NH_3 yields of different LDH NSs under visible light illumination ($\lambda > 400$ nm). Source: (g–i) Zhao et al. [34]. © 2017, John Wiley & Sons.

high photocatalytic NH_3 evolution rate of 104.2 μmol g^{-1} h^{-1} with the illumination of visible light (Figure 11.2f). Comparably, the ultrathin BiOBr NSs (4–8 nm in thickness) exposed with dominant {001} facets were synthesized for an effective promotion of N_2 reduction [37]. Prepared with the assistance of polyvinylpyrrolidone (PVP), desirable OVs were engineered onto the catalyst surface, which was mainly due to the combination of the negatively charged PVP molecules with the positive Bi atoms during the nucleation process. Ascribed to the narrowed bandgap and the lower energy required for the exciton generation, the BiOBr NSs with abundant OVs displayed a markedly improved photocatalytic NH_3 yield rate of 54.70 μmol g^{-1} h^{-1} (~10-fold that of BiOBr nanoplates).

Despite these attractive properties of bismuth oxyhalides in photocatalytic N_2 fixation, these materials usually suffer from the photo-corrosion issue, and OVs on the surface are easily to be oxidized, eventually losing the initial activities. To address this issue, developing robust photocatalysts with sustainable OVs is highly anticipated for feasible photocatalytic N_2 fixation. As confirmed in previous report, increasing the Bi/Br ratio in similar materials could not only enhance their optical absorption but also strengthen the electron migration and the separation of e–h pairs [38]. Moreover, the photostability could also be significantly increased with a high Bi/Br ratio. In this respect, Ye and coworkers reported a promising Bi_5O_7Br photocatalyst with a nanotube structure (Bi_5O_7Br–NT) to realize an excellent photo-reduction of atmospheric N_2 in pure H_2O [39]. The Bi_5O_7Br–NT with a suitable absorption edge and large specific area (>96 m^2 g^{-1}) possessed plentiful sustainable OVs on the surface. These OVs were firstly produced on the catalyst surface with the escape of partial O atoms under the light illumination. Interestingly, after the N_2 reduction reaction, the light-induced OVs were refilled via trapping O atoms of H_2O, effectively recovering to the pristine steady state. As a result, a high NH_3 formation rate of 1.38 mmol h^{-1} g^{-1}, accompanying with an impressive AQE of 2.3% (λ = 420 nm) was realized. More importantly, the Bi_5O_7Br–NT also exhibited a superior stability with subtle N_2 reduction activity loss after four cycles. Similarly, Xia and coworkers proposed an atomically thin Bi_3O_4Br NSs with a single-unit cell structure to boost photocatalytic NH_3 production efficiency, through enhancing the separation of generated charges in the bulk and on the catalyst surface simultaneously [40]. Via regulating the surface defect concentration of bismuth, the tunable OVs were engineered on the surface, realizing the efficient regulation of the electronic structure on Bi_3O_4Br, as well as a prompt charge separation rate. On account of the unique architecture and these available defects, a substantially enhanced photocatalytic N_2 fixation activity was obtained on the Bi_3O_4Br NSs (~30.9 times than bulk Bi_3O_4Br).

As another new kind of promising photocatalysts, LDHs have been extensively applied as catalysts or catalyst supports for heterogeneous catalysis, on account of their easily tunable metal cation composition and thickness [41]. As an attempt, Zhang and coworkers firstly reported efficient photocatalysts of ultrathin LDH NSs containing OVs for NH_3 synthesis with the N_2 and H_2O [34]. Through regulating the thickness and cation composition of fabricated LDHs, OVs as well as compressive strain were engineered into edges or lateral surfaces of LDH NSs (Figure 11.2g).

Among various LDHs, ultrathin CuCr–NS (~2.5 nm in thickness) with abundant OVs exhibited favorable photocatalytic performance toward N_2 reduction. The presence of Cu^{2+} resulted in the structural distortions (dotted section in Figure 11.2h) and introduced compressive strain, which stabilized OVs and produced a favorable interaction between the CuCr–NS and N_2 molecules. Under the visible light illumination, most of the as-prepared LDH NSs exhibited obvious N_2 photo-fixation activity, with CuCr–NS showing the optimal performance with the NH_3 production rate of 142.9 $\mu mol\,l^{-1}$ (Figure 11.2i).

Nitrogen Vacancies

Another attractive vacancy toward photocatalysis is nitrogen vacancies (NVs). Given the NVs displaying a comparable shape and dimension with the N atom, the NV-containing materials hold a great potential in the selective adsorption and activation of N_2. Moreover, defective NVs could also trap photo-excited electrons and further significantly increase the lifetime of the charge carriers [42].

To validate these hypothesis, Dong et al. firstly reported NV-incorporated graphitic carbon nitride (V–g-C_3N_4) for the photocatalytic N_2 reduction [43]. Through a second calcination under N_2 flow, abundant NVs were effectively engineered into the g-C_3N_4 (Figure 11.3a). As proven by electronspin resonance (ESR), a significantly enhanced Lorentzian line stemming from the unpaired electrons on carbon atoms was recorded on the V–g-C_3N_4 (Figure 11.3b), confirming the formation of abundant defect sites. Serving as the main reaction centers for the N_2 activation, the presence of indispensable chemical adsorption sites is critical to drive the photocatalytic N_2 reduction. In addition to an enhancement in the physical adsorption, the introduction of NVs substantially improved the chemical adsorption sites on V–g-C_3N_4 surface (Figure 11.3c). After the reaction for 15 h under the visible light, an impressive photocatalytic activity with the NH_3 generation of 2.4 mM was achieved on the V–g-C_3N_4, while no obvious amount of NH_3 was detected with pristine g-C_3N_4. Besides the mentioned approach, Wang and coworkers prepared the NVs enriched g-C_3N_4 with an etching method [45]. The etching with KOH resulted in the breakage of C=N in *s*-triazine, introducing plentiful NVs to strengthen the activation of N_2. Meanwhile, the grafted K^+ on the catalyst surface was favorable for NH_3 desorption. Serving as the proton source, methanol with a high solubility of N_2 was employed for the photocatalytic NH_3 production. Ascribed to these synthetic effects, a high NH_3 formation rate (~3.632 $mmol\,g^{-1}\,h^{-1}$) was observed on the designed photocatalyst.

Sulfur Vacancies

In view of the impressive performance of OVs, S-containing photocatalysts with sulfur vacancies (SVs) are also highly anticipated in the high-performance photocatalytic N_2 reduction, considering that S has similar chemical properties with O atom. As a paradigm, Hu et al. proposed a ternary metal sulfide (ZnSnCdS) with abundant SVs as an efficient N_2 reduction photocatalyst [46]. As anticipated, the presence of SVs generated plentiful chemical adsorption sites that were favorable

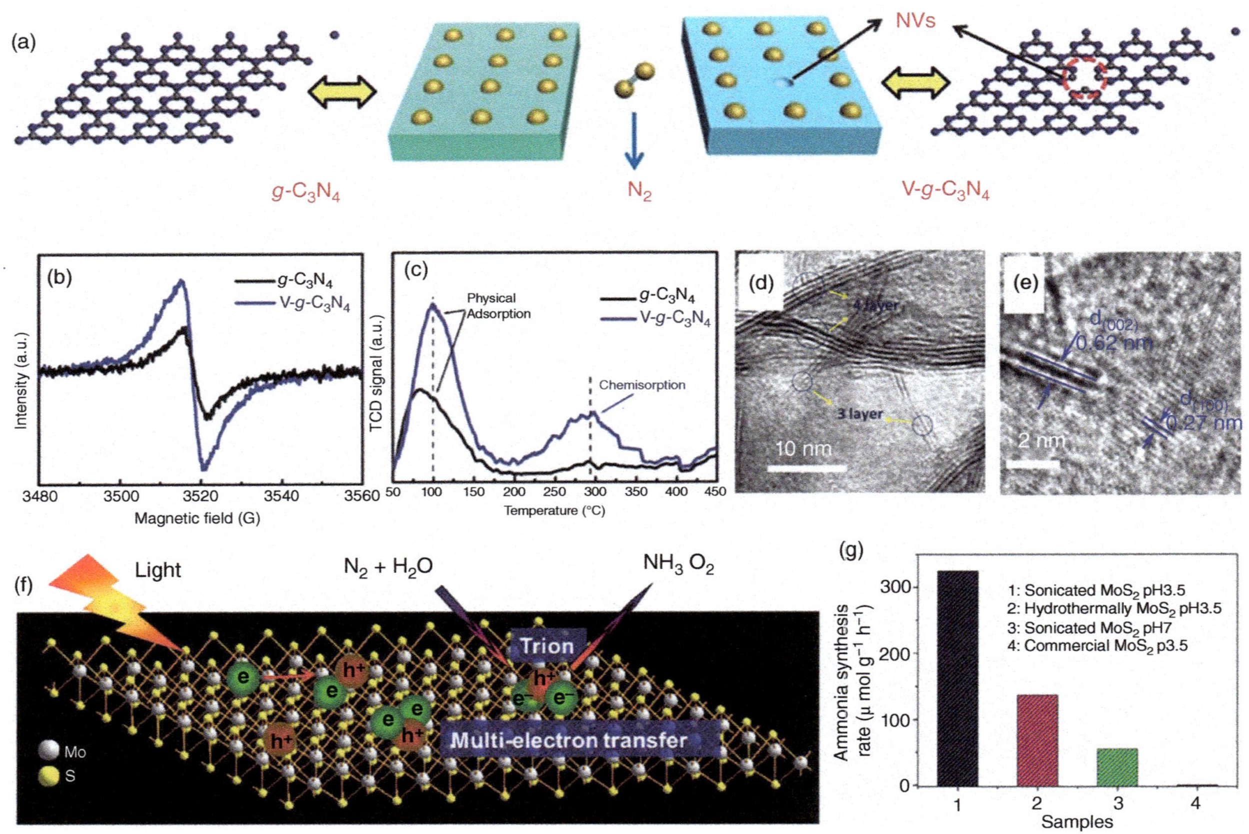

Figure 11.3 (a) Schematic of the synthesis process of engineering NVs into g-C_3N_4. (b) ESR and (c) TPD spectra of the prepared g-C_3N_4 and V–g-C_3N_4. Source: (a–c) Adapted from Dong et al. [43]. © 2015, Royal Society of Chemistry. (d) TEM, and (e) HRTEM images of prepared MoS_2 NSs. (f) Schematic of the Trion participated N_2 reduction reaction. (g) NH_3 evolution rates of the MoS_2 samples prepared under different conditions. Source: (d–g) Sun et al. [44]. © 2017, Elsevier.

for the activation of N_2 molecules. Meanwhile, the SVs could also serve as the electron trap, thus favorably facilitating the separation of photo-generated carriers. It was revealed that a higher concentration of SVs was beneficial for N_2 reduction. Possessing the most SVs among different catalysts, the photocatalyst of $Zn_{0.1}Sn_{0.1}Cd_{0.8}S$ exhibited the best performance in NH_3 generation.

As a classic transition metal dichalcogenide (TMD), MoS_2 shows superior electrical, optical, and optoelectronic properties [47]. The SV concentration within MoS_2 can be effectively regulated via controlling its layer thickness, and the ultrathin nanosheets with desirable atomic thickness are beneficial to introduce the SV defects. For this purpose, Wang and coworkers synthesized ultrathin SVs-enriched MoS_2 NSs with stoichiometry Mo/S ratio of 1 : 1.75 [44]. Prepared MoS_2 with few layers (3–5 layers) exhibited a typical lamellar morphology (Figure 11.3d), and these interplanar spacings (0.62 and 0.27 nm) were corresponded with the d spacing of (002) and (100) planes (Figure 11.3e). Under the light irradiation, the charged excitons (Trions) were generated from the free electrons on MoS_2 and situated around the Mo sites. Captured by the present SVs, the N_2 was then activated via donating and receiving electrons from the bonding and anti-bonding orbitals. Therefore, the charged excitons on the Mo sites cooperated with the center-adsorbed N_2, contributing to a six-electron transfer process for the photocatalytic N_2 reduction to NH_3 (Figure 11.3f). Accordingly, the MoS_2 NSs with sufficient SVs prepared in the sonication approach demonstrated a significantly improved NH_3 yield rate of 325 μmol h^{-1} g^{-1} (Figure 11.3g), while the bulk MoS_2 showing negligible activity toward N_2 reduction.

11.3.1.2 Heteroatom Doping

Despite the attractive properties of the vacancies in photocatalytic N_2 reduction, these defect generations within the lattice induce defect states within the electronic structures of catalysts. Instead of transferring to adsorbed N_2 molecules, most of photo-generated electrons may relax from CB to the formed defect band [48]. Due to the reduced energy of photo-excited electrons, the efficiencies in the N_2 activation and subsequent NH_3 formation would be severely impacted. In recent studies, the doping strategy has been demonstrated to effectively tackle this bottleneck to simultaneously improve N_2 chemisorption and electron transfer through refining these defect states of the photocatalysts [49–51]. The doping with various cation/anion heteroatoms could efficiently alter local electronic structures of the photocatalysts, thus inducing superior photocatalytic performance in N_2 reduction.

For the doping strategy, the selection of a proper element is of critical importance. Located on the on top of volcano diagrams, the Mo and Fe atoms have been theoretically estimated as the most active elements in N_2 reduction [52]. Moreover, Mo and Fe clusters are demonstrated as active sites in the nitrogenases for the natural biological N_2 fixation. Inspired by these discoveries, it is highly anticipated to explore the N_2 reduction ability after the doping with such kind of elements.

As a proof of concept, Xiong and coworkers firstly reported Mo-doped $W_{18}O_{49}$ (MWO-1) for photocatalytic N_2 reduction [53]. The designed photocatalyst possessed a desirable morphology of ultrathin nanowires (Figure 11.4a). Ascribed to an ultrahigh length/diameter ratio, a high specific area with sufficient exposed active sites would substantially boost the catalytic activity. As revealed in the theoretical simulations, the substitution of Mo atoms for W sites facilitated the chemisorption of N_2 molecules, giving a higher adsorption energy toward N_2 on MWO-1 (−2.48 eV) than the pristine $W_{18}O_{49}$ (−1.65 eV). Meanwhile, the charge difference was significantly increased from 0.45 to 0.58*e* within the two adsorbed N atoms (Figure 11.4b,c), indicating that the formation of Mo–W active centers was favorable for the polarization of adsorbed N_2 molecules. Furthermore, an enhanced M–O covalency with a higher peak (~533 eV) in the O K-edge XAS spectra indicated a promoted electron migration from metal active centers to chemisorbed N_2 (Figure 11.4d). Attributed to these favorable roles in the electron transfer, N_2 adsorption, and activation, as well as the elevated defect-band center plentiful energetic electrons were accumulated for subsequent N_2 fixation, following a multistep PCET process (Figure 11.4e). On account of these synergetic effects, a remarkably stable photocatalytic performance was acquired in N_2 reduction, with a high NH_3 evolution rate of 195.5 μmol g^{-1} h^{-1} (Figure 11.4f, ~7 fold that of $W_{18}O_{49}$), and an AQE of 0.33% at 400 nm.

In addition to Mo element, Fe atom is also widely adopted in the doping strategy, and several photocatalysts, such as Fe-doped TiO_2, Fe^{3+}-doped $g-C_3N_4$, and Fe-doped $BiMoO_6$, have been reported for the efficient N_2 fixation [55–57]. It is generally recognized that Fe doping could not only enhance the chemisorption of N_2 molecules, but accelerate the electron transfer from the photocatalysts to these adsorbed N_2 molecules. For instance, Zhao and coworkers recently synthesized a Fe-doped $SrMoO_4$ photocatalyst for efficient N_2 fixation [54]. Synthesized with a solvothermal method, the Fe atom was homogeneously distributed on the $SrMoO_4$ nanoparticles (Figure 11.4g). It was confirmed that Fe doping substantially narrowed the pristine bandgap of $SrMoO_4$ (3.98 to 2.93 eV), thus effectively extending the light absorption to the visible section. Retaining the thermodynamic activity toward N_2 reduction with suitable band energetics, the Fe doping also induced the OVs that could mitigate the quick recombination of photo-excited e–h pairs and therefore facilitate the interfacial charge transfer. With the light illumination, the excited electrons were transferred to the Fe–Mo centers and surface defects for the reduction of chemisorbed N_2 molecules, while the H_2O molecules were oxidized with the holes to produce O_2 (Figure 11.4h). Accordingly, the Fe-doped $SrMoO_4$ exhibited an enhanced NH_3 synthesis performance (93.1 μM g^{-1} h^{-1}) in ultrapure H_2O, compared with the pristine $SrMoO_4$ (~66.7 μM g^{-1} h^{-1}).

11.3.1.3 Amorphization

At the nanoscale, non-crystallinity, a metastable state with respect to its crystalline counterpart, is generally observed ascribed to disordered atomic arrangement on crystal lattice. Due to the presence of "dangling bonds" and unsaturated coordination sites, a mass of active sites in these defective amorphous phases could

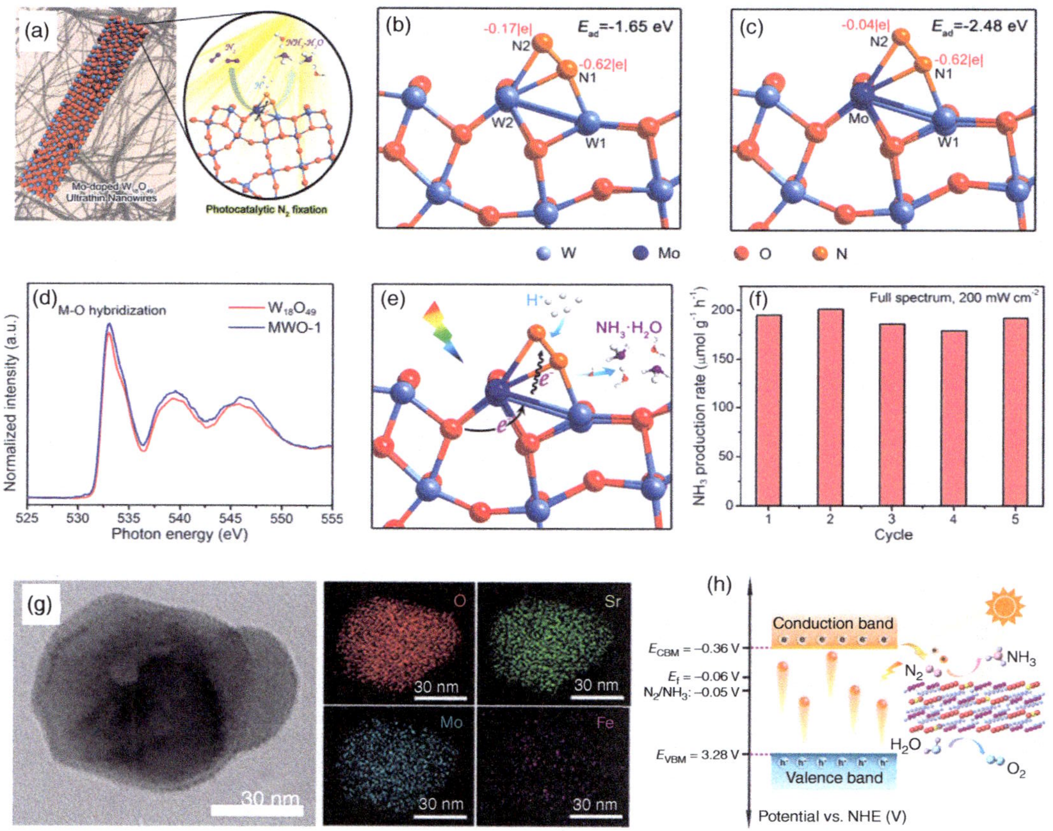

Figure 11.4 (a) Schematic photocatalytic N_2 reduction on ultrathin MWO-1 nanowires. Adsorption configurations of N_2 molecules with different charge dispersions on (b) defect-rich $W_{18}O_{49}$, and (c) Mo-doped $W_{18}O_{49}$. (d) O K-edge XAS spectra of different photocatalysts. (e) Schematic of the PCET process for photocatalytic N_2 fixation. (f) Photocatalytic NH_3 yield rates with MWO-1. Source: (a–f) Adapted from Zhang et al. [53]. © 2018, American Chemical Society. (g) TEM and corresponding EDS images of Fe doped $SrMoO_4$. (h) Mechanism illustration for photocatalytic N_2 fixation. Source: (g–h) Luo et al. [54]. © 2019, Elsevier.

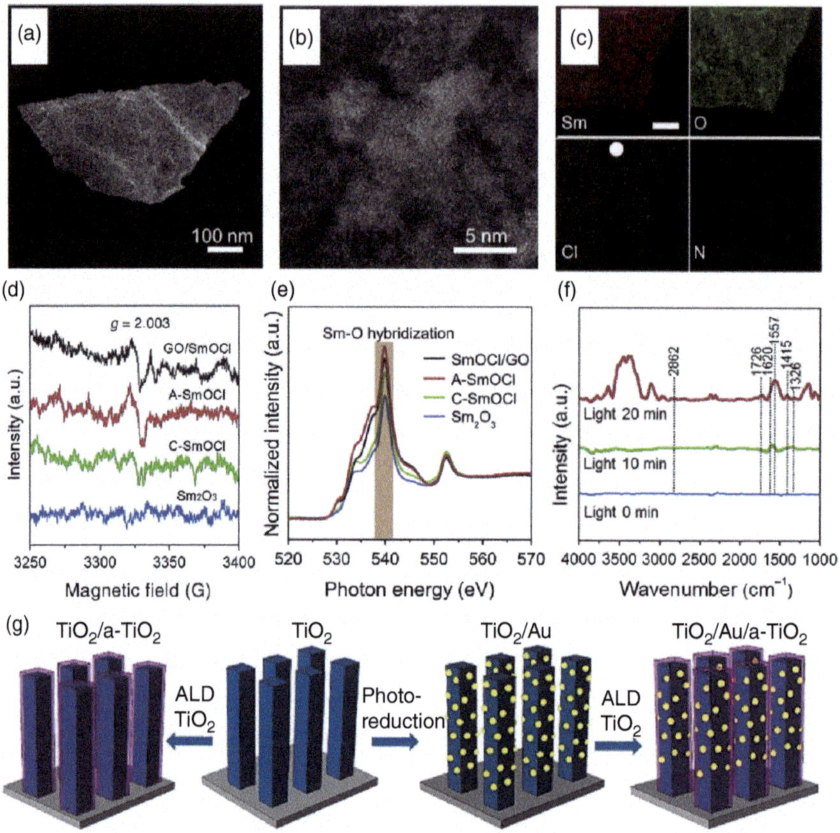

Figure 11.5 (a) and (b) HAADF–STEM images and (c) EDX mapping of A-SmOCl. (d) ESR signals and (e) O K-edge XAS spectra of different photocatalysts. (f) In situ DRIFT spectra recorded in the N_2 reduction process. Source: (a–f) Hou et al. [61]. © 2019, Elsevier. (g) Illustration for the preparation of various TiO_2-based photocatalysts. Source: (g) Li et al. [62]. © 2018, John Wiley & Sons.

significantly improve the activities toward the H_2O splitting and inert gas activation [58–60].

Recently, Wang and coworkers developed a photocatalyst of amorphous SmOCl NSs (A-SmOCl) with excellent performance for solar N_2 fixation [61]. Treated with a calcination at 550 °C, the two-dimensional nanosheets exhibited an amorphous structure in the absence of obvious lattice fringe (Figure 11.5a,b). As demonstrated in the energy dispersive X-ray (EDX) mapping, a homogeneous elemental distribution of Sm, O, and Cl was observed (Figure 11.5c). Accompanying with the formation of the amorphous phase, the presence of OVs was confirmed on the nanosheets, showing an obvious isotropic paramagnetic response with g value of 2.003 in ESR spectroscopy (Figure 11.5d). With an obviously enhanced intensity in the peak (~539.6 eV) in O K-edge XAS, the facilitated electron transfer was realized from active metal centers to chemisorbed N_2 (Figure 11.5e). Coupling with an efficient N_2 adsorption, fast electron delivery, and robust cleavage of N≡N bond, the efficient photocatalytic N_2 fixation with multistep reactions was successfully activated, showing several characteristic absorption peaks of intermediates with the in situ diffuse reflectance infrared Fourier transform (DRIFT, Figure 11.5f). The designed A-SmOCl exhibited a high NH_3 evolution rate of 426 μmol g^{-1} h^{-1} in the absence of any sacrificial agent, with an AQE of 0.32% at 420 nm.

As discussed in Section 11.3.1.1, the vacancies with coordinately unsaturated metal active sites played critical roles in the photocatalytic N_2 fixation. However, compared with the preferable surface vacancies, bulk vacancies would adversely act as electron traps, thus inducing charge recombination and further hindering the migration of photo-induced electrons [63]. Up to now, it still remains a great challenge to manipulate the vacancy positions and levels on the catalysts. Lately, as an attempt to introduce OVs without affecting bulk properties, Gong and coworkers reported a fine-tuning approach to introduce surface OVs by atomic layer deposition (ALD), through depositing an ultrathin amorphous TiO_2 (a-TiO_2) functional layer on TiO_2/Au (Figure 11.5g) [62]. ALD has been widely applied in the conformal growth of functional films on different kinds of nanostructures, and it is effective in controlling film thickness at the nanoscale level [64, 65]. With a thin thickness ~2.5 nm, the presence of uniformly deposited a-TiO_2 layer with OVs provided sufficient sites for N_2 adsorption and introduced extra energy bands to facilitate interfacial electron transfer. Utilizing the designed hybrid photocatalyst, a photoelectrochemical N_2 reduction was achieved with the NH_3 evolution rate of 13.4 nmol cm^{-2} h^{-1} (~2.6 times that of pristine TiO_2).

11.3.2 Structure Engineering

11.3.2.1 Morphology Regulation

Apart from the defect introduction into designed catalysts, the favorable morphology of nanostructured photocatalysts as well contributes to their photocatalytic activities. Versatile morphologies, including zero-dimensional (0D) quantum dots (QDs)/nanoparticles (NPs), one-dimensional (1D) nanowires/nanotubes, two-dimensional (2D) ultrathin nanosheets (NSs), and three-dimensional (3D) porous/array nanostructures, could not only provide a high specific area for reactant interaction and light absorption but also offer fast transfer paths for photo-excited e–h pairs to improve the charge separation [66–69].

Prepared with an oleate-modified hydrothermal method, Liu and coworkers reported the bismuth monoxide (BiO) QDs with excellent activity for photocatalytic NH_3 synthesis [70]. Possessing a small size of 2–5 nm, the BiO QDs provided a high specific area with sufficient active centers for N_2 fixation. Ascribed to the synergetic function of three surface Bi(II) species at low valence, the N_2 activation was dramatically enhanced through an efficient electron donation, eventually resulting in an impressive N_2 reduction performance. A high NH_3 production rate of ~1226 μmol g^{-1} h^{-1} was observed with no co-catalyst or sacrificial agent. Undergoing an alternative pathway with the 1N_2–3Bi side-on mode, a small amount of the by-product of N_2H_4 (~1.6%) was also observed after the reaction.

Meanwhile, applying the MoO_3 nanorods as the template, the hollow TiO_2 (B) nanotubes with a high specific surface area (~344 m^2 g^{-1}) were successfully fabricated for the highly efficient photocatalytic N_2 fixation [71]. On account of the multiple light scattering effect, the observed hollow structure was favorable for the light harvesting. Besides, the ultrathin shells endowed abundant defects, such as the Ti^{3+} and OVs into the nanotubes, which were demonstrated as critical electron trap centers for the rapid separation of the photo-generated carriers, as well as the active sites for N_2 adsorption and activation. As expected, a high NH_3 yield rate of 106.6 μmol g^{-1} h^{-1} (~1.61 times that of the TiO_2 NSs) was realized at ambient conditions.

As for the ultrathin 2D nanomaterials, the large lateral size with the tunable thickness imparts ultrahigh specific areas and plentiful exposed surface atoms, making them competent candidates for the photocatalytic N_2 reduction [35]. As a paradigm, Jiang and coworkers reported the ultrathin g-C_3N_4 NSs directly fabricated from urea polymerization through a facile one-step separation method [72]. Compared with the bulk g-C_3N_4 with a low specific area (~35.3 m^2 g^{-1}), the ultrathin structure (~2.0 nm in the thickness) imparted a significantly enhanced area of 90.2 m^2 g^{-1} on the g-C_3N_4 NSs. Besides, a high reduction potential, a fast interfacial electron migration, and an enhanced separation efficiency of charge carriers with increased lifetimes were realized on the synthesized g-C_3N_4 NSs. Attributed to the improved structural and optoelectronic properties of the g-C_3N_4 NSs, an impressive N_2 fixation rate of 60.5 μmol h^{-1} was acquired (~1.9 times that of bulk g-C_3N_4).

Compared with its counterparts, 3D micro-/nanostructures, including vertical arrays and porous structures, have also received great attention due to their unique merits in accelerating N_2 reduction and conversion efficiency [73]. For the 3D array nanostructures, the gallium nitride nanowire (GaN NW)-based photocatalysts were proposed for the NH_3 synthesis [74]. Fabricated by plasma-assisted molecular beam epitaxy, the as-synthesized GaN arrays consisting of vertically aligned nanowires were grown vertically on the Si substrate (Figure 11.6a). These nanowires with ~80 and 800 nm in diameter and length, respectively, grew along the *c*-axis direction (Figure 11.6b). Through doping with the Ge element, the n-type GaN NWs achieved an improved photocatalytic NH_3 production activity due to the altered Fermi-level position and carrier transfer properties. With the homogeneously dispersed Ru sub-nanoclusters (~0.8 nm in diameter, Figure 11.6c) as the electron reservoir, the photo-induced electrons were promptly immigrated from the nanowires to realize the cleavage of N≡N bonds, due to a barrier height (~0.94 eV) formed in the interfacial Schottky junction (Figure 11.6d). Furthermore, to extend the solar adsorption to the visible region, the bandgap of GaN NWs (3.4 eV) was reduced to 2.34 eV by introducing 25% In element, forming n-type InGaN on the GaN NWs (Figure 11.6e). Utilizing Ru as the co-catalyst, the n-InGaN/n-GaN showed a robust photocatalytic activity toward NH_3 synthesis ($\lambda > 400$ nm) (Figure 11.6f).

Another similar 3D array structure consisting of black silicon (bSi) structure decorated with Au NPs and a Cr layer (~50 nm in thickness) sputtered onto the reverse side of the Si wafer was reported for the photoelectrochemical N_2 reduction [75].

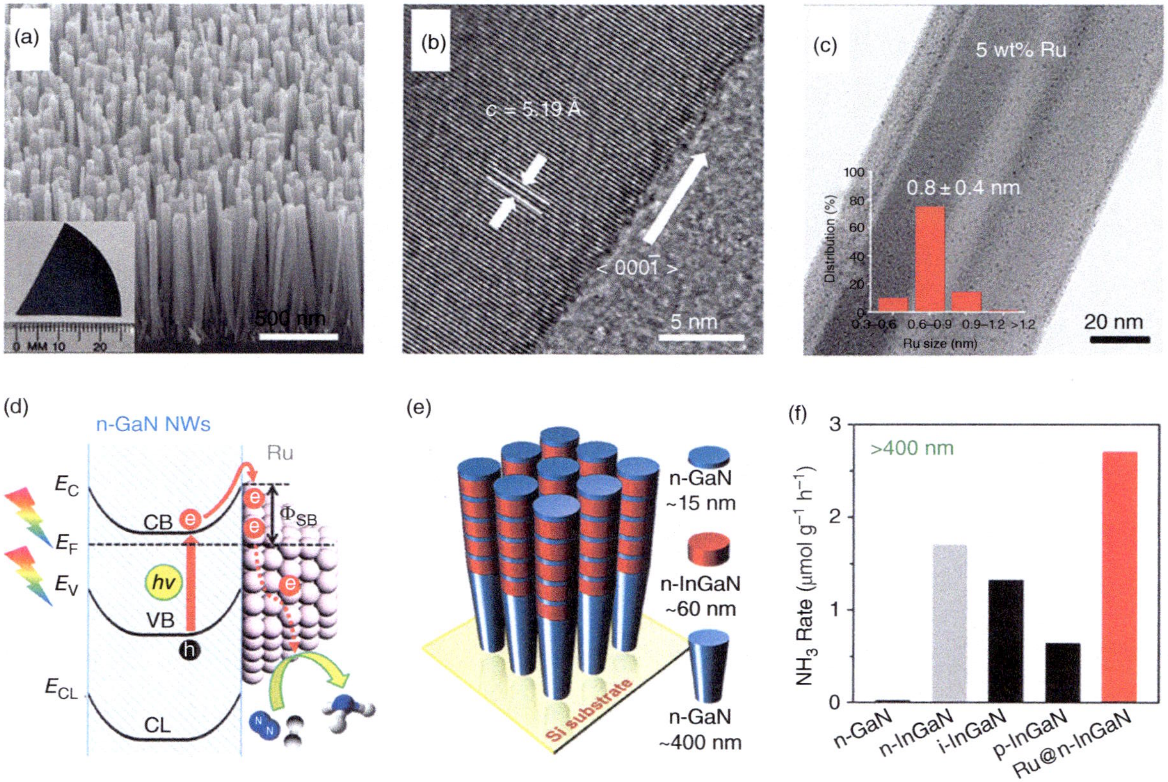

Figure 11.6 (a) SEM and (b) TEM images of the GaN NWs. (c) TEM image of Ru (5 wt%) decorated GaN NWs. Inset: the diameter distributions of Ru nanoclusters. (d) Schematic of the Schottky barrier between Ru clusters and n-type GaN NWs. (e) Schematic illustration for the InGaN/GaN with five segments of InGaN on the template of GaN nanowire. (f) NH_3 evolution rates of various photocatalysts under visible-light illumination. Source: (a–f) Li et al. [74]. © 2017, John Wiley & Sons.

Treated with a dry etching method, highly vertical bSi NWs not only enhanced the absorption of light but also offered a high specific area for the growth of Au NPs. The Cr layer was applied to collect the holes, further facilitating the electron transfer. Benefiting from the plasmon effect of Au NPs, the designed photocatalyst showed a superior performance in NH_3 production, with a high yield rate of 13.3 mg m^{-2} h^{-1} under two suns irradiation.

For the catalysts with porous or hollow structures, these special configurations not only facilitate the diffusion of reactants at both inner and outer surfaces but also increased the critical intermediate concentrations due the cage effect. In this respect, Chen's group developed a series of 3D macroporous graphene-based photocatalysts for the NH_3 production [76, 77]. Graphene, as a classic 2D carbon material, has received tremendous interests on account of its unique physicochemical properties, e.g. the large specific area, excellent charge mobility and thermal conductivity, efficient light absorption, and superior mechanical strength, all of which making it an ideal candidate in the highly efficient photocatalytic applications [78, 79]. Through the decoration of Fe_2O_3 NPs on the 3D graphene framework as the co-catalyst, photo-induced hot electrons on the graphene were quickly transferred to Fe_2O_3 to increase the electron density on the surface, achieving enhanced catalytic activity toward N_2 fixation [76]. Alternatively, these generated hot electrons might directly migrate into the antibond orbitals of adsorbed N_2 on Fe_2O_3 surface, realizing the activation of the inert molecules. With a particle size of ~6.6 nm, the optimized catalyst showed the best performance with a yield rate of 408 ± 68 μg g^{-1} h^{-1}. However, suffering from the gradual aggregation of Fe_2O_3 NPs during the reaction, the photocatalytic activity was decreased accordingly. To address this issue, the authors introduced a dispersing agent of Al_2O_3 NPs to prevent its aggregation on the graphene [77]. In addition to the function being a barrier, the Al_2O_3 was considered as a morphology controller that could significantly decreased the particle size of Fe_2O_3 catalyst. On account of these merits, the NH_3 formation rate of this promising catalyst reached to 430 μg g^{-1} h^{-1} with an excellent stability over 60 h.

As an extension of their previous work of hollow Au nanocages in electrochemical N_2 reduction [80, 81], El-Sayed and coworker recently reported the hybrid of Au–Ag_2O nanocages synthesized via a simple oxidation process for photocatalytic N_2 fixation (Figure 11.7a) [82]. As revealed in the HRTEM and the fast fourier transform (FFT) of the 100 oriented Au–Ag_2O hollow structures, the formed Ag_2O possessed the same orientation as Au (Figure 11.7b). Moreover, the observed lattice spacing of 2.23 Å and 2.68 Å belonging to the Ag–O and Au–Au atomic spacing confirmed the generation of Ag_2O on Au nanocages after the oxygen treatment (Figure 11.7c). The presence of Au facilitated the collection of photo-excited electrons from the CB of Ag_2O, therefore substantially enhancing charge separation efficiency. Meanwhile, Au NPs provided sufficient catalytic centers for N_2 reduction and H_2O oxidation, via producing hot electrons and holes upon the localized surface plasmon resonance (LSPR) excitation. Ascribed to the enhancement effect of the generated hot electrons and photo-induced electrons from Au NPs and Ag_2O, respectively, the highest activity was observed on the Au–Ag_2O in the photocatalytic N_2 reduction (Figure 11.7d). The designed photocatalyst possessed a robust and

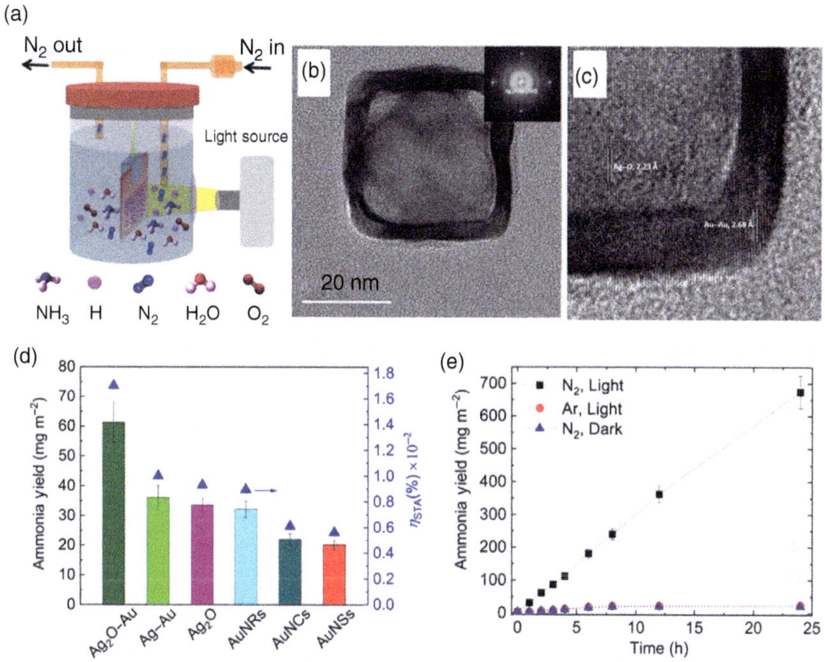

Figure 11.7 (a) Schematic illustration of photocatalytic cell for N_2 fixation. (b) and (c) HRTEM images of hollow Au–Ag_2O, the inset is the corresponding FFT of the nanoparticle. (d) NH_3 yields and solar-to-ammonia (STA) efficiencies of different photocatalysts. (e) NH_3 yields of Au–Ag_2O under various operating conditions. Source: (a–e) Nazemi et al. [82]. © 2019, Elsevier.

stable performance throughout 24 h, exhibiting a high NH_3 formation rate of 28.2 mg m^{-2} h^{-1} applying H_2O as an electron donor (Figure 11.7e).

11.3.2.2 Facet Control

Through the precise arrangement of active atoms, exposing specific crystal facets of designed catalysts has been considered as a crucial strategy to boost their catalytic ability [83–85]. As explored in theoretical calculations, the active centers on catalysts generally located on the stepped facets rather than flat terraces, and different facets exhibited distinct behaviors in the N_2 reduction [52, 86]. Inspired by these predictions, a promising strategy of crystal facet engineering is highly expected to prepare desirable photocatalysts for NH_3 synthesis.

Recently, Zhang and coworkers explored the effects of various exposed facets of BiOCl with OVs on the photocatalytic performance in N_2 reduction [87]. It was demonstrated that the N_2 adsorption showed different geometries on the (001) and (010) facets (BOC-001 and BOC-010). On the (001) surface, N_2 was bound with two adjacent Bi atoms, forming a terminal end-on bound configuration. A weak activation was observed with the slightly increased bond length to 1.137 Å, between the bond lengths of free N_2 (1.078 Å) and N_2H_2 (1.201 Å). In contrast, exhibiting a stable side-on bridging mode on the (010) surface, N_2 molecules were combined with two Bi atoms/one Bi atom in the outer layer/sublayer, respectively. In view of that the side-on mode was more efficient than the end-on mode, the lower bond order of adsorbed N_2 with a prolonged bond length (1.198 Å) was achieved on (010) surface. As a result, BOC-010 exhibited a staged NH_3 evolution behavior with the generation rates of 1.92 μmol h^{-1} (within 30 min) and 4.62 μmol h^{-1} (after 30 min), compared with the steady rate of 1.19 μmol h^{-1} on BOC-001. Following a symmetric alternating mode in BOC-010, a normalized N_2H_4 production rate of 4.14 μmol g h^{-1} m^{-2} was also observed in the first 30 min.

Besides, prepared by a calcination or hydrolysis approach, Ye and coworkers reported Bi_5O_7I NSs with exposed {001} and {100} facets (Bi_5O_7I-001 and Bi_5O_7I-100) and tested their photocatalytic performance in the NH_3 production [88]. Possessing a more negative CB position (−1.45 eV) than that of Bi_5O_7I-100 (−0.85 eV), Bi_5O_7I-001 with a higher reduction ability demonstrated a robust efficiency for the carrier separation, thereby creating more energetic electrons to promote N_2 reduction reaction. Upon light irradiation, the Bi_5O_7I-001 exhibited a higher formation rate of 111.5 μmol l^{-1} h^{-1} with the AQE of 5.1% (λ = 365 nm), while a lower rate of 47.6 μmol l^{-1} h^{-1} with AQE of 2.3% was observed on the Bi_5O_7I-100.

11.3.3 Interface Engineering

In addition to various active sites introduced at the atomic and nanoscales, another impressive strategy to boost the photocatalytic NH_3 production is the interface regulation at microscale. Through effectively restricting H_2O splitting and enhancing the accessibility of N_2 toward the catalyst surface, the production rate and the conversion efficiency of NH_3 would be improved simultaneously.

In the view of this point, a promising aerophilic–hydrophilic catalytic interface was firstly proposed for the photoelectrochemical synthesis of NH_3 [89]. Within the designed interface, the Si was introduced as the photo absorber, the porous poly(tetrafluoroethylene) (PTFE) was applied as the favorable N_2 diffusion layer, and Au NPs were served as the active centers (Au–PTFE/TS, Figure 11.8a). Considering the poor conductivity and hydrophobic property of PTFE, the deposition of homogeneously dispersed Au NPs provided an electric contact within the PTFE and Si, as well as improving the hydrophilicity of the interface to control proton activity. As a consequence, a smaller gas-bubble contact angle (CA_g, ~88°) accompanying with a contact angle of liquid (CA_l, ~125°) was acquired on the surface of Au–PTFE/TS (Figure 11.8b). As displayed in the interfacial water molecule spectra (Figure 11.8c), the intensive peaks at ~3400 and ~3600 cm^{-1} confirmed plentiful OHs straddling the interface.Based on the theoretical simulation, the formation of *NNH was indicated as the rate determining step, and the lower free energy change on Au–PTFE/TS (2.37 eV) suggested a facile hydrogenation step comparing with that on Au/TS (2.52 eV). Benefiting from these synergetic effects, a higher NH_3 production rate of 18.9 μg h^{-1} cm^{-2} with a remarkably increased FE of 37.8% was realized at −0.2 V (vs. RHE, Figure 11.8d).

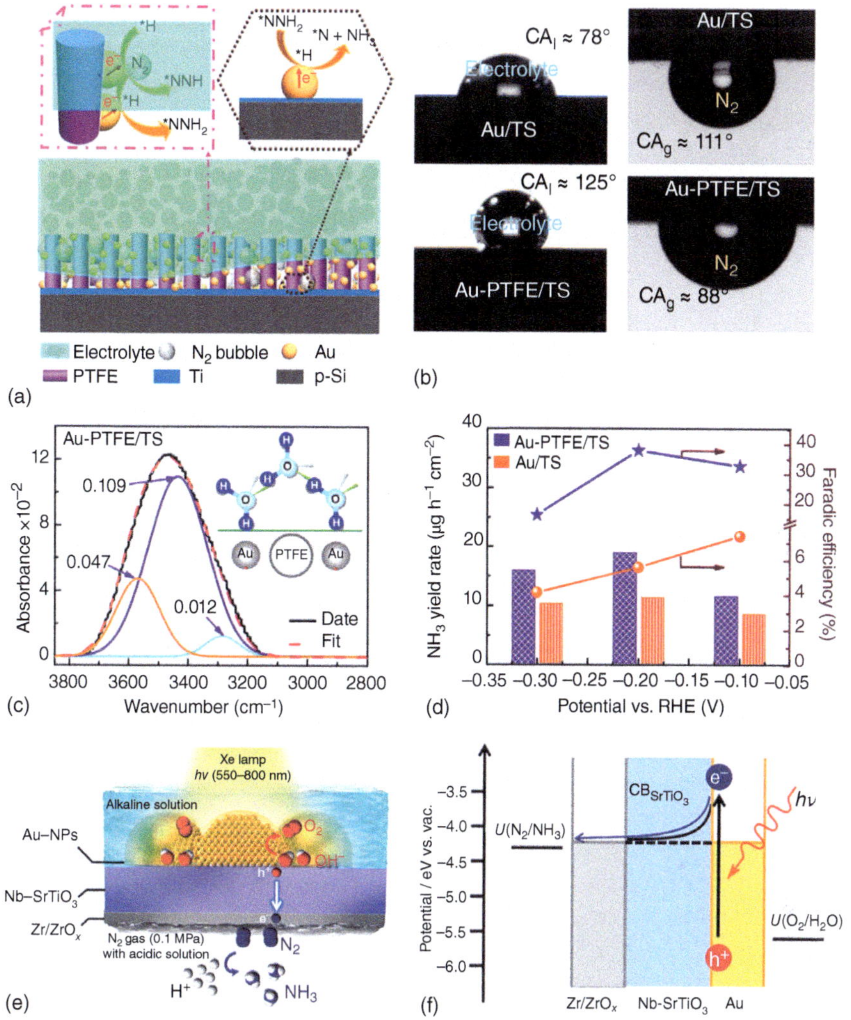

Figure 11.8 (a) Schematic of N_2 reduction on the aerophilic–hydrophilic interface. (b) Droplet shapes of the liquid and N_2 bubble on different interfaces. (c) Interfacial water molecule spectra on the surface of Au-PTFE/TS . (d) NH_3 yield rates and FEs on Au/TS and Au–PTFE/TS at different potentials. Source: (a–d) Zheng et al. [89]. © 2019, Elsevier. (e) Schematic of the Nb–$SrTiO_3$ interface modified with Au NPs and Zr/ZrOx thin film at two sides. (f) Energy-level diagram of the plasmon-induced NH_3 production. Source: (e–f) Oshikiri et al. [90]. © 2016, John Wiley & Sons.

Furthermore, Misawa's group developed a plasmon-induced NH_3 synthesis based on a niobium-doped strontium titanate (Nb–$SrTiO_3$) interface modified with plasmonic Au NPs and Ru co-catalyst on two sides, which divided the photoelectrochemical system into two cells for favorable separation of reaction products [91]. Under the visible light irradiation, plasmonic-induced charge separation occurred on the side of Au/Nb–$SrTiO_3$ in the anodic chamber, following by the electron transfer to the Ru side for the N_2 fixation in the cathodic chamber. However, suffering from a competitive H_2 generation, a low selectivity toward NH_3 production was observed within this reaction system. To address this issue, the authors substituted the Ru co-catalyst with the Zr/ZrO_x (Figure 11.8e), ascribed to their favorable affinity toward N adatoms than H adatoms [90]. The plasmon-induced charge separation occurred at the interface of Au NPs/Nb–$SrTiO_3$, and the formed electrons were transferred to the CB of Nb–$SrTiO_3$, which were subsequently participated in the N_2 reduction reaction at the Zr/ZrO_x surface (Figure 11.8f). Comparing with a main proton reduction on the Ru co-catalyst, the H_2 evolution was significantly reduced on the interface of Au NPs/Nb–$SrTiO_3/Zr/ZrO_x$. Accordingly, a greatly improved NH_3 production with high selectivity was acquired, achieving a NH_3 generation rate of 6.5 nmol h^{-1} cm^{-2} (6 times higher than that of the Ru co-catalyst).

11.3.4 Heterojunction Engineering

To suppress the fast recombination of e–h pairs and extend light absorption spectrum of photocatalysts, another most widely used strategy is to construct versatile heterojunctions, via combining with another metal, carbon, or semiconductor material. In the past few years, lots of efforts have been dedicated to the design and construction of heterojunctions to improve the photocatalytic and photoelectrochemical activity in NH_3 production, as summarized in the Table 11.2. According to the different components, the heterojunctions can be mainly classified into four categories, including the semiconductor–semiconductor (S–S) heterojunction, semiconductor–metal (S–M) heterojunction, the semiconductor–carbon (S–C) heterojunction, and the multicomponent (MC) heterojunction [108, 109].

For instance, Gao et al. proposed a 2D S–S heterojunction of AgCl/δ-Bi_2O_3 NSs for photocatalytic N_2 fixation, achieving a high yield rate of 606 μmol h^{-1} g^{-1} under visible light illumination [110]. Prepared with the hydrothermal precipitation approach, the 2D heterojunction exhibited an ultrathin morphology (~2.7 nm in thickness) with a length of ~250 nm. On the one hand, the construction of the heterojunction structure mitigated the recombination of e–h pairs, achieving a fast electron transfer and offering abundant electrons for N_2 fixation. On the other hand, the presence of OVs on the heterojunction facilitated the chemical adsorption and activation of N_2. Therefore, the photo-generated electrons were trapped by the active centers and promptly migrated to the adsorbed N_2, realizing the outstanding photocatalytic N_2 reduction.

Moreover, Zhang and coworkers reported a novel S–M heterojunction of Au/end–CeO_2 to realize the NIR-excited NH_3 production [111]. The CeO_2 was selectively grown at the terminals of gold nanorods (Au NRs) via a wet-chemistry route, in which the preferential adsorption of $PtCl_4^{2-}$ on the terminal sections of the nanorods and the subsequent redox reaction between $Ce(OH)_3$ and $PtCl_4^{2-}$ led to the nucleation and growth of CeO_2 (Figure 11.9a). This unique structure was confirmed by the HAADF–STEM image and the related elemental mapping. It was revealed that the prepared heterojunction contained the Au NRs at the center, while the Ce and O elements were homogeneously located at two ends (Figure 11.9b). After the absorption of NIR photon, the plasmon-excited hot electrons were formed

Table 11.2 Performance summary of various heterojunctions for photocatalytic and photoelectrochemical N_2 reduction.

Heterojunction	Light source	Reactant	Scavenger	Ammonia production rate	Year	References
g-C_3N_4/rGO	250 W high-pressure sodium lamp (400–800 nm)	Air/H_2O	Na–EDTA	9.276 mg l^{-1} g^{-1} h^{-1}	2016	[92]
g-C_3N_4/ZnMoCdS	250 W high-pressure sodium lamp (400–800 nm)	Air/H_2O	Ethanol	3.5 mg l^{-1} g^{-1} h^{-1}	2016	[93]
g-C_3N_4/ZnSnCdS	250 W high-pressure sodium lamp (400–800 nm)	N_2/H_2O	Ethanol	7.543 mg l^{-1} g^{-1} h^{-1}	2016	[94]
Carbon/$WO_3 \cdot H_2O$	Xe lamp (500 mW cm^{-2}, UV/Vis)	Air/H_2O	None	205 μmol g^{-1} h^{-1}	2016	[95]
g-C_3N_4/MgAlFeO	250 W high-pressure sodium lamp (400–800 nm)	N_2/H_2O	Ethanol	7.5 mg l^{-1} h^{-1} g^{-1}	2017	[96]
$W_{18}O_{49}$/g-C_3N_4	UV–vis–NIR	N_2/H_2O	Ethanol	2.6 mg l^{-1} h^{-1} g^{-1}	2017	[97]
Au/$(BiO)_2CO_3$	300 W Xe lamp	N_2/H_2O	None	38.23 μmol g^{-1} h^{-1}	2017	[98]
Ga_2O_3–DBD/g-C_3N_4	500 W Xe lamp (Vis)	N_2/H_2O	Methanol	112.5 μmol l^{-1} h^{-1}	2017	[99]
SiW_{12}/K–C_3N_4	Xe lamp (100 mW cm^{-2}, vis)	N_2/H_2O	None	353.2 μM g^{-1} h^{-1}	2018	[100]
GCN-γ-Ga_2O_3	100 W lamp (UV/vis)	N_2/H_2O	Ethanol	355.5 μmol l^{-1} h^{-1}	2018	[101]
MoS_2/C–ZnO	300 W Xe lamp ($\lambda > 420$ nm)	N_2/H_2O	Ethanol	245.7 μmol l^{-1} g^{-1} h^{-1}	2018	[102]
Bi_2MoO_6/BiOBr	300 W Xenon lamp	N_2/H_2O	None	90.7 μmol g^{-1} h^{-1}	2019	[103]
Au/P25	300 W Xe lamp (UV/vis)	N_2/H_2O	Methanol	1.02 mmol g^{-1} h^{-1}	2019	[104]
In_2O_3/In_2S_3	300 W Xe lamp	N_2/H_2O	None	40.04 μmol g^{-1} h^{-1}	2019	[105]
MoS_2/TiO_2	300 W Xe arc lamp (AM 1.5 G filter)	N_2/H_2O	None	1.42 μmol h^{-1} cm^{-2}	2019	[17]
TiO_2/$ZnFe_2O_4$	250 W Xe lamp (vis)	N_2/H_2O	Methanol	1.48 μmol l^{-1} min^{-1}	2019	[106]
Fe_2O_3/g-C_3N_4	300 W high pressure Xe lamp	N_2/H_2O	Ethanol	47.9 mg l^{-1} h^{-1}	2019	[107]

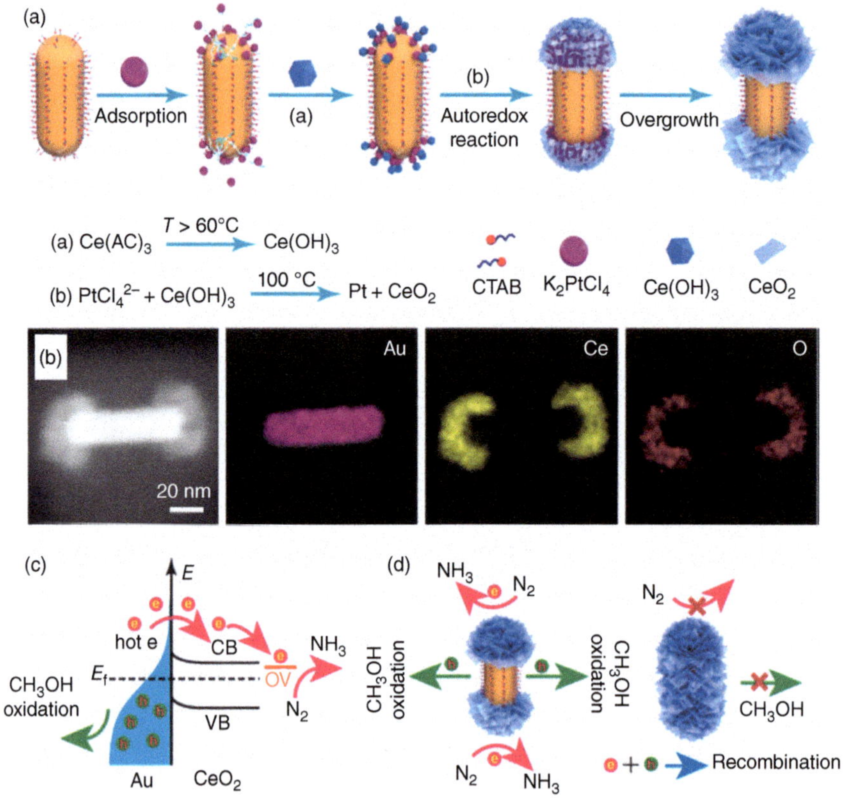

Figure 11.9 (a) Schematic of the synthesis steps for the Au/end–CeO_2 heterojunction. (b) HAADF–STEM image and the related elemental mapping. Mechanism for N_2 photo-fixation on the Au/end–CeO_2 heterojunction. The hot carrier separation behaviors on (c) the Au/end–CeO_2 and (d) the core@shell nanostructures. Source: (a–d) Jia et al. [111]. © 2019, American Chemical Society.

on the Au NRs and then transferred into the CB of CeO_2 (Figure 11.9c). Subsequently, accompanying with abundant OVs formed on the CeO_2, the adsorbed N_2 molecules were further activated to produce NH_3 on the heterojunction. Attributed to the unique design of the spatial separation, the Au/end–CeO_2 heterojunction demonstrated an excellent N_2 reduction rate of 114.3 μmol h^{-1} g^{-1} ($\lambda = 808$ nm). As a comparison, in the core@shell nanostructures with the Au NRs wrapped inside, the generated hot holes were hardly consumed by methanol, resulting in a severe recombination of e–h pairs (Figure 11.9d).

Synthesized via a facile solvothermal route, Lashgari and Zeinalkhani reported a S-C heterojunction of FeS_2/CNT as an effective and scavenger-free photocatalyst to convert N_2 molecules into NH_3 [112]. As one of the basic elements in the nitrogenase, Fe atom could preferably bind with N_2 molecules as well as adsorb H atoms. The FeS_2 was regarded as a stable narrow bandgap semiconductor to strongly absorb photons within a wide spectral region. Besides, it was confirmed that the introduction of CNT remarkably improved the specific area (121.2 m^2 g^{-1} compared with 15.3 m^2 g^{-1} of FeS_2), boosted the harvest of the incident light, retarded the charge recombination, decreased the interfacial impedance, and facilitated the charge transfer process. The superiority of FeS_2/CNT heterojunction endowed a high NH_3 production performance that was ~2 times that of FeS_2.

Lately, Jiang and coworkers reported a MC heterojunction of TiO_2@C/g-C_3N_4 and explored it property in the photocatalytic NH_3 formation [113]. The TiO_2 NPs decorated 2D carbon NSs coupling with g-C_3N_4 NSs were in situ thermally derived from a mixture of MXene $Ti_3C_2T_x$ and melamine. Supported on the 2D carbon NSs, TiO_2 with plentiful Ti^{3+} active species were tightly enwrapped by formed g-C_3N_4 NSs. Attributed to the synergetic effects, attractive properties with sufficient surface defects, suitable light adsorption ability, and excellent charge transfer were observed on the MC heterojunction. Thus, an impressive photo-activation for N_2 reduction, with the NH_3 evolution rate of 250.6 μmol g^{-1} h^{-1} was acquired under visible light illustration.

11.3.5 Co-catalyst Engineering

With a similar function of the heterojunction, the incorporation of co-catalyst, usually diverse metal catalysts, onto the photocatalyst has inspired an alternative route for the design of promising catalysts [114]. Utilizing the synergetic effect stemming from each component, for instance, the photocatalyst to produce the photo-induced electrons and the metal catalysts to activate the N_2 molecules, an enhanced efficiency in the photocatalytic NH_3 generation would be finally realized. More importantly, the presence of co-catalysts could also boost the e–h separation and facilitate the electron transfer [115]. In some cases, co-catalysts could be introduced to expand the light absorption range.

Playing a critical role in the overpotential for H_2 generation on different metals, the bond strength between metal–hydrogen is highly related to NH_3 production rate and selectivity. As depicted in the theoretical simulations, the Mo and Ru atoms on the top of volcano diagrams hold great potentials as efficient species for N_2 reduction [52]. Therefore, the incorporation of such metals as the co-calatysts has received enormous interests in the photocatalytic NH_3 generation. For instance, Sun and coworkers designed the TiO_2 NSs decorated with Ru atoms for photocatalytic N_2 reduction to NH_3 under Xe lamp illumination [116]. Stabilized by the OV sites of TiO_2 NSs, the isolated Ru atoms could not only substantially suppress H_2 evolution but also enhance N_2 chemisorption and promote charge carrier separation. Thus, an increased photocatalytic NH_3 yield rate of 56.3 μg h^{-1} g^{-1} was observed on the photocatalyst, which was approximately 2 fold than that of pure TiO_2 NSs.

In addition to function for the N_2 activation, the LSPR effect of Ru NPs could also contribute to the superior performance in photocatalytic N_2 reduction. For example, Zhang and coworkers demonstrated a photo-thermal synthesis strategy for the efficient NH_3 generation, based on a K promoted, Ru decorated $TiO_{2-x}H_x$ catalyst (K/Ru/$TiO_{2-x}H_x$) [117]. The introduction of K and the amorphous $TiO_{2-x}H_x$ support with abundant OVs effectively tuned electronic structure of Ru and further facilitated electron donation from the support of $TiO_{2-x}H_x$ toward Ru active sites.

More importantly, the reversibly incorporated H atoms in amorphous support were favorable to restrain the H_2 poisoning on Ru. Accompanying with the generation of electromagnetic field and regional heat due to the LSPR effect, the Ru clusters efficiently absorbed the light energy, and the temperature on the catalyst was heated up to 360 °C, finally achieving an impressive NH_3 yield rate of 112.6 μmol g^{-1} h^{-1}. As a comparison in the thermal process at the same temperature, the activity of K/Ru/$TiO_{2-x}H_x$ was only half of its photo-thermal catalysis. Interestingly, ascribed to the structure transformation of amorphous K/Ru/$TiO_{2-x}H_x$ to crystalline K/Ru/TiO_2, it was discovered that the reactivity of the photocatalyst could not be recovered in the thermal NH_3 synthesis. As an extension of this work, the same group recently proposed another promising $TiO_{2-x}H_y$/Fe composite for the dual-temperature photo-thermal NH_3 production (Figure 11.10a), effectively mitigating the adverse equilibrium shift toward NH_3 decomposition [118]. Prepared with a simple in situ reduction approach, $TiO_{2-x}H_y$/Fe was successfully obtained with the Fe nanospheres assembled into necklace nanostructures, and the $TiO_{2-x}H_y$ NPs uniformly dispersed on the Fe surface (Figure 11.10b). Instead of the weak physical interaction, it was revealed that $TiO_{2-x}H_y$ was strongly anchored to Fe surface via the lattice interaction with a protective layer of amorphous $Fe_2O_{3-\delta}$. Attributed to the plasmonic local heating effect, a high apparent temperature of 495 °C was observed on the hybrid photocatalyst, while a local temperature difference (LTD) was produced between the hot Fe zone and cooling $TiO_{2-x}H_y$ zone. The LTDs within $TiO_{2-x}H_y$/Fe catalyst was firstly theoretically simulated with the finite difference time domain (FDTD). It was confirmed that the LTDs were highly related to the size of Fe, and the maximum LTDs between $TiO_{2-x}H_y$ and Fe were estimated around 79–137 °C for d_{Fe} = 120 nm (Figure 11.10c,d). Furthermore, the experimental explorations were performed with surface enhanced Raman scattering (SERS) mapping to directly monitor the spatial temperature distributions between $TiO_{2-x}H_y$ and Fe, applying the adonitol as the temperature response probe. As shown in Figure 11.10e, the adonitol-rich area was overlapped with the $TiO_{2-x}H_y$-rich area, while away from the Fe-rich section, indicating that the high and low temperatures existed in the Fe and $TiO_{2-x}H_y$-rich area, respectively. Accordingly, a spatial temperature difference of 129 °C was calculated between hot Fe and cooling $TiO_{2-x}H_y$, which was consistent with the FDTD investigation. In view of the N_2 dissociation and subsequent hydrogenation occurred on the hot Fe and cooling of $TiO_{2-x}H_y$, respectively, the generation of the dual-temperature-zone effectively enhanced the NH_3 yield rate, delivering an impressive NH_3 concentration of 1939 ppm (Figure 11.10f). The anchoring of $TiO_{2-x}H_y$ significantly enhanced the stability of the Fe structure, and a high reactivity with an excellent reproducibility for five times was observed (Figure 11.10g). Interestingly, due to the favorable N_2 activation in the high-temperature area and facilitated NH_3 production in the low-temperature section, the acquired NH_3 concentrations exceeded the theoretical equilibrium limits at 495 °C (Figure 11.10h).

Recently, the Mo anchored polymeric carbon nitride (Mo–PCN) for enhanced photocatalytic NH_3 production was reported by Lu and coworkers, in which isolated Mo centers were stabilized via the formation of two-coordinated MoN_2

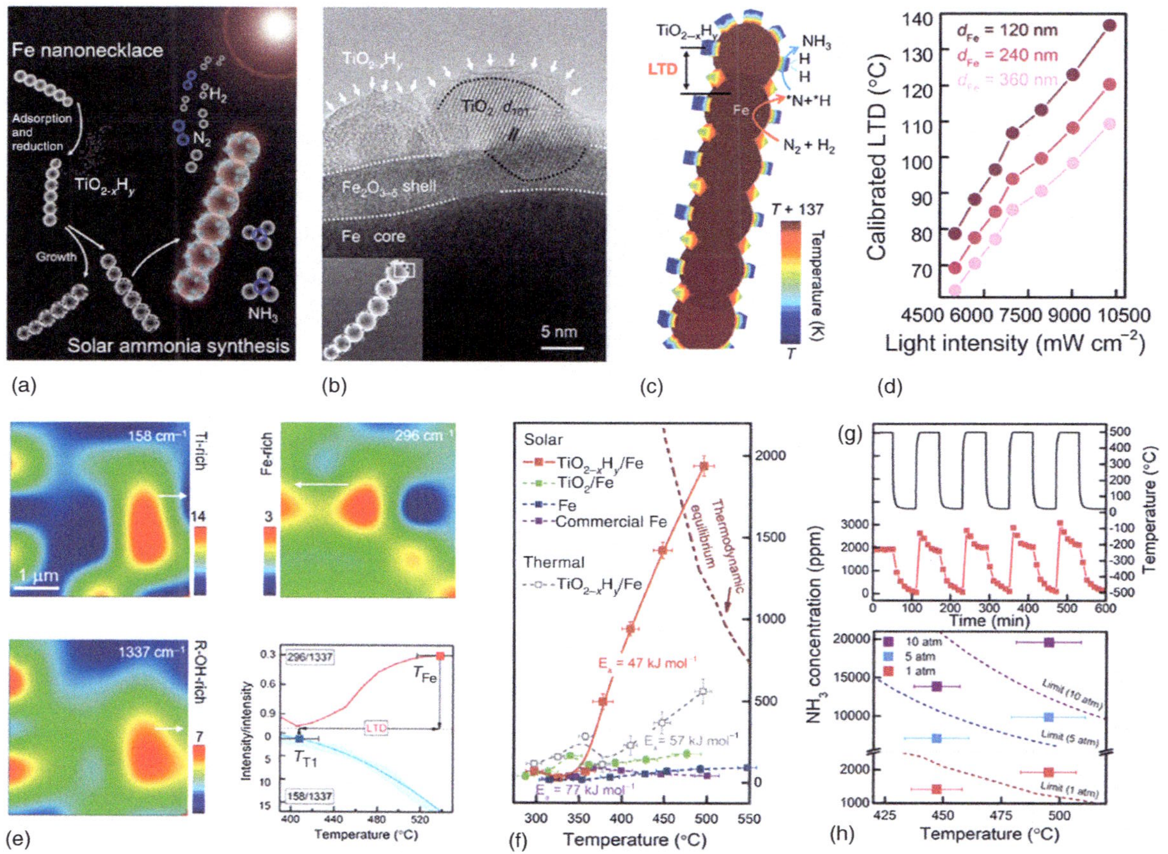

Figure 11.10 (a) Schematic of $TiO_{2-x}H_y$/Fe catalyst for dual-temperature-zone photo-thermal NH_3 synthesis. (b) High-resolution STEM image of $TiO_{2-x}H_y$/Fe catalyst. (c) Steady-state non-equilibrium temperature distribution and dual-temperature-zone NH_3 synthesis on $TiO_{2-x}H_y$/Fe. (d) LTDs between hot Fe and cooling $TiO_{2-x}H_y$. (e) SERS mapping of the spatial dispersion of $TiO_{2-x}H_y$, Fe, and the adonitol and the detected local LTD. (f) The NH_3 concentrations produced at different apparent catalyst temperatures. (g) Successive light-on/off measurement of photo-thermal NH_3 synthesis. (h) The impressive equilibrium-beyond reactivity of $TiO_{2-x}H_y$/Fe at different pressures. Source: (a–h) Mao et al. [118]. © 2019, Elsevier.

species [119]. Ascribed to the monodisperse of Mo on the PCN, no obvious nanoparticles were observed in Mo–PCN (Figure 11.11a), and the homogeneous atom distribution (e.g. Mo, C, and N) was confirmed in elemental mapping analysis (Figure 11.11b). As the direct evidence of Mo single atom, individual bright dots belonging to the active Mo species with a size less than 0.2 nm were captured in the HAADF–STEM (Figure 11.11c). Furthermore, the introduction of Mo center on the PCN substantially improved the electron transfer from PCN to Mo sites, thus enhancing the separation efficiency the photo-excited e–h pairs. Consequently, an obviously reduced radiative lifetime from 10.62 ns (PCN) to 5.06 ns (3-Mo–PCN) was observed (Figure 11.11d). Moreover, the presence of Mo sites also contributed to effectively enhanced N_2 adsorption on the PCN, with an obvious peak at 2197 cm^{-1} observed in the low-temperature Fourier transform infrared (LT–FTIR) spectra (Figure 11.11e). As further revealed in the density functional theory (DFT) simulations, the coordinative unsaturated metal center could strongly adsorb N_2 through an end-on configuration. Accordingly, the N≡N bond was elongated from 1.11 Å to 1.15 Å, following by the efficient N_2 fixation after receiving the photo-excited electrons under ambient conditions (Figure 11.11f). Owing to these impressive properties of Mo–PCN, an excellent photocatalytic activity with the NH_3 yield rate of 50.9 $\mu mol\,g^{-1}\,h^{-1}$ was achieved in H_2O. While introducing additional electron scavengers (e.g. ethanol), the NH_3 evolution rate was significantly improved to 830 $\mu mol\,g^{-1}\,h^{-1}$.

Apart from abovementioned metal atoms, the corporation of earth-abundant Cu into the ultrathin C_3N_4 NSs (Cu–CN) for photocatalytic NH_3 formation was recently reported by Xie's group [120]. Attributed to the interactions between highly dispersed Cu single atoms and the lone-pair electrons in p–CN, plentiful valence electron isolations were successfully yielded from its conjugated π-electron cloud. As confirmed in the ESR measurements, abundant free electrons were subsequently generated with the excitation of these isolated valence electrons under the light irradiation, which were participated in N_2 reduction and achieved a highly efficient NH_3 generation. The NH_3 yield rate of Cu–CN was calculated to be 186 $\mu mol\,g^{-1}\,h^{-1}$ (~7fold higher than that of pristine p-CN) under visible light illumination. Given that most conjugated polymers possess lone-pair electrons and π-electron cloud be within the structures, it is highly expected that more promising catalysts could be designed with the similar approach, via selecting proper metal atoms/small molecules to stimulate the isolation of single valence electrons, as well as the deformation of π electron cloud.

11.3.6 Biomimetic Engineering

As mentioned before, an easy transformation of N_2 toward NH_3 can be realized at ambient conditions with the assistance of the nitrogenases. Although the whole reaction is processed in the absence of light illumination, the ATP hydrolysis induced electron-transfer steps plays critical roles in N_2 reduction. Enlightened by this discovery, numerous efforts have been dedicated to the development of biomimetic photocatalysts for artificial NH_3 synthesis. King and coworkers

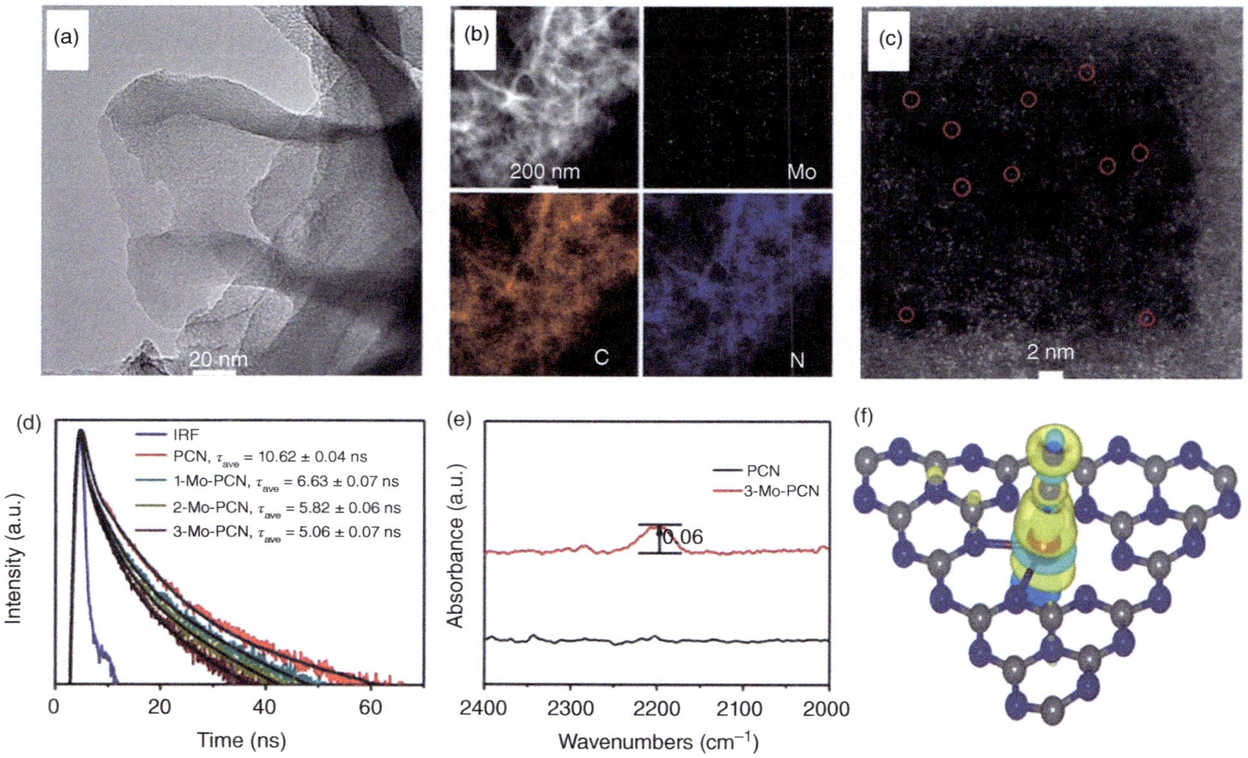

Figure 11.11 (a) HRTEM, (b) elemental mapping, and (c) HAADF–STEM images of Mo–PCN. (d) Time-resolved fluorescence kinetics. (e) LT–FTIR spectra for N_2 adsorption. (f) Charge density difference of Mo–PCN with adsorbed N_2. Source: (a–f) Guo et al. [119]. © 2019, Royal Society of Chemistry.

proposed highly photo-sensitive CdS NRs connecting to the MoFe-based protein to replace ATP hydrolysis for the biomimetic N_2 reduction [121]. With the visible-light response in the solar spectrum (E_g = 2.72 eV) and the suitable reduction potential of the first photo-induced state transition (−0.8 V vs. NHE), the introduction of CdS NRs was adequately negative for the reduction the MoFe protein (−0.31 V), which could supply electrons to promote the N_2 reduction and generate NH_3. Under the light irradiation (λ = 405 nm), this bio-photocatalyst displayed a superior activity for photocatalytic NH_3 generation, with a yield rate of 315 ± 55 nmol mg^{-1} min^{-1} and a turnover frequency (TOF) of 75 min^{-1} (~63% of biological reaction rate).

Instead of the partial mimic of the nitrogenases, Kanatzidis and coworkers prepared a biomimetic chalcogel of "$Mo_2Fe_6S_8$–Sn_2S_6" for the N_2 fixation using ultraviolet to visible light (Figure 11.12a) [122]. Consisted of double-cubane $Mo_2Fe_6S_8$ clusters connected by Sn_2S_6 ligands, the chalcogels with a random, amorphous network exhibited a strong optical absorption, in which the $Mo_2Fe_6S_8$ was served as the active center for N_2 fixation. The resultant chalcogels showed an impressive photocatalytic stability over 72 h. Furthermore, another biomimetic

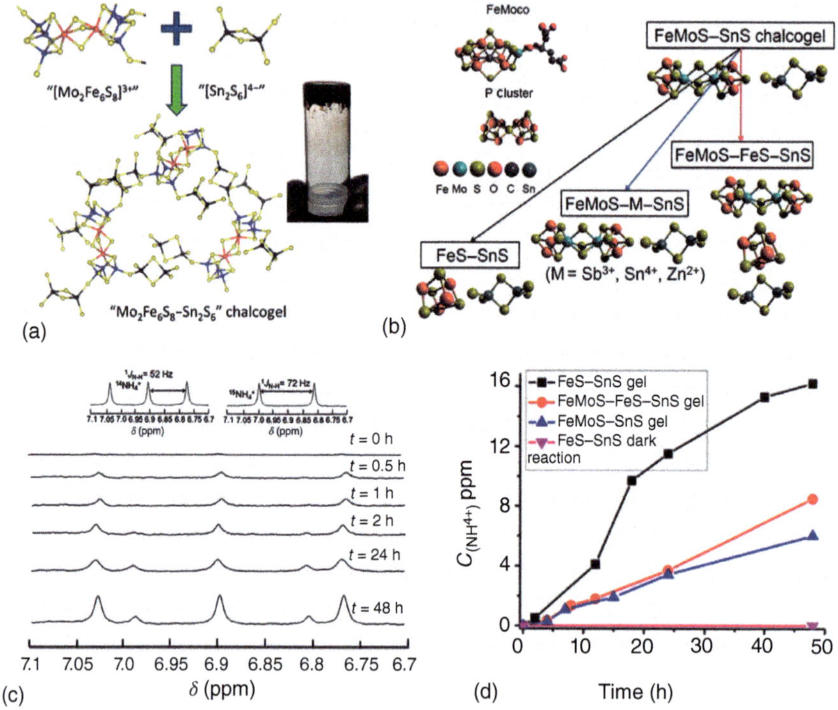

Figure 11.12 (a) Schemcatic of designed $Mo_2Fe_6S_8$–Sn_2S_6 biomimetic chalcogel. Source: (a) Adapted from Banerjee et al. [122]. © 2015, American Chemical Society. (b) The synthesis routes for the assembly of various nitrogenase-inspired biomimetic chalcogels. (c) Proton NMR spectra for the determination of $^{14}NH_4^+$ and $^{15}NH_4^+$ within the photocatalytic experiment. (d) NH_3 production performance of different chalcogels. Source: (b–d) Adapted from Liu et al. [123]. © 2016, National Academy of Sciences.

chalcogel consisting of $[Mo_2Fe_6S_8(SPh)_3]^{3+}$ and Fe_4S_4 clusters was reported by the same group for photocatalytic N_2 fixation (Figure 11.12b) [123]. It was revealed that observable photocatalytic activity for NH_3 formation was acquired even the Fe_4S_4 clusters were substituted by other ions (e.g. Sn^{4+}, Zn^{2+} or Sb^{3+}). The evidence for the photocatalytic production of NH_3 was confirmed by the 1H NMR of $^{14}NH_4^+$ and $^{15}NH_4^+$ samples, using the $^{15}N{\equiv}^{14}N$ as the feed gas (Figure 11.12c). Considering a lower N–N stretching mode frequency was observed on the Fe_4S_4 cluster than that of on $[Mo_2Fe_6S_8(SPh)_3]^{3+}$ cluster, the photo-activated Fe_4S_4 was beneficial for the weaken of the N≡N bond. Accordingly, Fe_4S_4-only chalcogel (FeS–SnS) achieved a higher NH_3 synthesis rate than that of $[Mo_2Fe_6S_8(SPh)_3]$-containing chalcogels (Figure 11.12d). Within the system, it was revealed that redox-active Fe sites were responsible for the N_2 activation in the NH_3 synthesis.

11.4 Conclusions and Outlook

Photocatalytic and photoelectrochemical N_2 reduction is an attractive route for the green and sustainable NH_3 synthesis, in which ideal catalysts can directly drive the conversion of N_2 and H_2O into NH_3 using the solar energy or an external bias, without the requirement of extra energy input and sacrificial reagents. Although significant research progresses related to the development of various photo- and photoelectro-catalysts have been achieved in the past few years, it still remains great challenges to be addressed before the large-scale applications of such NH_3 synthesis, due to the generally low yield rates and conversion efficiencies. Herein, versatile strategies including defect engineering, structure control, interface regulation, heterojunction construction, co-catalyst engineering and biomimetic engineering are systematically discussed in the design and preparation of promising catalysts toward effective N_2 reduction.

To develop more efficient routes for the design of desirable catalysts, additional attentions should be paid to several critical aspects in future research. (i) To confirm the presence, level, and position of defects within the catalysts, the advanced characterization techniques, e.g. the scanning tunnelling microscope (STM) and electron energy-loss spectroscopy (EELS), are highly anticipated and indispensable. For example, the existing vacancies on the catalyst surface instead of the bulk are preferable in highly efficient N_2 reduction. (ii) Developing promising approaches for catalyst synthesis is urgently required. It should be noted that the efficient introduction of various defects, the controllable regulation of the morphology and the construction of high-quality heterojunctions are closely dependent on the preparation methods. (iii) An efficient integration of multiple strategies. Through efficiently combining different design strategies, e.g. the vacancy introduction and morphology control, interface design, and heterojunction construction, it is extremely expected to tremendously boost the catalytic performance of the catalysts utilizing the resultant synergetic effects. (iv) Finally, a clear understanding of reaction mechanisms and pathways. With the assistance of theoretical explorations and advanced operando characterization methods (e.g. DRIFTS), in-depth insights into these fundamental details will enable the community to effectively design catalysts with high activity and selectivity.

In addition to the critical photocatalyst development, there are several other important aspects highly related to the performance evaluations of the constructed catalysts, such as the optimization of reaction systems (e.g. pH, counterion, and dissolved N_2 concentration), accurate methods for product determination and widely accepted benchmarks for the standardized assessment of performance. For instance, as the familiar colorimetric methods (e.g. Nessler's reagent and the indophenol blue methods) for the determination of NH_3, their accuracy is usually challenged with ion interference, pH variation, and some sacrificial reagents present in the reaction solution [124, 125]. Meanwhile, the ubiquitous NH_3 contaminations, either present in the atmosphere, human breath, or in the catalyst itself would seriously interfere with accurate performance quantification. Therefore, a rigorous test procedure would help to prevent false positives and enables researchers to reliably quantify the ammonia production [126–128].

With the persistent efforts and contributions from the research community in future, it is highly anticipated that a great prosperous development will be achieved within the photocatalytic and photoelectrochemical NH_3 synthesis.

References

1 Galloway, J.N., Townsend, A.R., Erisman, J.W. et al. (2008). *Science* 320: 889.

2 Valera-Medina, A., Xiao, H., Owen-Jones, M. et al. (2018). *Prog. Energy Combust. Sci.* 69: 63.

3 Galloway, J.N., Dentener, F.J., Capone, D.G. et al. (2004). *Biogeochemistry* 70: 153.

4 Bazhenova, T.A. and Shilov, A.E. (1995). *Coord. Chem. Rev.* 144: 69.

5 van der Ham, C.J.M., Koper, M.T.M., and Hetterscheid, D.G.H. (2014). *Chem. Soc. Rev.* 43: 5183.

6 Hoffman, B.M., Lukoyanov, D., Yang, Z.-Y. et al. (2014). *Chem. Rev.* 114: 4041.

7 Suryanto, B.H.R., Du, H.L., Wang, D.B. et al. (2019). *Nat. Catal.* 2: 290.

8 Wang, L., Xia, M.K., Wang, H. et al. (2018). *Joule* 2: 1055.

9 Schrauzer, G.N. and Guth, T.D. (1977). *J. Am. Chem. Soc.* 99: 7189.

10 Guo, C.X., Ran, J.R., Vasileff, A., and Qiao, S.Z. (2018). *Energy Environ. Sci.* 11: 45.

11 Zhang, S., Zhao, Y., Shi, R. et al. (2019). *EnergyChem* 1: 100013.

12 Wang, S.Y., Ichihara, F., Pang, H. et al. (2018). *Adv. Funct. Mater.* 28: 1803309.

13 Shipman, M.A. and Symes, M.D. (2017). *Catal. Today* 286: 57.

14 Li, J., Li, H., Zhan, G., and Zhang, L. (2017). *Acc. Chem. Res.* 50: 112.

15 Ithisuphalap, K., Zhang, H., Guo, L. et al. (2019). *Small Methods* 3: 1800352.

16 Li, H., Mao, C., Shang, H. et al. (2018). *Nanoscale* 10: 15429.

17 Ye, W., Arif, M., Fang, X.Y. et al. (2019). *ACS Appl. Mater. Interfaces* 11: 28809.

18 Chen, X., Li, N., Kong, Z. et al. (2018). *Mater. Horiz.* 5: 9.

19 Zhu, D., Zhang, L.H., Ruther, R.E., and Hamers, R.J. (2013). *Nat. Mater.* 12: 836.

20 Yang, J.H., Wang, D.G., Han, H.X., and Li, C. (2013). *Acc. Chem. Res.* 46: 1900.

21 Li, M., Huang, H., Low, J.X. et al. (2019). *Small Methods* 3: 1800388.

22 Shi, R., Zhao, Y.X., Waterhouse, G.I.N. et al. (2019). *ACS Catal.* 9: 9739.

23 Yan, D.F., Li, H., Chen, C. et al. (2019). *Small Methods* 3: 1800331.

24 Mao, C.L., Wang, J.X., Zou, Y.J. et al. (2019). *Green Chem.* 21: 2852.

25 Pan, X.Y., Yang, M.Q., Fu, X.Z. et al. (2013). *Nanoscale* 5: 3601.

26 Wang, Q., Liu, Z.Q., Liu, D.M. et al. (2018). *Appl. Catal., B* 236: 222.

27 Zhao, Y.F., Chen, G.B., Bian, T. et al. (2015). *Adv. Mater.* 27: 7824.

28 Li, H., Li, J., Ai, Z.H. et al. (2018). *Angew. Chem. Int. Ed.* 57: 122.

29 Hirakawa, H., Hashimoto, M., Shiraishi, Y., and Hirai, T. (2017). *J. Am. Chem. Soc.* 139: 10929.

30 Yang, J.H., Guo, Y.Z., Jiang, R.B. et al. (2018). *J. Am. Chem. Soc.* 140: 8497.

31 Jiang, R.B., Li, B.X., Fang, C.H., and Wang, J.F. (2014). *Adv. Mater.* 26: 5274.

32 Linic, S., Christopher, P., and Ingram, D.B. (2011). *Nat. Mater.* 10: 911.

33 Li, H., Shang, J., Ai, Z.H., and Zhang, L.Z. (2015). *J. Am. Chem. Soc.* 137: 6393.

34 Zhao, Y.F., Zhao, Y.X., Waterhouse, G.I.N. et al. (2017). *Adv. Mater.* 29: 1703828.

35 Zhao, Y.X., Zhao, Y.F., Shi, R. et al. (2019). *Adv. Mater.* 31: 1806482.

36 Di, J., Xia, J.X., Li, H.M. et al. (2017). *Nano Energy* 41: 172.

37 Xue, X.L., Chen, R.P., Chen, H.W. et al. (2018). *Nano Lett.* 18: 7372.

38 Jin, X.L., Ye, L.Q., Xie, H.Q., and Chen, G. (2017). *Coord. Chem. Rev.* 349: 84.

39 Wang, S.Y., Hai, X., Ding, X. et al. (2017). *Adv. Mater.* 29: 1701774.

40 Di, J., Xia, J.X., Chisholm, M.F. et al. (2019). *Adv. Mater.* 31: 1807576.

41 Fan, G.L., Li, F., Evans, D.G., and Duan, X. (2014). *Chem. Soc. Rev.* 43: 7040.

42 Niu, P., Yin, L.C., Yang, Y.Q. et al. (2014). *Adv. Mater.* 26: 8046.

43 Dong, G.H., Ho, W.K., and Wang, C.Y. (2015). *J. Mater. Chem. A* 3: 23435.

44 Sun, S.M., Li, X.M., Wang, W.Z. et al. (2017). *Appl. Catal., B* 200: 323.

45 Li, X.M., Sun, X., Zhang, L. et al. (2018). *J. Mater. Chem. A* 6: 3005.

46 Hu, S.Z., Chen, X., Li, Q. et al. (2016). *Catal. Sci. Technol.* 6: 5884.

47 Zhang, X., Lai, Z.C., Tan, C.L., and Zhang, H. (2016). *Angew. Chem. Int. Ed.* 55: 8816.

48 Pu, C., Qin, H., Gao, Y. et al. (2017). *J. Am. Chem. Soc.* 139: 3302.

49 Lv, X.S., Wei, W., Li, F.P. et al. (2019). *Nano Lett.* 19: 6391.

50 Cao, S.H., Fan, B., Feng, Y.C. et al. (2018). *Chem. Eng. J.* 353: 147.

51 Wang, K., Gu, G., Hu, S. et al. (2019). *Chem. Eng. J.* 368: 896.

52 Skulason, E., Bligaard, T., Gudmundsdottir, S. et al. (2012). *Phys. Chem. Chem. Phys.* 14: 1235.

53 Zhang, N., Jalil, A., Wu, D.X. et al. (2018). *J. Am. Chem. Soc.* 140: 9434.

54 Luo, J., Bai, X., Li, Q. et al. (2019). *Nano Energy* 66: 104187.

55 Zhao, W.R., Zhang, J., Zhu, X. et al. (2014). *Appl. Catal., B* 144: 468.

56 Hu, S.Z., Chen, X., Li, Q. et al. (2017). *Appl. Catal., B* 201: 58.

57 Meng, Q.Q., Lv, C.D., Sun, J.X. et al. (2019). *Appl. Catal., B* 256: 117781.

58 Smith, R.D.L., Prevot, M.S., Fagan, R.D. et al. (2013). *Science* 340: 60.

59 Chemelewski, W.D., Lee, H.C., Lin, J.F. et al. (2014). *J. Am. Chem. Soc.* 136: 2843.

60 Lv, C.D., Yan, C.S., Chen, G. et al. (2018). *Angew. Chem. Int. Ed.* 57: 6073.

61 Hou, T.T., Guo, R.H., Chen, L.L. et al. (2019). *Nano Energy* 65: 104003.

62 Li, C.C., Wang, T., Zhao, Z.J. et al. (2018). *Angew. Chem. Int. Ed.* 57: 5278.

63 Kong, M., Li, Y., Chen, X. et al. (2011). *J. Am. Chem. Soc.* 133: 16414.

64 Marichy, C., Bechelany, M., and Pinna, N. (2012). *Adv. Mater.* 24: 1017.

65 Johnson, R.W., Hultqvist, A., and Bent, S.F. (2014). *Mater. Today* 17: 236.

66 Wu, H.L., Li, X.B., Tung, C.H., and Wu, L.Z. (2019). *Adv. Mater.* 31: 1900709.

67 Ge, M.Z., Li, Q.S., Cao, C.Y. et al. (2017). *Adv. Sci.* 4: 1600152.

68 Di, J., Xiong, J., Li, H.M., and Liu, Z. (2018). *Adv. Mater.* 30: 1704548.

69 Zhang, P. and Lou, X.W. (2019). *Adv. Mater.* 31: 1900281.

70 Sun, S.M., An, Q., Wang, W.Z. et al. (2017). *J. Mater. Chem. A* 5: 201.

71 Wu, S.Q., Tan, X.J., Liu, K.D. et al. (2019). *Catal. Today* 335: 214.

72 Cao, S.H., Chen, H., Jiang, F., and Wang, X. (2018). *Appl. Catal., B* 224: 222.

73 Liu, M.X., Wang, Y.C., Kong, X.H. et al. (2019). *Iscience* 17: 208.

74 Li, L., Wang, Y.C., Vanka, S. et al. (2017). *Angew. Chem. Int. Ed.* 56: 8701.

75 Ali, M., Zhou, F.L., Chen, K. et al. (2016). *Nat. Commun.* 7: 11335.

76 Lu, Y.H., Yang, Y., Zhang, T.F. et al. (2016). *ACS Nano* 10: 10507.

77 Yang, Y., Zhang, T.F., Ge, Z. et al. (2017). *Carbon* 124: 72.

78 Geim, A.K. and Novoselov, K.S. (2007). *Nat. Mater.* 6: 183.

79 Yang, M.Q., Zhang, N., Pagliaro, M., and Xu, Y.J. (2014). *Chem. Soc. Rev.* 43: 8240.

80 Nazemi, M., Panikkanvalappil, S.R., and El-Sayed, M.A. (2018). *Nano Energy* 49: 316.

81 Nazemi, M. and El-Sayed, M.A. (2018). *J. Phys. Chem. Lett.* 9: 5160.

82 Nazemi, M. and El-Sayed, M.A. (2019). *Nano Energy* 63: 103886.

83 Jiang, J., Zhao, K., Xiao, X.Y., and Zhang, L.Z. (2012). *J. Am. Chem. Soc.* 134: 4473.

84 Yu, J.G., Low, J.X., Xiao, W. et al. (2014). *J. Am. Chem. Soc.* 136: 8839.

85 Yang, H.G., Sun, C.H., Qiao, S.Z. et al. (2008). *Nature* 453: 638.

86 Montoya, J.H., Tsai, C., Vojvodic, A., and Nørskov, J.K. (2015). *ChemSusChem* 8: 2180.

87 Li, H., Shang, J., Shi, J.G. et al. (2016). *Nanoscale* 8: 1986.

88 Bai, Y., Ye, L.Q., Chen, T. et al. (2016). *ACS Appl. Mater. Interfaces* 8: 27661.

89 Zheng, J.Y., Lyu, Y.H., Qiao, M. et al. (2019). *Chem* 5: 617.

90 Oshikiri, T., Ueno, K., and Misawa, H. (2016). *Angew. Chem. Int. Ed.* 55: 3942.

91 Oshikiri, T., Ueno, K., and Misawa, H. (2014). *Angew. Chem. Int. Ed.* 53: 9802.

92 Hu, S.Z., Zhang, W.D., Bai, J. et al. (2016). *RSC Adv.* 6: 25695.

93 Zhang, Q., Hu, S.Z., Fan, Z.P. et al. (2016). *Dalton Trans.* 45: 3497.

94 Hu, S., Li, Y., Li, F. et al. (2016). *ACS Sustainable Chem. Eng.* 4: 2269.

95 Li, X.M., Wang, W.Z., Jiang, D. et al. (2016). *Chem. Eur. J.* 22: 13819.

96 Wang, Y., Wei, W., Li, M. et al. (2017). *RSC Adv.* 7: 18099.

97 Liang, H., Zou, H., and Hu, S. (2017). *New J. Chem.* 41: 8920.

98 Xiao, C.L., Hu, H., Zhang, X.Y., and MacFarlane, D.R. (2017). *ACS Sustainable Chem. Eng.* 5: 10858.
99 Cao, S.H., Zhou, N., Gao, F.H. et al. (2017). *Appl. Catal., B* 218: 600.
100 Xiao, C.L., Zhang, L., Wang, K.F. et al. (2018). *Appl. Catal., B* 239: 260.
101 Devthade, V., Gupta, A., and Umare, S.S. (2018). *ACS Appl. Nano Mater.* 1: 5581.
102 Xing, P.X., Chen, P.F., Chen, Z.Q. et al. (2018). *ACS Sustainable Chem. Eng.* 6: 14866.
103 Xue, X.L., Chen, R.P., Yan, C.Z. et al. (2019). *Nanoscale* 11: 10439.
104 Bu, T.A., Hao, Y.C., Gao, W.Y. et al. (2019). *Nanoscale* 11: 10072.
105 Xu, H.C., Wang, Y., Dong, X.L. et al. (2019). *Appl. Catal., B* 257: 117932.
106 Rong, X.S., Chen, H.F., Rong, J. et al. (2019). *Chem. Eng. J.* 371: 286.
107 Liu, S.Z., Wang, S.B., Jiang, Y. et al. (2019). *Chem. Eng. J.* 373: 572.
108 Wang, H.L., Zhang, L.S., Chen, Z.G. et al. (2014). *Chem. Soc. Rev.* 43: 5234.
109 Shi, L., Yin, Y., Zhang, L.-C. et al. (2019). *Appl. Catal., B* 248: 405.
110 Gao, X.M., Shang, Y.Y., Liu, L.B., and Fu, F. (2019). *J. Catal.* 371: 71.
111 Jia, H.L., Du, A.X., Zhang, H. et al. (2019). *J. Am. Chem. Soc.* 141: 5083.
112 Lashgari, M. and Zeinalkhani, P. (2018). *Nano Energy* 48: 361.
113 Liu, Q.X., Ai, L.H., and Jiang, J. (2018). *J. Mater. Chem. A* 6: 4102.
114 Liu, H.M., Wu, P., Li, H.T. et al. (2019). *Appl. Catal., B* 259: 118026.
115 Qiu, P.X., Xu, C.M., Zhou, N. et al. (2018). *Appl. Catal., B* 221: 27.
116 Liu, S.Z., Wang, Y.J., Wang, S.B. et al. (2019). *ACS Sustainable Chem. Eng.* 7: 6813.
117 Mao, C.L., Yu, L.H., Li, J. et al. (2018). *Appl. Catal., B* 224: 612.
118 Mao, C.L., Li, H., Gu, H.G. et al. (2019). *Chem* 5: 2702.
119 Guo, X.W., Chen, S.M., Wang, H.J. et al. (2019). *J. Mater. Chem. A* 7: 19831.
120 Huang, P.C., Liu, W., He, Z.H. et al. (2018). *Sci. China-Chem.* 61: 1187.
121 Brown, K.A., Harris, D.F., Wilker, M.B. et al. (2016). *Science* 352: 448.
122 Banerjee, A., Yuhas, B.D., Margulies, E.A. et al. (2015). *J. Am. Chem. Soc.* 137: 2030.
123 Liu, J., Kelley, M.S., Wu, W.Q. et al. (2016). *Proc. Natl. Acad. Sci. U.S.A.* 113: 5530.
124 Zhao, Y.X., Shi, R., Bian, X.A.N. et al. (2019). *Adv. Sci.* 6: 1802109.
125 Andersen, S.Z., Colic, V., Yang, S. et al. (2019). *Nature* 570: 504.
126 Greenlee, L.F., Renner, J.N., and Foster, S.L. (2018). *ACS Catal.* 8: 7820.
127 Shi, L., Yin, Y., Wang, S.B., and Sun, H.Q. (2020). *ACS Catal.* 10: 6870.
128 Shi, L., Yin, Y., Wang, S.J. et al. (2020). *Appl. Catal., B* 278: 119325.

12

Photocatalytic Production of Hydrogen Peroxide Using MOF Materials

Xiaolang Chen[1], Yasutaka Kuwahara[1,2,3], Kohsuke Mori,[1,2] and Hiromi Yamashita[1,2]

[1] *Osaka University, Division of Materials and Manufacturing Science, Graduate School of Engineering, 2-1 Yamadaoka, Suita, Osaka, 565-0871, Japan*
[2] *Kyoto University, Elements Strategy Initiative for Catalysts and Batteries (ESICB), Katsura, Kyoto, 615-8520, Japan*
[3] *JST, PRESTO, 4-1-8 Honcho, Kawaguchi, Saitama, 332-0012, Japan*

12.1 Introduction

As an eco-friendly oxidant and reductant [1], hydrogen peroxide (H_2O_2) has been widely used in various chemical industries (like organic synthesis and paper bleaching) and environmental treatment [2] (such as disinfection and water treatment) [3]. In addition, H_2O_2 is a green and environmental friendly energy alternative to hydrogen energy (H_2) [4] and can be used in fuel cells [1]. Liquid H_2O_2 is more convenient and safer for being stored and transported than gaseous H_2 due to its good water solubility [5]. As a potential energy carrier, the theoretical potential of aqueous H_2O_2 is 1.09 V, which is close to that of a traditional H_2/O_2 fuel cell (1.23 V) [1]. On the other hand, some disadvantages (for example, high costs, complicated industrial routes, and substantial toxic by-products) occur to the conventional anthraquinone method for H_2O_2 production [6]. Moreover, both electrocatalytic oxygen reduction reaction (ORR) route and the direct synthesis of H_2O_2 from H_2 and O_2 using metals show disadvantages such as high energy consumption or potentially explosive danger [7]. Thus, it is necessary to develop a safer, environmental-friendly, and low-cost technology for H_2O_2 production.

Photocatalytic H_2O_2 production has recently captured wide interests because of its safety, consumption of low energy, and pollution free [1, 6, 8]. So far, many efforts have been devoted to developing effective photocatalysts for the photocatalytic H_2O_2 production by the reduction of O_2. Generally, it can be divided into two categories by the nature of the photocatalysts: the one is molecule-based photocatalyst, and the other is semiconductor-based photocatalyst. In the molecule-based photocatalysts, Fukuzumi and coworkers [9] in 2013 reported the utilization of $[Ru(Me_2phen)_3)]^{2+}$ ($Me_2phen = 4, 4'$-dimethyl-1,10-phenanthroline) as a visible-light-responsive photocatalyst in an O_2-saturated H_2SO_4 aqueous solution. One of its disadvantages is that the photocatalyst is not recyclable due to its water solubility. In the semiconductors,

Solar-to-Chemical Conversion: Photocatalytic and Photoelectrochemcial Processes, First Edition. Edited by Hongqi Sun.

some photocatalysts such as TiO_2-based [10, 11], CdS-based [8], and carbon nitride (g-C_3N_4)-based [1, 6] materials were reported. In 2013, Shiraishi et al. [11] reported selective formation of H_2O_2 by TiO_2 photocatalyst with benzylic alcohols and O_2 in water, and in 2014, Choi and coworkers [10] reported the production of H_2O_2 on reduced graphene oxide-TiO_2 hybrid photocatalysts consisting of earth-abundant elements. However, wide band gap of TiO_2 has limited its application under solar light irradiation, as only 5% of solar energy is composed of UV light to generate carriers in TiO_2. In 2016, Kim et al. [8] reported the harnessing low energy photons (635 nm) for the production of H_2O_2 using upconversion CdS photocatalyst. Unfortunately, their system is complex, and the photocatalytic activity was still low, just producing only micromolar levels of H_2O_2. Then, Shiraishi and coworkers [1] in 2016 reported carbon nitride–aromatic diimide–graphene nanohybrids for H_2O_2 production from pure water and O_2. In 2018, Zhu and coworkers [6] reported efficient visible-light-driven selective O_2 reduction to H_2O_2 by oxygen-enriched g-C_3N_4 polymers. Nonetheless, their catalytic efficiencies were not enough, which are suffered from the low surface area of g-C_3N_4. Therefore, it is necessary to develop some new materials to meet these deficiencies.

Metal–organic frameworks (MOFs), hybrid porous materials consist of organic linkers and metal oxide clusters, have been widely employed in photocatalytic H_2 evolution [12], CO_2 reduction [13], dye degradation [14], and organic transformations [15] due to their advantages such as large surface area, good chemical stability, tailorable structural diversity, etc. [16] Moreover, some MOFs have the characteristic of visible-light response originating from their amine-functionalized linkers [17], which can effectively improve the utilization rate of visible light. Therefore, it is very promising to apply MOFs into the field of photocatalytic H_2O_2 production.

MIL-125-NH_2 composed of $Ti_8O_8(OH)_4$ clusters and 2-aminoterephthalic acid linkers is one of the amino-functionalized MOFs often used in photocatalysis [18, 19]. Specially, when MIL-125-NH_2 is irradiated under visible light, the photogenerated electrons originating from linkers reduce Ti^{IV} in the $Ti_8O_8(OH)_4$ clusters to Ti^{III} through linker-to-cluster charge transfer (LCCT) mechanism, and the photogenerated holes are located in the organic linker [19]. On basis of this, photocatalytic H_2O_2 production through visible-light-excited O_2 reduction coupled with oxidation reactions was designed (Scheme 12.1). The formed Ti^{III} is expected to reduce O_2 to super oxide radical ($O_2{\cdot}^-$), which forms H_2O_2, while another oxidation product is produced by photogenerated holes.

12.2 Photocatalytic H_2O_2 Production Through Selective Two-Electron Reduction of O_2 Utilizing NiO/MIL-125-NH_2

MIL-125-NH_2 was synthesized and deposited with NiO according to a reported method to obtain NiO/MIL-125-NH_2 [20]. Ultraviolet visible diffuse reflectance spectra (UV-vis DRS) of the as-synthesized MIL-125-NH_2 and NiO/MIL-125-NH_2 in Figure 12.1a showed only insignificant differences, demonstrating that MIL-125-NH_2 sustained its original optical absorption characteristics after the

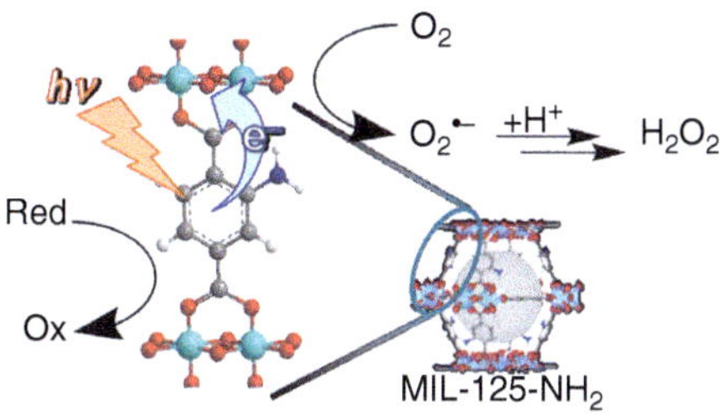

Scheme 12.1 Photocatalytic H_2O_2 production through O_2 reduction coupled with oxidation reactions under photoirradiation. Source: Isaka et al. [24]. © 2018, Royal Society of Chemistry.

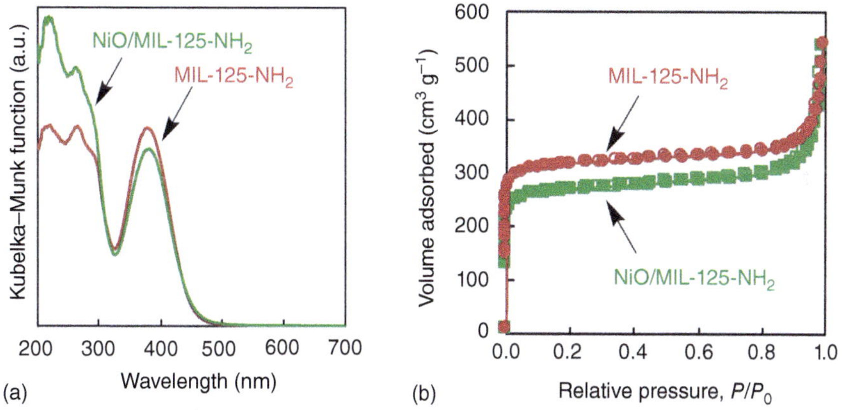

Figure 12.1 (a) UV–vis DRS and (b) N_2 adsorption isotherms at 77 K for MIL-125-NH$_2$ and NiO/MIL-125-NH$_2$. Source: Isaka et al. [24]. © 2018, Royal Society of Chemistry.

deposition of NiO nanoparticles. The absorption bands in the high energy region ($\lambda < 320$ nm) is ascribed to the $\pi \rightarrow \pi^*$ transition of the linkers, while the absorption band extending to the visible region from 320 to 500 nm originates from the LCCT of amino-functional groups (–NH$_2$) on the organic linkers [21]. Those results indicate that both MIL-125-NH$_2$ and NiO/MIL-125-NH$_2$ have visible-light-responsive property, which are helpful for the photochemical reaction excited by visible light.

Besides, the deposition of NiO nanoparticle resulted in a slight decrease (from 1498 m^2 g^{-1} for MIL-125-NH$_2$ to 1192 m^2 g^{-1} for NiO/MIL-125-NH$_2$) in Brunauer–Emmett–Teller surface area (S_{BET}) in Figure 12.1b.

Fourier transform extended X-ray absorption fine structure (FT-EXAFS) spectra (Figure 12.2) and energy dispersive X-ray (EDX) mappings (Figure 12.3c-f) indicated that the valence state of Ni element in deposited NiO nanoparticles for NiO/MIL-125-NH$_2$ was divalent [22]. The content of Ni species in NiO/MIL-125-NH$_2$ was determined to be 0.77 wt%. Transmission electron microscope (TEM) and high-angle annular dark field-scanning transmission electron microscope (HAADF-STEM) images of NiO/MIL-125-NH$_2$ in Figure 12.3a and b showed the deposition of the NiO nanoparticles with diameters about 7 nm.

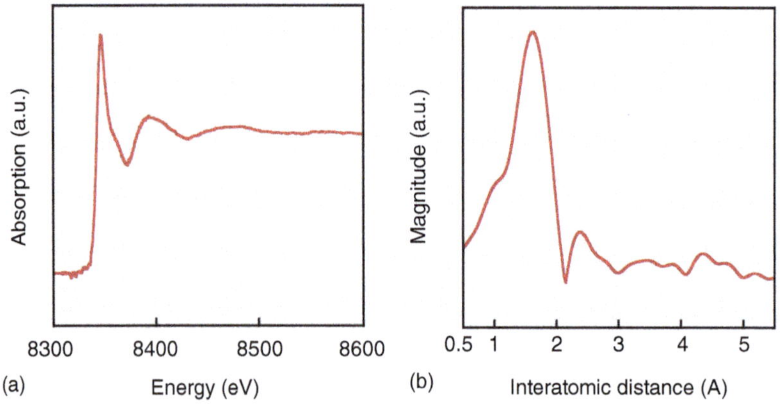

Figure 12.2 Ni K-edge (a) XAFS and (b) FT-EXAFS spectra of NiO/MIL-125-NH_2. Source: Isaka et al. [24]. © 2018, Royal Society of Chemistry.

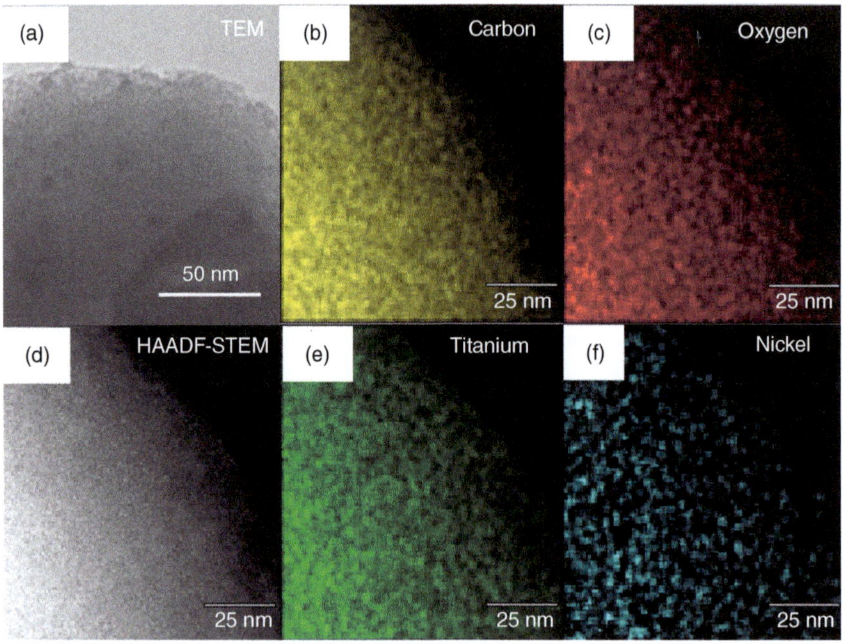

Figure 12.3 (a) TEM image, (b) HAADF-STEM image, and (c-f) EDX mappings of NiO/MIL-125-NH_2. Source: Reproduced with permission from Isaka et al. [24].

Photocatalytic H_2O_2 production was carried out when MIL-125-NH_2 dispersed in an O_2-saturated acetonitrile solution using triethanolamine (TEOA) as the electron donor under visible-light ($\lambda > 420$ nm) irradiation (Figure 12.4). No H_2O_2 production was observed in the absence of light irradiation or TEOA (Figure 12.4a). Both the rate and activity of H_2O_2 production were enhanced when NiO/MIL-125-NH_2 or Pt/MIL-125-NH_2 was used as the photocatalyst as compared with that of

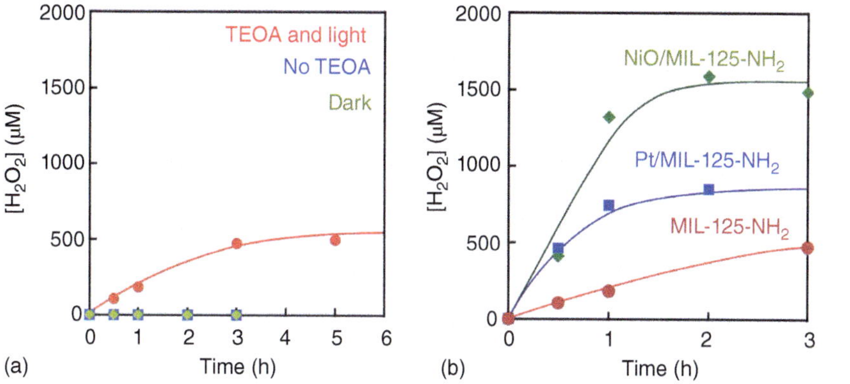

Figure 12.4 (a) Time course of photocatalytic H_2O_2 production catalyzed by MIL-125-NH_2 in the presence and absence of TEOA and visible-light irradiation ($\lambda > 420$ nm). (b) Time courses of photocatalytic H_2O_2 production catalyzed by MIL-125-NH_2, NiO/MIL-125-NH_2, and Pt/MIL-125-NH_2. Source: (a) Isaka et al. [24]. © 2018, Royal Society of Chemistry.

the MIL-125-NH_2 (Figure 12.4b). Moreover, the activity of H_2O_2 production for NiO/MIL-125-NH_2 was even higher than that for Pt/MIL-125-NH_2.

Oxidation product of TEOA was determined by GC-MS analysis of the reaction solution (Figure 12.5). A fragment with $m/z = 116$ was detected from the reaction solution. Some reported oxidation products of TEOA in literatures were shown in Figure 12.5c,d [23]. Aldehyde and successive hydration are the oxidation products of alcohol moiety in TEOA. In this work, complete oxidation of an alcohol moiety in TEOA to carboxylic acid and hydration is expected to form the molecule shown in Figure 12.5e. This molecule is suggested to be the oxidation product of TEOA in the solution because it can coincidentally form a fragment with $m/z = 116$ after deprotonation:

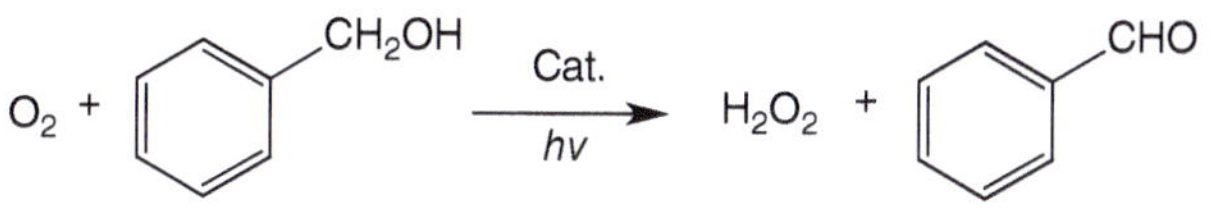

(12.1)

However, when TEOA was used as the electron donor, its oxidation product is not beneficial for analysis of the chemical stoichiometric relationship between the produced H_2O_2 and electron donor due to the multiple alcohol moieties in TEOA. Thus, photocatalytic H_2O_2 production through visible-light-excited O_2 reduction coupled with benzyl alcohol oxidation was performed (Eq. (12.1)). An observable activity for pure MIL-125-NH_2 is shown in Figure 12.6a, indicating that this reaction (Eq. (12.1)) can be effectively driven by MIL-125-NH_2 in an O_2-saturated acetonitrile solution (Figure 12.6a) [24]. Furthermore, the catalytic activity could be greatly enhanced by deposition of NiO nanoparticles onto the MOF (Figure 12.6b). It is worth noting that the concentration of H_2O_2 produced was only 5.7% of the produced benzaldehyde after 8 h of photoirradiation when the reaction was catalyzed by MIL-125-NH_2. Then, the yield surprisingly reached 93% when it was catalyzed by NiO/MIL-125-NH_2. On the other hand, the formation rate

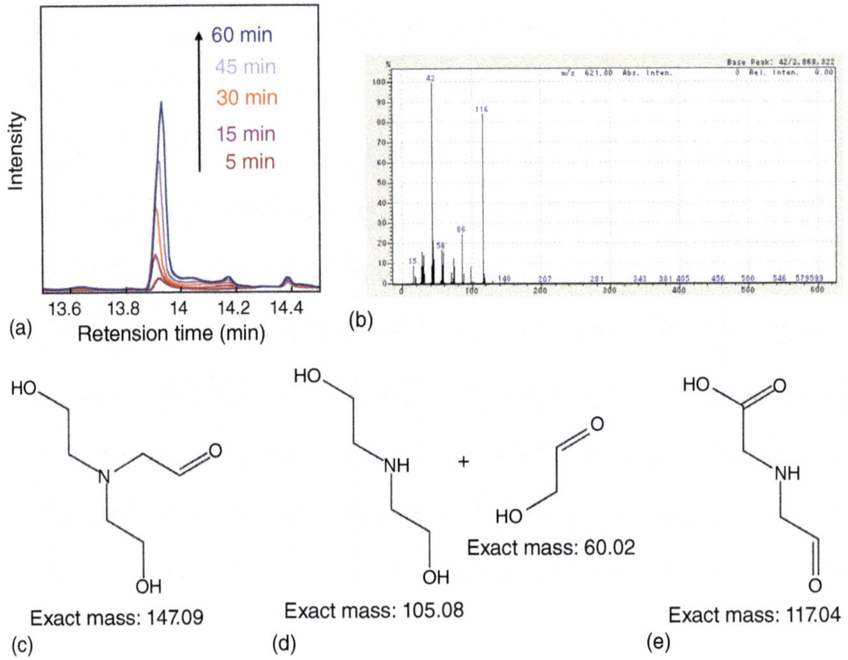

Figure 12.5 (a) Time course of TEOA oxidation product formation under visible-light ($\lambda > 420$ nm) irradiation for MIL-125-NH_2. (b) Mass spectrum of the peak after photoirradiation. (c,d) Reported oxidation products of TEOA in literatures. (e) Plausible product of TEOA oxidation. Source: Isaka et al. [24]. © 2018, Royal Society of Chemistry.

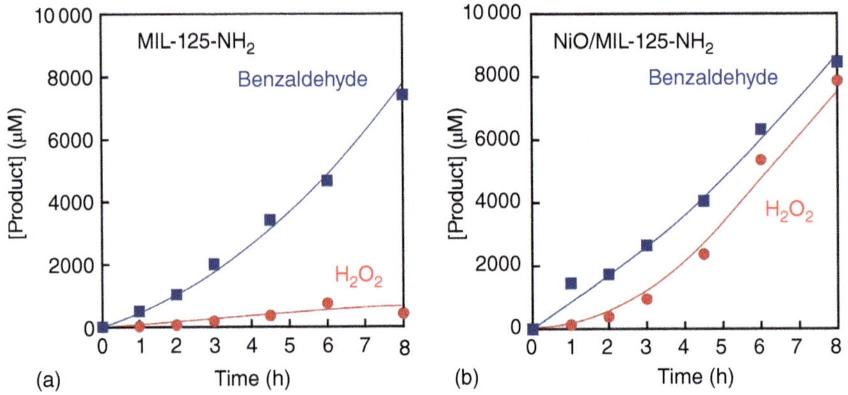

Figure 12.6 Time courses of H_2O_2 and benzaldehyde production of (a) MIL-125-NH_2 and (b) NiO/MIL-125-NH_2 dispersed in an acetonitrile solution of benzyl alcohol under visible-light irradiation ($\lambda > 420$ nm). Source: Isaka et al. [24]. © 2018, Royal Society of Chemistry.

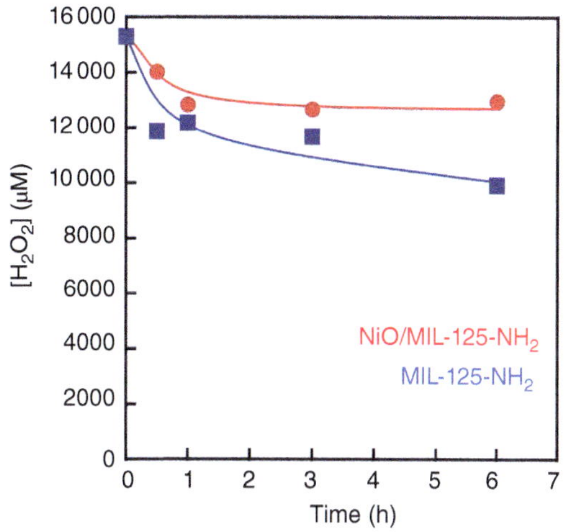

Figure 12.7 Time courses of H_2O_2 (15 mM) decomposition dissolved in 5.0 ml of acetonitrile suspension of 5.0 mg of MIL-125-NH_2 and NiO/MIL-125-NH_2 at 313 K. Source: Isaka et al. [24]. © 2018, Royal Society of Chemistry.

of benzaldehyde remained similar, which means the remarkable increase in H_2O_2 production for NiO/MIL-125-NH_2 was not due to an accelerated reaction rate. The enhancement of selectivity for the two-electron reduction of O_2 to H_2O_2 needs to be considered by inhibiting the decomposition of the produced H_2O_2.

Two-electron reduction of H_2O_2 to H_2O and disproportionation of H_2O_2 are considered as two possible pathways for the decomposition of H_2O_2. The decomposition rate of H_2O_2 for NiO/MIL-125-NH_2 measured at high H_2O_2 concentration was about half that for MIL-125-NH_2 (Figure 12.7) However, the drop in the decomposition of H_2O_2 through disproportionation is not sufficient to explain the observed remarkable enhancement of selectivity. Therefore, suppression of the two-electron reduction of H_2O_2 by NiO was more reasonable to explain the enhanced selectivity.

To further verify this point of selectivity for the two-electron reduction of O_2, electron paramagnetic resonance (EPR) experiments were performed to detect the $O_2{\cdot}^-$ species in the reaction using 5,5-dimethyl-1-pyrroline N-oxide (DMPO) as a capture agent. The results assuredly confirmed the formation of the DMPO–$O_2{\cdot}^-$ under visible-light irradiation catalyzed by MIL-125-NH_2 (Figure 12.8a) [25]. It should be noted that the formation rate of the DMPO–$O_2{\cdot}^-$ was significantly decreased when it was catalyzed by NiO/MIL-125-NH_2 (Figure 12.8b). These results suggest that O_2 rapidly formed H_2O_2 in the presence of deposited NiO, which is consistent with some previous reports that Ni species is beneficial for promoting the decomposition of intermediate species for H_2O_2 production [26]. Faster decomposition of $O_2{\cdot}^-$ by NiO/MIL-125-NH_2 is reasonable to explain the high H_2O_2 selectivity.

The photocatalytic H_2O_2 production from O_2 with benzyl alcohol as an electron donor catalyzed by NiO/MIL-125-NH_2 is proposed in Scheme 12.2. When MIL-125-NH_2 is irradiated under visible light, the photogenerated electrons originating from linkers reduce Ti^{IV} in the $Ti_8O_8(OH)_4$ clusters to Ti^{III} through LCCT process, resulting in the formation of $Ti_8O_8(OH)_4{\cdot}^-$. Meanwhile, the photogenerated holes are formed in the organic linkers, which are reduced by benzyl alcohol

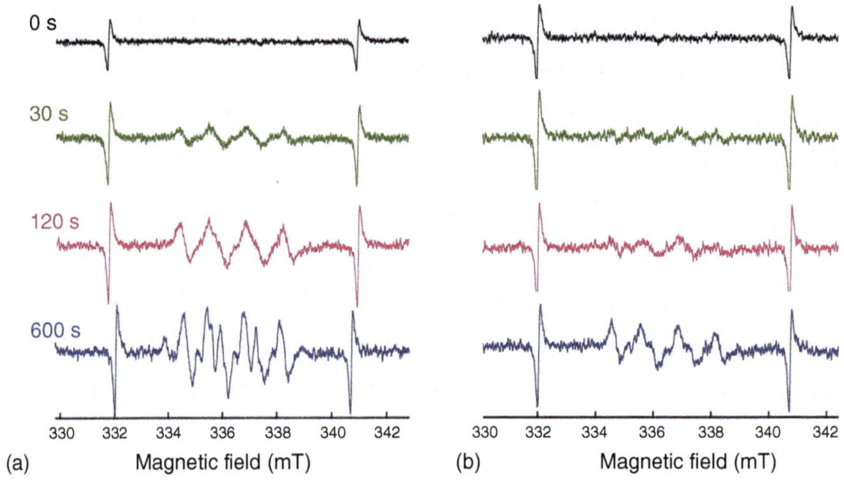

Figure 12.8 EPR spectral of suspensions containing DMPO and (a) MIL-125-NH_2 or (b) NiO/MIL-125-NH_2 under visible-light irradiation ($\lambda > 420$ nm). Source: (a) Isaka et al. [24]. © 2018, Royal Society of Chemistry.

followed by O_2 reduction by $Ti_8O_8(OH)_4{\cdot}^-$ to form $O_2{\cdot}^-$. Eventually, $O_2{\cdot}^-$ forms H_2O_2 through a disproportionation process, which can be accelerated by NiO.

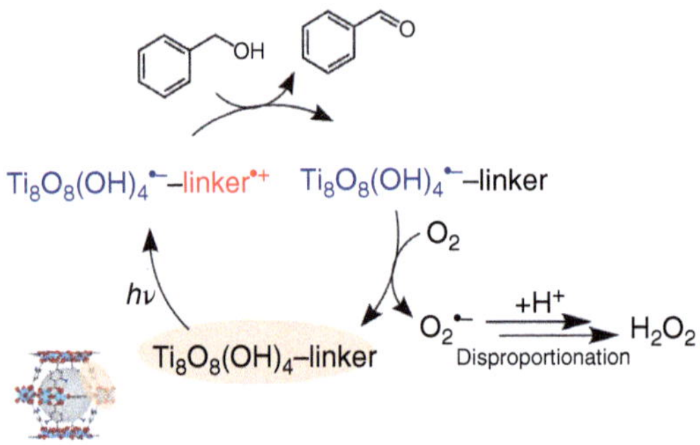

Scheme 12.2 Reaction mechanism of photocatalytic H_2O_2 production catalyzed by MIL-125-NH_2. Source: Isaka et al. [24]. © 2018, Royal Society of Chemistry.

12.3 Two-Phase System Utilizing Linker-Alkylated Hydrophobic MIL-125-NH_2 for Photocatalytic H_2O_2 Production

While NiO/MIL-125-NH_2 could effectively catalyze the production of H_2O_2 in the single-phase system in the previous work, its activity was not maintained when the used NiO/MIL-125-NH_2 was centrifugally collected for the recycling tests. This is plausibly resulted from the unstable structure of the MIL-125-NH_2 as generally reported for Ti-based MOFs [27]. On the other hand, the mixture of H_2O_2 and

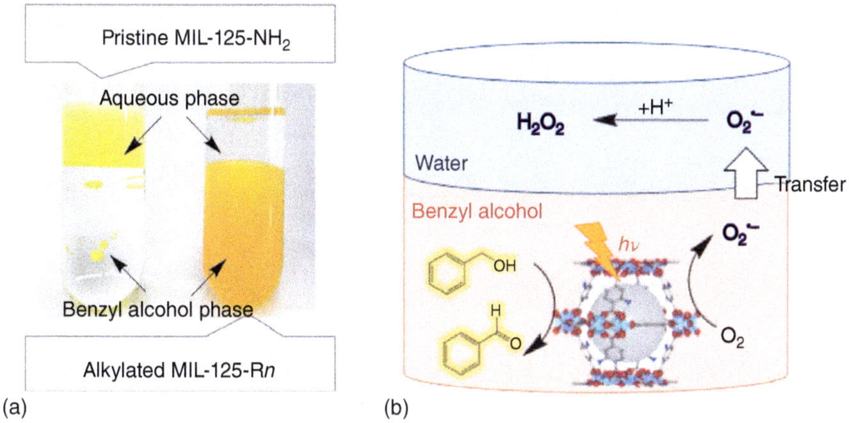

Figure 12.9 (a) Digital photographs of two-phase systems composed of an aqueous phase and a benzyl alcohol phase containing MIL-125-NH_2 (left) and MIL-125-R*n* ($n = 4$ and 7, right). (b) Photocatalytic H_2O_2 production in the two-phase system. Source: Isaka et al. [28]. © 2019, John Wiley & Sons.

catalysts in the single-phase system can not only result in the uneasy collection of the formed H_2O_2 but also lead to its catalytic decomposition by the catalysts. Thus, extra separation process is needed to separate the H_2O_2 and MOFs.

To solve the mentioned problems caused by the mixing of H_2O_2 and MOFs, photocatalytic H_2O_2 production in a two-phase system composed of benzyl alcohol and water (BA/water) (Figure 12.9) was designed in this part [28]. This two-phase system realized spontaneous separation of the MOFs in the BA phase (Figure 12.9a, right) and the H_2O_2 in the aqueous phase through the hydrophobic treatment of MIL-125-NH_2.

Post-synthetic modification (PSM) was performed to modify the MOFs with alkyl chains (Scheme 12.3). The linker-alkylated MOFs are denoted as MIL-125-R*n* ($n = 4$ and 7, it is the carbon number in the alkyl chain, representing the length of the alkyl chain).

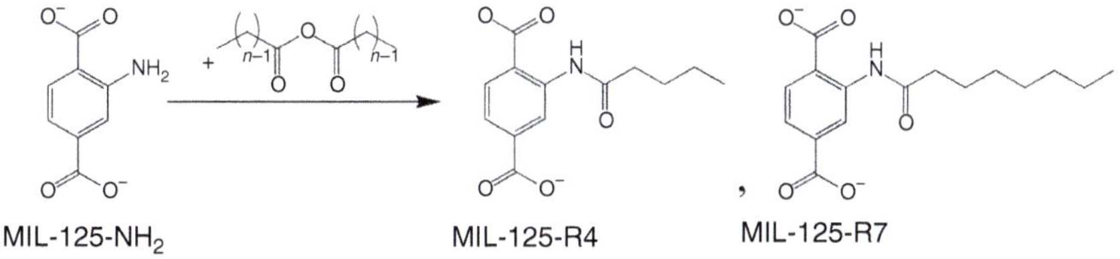

Scheme 12.3 Alkylation of linkers of MIL-125-NH_2 to form MIL-125-R*n* ($n = 4$ and 7). Source: Isaka et al. [28]. © 2019, John Wiley & Sons.

The X-ray diffraction (XRD) patterns of the as-synthesized samples in Figure 12.10a showed that the linker-alkylated MIL-125-R*n* ($n = 4$ and 7) totally maintained the original crystal structure of the pristine MIL-125-NH_2. Then, MIL-125-R*n* ($n = 4$ and 7) were further characterized by the UV–vis DRS

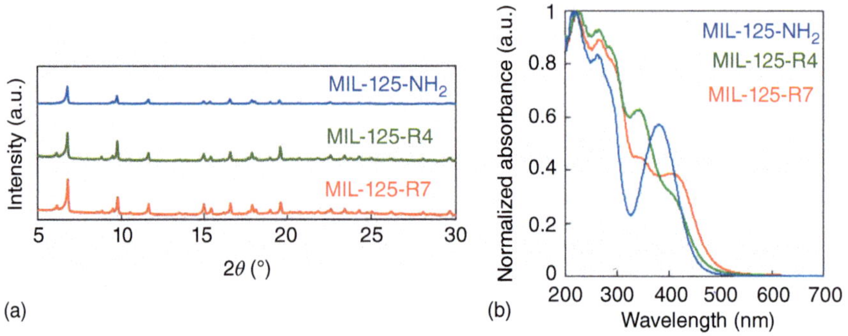

Figure 12.10 (a) XRD patterns and (b) UV–vis DRS for MIL-125-NH_2, MIL-125-R4, and MIL-125-R7. Source: Isaka et al. [28]. © 2019, John Wiley & Sons.

in Figure 12.10b. The absorption peak at around λ = 390 nm that originates from the LCCT of the –NH_2 in MIL-125-NH_2 was slightly weaken in MIL-125-R*n* (n = 4 and 7), while new peaks appeared at around λ = 350 nm. Since highest occupied molecular orbital (HOMO) level of the alkylated linkers is beneficial for the LCCT process, this partial blueshift in absorbance is possibly ascribed to the linker alkylation in the MIL-125-R*n* (n = 4 and 7), which leads to lowered HOMO levels. Fortunately, overall absorption in the visible region still maintained even after the linker alkylation.

Field emission scanning electron microscope (FE-SEM) (Figure 12.11a,b) and TEM (Figure 12.12a,b) images of MIL-125-NH_2 and MIL-125-R7 demonstrated that the morphology of cuboid shaped MIL-125-NH_2 nanoparticles was almost no changed after linker alkylation.

Furthermore, thermogravimetric analysis–differential thermal analysis (TG-DTA) measurements in Figure 12.13a–f were performed to calculate the amount of alkylated linker in MIL-125-R*n* (n = 4 and 7).

For MIL-125-NH_2, the unit cell of MIL-125-NH_2 is $Ti_8O_8(OH)_4$-L_6, in which L represents 2-aminoterephthalic acid linker (Figure 12.13g). After combustion, this unit cell will theoretically yield 8TiO_2. The theoretical value of the weight of the 6L (1074 g mol^{-1}) to 8TiO_2 (638.9 g mol^{-1}) is calculated to be 1.682. In Figure 12.13a, the endothermal lost weights showed in a-1 and a-2 were ascribed to the H_2O and residual DMF molecules from synthesis process, respectively. Exothermal lost weights in a-3 and a-4 were attributed to the combustion of organic linkers and residual weight of the produced TiO_2, respectively. The experimental ratio of lost weights due to combustion of organic linkers (41.8%) to 8TiO_2 (25.8%) was determined to be 1.62, which is similar with the expected theoretical value (1.682).

For MIL-125-R*n*, its unit cell can be expressed as $Ti_8O_8(OH)_4$-L_{6-n}-R_n, in which R is the alkylated linker and n is an average number of linkers that are alkylated per unit cell. Its structure is showed in Figure 12.13h,i for MIL-125-R4 and MIL-125-R7, respectively. When combusted, 1 mol of this unit cell loses $[179.1*(6-n)+Mn]$ g (M is molecular weight of an alkylated linker and equals to 263.3 for MIL-125-R4 and 305.3 for MIL-125-R7) due to combustion of organic linkers and leaves 638.9 g of residual 8TiO_2. The lost weights resulted from combustion of organic linkers of

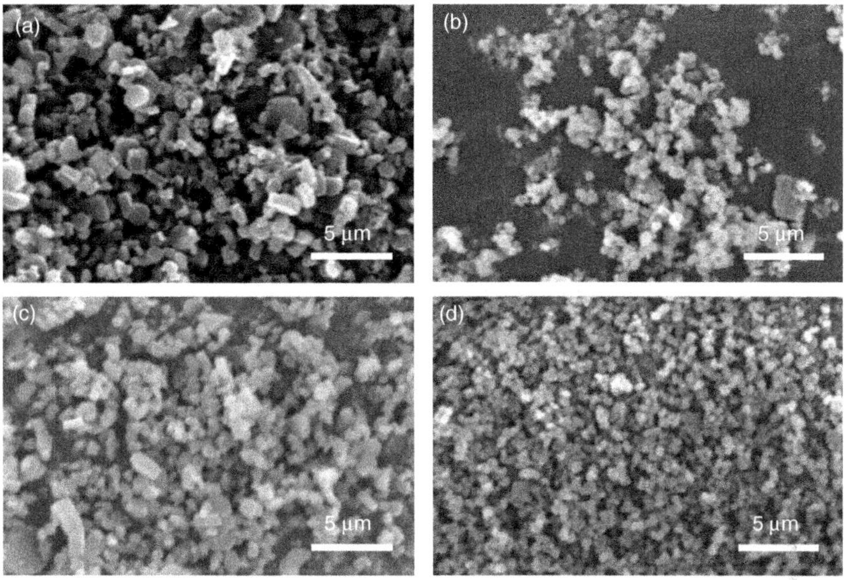

Figure 12.11 FE-SEM images of (a,c) MIL-125-NH_2 and (b,d) MIL-125-R7 (a,b) before and (c,d) after three hours of photoirradiation ($\lambda > 420$ nm) reaction in the two-phase system composed of benzyl alcohol (5.0 ml) and water (2.0 ml). Source: Reproduced with permission from Isaka et al. [28].

MIL-125-R4 and MIL-125-R7 are showed in c-2 (57.8%) and e-2 (63.5%), respectively. The weight of residual $8TiO_2$ of MIL-125-R4 and MIL-125-R7 is showed in c-3 (26.8%) and e-3 (26.6%), respectively. By comparing the ratios of these values with the theoretically calculated values, n can be calculated to be 3.63 for MIL-125-R4 and 3.56 for MIL-125-R7. Thus, the amount of alkylated linker in terephthalic linker can be calculated to be 61% in MIL-125-R4 and 59% in MIL-125-R7, respectively.

Hydrophobicity of MIL-125-NH_2 and MIL-125-Rn ($n = 4$ and 7) was examined by water adsorption (Figure 12.14a). For all samples, the amount of adsorbed water linearly increased at low pressure. However, steep increase of the water adsorption at $P/P_0 > 0.2$ due to capillary condensation was observed only for MIL-125-NH_2, while it was not observed in the case of linker-alkylated MIL-125-Rn (n = 4 and 7) [29]. The total amount of the adsorbed water decreased by 63% for MIL-125-R4 [267 cm^3 (STP) g^{-1}] and 72% for MIL-125-R7 [206 cm^3 (STP) g^{-1}], respectively, as compared with that of the pristine MIL-125-NH_2 [726 cm^3 (STP) g^{-1}]. To further examine the hydrophobic nature of the MOFs, water contact angles were measured for MIL-125-NH_2, MIL-125-R4, and MIL-125-R7 (Figure 12.14b–d). The water contact angle for MIL-125-NH_2 was 30°. The MOFs showed hydrophobic property after linker alkylation with contact angles exceeding 90° (101° for MIL-125-R4 and 124° for MIL-125-R7, respectively). These results indicate the hydrophobic nature of MIL-125-Rn ($n = 4$ and 7).

When MIL-125-NH_2 was added into two-phase system (BA/water), it dispersed selectively to the water phase (Figure 12.15a, left). H_2O_2 production was observed

Figure 12.12 TEM images of (a,c) MIL-125-NH_2 and (b,d) MIL-125-R7 (a,b) before and (c,d) after three hours of photoirradiation ($\lambda > 420$ nm) reaction in the two-phase system composed of benzyl alcohol (5.0 ml) and water (2.0 ml). Source: Reproduced with permission from Isaka et al. [28].

under visible-light irradiation ($\lambda > 420$ nm) (Figure 12.15b), which only appeared in aqueous phase.

When MIL-125-Rn ($n = 4$ and 7) was added into the two-phase system, it dispersed selectively in the BA phase (Figure 12.15a, right). Formation of H_2O_2 was observed (Figure 12.15b) when the system was irradiated under visible light ($\lambda > 420$ nm). Similar as MIL-125-NH_2, the H_2O_2 was only appeared in aqueous phase. It is worth noting that the rate and the total concentration of produced H_2O_2 were both obviously enhanced when MIL-125-Rn (n = 4 and 7) were utilized as compared with that of MIL-125-NH_2. Besides, the time courses of benzaldehyde formation were similar between MIL-125-NH_2 and linker-alkylated MIL-125-Rn (n = 4 and 7) in Figure 12.16 [28], meaning that their oxidation rates of benzyl alcohol were similar. Thus, what can be inferred is that rather selective H_2O_2 production has resulted in the observed increased activity. When H_2O_2 and photocatalyst both appears in the same phase (water), photocatalytic decomposition of H_2O_2 to OH^- and $\cdot OH$ by the catalyst happens, which finally resulted in the suppression of H_2O_2 production. Such phenomena can be prevented by spatial separation of the H_2O_2 and photocatalysts (hydrophobic MIL-125-Rn [n = 4 and 7]) in this two-phase system.

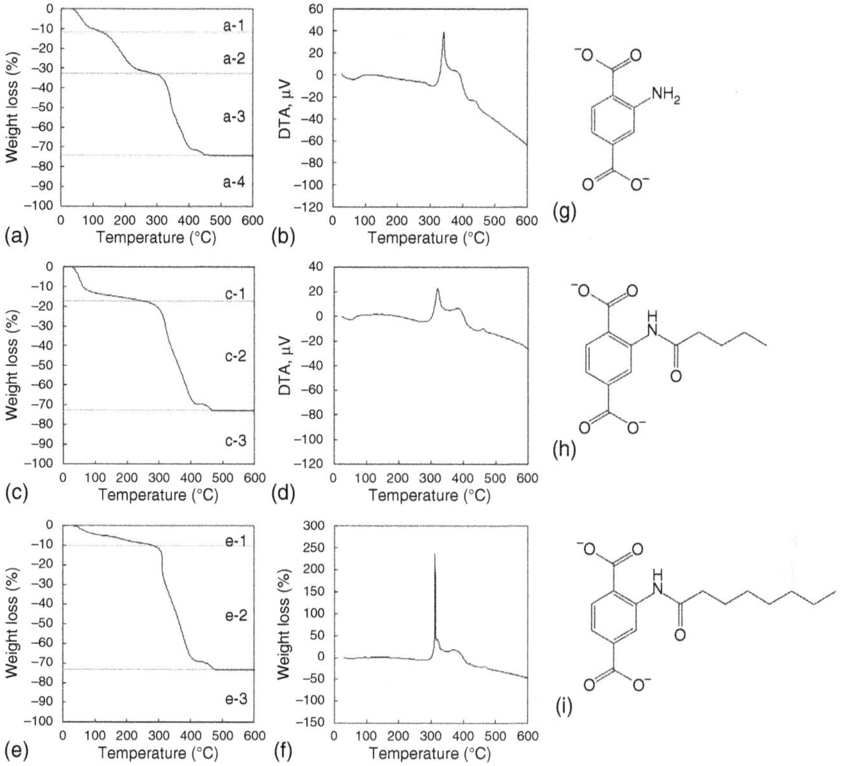

Figure 12.13 TG-DTA measurement of (a,b) MIL-125-NH_2, (c,d) MIL-125-R4, and (e,f) MIL-125-R7. Chemical structures of linkers in (g) MIL-125-NH_2, alkylated linker in (h) MIL-125-R4, and (i) MIL-125-R7. Source: Isaka et al. [28]. © 2019, John Wiley & Sons.

As shown in Figure 12.17, the XRD patterns of MIL-125-NH_2 and MIL-125-Rn (n = 4 and 7) after reaction indicate that the crystal structures of MIL-125-NH_2 and MIL-125-R7 were retained. Besides, the morphologies of the MIL-125-NH_2 and MIL-125-R7 were also hardly changed after reaction as shown in SEM and TEM images (Figures 12.11 and 12.12).

Recycling tests of H_2O_2 production catalyzed by MIL-125-NH_2 and MIL-125-R7 were performed in Figure 12.18 to further test the stability of the catalytic activities. Both MIL-125-NH_2 and MIL-125-R7 were active for at least three recycling tests. However, MIL-125-R7 still sustained relatively higher activity after three recycling tests (78%) as compared with MIL-125-NH_2 (63%), indicating the positive effect of hydrophobization on increasing the stability of the MOF.

Obtaining highly concentrated H_2O_2 solution is the key to efficient storage of the solar energy under high volumetric energy density. To avoid the energy consuming during the concentration process of H_2O_2, concentration of H_2O_2 at the production step is believed to be preferable; it is expected that reduced volume of the aqueous phase in the two-phase system will obtain such a concentrated H_2O_2 solution. When H_2O_2 production in various volumes of aqueous phase was performed, the activity of

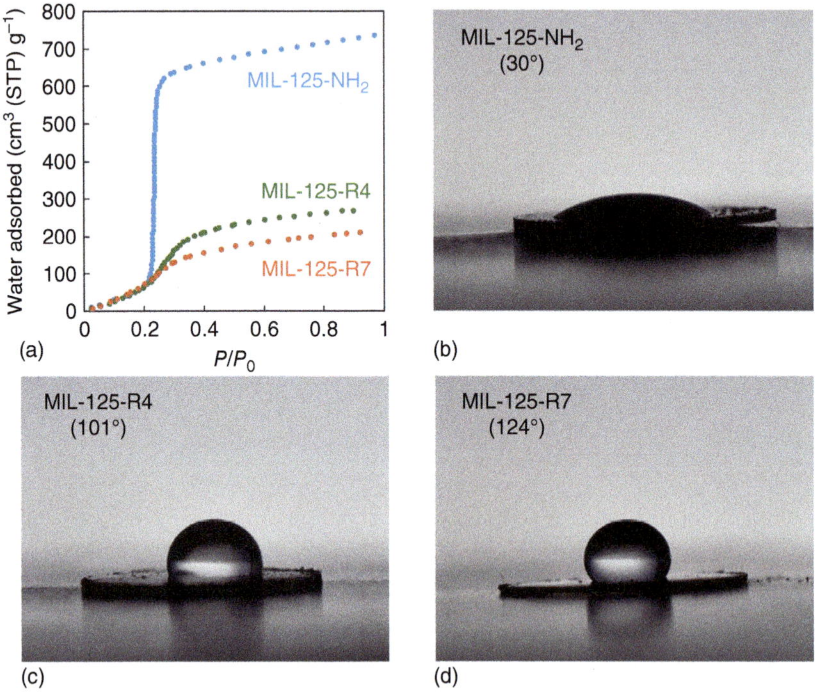

Figure 12.14 (a) Water adsorption isotherms at 298 K for MIL-125-NH_2, MIL-125-R4, and MIL-125-R7. The water contact angles of (b) MIL-125-NH_2, (c) MIL-125-R4, and (d) MIL-125-R7. Source: (a) Isaka et al. [28]. © 2019, John Wiley & Sons; (b–d) Reproduced with permission from Isaka et al. [28].

H_2O_2 was inversely proportional to the volume of aqueous phase after three hours of photoirradiation in Figure 12.19a. This result indicates that an effective concentration process of H_2O_2 at its production step is realized by application of the two-phase system.

The result that MIL-125-NH_2 produces H_2O_2 through disproportionation of $O_2{\cdot}^-$ has been studied in the single phase in Section 12.2 [24]. $O_2{\cdot}^-$ species can be stabilized under low pH conditions to accelerate the formation of H_2O_2 [9, 30]. However, it is not easy to use MIL-125-NH_2 in extremely low pH aqueous solutions because protonation of the 2-aminoterephthalate linkers leads to demetallation of the linker from $Ti_8O_8(OH)_4$ clusters, which will badly destroy the structure of MOF. Actually, the pure MIL-125-NH_2 was dissolved after a few seconds when it dispersed in aqueous solution at pH = 0.3. This limitation in the stability of MOF structure could be overcome by using the linker-alkylated MIL-125-R*n* (n = 4 and 7) because it only dispersed in the BA phase that is separated from low pH aqueous phase. In order to illustrate this point, photocatalytic H_2O_2 production in two-phase system was performed under different pH conditions (0.3, 1.3, 2.1, and 6.7) of the aqueous phase. The pH-dependent increase for activity of H_2O_2 production was shown in Figure 12.19b, in which positive correlation appeared between activity and pH value. This remarkably enhanced activity under low pH conditions of the aqueous

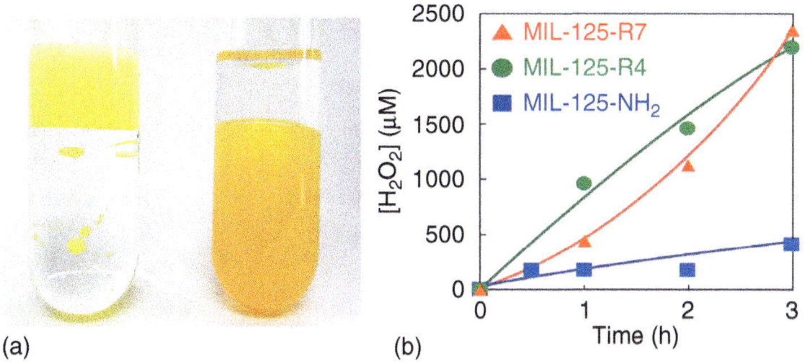

Figure 12.15 (a) Digital photograph of MIL-125-NH_2 (left) and MIL-125-R7 (right) dispersed into the two-phase system; aqueous phase (2.0 ml) was observed on top of the benzyl alcohol phase (5.0 ml). (b) Time courses of H_2O_2 production under photoirradiation ($\lambda > 420$ nm) in the two-phase system catalyzed by 5.0 mg of catalysts. Source: (a) Reproduced with permission from Isaka et al. [28]; (b) Isaka et al. [28]. © 2019, John Wiley & Sons.

Figure 12.16 Time courses of benzaldehyde formation under photoirradiation ($\lambda > 420$ nm) of the two-phase system composed of benzyl alcohol (5.0 ml) and water (2.0 ml) catalyzed by 5.0 mg of catalysts. Source: Isaka et al. [28]. © 2019, John Wiley & Sons.

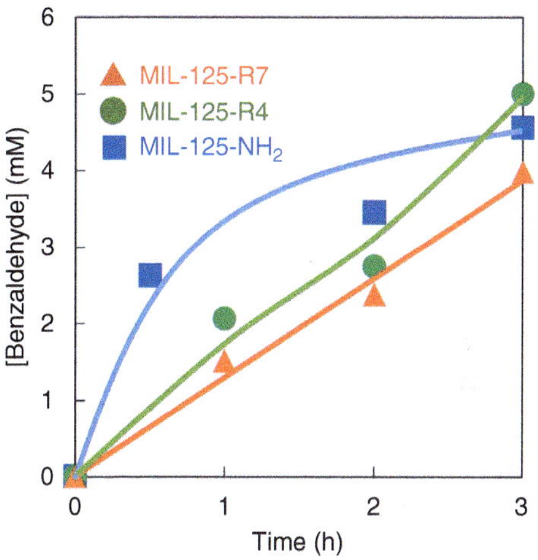

Figure 12.17 XRD patterns of MIL-125-NH_2 and MIL-125-R7 after three hours of photoirradiation ($\lambda > 420$ nm) reaction in the two-phase system composed of benzyl alcohol (5.0 ml) and water (2.0 ml). Source: Isaka et al. [28]. © 2019, John Wiley & Sons.

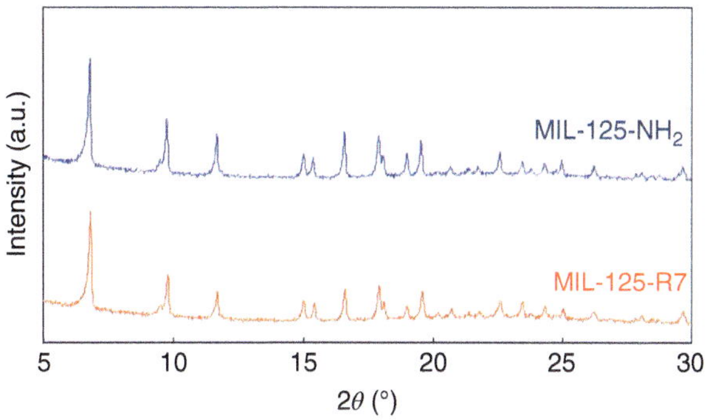

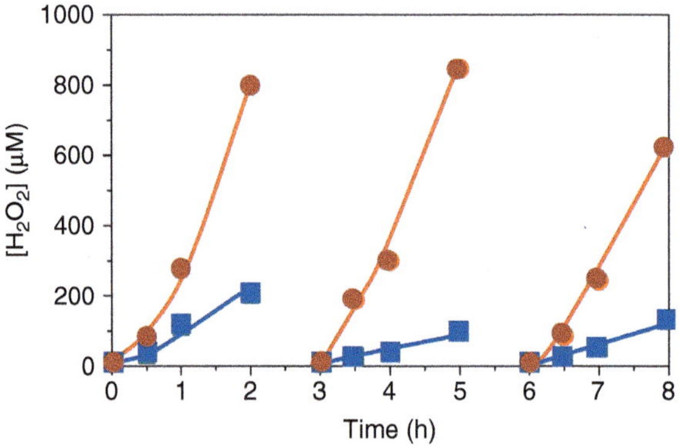

Figure 12.18 Recycling tests of H_2O_2 production under photoirradiation ($\lambda > 420$ nm) of the two-phase system composed of benzyl alcohol (10.0 ml) and water (4.0 ml) catalyzed by 10.0 mg of MIL-125-NH_2 (blue) and MIL-125-R7 (orange). Source: Isaka et al. [28]. © 2019, John Wiley & Sons.

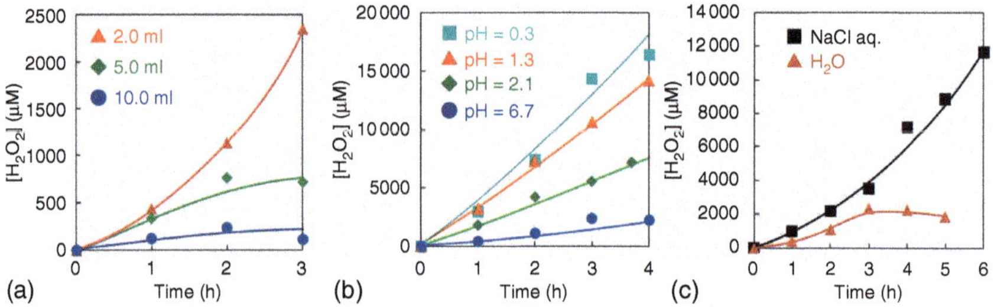

Figure 12.19 (a) Time courses of H_2O_2 production under photoirradiation ($\lambda > 420$ nm) of the two-phase system composed of benzyl alcohol (5.0 ml) and water (2.0, 5.0, and 10.0 ml) catalyzed by 5.0 mg of MIL-125-R7. (b) Time courses of H_2O_2 production under photoirradiation ($\lambda > 420$ nm) of the two-phase system composed of benzyl alcohol (5.0 ml) and water (2.0 ml) at different pH values (0.3, 1.3, 2.1, and 6.7) catalyzed by 5.0 mg of MIL-125-R7. (c) Time courses of H_2O_2 production under photoirradiation ($\lambda > 420$ nm) of the two-phase system composed of benzyl alcohol (5.0 ml) and an aqueous phase (2.0 ml, deionized water or saturated NaCl aqueous solution) catalyzed by 5.0 mg MIL-125-R7. Source: Isaka et al. [28]. © 2019, John Wiley & Sons.

phase indicates that $O_2{\cdot}^-$ formed in BA phase is transferred to aqueous phase where efficient disproportionation of $O_2{\cdot}^-$ to H_2O_2 occurs.

Furthermore, utilization of seawater instead of deionized water is thought to be more valuable because seawater is the largest source of water on the Earth [30]. Therefore, saturated NaCl aqueous solution was used as the aqueous phase. As shown in Figure 12.19c, the enhanced activity of H_2O_2 was obviously observed in the saturated NaCl aqueous phase. Stabilization of $O_2{\cdot}^-$ through its complexation with a Lewis acid (Na^+) inhibited reaction between $O_2{\cdot}^-$ and H_2O_2, leading to undesired decomposition of H_2O_2 to H_2O. Besides, the activity of H_2O_2 was continually enhanced within prolonged photoirradiation time for saturated NaCl aqueous phase as compared with the reaction system containing deionized water. It can be explained by the increased ionic strength of the aqueous phase resulting in the lower concentration of water in the BA phase.

Based on the above discussions, the reaction mechanism in the two-phase reaction system is proposed in Scheme 12.4. When the hydrophobic MIL-125-R*n* ($n = 4$ and 7)

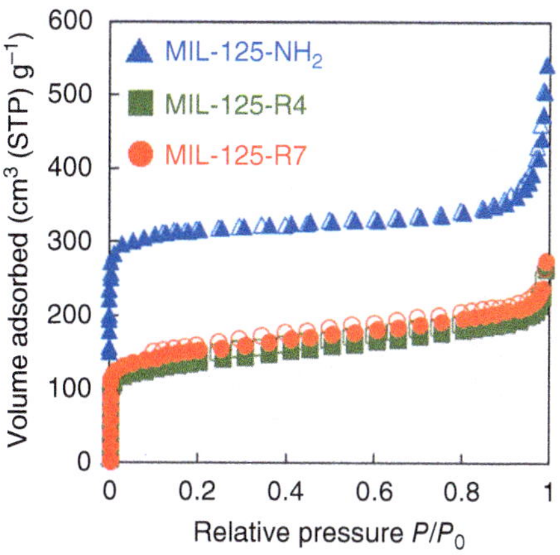

Figure 12.20 N_2 adsorption isotherms at 77 K for MIL-125-NH_2, MIL-125-R4, and MIL-125-R7. Source: Isaka et al. [28]. © 2019, John Wiley & Sons.

is irradiated under visible light, the photogenerated electrons originating from linkers reduce Ti^{IV} in the $Ti_8O_8(OH)_4$ clusters to Ti^{III} through LCCT process, resulting in the formation of $Ti_8O_8(OH)_4{\cdot}^-$. Meanwhile, the photogenerated holes are formed in the organic linkers, which are reduced by benzyl alcohol followed by O_2 reduction by $Ti_8O_8(OH)_4{\cdot}^-$ to form $O_2{\cdot}^-$. $O_2{\cdot}^-$ is considered to transfer to aqueous phase and then formed H_2O_2 through disproportionation process in aqueous phase, which was accelerated by H^+ or Na^+.

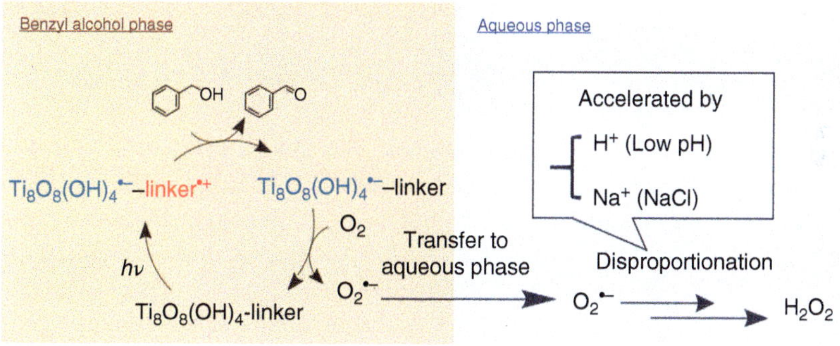

Scheme 12.4 Reaction mechanism of H_2O_2 production in the two-phase system. Source: Isaka et al. [28]. © 2019, John Wiley & Sons.

Although the photocatalytic activity of H_2O_2 production for MIL-125-R7 is four times higher than that for MIL-125-NH_2 (Figure 12.15b), the activity in this system was still not sufficient. We try to attribute this insufficiency to the reduced S_{BET} surface area. The surface area of MIL-125-R7 (560.7 $m^2\,g^{-1}$) was decreased dramatically, compared with that of MIL-125-NH_2 (1498 $m^2\,g^{-1}$; Figure 12.20, Table 12.1), which is resulted from smaller pore volume due to the linker alkylation (Figure 12.21a). It could be assumed that hydrophobization with larger pore volume would enhance photocatalytic activity of H_2O_2 production.

Table 12.1 Structure parameters of MIL-125-NH_2, MIL-125-R7, and OPA/MIL-125-NH_2.

Sample	S_{BET} ($m^2\ g^{-1}$)	V_p ($cm^3\ g^{-1}$)	D_p (nm)
MIL-125-NH_2	1498	0.58	0.57
MIL-125-R7	561	0.23	0.55
OPA/MIL-125-NH_2	1242	0.48	0.54
OPA/MIL-125-NH_2 (used)	1038	0.65	0.39

Source: Kawase et al. [31]. © 2019, Royal Society of Chemistry.

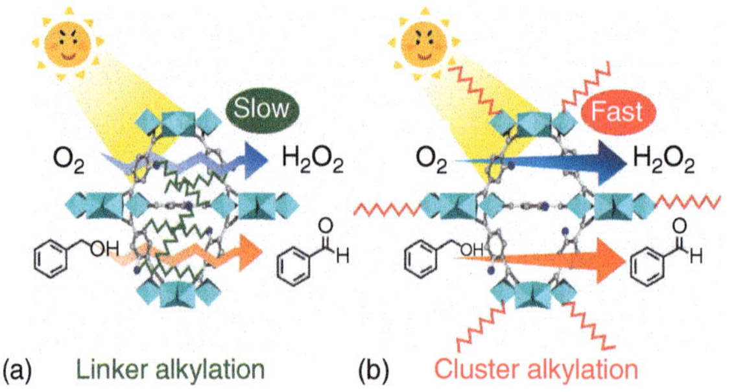

Figure 12.21 Photocatalytic H_2O_2 production with (a) linker-alkylated MOF: MIL-125-R7 and (b) cluster-alkylated MOF: OPA/MIL-125-NH_2. Source: Kawase et al. [31]. © 2019, Royal Society of Chemistry.

12.4 Ti Cluster-Alkylated Hydrophobic MIL-125-NH_2 for Photocatalytic H_2O_2 Production in Two-Phase System

In order to solve the problem of the decrease in surface area of MIL-125-NH_2 resulted from the linker alkylation, another hydrophobic MOF, OPA/MIL-125-NH_2, whose Ti cluster was alkylated with octadecylphosphonic acid (OPA), was designed [31]. For MIL-125-R7, the alkyl chains grafted onto the linkers filled its pores (Figure 12.21a). As for OPA/MIL-125-NH_2, the OPA only modified the $Ti_8O_8(OH)_4$ cluster with open coordination sites located on the outermost surface of MIL-125-NH_2 (Figure 12.21b). That means the outermost surface of MIL-125-NH_2 was selectively modified, and most of the pores could be maintained well. Thus, higher photocatalytic activity for H_2O_2 production is expected due to the faster diffusion of reactants and products from the maintained pores of the MOFs.

OPA/MIL-125-NH_2 was synthesized by cluster alkylation of MIL-125-NH_2 with OPA according to the process in Scheme 12.5 [32]. MIL-125-NH_2 was simply immersed in ethanol solution of OPA for 24 hours. Then, it was centrifuged and washed several times with ethanol to remove free OPA.

The XRD patterns in Figure 12.22a showed that the synthesized OPA/MIL-125-NH_2 still sustained the original crystal structure of MIL-125-NH_2. The obtained UV–vis DRS of OPA/MIL-125-NH_2 are shown in Figure 12.22b; the peak at $\lambda = 390$ nm derived from LCCT in MIL-125-NH_2 was still remained.

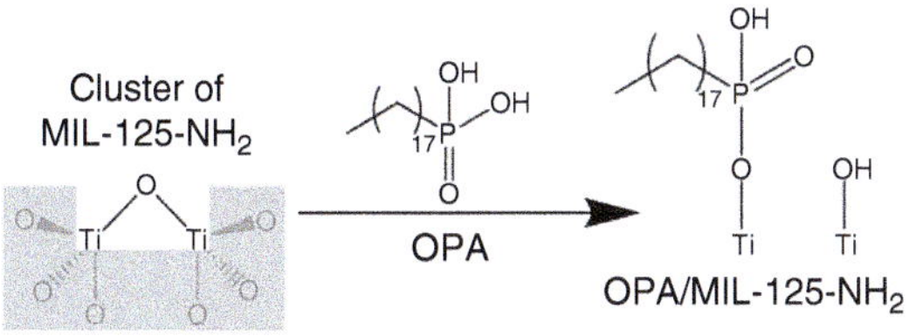

Scheme 12.5 Alkylation of a $Ti_8O_8(OH)_4$ cluster in MIL-125-NH_2 to form OPA/MIL-125-NH_2. Source: Kawase et al. [31]. © 2019, Royal Society of Chemistry.

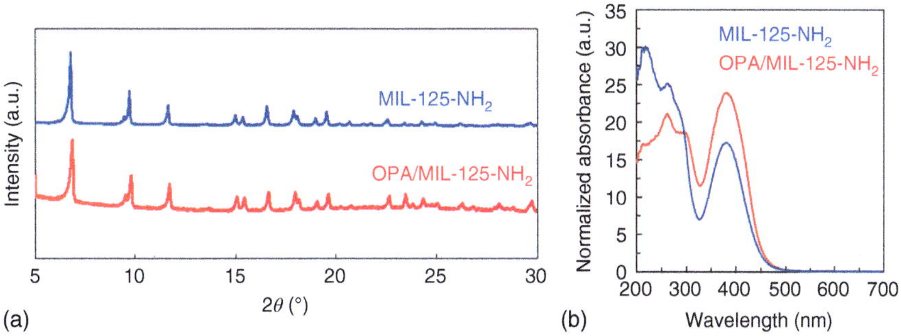

Figure 12.22 (a) XRD patterns and (b) UV–vis DRS of MIL-125-NH_2 and OPA/MIL-125-NH_2. Source: Kawase et al. [31]. © 2019, Royal Society of Chemistry.

Figure 12.23 N_2 adsorption isotherms at 77 K for MIL-125-NH_2 and OPA/MIL-125-NH_2. Source: Kawase et al. [31]. © 2019, Royal Society of Chemistry.

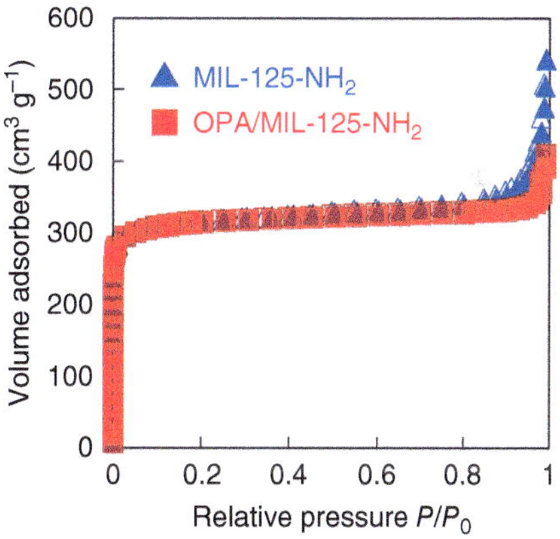

The S_{BET} of OPA/MIL-125-NH_2 was determined to be 1242 $m^2\,g^{-1}$ (Figure 12.23, Table 12.1). It revealed that most of the surface area was retained after cluster alkylation of MIL-125-NH_2, maintaining its original pore diameter (D_p) and volume (V_p) (Table 12.1).

TG-DTA measurements were carried out in Figure 12.24a,b to determine the amount of the alkylated Ti atoms in MIL-125-NH_2 by OPA.

For MIL-125-NH_2, corresponding calculation process was already showed in the part 12.3. The experimental ratio of lost weight resulted from combustion of organic

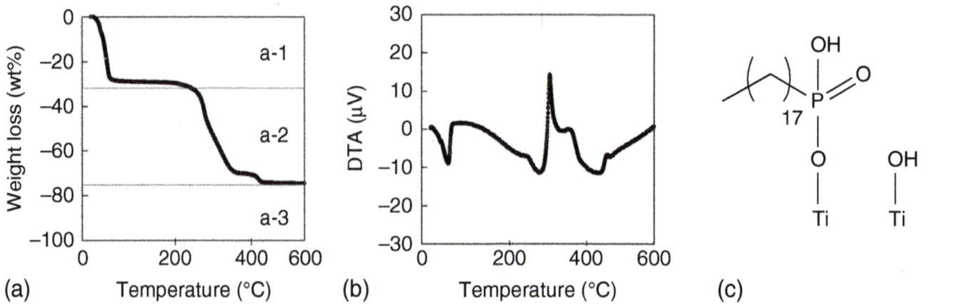

Figure 12.24 (a) TG and (b) DTA profiles of OPA/MIL-125-NH_2. (c) Chemical structures of alkylated clusters in OPA/MIL-125-NH_2. Source: Kawase et al. [31]. © 2019, Royal Society of Chemistry.

linkers was determined to be 1.62, which is similar with the expected theoretical value (1.682).

For OPA/MIL-125-NH_2, its unit cell can be expressed as $OPA_nTi_8O_8(OH)_4$-L_6, in which n is an average number of OPA that modify the clusters per unit cell. The structure of OPA in OPA/MIL-125-NH_2 is showed in Figure 12.24c. When combusted, assuming the monodentate species (Figure 12.24c), 1 mol of this unit cell loses [179.1*6 + 334.5*n] g due to combustion of organic linkers and OPA and leaves 638.9 g of residual 8TiO_2. The lost weight resulted from combustion of organic linkers and OPA of OPA/MIL-125-NH_2 is showed in a-2 (64.7%). The weight of residual 8TiO_2 of OPA/MIL-125-NH_2 is showed in a-3 (35.3%). Thus, n can be calculated to be 0.283. This indicates the amount of the alkylated Ti atoms in OPA/MIL-125-NH_2 is 3.54%.

This alkylation ratio of 3.54% demonstrates that only the outermost surface of MIL-125-NH_2 was modified by cluster alkylation. The ratio was much smaller than that observed for the linker alkylation of MIL-125-R7 (59%).

To further verify the success of cluster alkylation by OPA, Fourier transform infrared (FTIR) spectroscopy and X-ray photoelectron spectroscopy (XPS) of Ti 2p in MIL-125-NH_2 and OPA/MIL-125-NH_2 were obtained in Figures 12.25 and 12.26. Two peaks ascribe to $\nu(-CH_2-)$ band in OPA appeared at around 2900 cm^{-1}

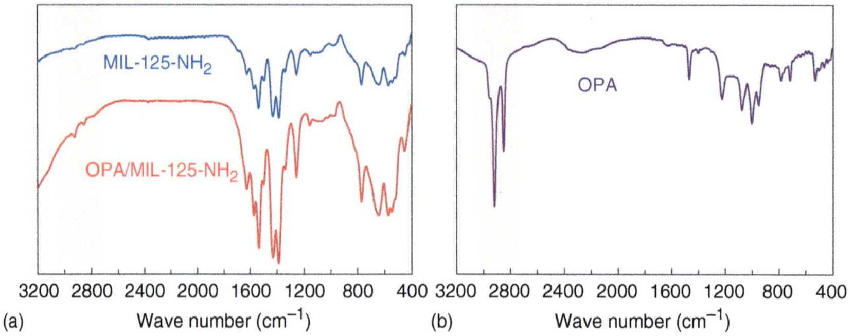

Figure 12.25 FTIR spectra of (a) MIL-125-NH_2, OPA/MIL-125-NH_2, and (b) OPA. Source: Kawase et al. [31]. © 2019, Royal Society of Chemistry.

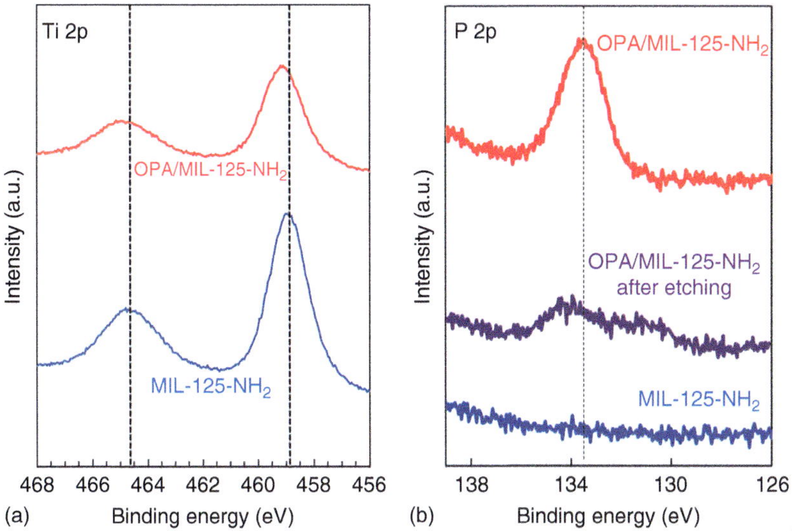

Figure 12.26 (a) XPS spectra of Ti 2p in MIL-125-NH_2 and OPA/MIL-125-NH_2. (b) XPS spectra of P 2p in MIL-125-NH_2 and OPA/MIL-125-NH_2 before and after etching. Source: Kawase et al. [31]. © 2019, Royal Society of Chemistry.

in the FTIR spectra of OPA/MIL-125-NH_2, while these peaks were not observed in the spectra of MIL-125-NH_2 (Figure 12.25a), indicating the presence of OPA in OPA/MIL-125-NH_2. Moreover, the binding energies of Ti 2p in OPA/MIL-125-NH_2 both show positive shifts of 0.3 eV, compared with those in pristine MIL-125-NH_2 as shown in Figure 12.26a. Those result indicates an interaction between OPA and $Ti_8O_8(OH)_4$ clusters, supporting the successful cluster alkylation by OPA [33]. To further confirm that cluster alkylation modified only the outermost surface of MIL-125-NH_2, XPS spectra of P 2p in OPA/MIL-125-NH_2 before and after etching were obtained in Figure 12.26b. A peak at around 133 eV assigned to P 2p was observed for OPA/MIL-125-NH_2 before etching, verifying the existence of OPA. However, this peak almost disappeared after etching. This result indicates that OPA molecules locate on only the outermost surface of OPA/MIL-125-NH_2. This is because only the $Ti_8O_8(OH)_4$ cluster that locates on the outermost surface has open coordination sites modified by OPA.

The water contact angle of 110° for OPA/MIL-125-NH_2 indicates that OPA/MIL-125-NH_2 has a hydrophobic nature in Figure 12.27a.

In the two-phase system of BA/water, OPA/MIL-125-NH_2 dispersed in the BA phase selectively, and H_2O_2 production was observed in the water phase under visible-light irradiation ($\lambda > 420$ nm). The time courses of H_2O_2 in Figure 12.27b show that OPA/MIL-125-NH_2 has three times higher activity of H_2O_2 than MIL-125-R7 after three hours of photoirradiation. The activity of BA using OPA/MIL-125-NH_2 was 1.7 times higher than that using MIL-125-R7 (Figure 12.28) because of the faster diffusion of BA passing unblocked pores in OPA/MIL-125-NH_2. These results confirm that cluster-alkylated OPA/MIL-125-NH_2 is more efficient for photocatalytic H_2O_2 production.

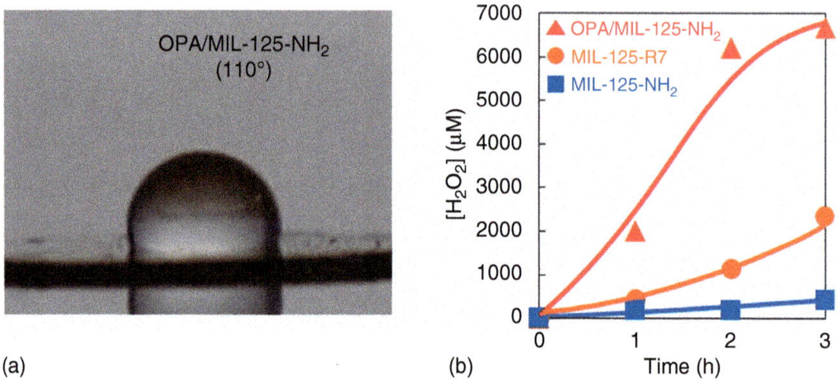

Figure 12.27 (a) Water contact angle of OPA/MIL-125-NH$_2$. (b) Time courses of H_2O_2 production of OPA/MIL-125-NH$_2$ under photoirradiation ($\lambda > 420$ nm) in the two-phase system. Source: (a) Kawase et al. [31]. © 2019, Royal Society of Chemistry; (b) Reproduced with permission from Kawase et al. [31].

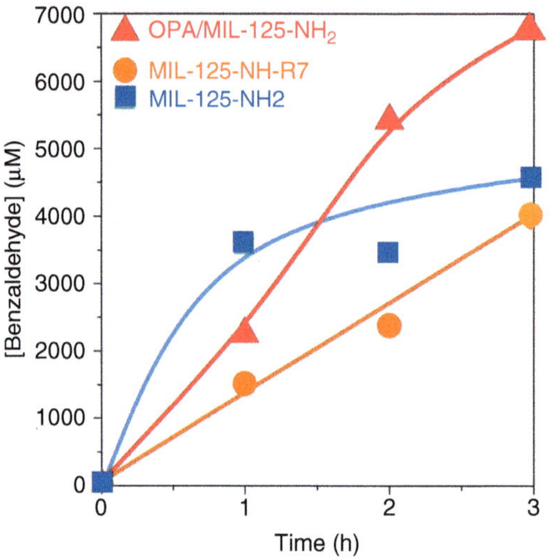

Figure 12.28 Time courses of benzaldehyde production of OPA/MIL-125-NH$_2$ under photoirradiation ($\lambda > 420$ nm) in the two-phase system. Source: Kawase et al. [31]. © 2019, Royal Society of Chemistry.

In order to exclude the possibility that the grafted OPA in OPA/MIL-125-NH$_2$ contributes to the improvement of activity, photocatalytic H_2O_2 production was carried out in a single-phase reaction system using MOFs (Figure 12.29). The activity of OPA/MIL-125-NH$_2$ was similar to that of MIL-125-NH$_2$, verifying that the enhanced activity is not caused by the grafted OPA. Therefore, the open pores in OPA/MIL-125-NH$_2$ were considered as the main reason of the enhancement in activity.

To evaluate the reusability of photocatalysts, recycling tests of MIL-125-NH$_2$, MIL-125-R7, and OPA/MIL-125-NH$_2$ were carried out in Figure 12.30. Unlike MIL-125-NH$_2$ and MIL-125-R7, the photocatalytic activity of OPA/MIL-125-NH$_2$ for H_2O_2 production was maintained even after three cycles.

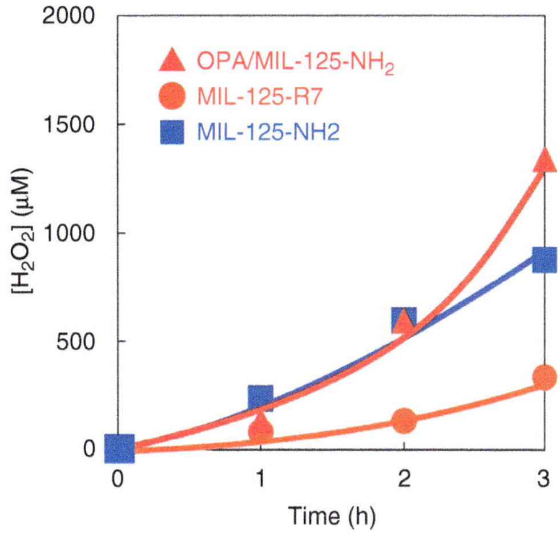

Figure 12.29 Time courses of H_2O_2 production under photoirradiation ($\lambda > 420$ nm) of a single-phase system composed of an acetonitrile solution (5.0 ml) of BA (1.0 ml) catalyzed by 5.0 mg of catalysts. Source: Kawase et al. [31]. © 2019, Royal Society of Chemistry.

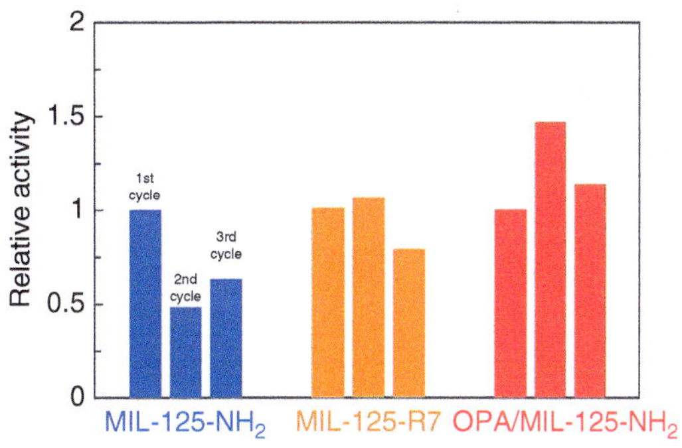

Figure 12.30 Recycling tests of MIL-125-NH_2, MIL-125-R7, and OPA/MIL-125-NH_2. Source: Kawase et al. [31]. © 2019, Royal Society of Chemistry.

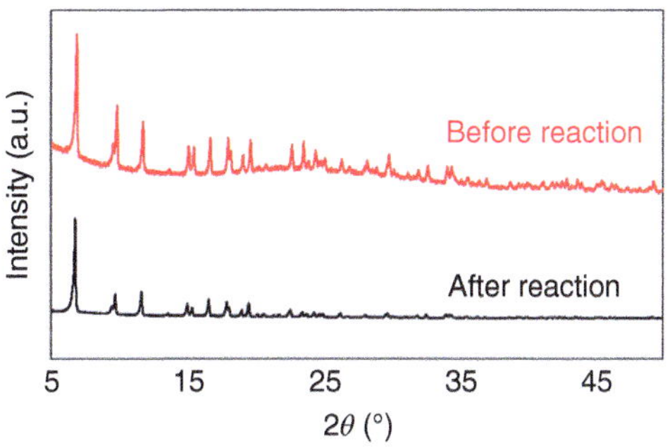

Figure 12.31 XRD patterns of OPA/MIL-125-NH_2 before and after reaction. Source: Kawase et al. [31]. © 2019, Royal Society of Chemistry.

According to the XRD patterns and S_{BET} of OPA/MIL-125-NH_2 after reaction in Figures 12.31 and 12.32 and Table 12.1, there is reason to believe that the crystal structure and surface area of OPA/MIL-125-NH_2 were retained. It can be assumed that the high stability of OPA/MIL-125-NH_2 in the two-phase system is conducive to the retention of the activity enhancement. Therefore, cluster-alkylated OPA/MIL-125-NH_2 as a catalyst is highly efficient for photocatalytic H_2O_2 production.

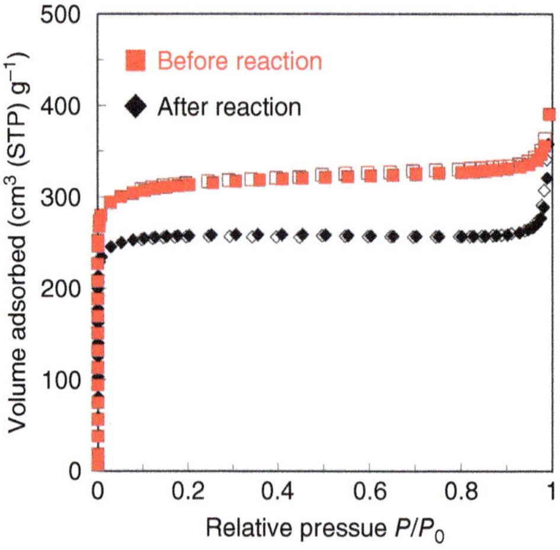

Figure 12.32 N_2 adsorption isotherms at 77 K for OPA/MIL-125-NH_2 before and after the reaction. Source: Kawase et al. [31]. © 2019, Royal Society of Chemistry.

12.5 Conclusion and Outlooks

The MOF, MIL-125-NH_2, was used for photocatalytic H_2O_2 production in a single-phase system composed of acetonitrile and benzyl alcohol. The activity was highly enhanced in the presence of deposited NiO nanoparticles. To better separate the formed H_2O_2 and protect the structure of MOF, a hydrophobic linker-alkylated MIL-125-NH_2 dispersed selectively into the organic phase of a two-phase system composed of benzyl alcohol and water and catalyzed photocatalytic H_2O_2 production rather efficiently. Furthermore, the catalytic activity was even more enhanced when using the hydrophobic cluster-alkylated MIL-125-NH_2 due to its maintained surface area. Those works creatively advanced the application of MOFs in the field of H_2O_2 production.

The future development of other types of MOF-based materials such as Zr, Fe, and Cr-based MOFs for photocatalytic H_2O_2 production is promising. It still needs deeper understanding, and many challenges remain in this field. We believe that the higher concentration of H_2O_2 that are available on the market will be obtained using MOF-based materials with photocatalysis in the near future.

Reference

1 Kofuji, Y., Isobe, Y., Shiraishi, Y. et al. (2016). *J. Am. Chem. Soc.* 138: 10019.

2 Campos-Martin, J.M., Blanco-Brieva, G., and Fierro, J.L.G. (2006). *Angew. Chem. Int. Ed.* 45: 6962.

3 (a) Seh, Z.W., Kibsgaard, J., Dickens, C.F. et al. (2017). eaad4998. *Science* 355. (b) Siahrostami, S., Verdaguer-Casadevall, A., Karamad, M. et al. (2013). *Nat. Mater.* 12: 1137.

4 (a) Zheng, Y., Yu, Z., Ou, H. et al. (2018). *Adv. Funct. Mater.* 28: 1705407. (b) Zhao, S., Guo, T., Li, X. et al. (2018). *Appl. Catal., B* 224: 725. (c) Yin, H., Kuwahara, Y., Mori, K., and Yamashita, H. (2018). *J. Mater. Chem. A* 6: 10932.

5 (a) Zhao, S. and Zhao, X. (2018). *J. Catal.* 366: 98. (b) Zhao, S. and Zhao, X. (2019). *Appl. Catal., B* 250: 408.

6 Wei, Z., Liu, M., Zhang, Z. et al. (2018). *Energy Environ. Sci.* 11: 2581.

7 (a) Zheng, Z., Ng, Y.H., Wang, D.-W., and Amal, R. (2016). *Adv. Mater.* 28: 9949. (b) Pritchard, J.C., He, Q., Ntainjua, E.N. et al. (2010). *Green Chem.* 12: 915. (c) Wilson, N.M. and Flaherty, D.W. (2016). *J. Am. Chem. Soc.* 138: 574.

8 Kim, H.-i., Kwon, O.S., Kim, S. et al. (2016). *Energy Environ. Sci.* 9: 1063.

9 Kato, S., Jung, J., Suenobu, T., and Fukuzumi, S. (2013). *Energy Environ. Sci.* 6: 3756.

10 Moon, G.-h., Kim, W., Bokare, A.D. et al. (2014). *Energy Environ. Sci.* 7: 4023.

11 Shiraishi, Y., Kanazawa, S., Tsukamoto, D. et al. (2013). *ACS Catal.* 3: 2222.

12 Meyer, K., Ranocchiari, M., and van Bokhoven, J.A. (2015). *Energy Environ. Sci.* 8: 1923.

13 Chen, Y., Wang, D., Deng, X., and Li, Z. (2017). *Catal. Sci. Technol.* 7: 4893.

14 Wang, C.-C., Li, J.-R., Lv, X.-L. et al. (2014). *Energy Environ. Sci.* 7: 2831.

15 Deng, X., Li, Z., and García, H. (2017). *Chem. Eur. J.* 23: 11189.

16 (a) Wu, Z.-L., Wang, C.-H., Zhao, B. et al. (2016). *Angew. Chem. Int. Ed.* 55: 4938. (b) Zhang, T. and Lin, W. (2014). *Chem. Soc. Rev.* 43: 5982. (c) Wang, W., Xu, X., Zhou, W., and Shao, Z. (2017). *Adv. Sci.* 4: 1600371.

17 Wen, M., Mori, K., Kuwahara, Y., and Yamashita, H. (2017). *ACS Energy Lett.* 2: 1.

18 (a) Dan-Hardi, M., Serre, C., Frot, T. et al. (2009). *J. Am. Chem. Soc.* 131: 10857. (b) Logan, M.W., Ayad, S., Adamson, J.D. et al. (2017). *J. Mater. Chem. A* 5: 11854.

19 (a) Horiuchi, Y., Toyao, T., Saito, M. et al. (2012). *J. Phys. Chem. C* 116: 20848. (b) Fu, Y., Sun, D., Chen, Y. et al. (2012). *Angew. Chem. Int. Ed.* 51: 3364.

20 (a) Martis, M., Mori, K., Fujiwara, K. et al. (2013). *J. Phys. Chem. C* 117: 22805. (b) Puthiaraj, P. and Ahn, W.-S. (2015). *Catal. Commun.* 65: 91.

21 (a) Sun, D., Fu, Y., Liu, W. et al. (2013). *Chem. Eur. J.* (19): 14279. (b) Gomes Silva, C., Luz, I., Llabrés i Xamena, F.X. et al. (2010). *Chem. Eur. J.* 16: 11133.

22 Nietubyć, R., Czerwosz, E., Diduszko, R. et al. (2009). *J. Alloys Compd.* 484: 896.

23 (a) Pellegrin, Y. and Odobel, F. (2017). *C. R. Chim.* 20: 283. (b) Kalyanasundaram, K., Kiwi, J., and Grätzel, M. (1978). *Helv. Chim. Acta* 61: 2720.

24 Isaka, Y., Kondo, Y., Kawase, Y. et al. (2018). *Chem. Commun.* 54: 9270.

25 (a) Sun, D., Ye, L., and Li, Z. (2015). *Appl. Catal., B* 164: 428. (b) Konaka, R., Kasahara, E., Dunlap, W.C. et al. (1999). *Free Radical Biol. Med.* 27: 294.

26 (a) Ibrahim, M.M., Ramadan, A.E.-M.M., Shaban, S.Y. et al. (2017). *J. Inorg. Organomet. Polym.* 27: 1252. (b) Pap, J.S., Kripli, B., Váradi, T. et al. (2011). *J. Inorg. Biochem.* 105: 911. (c) Barondeau, D.P., Kassmann, C.J., Bruns, C.K. et al. (2004). *Biochemistry* 43: 8038.

27 (a) Burtch, N.C., Jasuja, H., and Walton, K.S. (2014). *Chem. Rev.* 114: 10575. (b) Canivet, J., Fateeva, A., Guo, Y. et al. (2014). *Chem. Soc. Rev.* 43: 5594.

28 Isaka, Y., Kawase, Y., Kuwahara, Y. et al. (2019). *Angew. Chem.* 131: 5456.

29 Oh, J.S., Shim, W.G., Lee, J.W. et al. (2003). *J. Chem. Eng. Data* 48: 1458.
30 Mase, K., Yoneda, M., Yamada, Y., and Fukuzumi, S. (2016). *Nat. Commun.* 7: 11470.
31 Kawase, Y., Isaka, Y., Kuwahara, Y. et al. (2019). *Chem. Commun.* 55: 6743.
32 Sun, Y., Sun, Q., Huang, H. et al. (2017). *J. Mater. Chem. A* 5: 18770.
33 Ansari, S.A. and Cho, M.H. (2016). *Sci. Rep.* 6: 25405.

13

Photocatalytic and Photoelectrochemical Reforming of Methane

Jinqiang Zhang and Hongqi Sun

Edith Cowan University, School of Engineering, 270 Joondalup Drive, Joondalup, WA 6027, Australia

13.1 Introduction

Energy shortage and environmental deterioration are the two main issues that impede the speed of human development [1]. With the discovery of large amount of natural gas, shale gas, and flammable ice, methane as their main component appeals much more attention, which is deemed as the solution for the problem of energy shortage [2–4]. However, as one kind of greenhouse gas, the global warming potential of methane is over 20 times higher than that of CO_2 with same mass [5]. Besides, its transportation is also a Gordian knot despite of its large reserve. Therefore, the current utilization of methane is mostly limited with the heat gained through direct combustion. Nevertheless, simple combustion process cannot meet the requirements of sustainable development and essentially cannot solve the shortage problem of fossil fuels. How to make full use of methane at a large scale is a big challenge.

Fortunately, as the simplest molecule structure, methane could be converted into C1 oxygenates, other alkanes and olefins with long chains and aromatics, which are deemed as promising alternations to the fossil fuels. However, four strong and localized C—H bonds (with a bond energy of 434 kJ mol^{-1}) [3] and the absence of low-energy empty orbitals and high-energy filled orbitals in methane molecule make it difficult for methane to readily participate in a chemical reaction. Therefore high temperatures and rigorous reaction conditions are required to overcome the reaction barriers [6]. This means additional energy inputs and specific reactors will be needed, which are not favorable for the safety requirements during practical industrial processes [7]. Furthermore, reactions conducted at a high temperature would easily result in the formation of carbon or coke on the catalysts, which rapidly leads to the catalyst deactivation [8, 9]. As a result, finding a mild energy input with a high efficiency for methane activation is the key to take advantage of the luxuriant methane energy.

Photo-assisted catalysis is the process that can make full use of the merits of wide distribution, large amount reserve, and environmentally benign nature of

Solar-to-Chemical Conversion: Photocatalytic and Photoelectrochemcial Processes, First Edition.
Edited by Hongqi Sun.

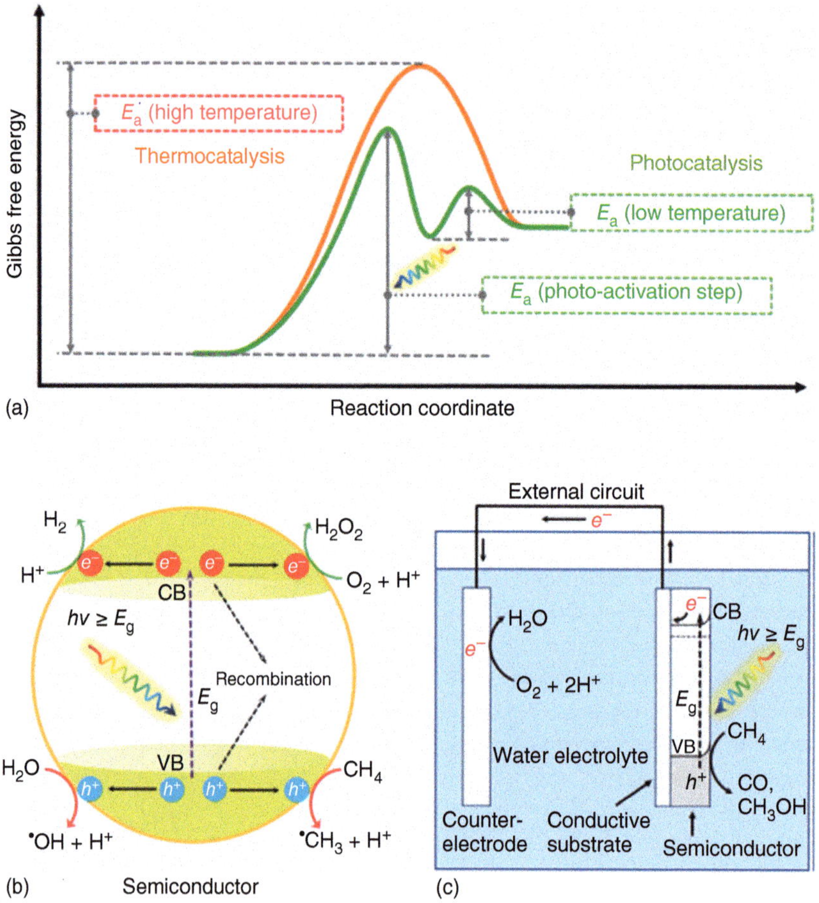

Figure 13.1 (a) The kinetic progress of methane activation on thermocatalysis and photocatalysis methods, (b) reaction mechanism of photocatalysis, and (c) reaction mechanism of photoelectrochemical catalysis. Source: Song et al. [2].

solar energy [10, 11]. Importantly, as shown in Figure 13.1a, this novel "darling" could lower the reaction activation energy in comparison with thermocatalysis and even could activate the reactions that could not be realized under thermocatalysis [12–14]. As a result, solar energy-mediated catalytic processes have been widely involved into water splitting for hydrogen/oxygen evolution and CO_2 or CO reduction for hydrocarbons [15]. Considering the difficulties on the methane activation in thermocatalysis, recently solar energy was introduced into a variety of methane activation reactions, and encouraging results were obtained, thus at present is recognized as a promising way for solar energy utilization and methane conversion [16, 17]. As such, this chapter timely gives introduction on the methane activation processes in which light is involved, including photocatalysis and photoelectrochemical (PEC) catalysis processes. Meanwhile, the differences of reaction mechanisms between thermocatalysis and photo-assisted catalysis will be discussed in detail. Then recent research progresses on the light-initiated methane activation reactions with related catalysts are provided, including methane dry reforming, methane steam reforming for syngas or hydrogen, methane coupling, methane oxidation, and methane dehydroaromatization reactions. Besides, their corresponding reaction mechanisms using photo-mediated catalysis are summarized. At last, the promising perspectives on the methane upgrading via solar energy excitation are proposed. This chapter is anticipated to provide a compact source of relevant information and is of great interest to all chemists, materials scientists, engineers, and students entering or already working in the field of photo-assisted methane conversion.

13.2 Photo-Mediated Processes

As reported, photo-mediated catalytic processes for methane activation are mainly dependent on the means of photocatalysis and photoelectrocatalysis (PEC). Besides, in the past five years, a novel kind of photo-assisted method, namely, plasma-based photocatalysis, was brought up and satisfied quantum efficiencies were acquired [18]. Here in this part, photocatalysis and PEC processes will be emphatically introduced. At the same time, the principle of plasma-based photocatalysis for methane conversion will also be provided.

Photocatalysis and photoelectrocatalysis are two basic methods in photo-assisted catalysis, which could offer the possibility to initiate even inflexible processes at ambient temperatures [19]. Especially at low temperature such as room temperature, some advantages such as lowering the energy consumption and relax restrictions on the reaction conditions, prohibiting the deactivation and carbon coking on catalyst at critical conditions and importantly improving the safety of the reactor, would be shown up [20]. The reason why solar energy could offer these advantages is that solar energy can exceed the activation energy for various chemical processes. As calculated by Planck constant, the energy of one photon from a certain wavelength of light photons of 300 nm ultraviolet (UV) light has the equal energy to 399 kJ mol^{-1}. Although this energy is too low to directly break the C–H bonding in methane, it is anticipated that using a suitable photocatalyst would let the methane activation be feasible. Thus, together with the photocatalyst, this high energy of the photon would enable thermodynamically unfavorable reactions ($\Delta G < 0$) to proceed photocatalytically in mild conditions [21]. As a result, since solar energy was firstly brought into catalysis, photocatalysis became the hot spot in the field of energy preparation, and then large amount of researches burst out. Till now, the principal of photocatalysis for methane conversion has been investigated in detail. As shown in Figure 13.1b, when the light irradiates on the semiconductor, electron and hole pairs will generate and separate for the redox reactions [22]. Basically, photoinduced electrons move to the surface of photocatalyst to reduce the water to hydrogen, while holes activate the C—H bond and initiate a series of reactions for methane conversion [23]. As calculated by finite element method, when the photocatalyst is irradiated with light, an enhanced electric field will form around and between the nanoparticle of photocatalyst [24]. However, within a semiconductor, only the electric field is not strong enough to force the separation of electron–hole pairs, which possibly cannot effectively initiate the methane activation reaction. Therefore, an external potential is innovatively applied into the photocatalysis system to create an extra electric field [25]. With the extra electric field, the separated speed of photogenerated electron–hole pairs can be significantly accelerated. The mechanism scheme can be inferred as Figure 13.1c, in which a photocatalyst is prepared into electrode and used as a working electrode. Same phenomenon with photocatalysis would occur when the photocatalyst is irradiated with light. Different from the low separation efficiency of electron–hole pairs in photocatalysis, the applied bias voltage would drive the photogenerated electrons to the counter electrode, prohibiting the recombination of charge carriers. The holes remain on the surface of

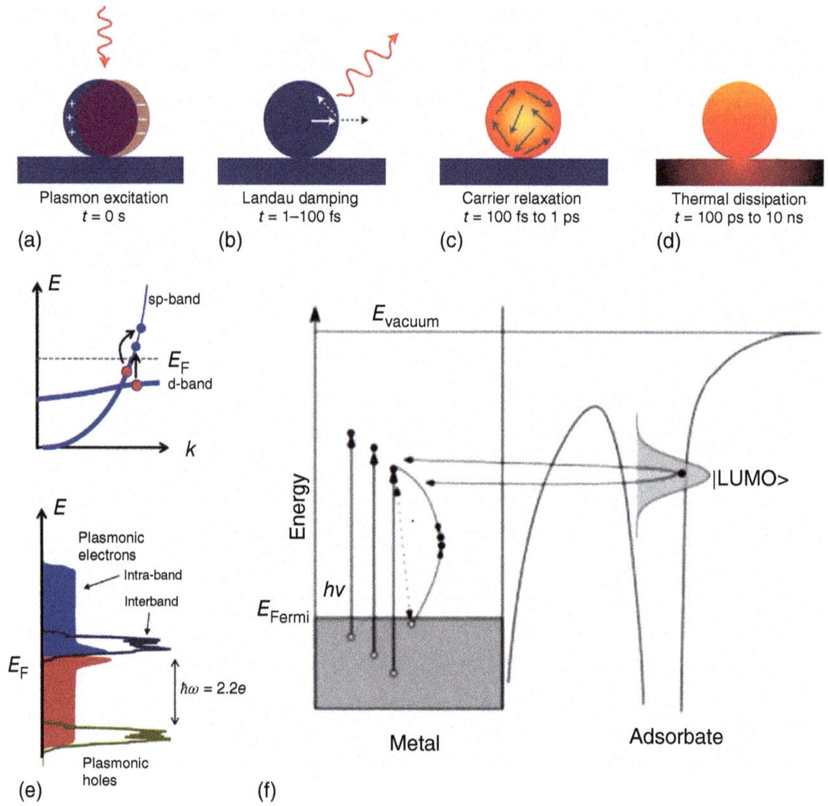

Figure 13.2 (a–d) Photoexcitation and subsequent relaxation processes following the illumination of a metal nanoparticle with a laser pulse and characteristic time scales. Source: Brongersma et al. [28]. (e) Simplified diagrams showing the interband and intra-band transitions in gold and distributions of plasmonic carriers created via the intra-band transitions (blue and red bands) and the distributions of carriers excited through the interband transitions (solid lines). Source: Govorov et al. [29]. (f) Schematic illustration of hot-electron transfer from a metal to an unoccupied molecular orbital at the surface. Source: Govorov et al. [29].

working electrode to activate the C–H bonding, while induced electrons transfer to the counter electrode to initiate the reduction reaction [26]. Apart from these two methane activation methods, plasma-based photocatalysis as one special photocatalysis has received satisfied efficiency during the methane activation and thus is regarded as a promising method in the field of energy preparation [27]. As shown in Figure 13.2, when the light irradiates the plasmonic nanometals and at the same time the minimum energy of a photon is equal to the metal's work function, the absorbed energy will be stored in the collective resonant oscillations of conduction electrons, which is called the surface plasmon resonance (SPR) effect (shown in Figure 13.2a). At this stage, hot carriers generated by intra-band or interband extinction (also called Landau damping) (shown in Figure 13.2b,e) will be consumed to activate methane molecule if the energy of electron is large enough to overcome the lowest unoccupied molecular orbital (LUMO) energy of CH_4 molecule (Figure 13.2f). While when the energy of hot carriers cannot exceed the LUMO energy of CH_4 molecule, relaxation (Figure 13.2c) due to the ohmic damping would occur, and more heat would be generated (Figure 13.2d). Fortunately, the generated heat is also an important factor for not only methane activation but also boosting the mobility of charge carriers. Thus recently plasma-based photocatalysis was frequently employed into methane conversion reactions. For example, Au, Ag, and Cu are widely reported as localized surface plasmon resonance (LSPR) nanometals that could convert and store solar energy and generate more hot electrons for photocatalytic reactions. While metals locate at VIII group can also be initiated and generate hot electrons by solar energy, however the energy stored in the hot electrons is relatively low, which would be easily dissipated into heat for improving the methane conversion efficiency.

13.3 Differences Between Photo-Assisted Catalysis and Thermocatalysis

It is essential to point out the differences between photo-mediated catalysis and thermocatalysis as this could provide guidance for experiments and promote their further developments. Compared with traditional thermocatalysis, special requirements should be considered on photocatalysis and photoelectrocatalysis processes, including the way to introduce light through reactor and the nature of catalyst involved in photo-assisted processes [17]. Critically, with the participation of solar energy, the reaction pathway in photo-assisted catalytic processes is also different from that of thermocatalysis. Therefore, this will be also mentioned so as to help design the reaction, enhance the methane conversion, and improve the selectivity of target products.

13.3.1 Catalyst Involved

Thermocatalysis for methane conversion as one kind of traditional C–H activation method has already been mature with large amounts of researches reported [30, 31], and that some activation reactions have reached to the industrialization level. Accordingly, catalysts with excellent CH_4 activation performances have been exploited. Among these reported catalysts, noble metals such as Pt [32], Pd [33], Rh [34], etc. loaded on the substrate are accepted as the most efficient catalysts. However, the high price and hazards to environment impede their practical applications. Thus, non-noble metals such as Fe [35], Ni [8], and Co [36] are subsequently developed, and their catalytic efficiencies have been improved to a comparable level with noble metals. As studied, the electron orbits of noble metals are responsible for the C–H activation over thermocatalysis. While in the field of photo-mediated catalysis, semiconductors that can be excited by solar light are indispensable. As reported, metal oxides (TiO_2 [37], WO_3 [38], ZnO [5], etc.), GaN [39] as well as Bi-based catalysts [40], were employed into photocatalytic and photoelectrocatalytic methane activation. Unfortunately, the quantum efficiencies of these pristine semiconductors are still unsatisfied due to the low light harvesting, inferior electrical conductivity, and fast electron–hole recombination [41]. Thus, inspired by the catalyst modification method from photocatalytic water splitting, some semiconductor heterojunctions were fabricated to promote the separation rate of electron–hole pairs. Tang and coworkers [42] recently prepared FeO_x/TiO_2 heterojunction, which was used for photocatalytic methane activation; a conversion rate of 15% was obtained under one sun irradiation and mild condition. Importantly, the selectivity of alcohol was above 97%. Although the quantum efficiency of photocatalysis for methane activation has been improved via the modification of pristine semiconductor, a large gap to the requirements of industrialization remain.

Based on this, metals with LSPR effect are loaded on the substrates and applied into plasma-based photocatalysis processes for methane activation. Metals involved in plasma-based photocatalysis are currently limited to the metals locate in VIII [43] and IB [44] groups, which have been proven to exhibit the LSPR effect under light irradiations. With the emerge of plasma-based photocatalysis, the quantum efficiency is significantly improved. For example, Ye and coworkers [45] prepared Ni-loaded Al_2O_3 catalyst in which the Ni nanoparticles served as both active sites for dry methane reforming and the plasmonic promoter to enhance the electrical field. Using this photocatalyst into methane dry reforming reaction, an apparent quantum efficiency (AQE) value of 19.0% was acquired on 10% Ni/Al_2O_3, showing limitless application prospective.

13.3.2 Reactors

In the field of thermocatalysis, the design principal of reactor mainly places emphasis on its endurance of high temperature and critical pressure. The reactor requirements in photocatalysis and photoelectrocatalysis are different. As reported, there are two categories of photocatalysis reactor designed, including the sealed quartz photo-reactor also named fixed bed reactor [46] and the flow photo-reactor [47]. As shown in Figure 13.3A, the sealed reactor is replaced by reacting gas (including methane), and catalyst is fixed on a membrane that could receive the light from a light source. After the light is turned on, the reaction will be initiated, and products will be detected by GC or GC-MS dependent on the reaction time. Design principle

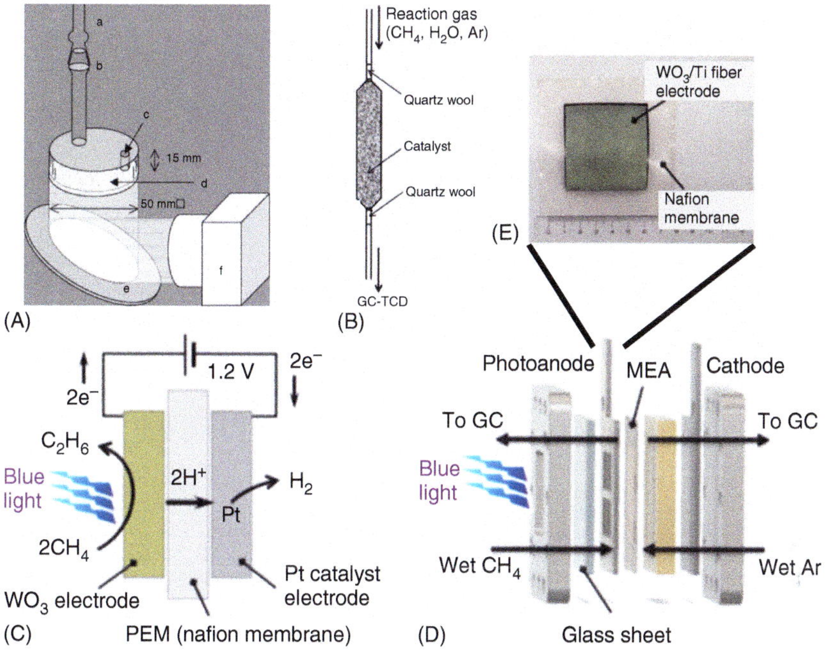

Figure 13.3 (A) Schematic (a) a conventional vacuum line equipped with pressure gage; (b) joint; (c) small hole for thermo-couple; (d) catalyst bed; (e) UV-reflection mirror; (f) Xe lamp. drawing of the fixed bed photo-reactor. Source: Yoshida et al. [46], (B) photocatalytic reaction cell made of quartz for the flow reactor. Source: Yoshida et al. [47], and (C–E) photoelectrochemical (PEC) system for gas-phase CH_4 activation. Source: Amano et al. [48].

of a flow reactor is similar with that of sealed reactor, in which as observed from Figure 13.3B a heterotype window is created on the quartz tube. The window is filled with catalyst that is irradiated by a light source, and the effluent product is analyzed by the online GC. Different with the photocatalysis reactor, a special PEC cell is designed and successfully employed for PEC methane activation [48]. As illustrated from Figure 13.3C,D, the cell is assembled with a working electrode and Pt electrode that are separated by a proton exchange membrane (PEM). On the side of the PEC membrane flow reactor, an optical window is designed to allow the irradiation of visible light, and the product gas is connected to a GC analyzer. Amano et al. [48] applied this PEC cell for photoelectrolysis of CH_4 into C_2H_6 and H_2, and the conversion efficiency of incident photon-to-current can reach as high as 11% under blue light with an applied voltage of 1.2 V.

13.3.3 Mechanism

As different reaction mechanisms could obtain different products, it is crucial to investigate and manage to control the reaction pathways to improve the selectivity of target product. It is demonstrated from Figure 13.4a that the activation energy could be lowered after the introduction of solar energy. Similarly, the reaction mechanism of thermocatalysis is also different from that with solar light. As previously

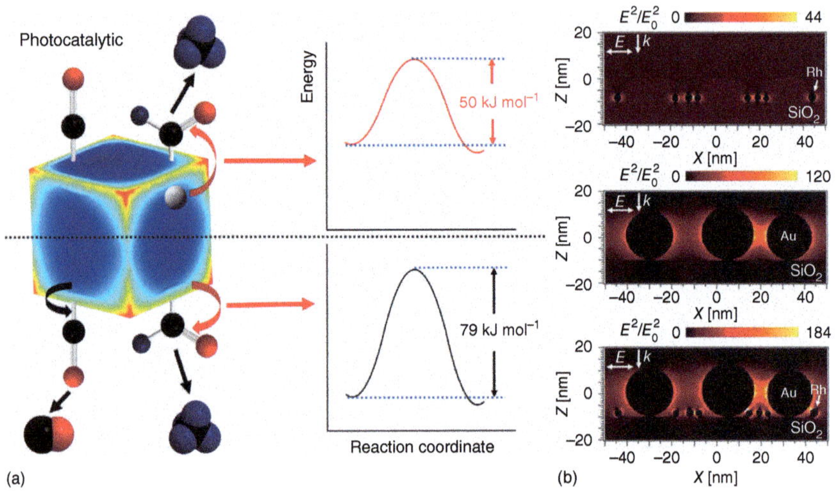

Figure 13.4 (a) Difference on activation energy between photocatalysis and thermocatalysis. Source: Chen et al. [5], licensed under CC BY 4.0 and (b) electrical field of Rh/SiO_2, Au/SiO_2, and Au–Rh/SiO_2 catalysts calculated by FDTD method. Source: Liu et al. [16].

reported, thermocatalysis is mainly dependent on the electron orbit of metal or the vacancy of metal oxide [49], whereas photo- assisted catalysis is principally initiated by the photogenerated charge carriers [50]. As a result, the selectivity of product varies after the introduction of solar light. Indeed, Zhang et al. [12] found that the catalytic performance of rhodium nanoparticles was profoundly improved after the addition of solar light. The reaction activation energy was significantly reduced, and the selectivity of desired product was enhanced, but unfavorable product was kinetically suppressed. Taking methane dry reforming with carbon dioxide as example, here the difference of reaction mechanism between thermocatalysis and photocatalysis is compared in detail. The thermocatalysis reaction mechanism of methane dry reforming is widely accepted, which is shown from Eqs. (13.1)–(13.7):

$$CH_4 + * \rightleftharpoons CH_3{}^* + H \rightleftharpoons CH_2{}^* + 2H \rightleftharpoons CH^* + 3H \rightleftharpoons C^* + 4H \tag{13.1}$$

$$CO_2 + \& \rightleftharpoons CO + O^{\&} \tag{13.2}$$

$$H + H \rightleftharpoons H_2 \tag{13.3}$$

$$CH_x{}^* + O^{\&} \rightleftharpoons CO + 0.5 \times H_2 + * + \& \tag{13.4}$$

$$C^* + CO_2 \rightleftharpoons 2CO + * \tag{13.5}$$

$$CH_x{}^* + O^{\&} \rightleftharpoons CH_{x-1}{}^* + HO^{\&} \tag{13.6}$$

$$HO^{\&} + H \rightleftharpoons H_2O + \& \tag{13.7}$$

where * and & represent the active sites on the catalyst for the activation of CH_4 and CO_2, respectively.

For the reaction pathway using photocatalysis, Ye and coworkers [16] investigated and proposed a mechanism that is shown from Eqs. (13.8) and (13.9). Compared with more intermediates in thermocatalysis, reaction pathway of photocatalysis seems simple, which should be the reason that the selectivity of specific product is relatively high.

$$e\,(\text{with high energy}) + CO_2 \rightarrow CO + O + e \tag{13.8}$$

$$e\,(\text{with high energy}) + CH_4 \rightarrow CH_x + (4 - x)H + e \tag{13.9}$$

13.3.4 Equations for Quantum Efficiency

It is of vital importance to standardize the calculations for the efficiencies of solar energy to chemical fuels. Till now, three equations (Eqs. (13.10)–(13.12)) were reported and summarized below, including solar-to-fuel (STF) efficiency, apparent quantum yield (AQY), and incident photon-to-electron conversion efficiency (IPCE). STF equation is merely suitable for uphill reactions, while AQY and IPCE can be used for evaluating the efficiency of photocatalytic and PEC methane conversion, respectively [2]:

$$\text{STF}\,(\%) = \frac{\text{Stored chemical energy}}{\text{Incident solar energy}} \times 100 \tag{13.10}$$

$$\text{AQY}\,(\%) = \frac{\text{Number of reacted electrons}}{\text{Number of incident photons}} \times 100 \tag{13.11}$$

$$\text{STF}\,(\%) = \frac{1240 \times J}{\lambda \times P_\lambda} \times 100 \tag{13.12}$$

where J, λ, and P_λ in Eq. (13.12) represent the measured photocurrent density, the wavelength of the incident light, and the measured irradiance at the specific wavelength, respectively.

13.4 Reactions of Methane Conversion via Photo-Assisted Catalysis

In the field of traditional thermocatalysis, various methane conversion methods including C1 conversion, carbon chain growth, and aromatization reactions have been exploited and received the conversion efficiencies to the level of industrialization. However, as a novel hot spot, photocatalysis is still confronted with the unsatisfied quantum efficiencies because of the low mobilities of photoinduced charge carriers of photocatalysts [51]. Since the first report of PEC catalysis on the year of 1972 [11], photo-assisted catalysis was widely employed into water splitting [52], carbon dioxide reduction [13], and degradation of contamination [53]. Recently, considering the huge reserves of natural gas, activation of methane and its conversion to high-value-added products appeal much more attention of scientists. Although the conversion efficiencies of methane are extremely low due to the strong chemical bonding in methane, fortunately its activation at relatively low temperature is also regarded as a hope compared with the critical conditions using thermocatalysis method. Till now, photo-assisted catalytic means are involved into several methane conversion reactions including methane dry reforming reaction [54], methane steam reforming reaction [55], methane coupling reaction [56], the oxidation reaction of methane [5, 57], and conversion of methane to benzene [39]. Here in this section, these reactions will be introduced; meanwhile, the reported catalysts and related conversion efficiencies will be summarized. Importantly, the reaction mechanism will be analyzed, which could provide guidance to further improve the efficiencies of methane conversion.

13.4.1 Methane Dry Reforming

Methane reforming reaction mainly includes methane dry reforming with carbon dioxide and methane steam reforming depending on the oxidizing agents. As shown in Table 13.1, methane dry reforming reaction is thermodynamically endothermic reaction that hardly occurs at a low temperature (<450 °C). Therefore, critical requirements on reactor and energy-extensive consumption impede its practical application despite its high conversion efficiency. Reducing the reaction temperature to room temperature becomes the aim of scientists all over the world. Recently, photo-assisted catalysis was used into methane dry reforming reaction, and satisfied results were obtained. Considering the difficulty on methane activation, semiconductors with wide band gaps are initially involved in photocatalytic methane dry reforming reaction that have enough oxidation abilities to break the stubborn C–H bonding in methane. Based on this, Yoshida and coworkers [58] used Ga_2O_3 into photocatalytic methane dry reforming reaction. Prior to the introduction of UV light, no product was detected in methane dry reforming reaction using thermal only. However, after irradiation with UV light, 2.48 μmol H_2 and 1.08 μmol CO were acquired. Meanwhile, the yield of CO increased with the reaction temperature rose from 473 up to 673 K, attributing to the boosted mobility

Table 13.1 Reaction categories involved in photo-mediated catalysis and related Gibbs free energy.

Reaction category	Reaction equation	$\Delta G^{\theta}_{\text{free energy}}$
Methane dry reforming	$CH_4 + CO_2 = 2CO + 2H_2$	171 kJ mol^{-1}
Methane steam reforming	$CH_4 + 2H_2O = CO_2 + 4H_2$	113 kJ mol^{-1}
	$CH_4 + H_2O = CH_3OH + H_2$	117 kJ mol^{-1}
Methane coupling	$2CH_4 = C_2H_6 + H_2$	68.6 kJ mol^{-1}
Methane oxidation	$CH_4 + 0.5O_2 = CH_3OH$	−223 kJ mol^{-1}
	$CH_4 + 2O_2 = CO_2 + 2H_2O$	−801 kJ mol^{-1}
Methane dehydroaromatization	$6CH_4 = C_6H_6 + 9H_2$	434 kJ mol^{-1}

of photoinduced charge carriers. Therefore, it can be concluded that both thermal and solar energy played important roles on initialing the photocatalytic methane dry reforming reaction. The reaction mechanism of photocatalytic methane dry reforming over semiconductors was also investigated by Tanaka and coworkers [59]. As revealed by electron paramagnetic resonance (EPR) and Fourier transform infrared spectroscopy (FTIR) spectra, CO_2 molecule absorbed on the surface of MgO was excited to CO_2^- radicals that anchored on the surface of MgO could build up a new acceptor level between the valence band and the conduction band of MgO on CO_2-adsorbed MgO. Furthermore, using ^{13}C-labeled method, the obtained CO was confirmed to come from CO_2, rather than from the methane.

Addressing the endothermic methane activation, plasma nanometal-loaded catalysts are employed to initiate the photocatalytic methane dry reforming reaction. Because of appreciable heat generated under light irradiations and more active sites on the surface of plasma nanometals, the conversion rate of this reaction could be significantly enhanced [29, 60]. Ni-based catalysts are widely used in the thermocatalytic methane dry reforming reaction owing to the low price and high performance [61]. However, sintering deactivation and carbon coking on Ni-based catalysts at high temperatures limited its further developments [20]. Fortunately, Ni as one kind of plasma metal could produce more heat under the light irradiations, which could avoid the critical reaction conditions [62]. Ye and coworkers [45] reported Ni/Al_2O_3 catalyst with varied Ni loading could effectively initiate methane dry reforming at 773 K. Compared with the photocatalytic abilities using traditional thermocatalysis (38.5 μmol g^{-1} min^{-1}), the performance of methane dry reforming after the introduction of light was greatly enhanced with the methane conversion above 60 μmol g^{-1} min^{-1}. Meanwhile, the photocatalytic capabilities were almost linearly improved with the light intensity increasing, even at the light intensity as high as 1.07 W cm^{-2}, proving that Ni nanoparticles were stable and not severely aggregated under the irradiations. Combined with the finite-difference time-domain (FDTD) method, the reason of the significant improvement on methane dry reforming performance was attributed to the enhanced electrical field excited from the LSPR of Ni nanoparticles. The reaction mechanism was also investigated, which can be referred below (Eqs. (13.13–13.20)). Meanwhile, Ye's group [14] studied the photocatalytic performances on all the metals in VIII group. Results revealed that the surface temperature of Ni/Al_2O_3 could rise to approximately 400 °C, which is more favorable to the endothermic reaction. Bimetallic catalysts exhibit further enhancement in comparison with monometallic catalysts on catalysis because of the synergistic effect between different metals. Based on this, Ye and coworkers, respectively, prepared Rh/Au- [16] and Pt/Au- [63] loaded bimetallic catalysts, which were employed into photocatalytic methane dry reforming reaction. As calculated by FDTD method (shown in Figure 13.4b), the electrical field was improved by 1.5 and more than twice on Rh/Au and Pt/Au bimetallic catalysts, respectively, to Au-loaded monometallic catalyst. As a result, the methane conversions on Rh/Au- and Pt/Au-loaded bimetallic catalysts were correspondingly improved to 3200 μmol g^{-1} s^{-1} (0.3 W cm^{-2}) and 55 μmol g^{-1} min^{-1} (0.6 W cm^{-2}), respectively. For the sake of avoiding the sintering deactivation and carbon coking of nanometals, different morphologies of catalysts were designed. For example, Li and coworkers [64] designed and successfully prepared a catalyst named SCM-Ni/SiO_2 in which the surface of Ni nanoparticles was surrounded with silica clusters. Since the Ni nanoparticles were protected by the silica clusters that had no effect on the transportation of light, the STF efficiency of the catalyst can be improved to as high as 12.5%. Importantly, the catalyst still exhibited high performance and selectivity on photocatalytic methane dry reforming reaction after a 100-hours test, showing promising application prospect:

$$CH_4 + * = CH_x{}^* + (4 - x)H \text{ (fast)} \tag{13.13}$$

$$CO_2 + *' = CO + O^{*\prime} \text{ (fast)} \tag{13.14}$$

$$CH_x{}^* + O^{*\prime} = CO + 0.5 \times H_2 + * + *' \text{ (slow)} \tag{13.15}$$

$$CO_2 = CO + 0.5O_2 \tag{13.16}$$

$$e \text{ (energetic electron)} + CO_2 = CO + O^{\cdot} + e \tag{13.17}$$

$$e \text{ (energetic electron)} + CH_4 = CH_x \cdot + (4 - x)H \cdot + e \tag{13.18}$$

$$CH_x{}^* + O\cdot = CO + 0.5 \times H_2 + * + \cdot \tag{13.19}$$

$$CH_x \cdot + O^{*\prime} = CO + 0.5 \times H_2 + *' + \cdot \tag{13.20}$$

13.4.2 Methane Steam Reforming

Different from CO_2 as an oxidative agent in methane dry reforming reaction, methane steam reforming process mainly depends on steam for the methane oxidation. As reported photocatalysis for methane steam reforming reaction, it can be divided into methane steam reforming for hydrogen [65, 66] or syngas [26] and methane steam reforming for oxygenated products [67, 68] (mainly methanol). Here in this section, methane steam reforming process to hydrogen and methane steam reforming process to methanol will be emphatically introduced. As shown in Table 13.1, both of these chemical processes are endothermic reactions; therefore, high temperature could also promote the reaction conversion of these two kinds of reforming reactions. Although the methane steam reforming reaction has realized its practical application, improvements still need to be done to lower the requirements of reaction conditions and reactor. Yoshida's group pioneeringly studied this reaction using varieties of noble metal-loaded photocatalysts. In their researches, Pt loaded on different substrates including TiO_2 [69], $CaTiO_3$ [66], and Ga_2O_3 [70] was prepared and employed into photocatalytic methane steam reforming for hydrogen energy. Corresponding results are summarized, which can be referred in Table 13.2. Results indicated that introduction of light with full spectrum could initiate the methane steam reforming reaction in mild conditions. The addition of light not only excited the photocatalysts generating charge carriers but also made the reaction temperature increase up to roughly 348 K that also promoted the reaction conversion. It should be noted here that the effluents in Yoshida's work only contained hydrogen and carbon dioxide, in which other products such as CO, CH_3OH, and C_2H_6 were not detected, confirming that this chemical process could be selectively controlled with the light irradiation. Semiconductor substrates in photocatalysts also play important roles in photocatalytic methane steam reforming reactions. Based on this, Yoshida and coworkers [65] further chose $NaTaO_3$ as target to study the influence of substrate on the reaction conversion. After an optimized amount of La^{3+} was inserted into the bulk $NaTaO_3$ photocatalyst, the highest activity of hydrogen production was observed on Pt/$NaTaO_3$:La photocatalyst, with the hydrogen production of rate 4.5 μmol min^{-1} and the methane conversion of 0.6%. The doping of La^{3+} into $NaTaO_3$ improved the density and boosted the mobility rate of the photoexcited charge carriers, while the loading of Pt enhanced the separation of electron–hole pairs, both leading to the relatively high photocatalytic methane steam reforming performance. Besides, Rh nanometal-loaded $K_2Ti_6O_{13}$ photocatalyst was also selected to activate methane molecule with water vapor under light irradiation [55]. Compared with Pt-loaded $K_2Ti_6O_{13}$ catalyst, twice higher performance was acquired on Rh/$K_2Ti_6O_{13}$, and the catalyst was more stable without deactivation after many hours. Furthermore, increasing the reaction temperature can thermodynamically improve the methane conversion due to the endothermic property of reaction; meanwhile the mobility of photoexcited charge carriers can also be accelerated by the elevated temperature. Indeed as studied by Yoshida and coworkers [75], the hydrogen evolution rate on Pt/$CaTiO_3$ can be improved from 0.38 μmol min^{-1} at 313 K to 0.40 μmol min^{-1} at 323 K. Reported hydrogen generation rate on various photocatalysts and different reaction conditions are summarized in Table 13.2.

Similar with the reaction of methane steam reforming to hydrogen, methane steam reforming to methanol in traditional thermocatalytic field is also of vital importance as methanol is one kind of chemical feedstock to synthesize rubber, dimethyl terephthalate (DMT), and other organic chemicals [31]. As novel catalytic method, photocatalysis has also been applied into this reaction. Thus, here some advanced researches are reviewed. The widely used photocatalysts are WO_{3-} [67] and Bi [72]-based catalysts. Katherine et al. [73] firstly inserted La into mesoporous WO_3, which was subsequently applied into photocatalytic methane steam reforming reaction. Results revealed that the addition of La could increase the amount of OH groups on the surface of mesoporous WO_3, leading to the enhanced yield and selectivity to methanol. Then Villa et al. [71] also investigated the effects of electron scavengers (Fe^{3+}, Cu^{2+}, Ag^+) and H_2O_2 species in the mesoporous WO_3 system for methane steam reforming. As shown in Table 13.2, the presence of Fe^{3+} promoted the methanol production about twice higher than that using unmodified mesoporous WO_3, for which the significantly facilitated electron–hole separation in this system was responsible. Meanwhile, the selectivity to methanol was also enhanced because of the presence of electron scavengers. Besides, $BiVO_4$ photocatalyst is also active in the reaction of photocatalytic methane steam reforming for methanol. As

Table 13.2 Hydrogen or methanol yield in methane steam reforming reaction using various photocatalysts.

Catalyst	Reactant gas	Hydrogen production (μmol min^{-1})	Methanol production (μmol g^{-1} h^{-1})	References
Pt/TiO_2	$CH_4 + H_2O(g)$	0.56	—	[55]
$Pt/NaTaO_3$:La	$CH_4 + H_2O(g)$	4.5	—	[65]
Rh/TiO_2	$CH_4 + H_2O(g)$	0.49	—	[55]
$Pt/CaTiO_3$ a)	$CH_4 + H_2O(g)$	0.38	—	[66]
$Pt/CaTiO_3$ b)	$CH_4 + H_2O(g)$	0.40	—	[66]
Pt/Ga_2O_3	$CH_4 + H_2O(g)$	0.65	—	[55]
Rh/Ga_2O_3	$CH_4 + H_2O(g)$	0.52	—	[55]
$Pt/K_2Ti_6O_{13}$	$CH_4 + H_2O(g)$	0.78	—	[55]
$Rh/K_2Ti_6O_{13}$	$CH_4 + H_2O(g)$	1.5	—	[55]
$Pt/Na_2Ti_6O_{13}$	$CH_4 + H_2O(g)$	0.56	—	[55]
$Rh/Na_2Ti_6O_{13}$	$CH_4 + H_2O(g)$	0.80	—	[55]
WO_3	$CH_4 + H_2O$	—	27.1	[71]
$BiVO_4$	$CH_4 + H_2O$	—	>100	[72]
Fe^{3+}/WO_3	$CH_4 + H_2O$	—	67.5	[71]
La-WO_3	$CH_4 + H_2O$	—	~30	[73]
HBEA	$CH_4 + H_2O$	—	10	[74]
V-HBEA	$CH_4 + H_2O$	—	11.3	[74]
Bi-V-HBEA	$CH_4 + H_2O$	—	10.7	[74]
V2O5	$CH_4 + H_2O$	—	2	[74]

a) Reaction temperature of 313 K.
b) Reaction temperature of 323 K.

reported by Sadtler and coworkers [72], after controlling different facets on $BiVO_4$ photocatalyst and employing them into photocatalytic methane steam reforming reaction, bipyramidal microcrystals were found to exhibit the highest capability for CH_3OH formation with an activity above 100 μmol g^{-1} h^{-1} and a selectivity above 80% during the first two hours of CH_4 oxidation. Furthermore, the reaction mechanism was also proposed, which can be referred in Eqs. (13.21–13.30):

$$h^+_{VB} + H_2O \rightarrow OH^{\cdot} + H^+ \tag{13.21}$$

$$OH^{\cdot} + CH_4 \rightarrow CH_3^{\cdot} + H_2O \tag{13.22}$$

$$CH_3^{\cdot} + H_2O \rightarrow CH_3OH + H^{\cdot} \tag{13.23}$$

$$e^-_{CB} + H^{\cdot} + H_2O \rightarrow H_2 + OH^- \tag{13.24}$$

$$e^-_{CB} + H^+ \rightarrow H_{ads}\cdot \tag{13.25}$$

$$OH\cdot + OH\cdot \rightarrow H_2O_2 \tag{13.26}$$

$$2e^-_{CB} + H_2O_2 + 2H^+ \rightarrow 2H_2O \tag{13.27}$$

$$CH_3\cdot + CH_3\cdot \rightarrow C_2H_6 \tag{13.28}$$

$$CH_4 + H_2O \rightarrow CH_3OH + H_2 \tag{13.29}$$

$$CH_4 + 2H_2O \rightarrow CO_2 + 4H_2 \tag{13.30}$$

Photocatalysis involved in the methane steam reforming is mainly reported for the products of hydrogen or methanol. Selective conversion of methane to highly value-added chemicals including CO and CH_3OH is still a challenge. Based on this, Wang and coworkers [26] exploited a PEC system for the reforming of methane. Using the TiO_2 prepared by atomic layer deposition (ALD) method, Pt wire and NaOH solution (pH = 13.6) worked as the photoelectrode, counter electrode and electrolyte, respectively. A high CO yield of 81.9% was obtained at a moderately positive applied potential (between 0.4 and 1.2 V vs. reversible hydrogen electrode [RHE]). Meanwhile, as studied the positive applied potentials lead to the high selectivity for CO. As a result, the combination of solar energy and electricity not only accelerated the mobility of photoinduced charge carriers but also could force charge carriers involving into target free radical reaction. The reaction mechanism using PEC method is shown in Figure 13.5a.

13.4.3 Methane Coupling

Photocatalysis was firstly involved into methane coupling reaction on the year of 1998 (Table 13.1, Entry 4) [21]. As revealed from the reaction free energy, the photocatalytic methane coupling reaction is endothermic; thus the reaction is extremely difficult to be initiated at room temperature, even under the UV irradiations. Lowering the reaction temperature to room temperature is a formidable task. Based on this, Yoshida et al. [77] successfully fabricated a SiO_2–Al_2O_3–TiO_2 ternary photocatalyst and applied it into photocatalytic methane coupling reaction at room temperature. As shown in Table 13.3, results indicated that SiO_2–Al_2O_3–TiO_2 catalyst with optimized Al and Ti concentrations exhibited about 20 times photocatalytic capability higher than that of the SiO_2–Al_2O_3 binary system. Subsequently, zeolite materials were found to be active and then used in photocatalytic methane coupling reaction. Chen and coworkers successively used Zn^{2+} [76] and Ga^{3+} [79] to modify the zeolite materials and then utilized them into photocatalytic methane coupling reaction. Results in Table 13.3 showed that both the insertion of these two ions could largely improve the methane conversion to ethane of which the enhancements were attributed to the longer lifetime of photoexcited charge carriers. The corresponding reaction mechanism and schematic energy diagram for the processes of the

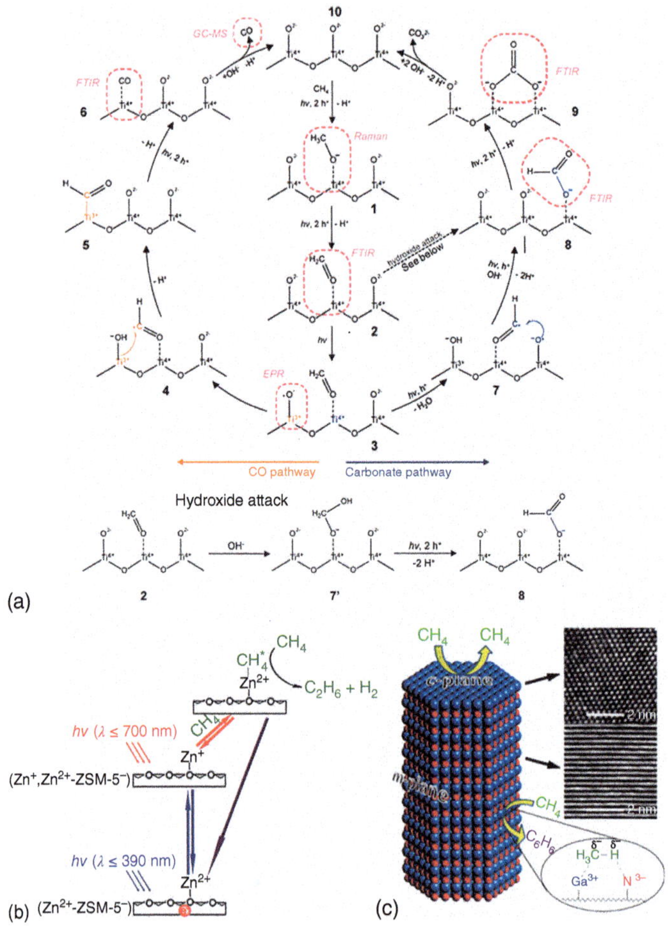

Figure 13.5 (a) Proposed mechanisms of photooxidation of CH_4 on TiO_2. Source: Li et al. [26] © 2018, American chemical society, (b) photocatalytic mechanism of methane coupling reaction over (Zn^+,Zn^{2+})-ZSM-5^- catalyst. Source: Li et al. [76], and (c) photocatalytic mechanism of methane dehydroaromatization reaction over GaN catalyst. Source: Li et al. [39].

photocatalytic reaction can be referred in Eqs. (13.31)–(13.34) and in Figure 13.5b.:

$$h^+_{VB} + CH_4 \rightarrow CH_3 \cdot + H^+ \tag{13.31}$$

$$e^-_{CB} + H^+ \rightarrow H_{ads}\cdot \tag{13.32}$$

$$CH_3 \cdot + CH_3\cdot \rightarrow C_2H_6 \tag{13.33}$$

$$H_{ads} \cdot + H_{ads}\cdot \rightarrow H_2 \tag{13.34}$$

Except the photocatalysis process for methane coupling reaction, PEC method is also reported to be used into methane activation to ethane. Amano et al. [48]

Table 13.3 Methane coupling reaction with photo-assisted catalysis.

Catalyst	Reactant gas	Ethane yield (%)[a)]	References
SiO_2	CH_4	0.02	[77]
Al_2O_3	CH_4	0.05	[77]
TiO_2	CH_4	0.11	[77]
SiO_2–Al_2O_3	CH_4	0.10	[77]
SiO_2–TiO_2	CH_4	0.09	[77]
Al_2O_3–TiO_2	CH_4	0.43	[77]
SiO_2–Al_2O_3–TiO_2	CH_4	2.07/1.3 $\mu mol\, g^{-1}\, h^{-1}$[b)]	[77]
MgO/SiO_2	CH_4	2.72	[78]
ZrO_2/SiO_2	CH_4	2.14	[78]
Ga_2O_3	CH_4	0.4	[56]
Ga_2O_3/SiO_2	CH_4	0.062	[56]
Zn^{2+}-ZSM-5^-	CH_4	9.8 $\mu mol\, g^{-1}\, h^{-1}$[b)]	[76]
Ga^{3+}-ETS-10	CH_4	29.8 $\mu mol\, g^{-1}\, h^{-1}$[b)]	[79]

a) Ethane concentration in effluent products.
b) Methane conversion rate.

recently prepared tungsten trioxide coated with a proton-conducting ionomer as the gas-diffusion photoanode that was used into PEC methane coupling reaction. The gas-phase PEC system revealed that incident photon-to-current conversion efficiency can reach as high as 11% under blue light at an applied voltage of 1.2 V. Meanwhile, the selectivity of homocoupling of CH_4 to form ethane (C_2H_6) can reach 54% of which the reason was attributed to the induced photogenerated holes.

13.4.4 Methane Oxidation

Methane oxidation reaction in the presence of light was firstly reported on the year of 1978 [21]. It was proposed that the photogenerated holes from metal oxides were responsible for the methane activation. As observed from Table 13.1, till now the reported works on photocatalytic methane oxidation mainly include total oxidation of methane and partial oxidation of methane to methanol processes. Reaction free energy reveals that both of these oxidation processes are exothermic. As the greenhouse effect caused by CH_4 is over 20 times than that caused from CO_2, it is necessary to convert methane to carbon dioxide via total oxidation process with the increasing amount of natural gas and shale gas [80]. Yi and coworkers [5] reported a Ag-decorated ZnO photocatalyst to convert methane under light irradiations. As studied, this novel nanocatalyst exhibited a high performance for methane oxidation under simulated sunlight illuminations, and the decoration of silver nanoparticle on the ZnO surface further improved the activity due to the SPR effect. The high quantum yield of 8% at wavelengths of below 400 nm and over

0.1% at wavelengths roughly 470 nm were obtained, showing a great promise for photocatalytic methane total oxidation. Meanwhile, the reaction mechanism was also investigated and referred in Eqs. (13.35)–(13.38). The intermediate product of formaldehyde can eventually be oxidized to CO_2 and H_2O with the assistance of active oxygen species:

$$e^-_{CB} + O_2 \rightarrow O_2^- \tag{13.35}$$

$$CH_3 \cdot + O_2^- \rightarrow CH_2O + OH^- \tag{13.36}$$

$$h^+_{VB} + O_2^- \rightarrow O^{2-} + HO_2\cdot \tag{13.37}$$

$$CH_3 \cdot + HO_2\cdot \rightarrow CH_3OOH \rightarrow HCHO + H_2O \tag{13.38}$$

Competing with the total oxidation of methane to carbon dioxide, partial oxidation of methane to methanol has more practical application value as the greenhouse gas can be converted to chemical feedstock. Recently β zeolite catalyst with a large amount of internal silanol groups was employed by Garcia and coworkers [57] to directly convert methane to methanol and other liquid C1 oxygenated products in the presence of air and water. With the assistance of deep UV light, a 13% conversion was achieved, and the reaction mechanism was proposed in Eqs. (13.39)–(13.41). Except air and O_2, other oxidizing agents were also exploited for prohibiting the over-oxidation of methane. Anpo and coworkers [81] took V-MCM-41 mesoporous molecular sieves as a catalyst and nitric oxide as the oxidized agent to partial oxidize methane to methanol. The methanol yield was in accord with the photoluminescence yield of the isolated tetrahedrally coordinated V-oxide species, proving that the active sites in this partial oxidation reaction were triplet state of these species excited by charge transfer. Furthermore, Ohkubo and Hirose [82] used chlorine dioxide radical as the oxidizing agent aerobic oxygenation of methane to methanol and formic acid under photoirradiation. The methanol yield was 14%, with a methane conversion of 99% without formation of the further oxygenated products such as CO_2 and CO:

$$\text{catalyst} - OH + \text{UV light} \rightarrow \text{catalyst} - O \cdot + H\cdot \tag{13.39}$$

$$\text{catalyst} - O \cdot + CH_3\cdot \rightarrow \text{catalyst} - OCH_3 \tag{13.40}$$

$$\text{catalyst} - OCH_3 + H_2O \rightarrow \text{catalyst} - OH + CH_3OH \tag{13.41}$$

13.4.5 Methane Dehydroaromatization

Most above involved reactions for photo-assisted methane conversion are limited to C1 conversion in which the products are mainly CO, CO_2, and CH_3OH, except for ethane from the methane coupling reaction. Methane dehydroaromatization reaction is more difficult not only on the break of C–H bonding but also on the growth and cyclization of C chain. Indeed, as shown in Table 13.1, the reaction free energy of methane dehydroaromatization reaction is 434 kJ mol^{-1}, proving the hardship of feasibility on this reaction. Fortunately, Li et al. [39] recently discovered that Si-doped GaN nanowires (NWs) with a 97% rationally constructed m-plane can directly convert methane into benzene and hydrogen under UV illumination at room temperature. As illustrated from Figure 13.5c, the exposed m-plane of GaN and light irradiation were responsible for methane C—H bond activation and the formation of benzene.

13.5 Conclusions and Perspectives

To conclude, this chapter firstly introduces the categories of photo-assisted catalysis for methane activation, including photocatalysis, plasma-based photocatalysis, and PEC catalysis processes. Then the differences between photocatalysis and thermo-catalysis in terms of catalyst, reactor, and reaction mechanism are also analyzed in detail. Then, recent developments in employing photo-assisted catalysis for methane conversion reactions including methane dry reforming, methane steam reforming, methane coupling, methane oxidation, and methane dehydroaromatization reactions are summarized. Through this study, we can conclude that solar energy can initiate the methane conversion reactions even at room temperature where thermo-catalysis cannot. However, the low quantum efficiency using photo-assisted catalysis indicates that further endeavors are still required; thus at last several strategies are proposed. This chapter aims to provide a compact source of relevant information to a wide readership.

As a new-fashioned mean of catalysis, photocatalysis has made itself conspicuous on the methane activation to C1, C2, and even aromatic hydrocarbons, and relatively satisfied results have been achieved at the present. However, its low quantum efficiency is always a criticism that is still far from the requirements of practical application. It is demanding to elevate the quantum efficiency of photocatalysis despite the high cost it will take. As a result, endeavors are still required, and several strategies are here proposed for the further studies:

(1) For sake of promoting the quantum efficiency of photocatalytic methane conversion, it is widely accepted that exploitation of photocatalysts with a high performance is the key to realize the practical application of photocatalysis. Now, most works mainly focus on enhancing the visible-light absorption, boosting the separation of photogenerated charge carriers, and improving electrical conductivity of photocatalyst. Besides, discovering and exploiting novel semiconductors with ultrahigh photocatalytic capabilities are still crucial. Furthermore, recently plasma-based photocatalysis for energy preparation is a hot spot, which seems a sally port for the conversion of solar energy to chemical fuels.
(2) Except the development of photocatalyst, the reactor categories also determine the final quantum efficiency. As reported, the reactor could influence the receiving of the solar light and light-to-heat conversion efficiency. As a result, design and preparation of new photocatalytic reactor are also important for methane

conversion. Besides, optimization of reaction conditions such as temperature and pressure could also promote the quantum efficiency of photocatalysis.

(3) Previously reported photocatalytic processes for methane activation are limited to the reactions mentioned above. Therefore, the introduction of solar energy into novel processes for methane conversion would also be necessary especially for some tough reactions even using thermocatalysis.

(4) Multidiscipline such as PEC catalysis and photothermal catalysis can be used for methane conversion reaction as the addition of thermal or electrical energy into photocatalysis can effectively facilitate the mobility of charge carriers, improving the quantum efficiency. Besides, combination of several reactions into one system could also promote the consumption of by-products and correspondingly improve the yield of target products.

Acknowledgment

This work was partially supported by the Australian Research Council (DP170104264).

References

1 (a) De Smit, E. and Weckhuysen, B.M. (2008). *Chem. Soc. Rev.* 37: 2758. (b) Li, S. and Gong, J. (2014). *Chem. Soc. Rev.* 43: 7245. (c) Pakhare, D. and Spivey, J. (2014). *Chem. Soc. Rev.* 43: 7813. (d) Rodriguez, J.A., Grinter, D.C., Liu, Z. et al. (2017). *Chem. Soc. Rev.* 46: 1824. (e) Wang, W., Wang, S., Ma, X., and Gong, J. (2011). *Chem. Soc. Rev.* 40: 3703.

2 Song, H., Meng, X., Wang, Z.-j. et al. (2019). *Joule* 3: 1606.

3 Guo, X., Fang, G., Li, G. et al. (2014). *Science* 344: 616.

4 Hashiguchi, B.G., Konnick, M.M., Bischof, S.M. et al. (2014). *Science* 343: 1232.

5 Chen, X., Li, Y., Pan, X. et al. (2016). *Nat. Commun.* 7: 12273.

6 (a) Dietl, N., Schlangen, M., and Schwarz, H. (2012). *Angew. Chem. Int. Ed.* 51: 5544. (b) Schwach, P., Pan, X., and Bao, X. (2017). *Chem. Rev.* 117: 8497. (c) Tang, P., Zhu, Q., Wu, Z., and Ma, D. (2014). *Energy Environ. Sci.* 7: 2580.

7 (a) Alvarez-Galvan, M.C., Mota, N., Ojeda, M. et al. (2011). *Catal. Today* 171: 15. (b) Schulz, H. (1999). *Appl. Catal., A* 186: 3.

8 Che, F., Gray, J.T., Ha, S., and McEwen, J.-S. (2016). *ACS Catal.* 7: 551.

9 Kechagiopoulos, P.N., Angeli, S.D., and Lemonidou, A.A. (2017). *Appl. Catal., B* 205: 238.

10 (a) Asahi, R., Morikawa, T., Ohwaki, T. et al. (2001). *Science* 293: 269. (b) Kubacka, A., Fernandez-Garcia, M., and Colon, G. (2012). *Chem. Rev.* 112: 1555. (c) Liu, J., Liu, Y., Liu, N.Y. et al. (2015). *Science* 347: 970.

11 Fujishima, A. and Honda, K. (1972). *Nature* 238: 37.

12 Zhang, X., Li, X., Zhang, D. et al. (2017). *Nat. Commun.* 8: 14542.

13 Zhang, H., Wang, T., Wang, J. et al. (2016). *Adv. Mater.* 28: 3703.

14 Meng, X., Wang, T., Liu, L. et al. (2014). *Angew. Chem. Int. Ed.* 53: 11478.

15 (a) Sastre, F., Puga, A.V., Liu, L. et al. (2014). *J. Am. Chem. Soc.* 136: 6798. (b) Kho, E.T., Tan, T.H., Lovell, E. et al. (2017). *Green Energy Environ.* 2: 204. (c) Chen, G., Gao, R., Zhao, Y. et al. (2018). *Adv. Mater.* 30: 1704663. (d) Guo, X.-N., Jiao, Z.-F., Jin, G.-Q., and Guo, X.-Y. (2015). *ACS Catal.* 5: 3836.

16 Liu, H., Meng, X., Dao, T.D. et al. (2015). *Angew. Chem. Int. Ed.* 54: 11545.

17 Han, B., Wei, W., Chang, L. et al. (2015). *ACS Catal.* 6: 494.

18 (a) Kale, M.J., Avanesian, T., and Christopher, P. (2013). *ACS Catal.* 4: 116. (b) Christopher, P., Xin, H., and Linic, S. (2011). *Nat. Chem.* 3: 467.

19 (a) Jia, J., O'Brien, P.G., He, L. et al. (2016). *Adv. Sci.* 3: 1600189. (b) Hoch, L.B., O'Brien, P.G., Jelle, A. et al. (2016). *ACS Nano* 10: 9017. (c) Meng, X., Liu, L., Ouyang, S. et al. (2016). *Adv. Mater.* 28: 6781.

20 Bitter, J.H., Seshan, K., and Lercher, J.A. (1999). *J. Catal.* 183: 336.

21 Yuliati, L. and Yoshida, H. (2008). *Chem. Soc. Rev.* 37: 1592.

22 Hisatomi, T., Kubota, J., and Domen, K. (2014). *Chem. Soc. Rev.* 43: 7520.

23 Chen, X., Shen, S., Guo, L., and Mao, S.S. (2010). *Chem. Rev.* 110: 6503.

24 (a) Baffou, G. and Quidant, R. (2014). *Chem. Soc. Rev.* 43: 3898. (b) Wang, C., Ranasingha, O., Natesakhawat, S. et al. (2013). *Nanoscale* 5: 6968.

25 Kuang, Y., Yamada, T., and Domen, K. (2017). *Joule* 1: 290.

26 Li, W., He, D., Hu, G. et al. (2018). *ACS Cent. Sci.* 4: 631.

27 (a) Linic, S., Christopher, P., and Ingram, D.B. (2011). *Nat. Mater.* 10: 911. (b) Linic, S., Aslam, U., Boerigter, C., and Morabito, M. (2015). *Nat. Mater.* 14: 567. (c) Zhang, Y., He, S., Guo, W. et al. (2018). *Chem. Rev.* 118: 2927.

28 Brongersma, M.L., Halas, N.J., and Nordlander, P. (2015). *Nat. Nanotechnol.* 10: 25.

29 Govorov, A.O., Zhang, H., Demir, H.V., and Gun'ko, Y.K. (2014). *Nano Today* 9: 85.

30 (a) Kwon, Y., Kim, T.Y., Kwon, G. et al. (2017). *J. Am. Chem. Soc.* 139: 17694. (b) Zuo, Z., Ramirez, P.J., Senanayake, S.D. et al. (2016). *J. Am. Chem. Soc.* 138: 13810. (c) Cui, X., Li, H., Wang, Y. et al. (2018). *Chem* 4: 1902. (d) Ikuno, T., Zheng, J., Vjunov, A. et al. (2017). *J. Am. Chem. Soc.* 139: 10294.

31 Ravi, M., Ranocchiari, M., and van Bokhoven, J.A. (2017). *Angew. Chem. Int. Ed.* 56: 16464.

32 Xie, P., Pu, T., Nie, A. et al. (2018). *ACS Catal.* 8: 4044.

33 Ab Rahim, M.H., Forde, M.M., Jenkins, R.L. et al. (2013). *Angew. Chem. Int. Ed.* 52: 1280.

34 Tang, Y., Li, Y., Fung, V. et al. (2018). *Nat. Commun.* 9: 1231.

35 Osadchii, D.Y., Olivos-Suarez, A.I., Szécsényi, Á. et al. (2018). *ACS Catal.* 8: 5542.

36 Khodakov, A.Y., Chu, W., and Fongarland, P. (2007). *Chem. Rev.* 107: 1692.

37 Zhao, W., Ma, W.H., Chen, C.C. et al. (2004). *J. Am. Chem. Soc.* 126: 4782.

38 Wang, L., Wang, Y., Cheng, Y. et al. (2016). *J. Mater. Chem. A* 4: 5314.

39 Li, L., Fan, S., Mu, X. et al. (2014). *J. Am. Chem. Soc.* 136: 7793.

40 Pan, C., Xu, J., Wang, Y. et al. (2012). *Adv. Funct. Mater.* 22: 1518.

41 Li, X.H., Chen, J.S., Wang, X.C. et al. (2011). *J. Am. Chem. Soc.* 133: 8074.

42 Xie, J., Jin, R., Li, A. et al. (2018). *Nat. Catal.* 1: 889.

43 Ren, J., Ouyang, S., Xu, H. et al. (2017). *Adv. Energy Mater.* 7: 1601657.

44 Liu, L., Dao, T.D., Kodiyath, R. et al. (2014). *Adv. Funct. Mater.* 24: 7754.

45 Liu, H., Dao, T.D., Liu, L. et al. (2017). *Appl. Catal., B* 209: 183.

46 Yoshida, H., Chaskar, M.G., Kato, Y., and Hattori, T. (2003). *J. Photochem. Photobiol., A* 160: 47.

47 Yoshida, H., Hirao, K., Nishimoto, J.I. et al. (2008). *J. Phys. Chem. C* 112: 5542.

48 Amano, F., Shintani, A., Tsurui, K. et al. (2019). *ACS Energy Lett.* 4: 502.

49 Kaliaguine, S.L., Shelimov, B.N., and Kazansky, V.B. (1978). *J. Catal.* 55: 384.

50 (a) Du, A.J., Sanvito, S., Li, Z. et al. (2012). *J. Am. Chem. Soc.* 134: 4393. (b) Wu, W., Zhang, J., Fan, W. et al. (2016). *ACS Catal.* 6: 3365.

51 (a) Wang, X., Maeda, K., Thomas, A. et al. (2009). *Nat. Mater.* 8: 76. (b) Sun, J., Zhang, J., Zhang, M. et al. (2012). *Nat. Commun.* 3: 1139.

52 Kudo, A. and Miseki, Y. (2009). *Chem. Soc. Rev.* 38: 253.

53 Chen, C., Ma, W., and Zhao, J. (2010). *Chem. Soc. Rev.* 39: 4206.

54 Liu, H., Li, M., Dao, T.D. et al. (2016). *Nano Energy* 26: 398.

55 Shimura, K., Kawai, H., Yoshida, T., and Yoshida, H. (2012). *ACS Catal.* 2: 2126.

56 Yuliati, L., Hattori, T., Itoh, H., and Yoshida, H. (2008). *J. Catal.* 257: 396.

57 Sastre, F., Fornes, V., Corma, A., and Garcia, H. (2011). *J. Am. Chem. Soc.* 133: 17257.

58 Yuliati, L., Itoh, H., and Yoshida, H. (2008). *Chem. Phys. Lett.* 452: 178.

59 Teramura, K., Tanaka, T., Ishikawa, H. et al. (2004). *J. Phys. Chem. B* 108: 346.

60 (a) Luther, J.M., Jain, P.K., Ewers, T., and Alivisatos, A.P. (2011). *Nat. Mater.* 10: 361. (b) Abe, T., Tanizawa, M., Watanabe, K., and Taguchi, A. (2009). *Energy Environ. Sci.* 2: 315.

61 (a) Kawi, S., Kathiraser, Y., Ni, J. et al. (2015). *ChemSusChem* 8: 3556. (b) Wang, Y., Yao, L., Wang, S. et al. (2018). *Fuel Process. Technol.* 169: 199.

62 Aramouni, N.A.K., Touma, J.G., Tarboush, B.A. et al. (2018). *Renewable Sustainable Energy Rev.* 82: 2570.

63 Song, H., Meng, X., Dao, T.D. et al. (2018). *ACS Appl. Mater. Interfaces* 10: 408.

64 Huang, H., Mao, M., Zhang, Q. et al. (2018). *Adv. Energy Mater.* 8: 1702472.

65 Shimura, K., Kato, S., Yoshida, T. et al. (2010). *J. Phys. Chem. C* 114: 3493.

66 Shimura, K. and Yoshida, H. (2010). *Energy Environ. Sci.* 3: 615.

67 (a) Gondal, M.A., Hameed, A., and Suwaiyan, A. (2003). *Appl. Catal., A* 243: 165. (b) Gondal, M.A., Hameed, A., Yamani, Z.H., and Arfaj, A. (2004). *Chem. Phys. Lett.* 392: 372.

68 Taylor, C.E. (2003). *Catal. Today* 84: 9.

69 Yu, L., Shao, Y., and Li, D. (2017). *Appl. Catal., B* 204: 216.

70 Shimura, K., Yoshida, T., and Yoshida, H. (2010). *J. Phys. Chem. C* 114: 11466.

71 Villa, K., Murcia-López, S., Andreu, T., and Morante, J.R. (2015). *Appl. Catal., B* 163: 150.

72 Zhu, W., Shen, M., Fan, G. et al. (2018). *ACS Appl. Nano Mater.* 1: 6683.

73 Villa, K., Murcia-López, S., Morante, J.R., and Andreu, T. (2016). *Appl. Catal., B* 187: 30.

74 Murcia-López, S., Bacariza, M.C., Villa, K. et al. (2017). *ACS Catal.* 7: 2878.

75 Shimura, K., Maeda, K., and Yoshida, H. (2011). *J. Phys. Chem. C* 115: 9041.
76 Li, L., Li, G.D., Yan, C. et al. (2011). *Angew. Chem. Int. Ed.* 50: 8299.
77 Yoshida, H., Matsushita, N., Kato, Y., and Hattori, T. (2003). *J. Phys. Chem. B* 107: 8355.
78 Yuliati, L., Hattori, T., and Yoshida, H. (2005). *Phys. Chem. Chem. Phys.* 7: 195.
79 Li, L., Cai, Y.Y., Li, G.D. et al. (2012). *Angew. Chem. Int. Ed.* 51: 4702.
80 Tavasoli, A. and Ozin, G. (2018). *Joule* 2: 571.
81 Hu, Y., Higashimoto, S., Takahashi, S. et al. (2005). *Catal. Lett.* 100: 35.
82 Ohkubo, K. and Hirose, K. (2018). *Angew. Chem. Int. Ed.* 57: 2126.

14

Photocatalytic and Photoelectrochemical Reforming of Biomass

Xiaoqing Liu, Wei Wei and Bing-Jie Ni

University of Technology Sydney, Centre for Technology in Water and Wastewater, School of Civil and Environmental Engineering, Sydney, NSW 2007, Australia

14.1 Introduction

The depletion of fossil fuel resources and the resulting adverse effects on the global environment and climate are of major academic, economic, and political concerns worldwide [1]. The demands for energy and commodities increase steadily due to the population and economic growth, as well as the advances in lifestyle. To alleviate these issues and reduce our dependence on fossil fuels, it is urgent and necessary to exploit renewable feedstocks, typically biomass to obtain energy and valuable chemicals [2–5].

Every year, 170 billion tons of biomass are produced globally [6]. Only 3% of the global biomass is cultivated and employed in food and non-food applications [5]. Carbohydrates (75%) and lignin (20%) represent the two largest fractions of biomass and consequently play a major role in the industrial utilization of biomass (Figure 14.1). The remaining 5% comprise triglycerides (fats and oils), proteins, and terpenes. The formation of biodiesel from triglycerides is an exemplary real-world application for biomass conversion [7]. Glycerol is the main product in the biodiesel production process [5]. Carbohydrates can be divided into storage carbohydrates, including starch, inulin, and sucrose, and structural polysaccharides, such as cellulose, hemicellulose, and chitin. In addition, aquatic carbohydrates derived from micro- and macroalgae, consist of a variety of polysaccharides that differ in structure from those of terrestrial biomass.

Lignocellulose, the fibrous material that constitutes the cell walls of plants, is available in very large quantities. Lignocellulose is much more difficult to process than the first-generation renewable feedstocks (sugars, starches, and vegetable oils). Nevertheless, it has the advantage that it is not edible, not competing with food supply. Lignocellulose is the most abundant renewable biomass on earth [8]. The major components of lignocellulose are cellulose (35–50%), hemicellulose (20–30%), and lignin (10–25%). Cellulose is a linear crystalline polymer composed of glucose units. Hemicellulose is an amorphous branched polymer. Lignin is an amorphous polymer

Solar-to-Chemical Conversion: Photocatalytic and Photoelectrochemcial Processes, First Edition.
Edited by Hongqi Sun.

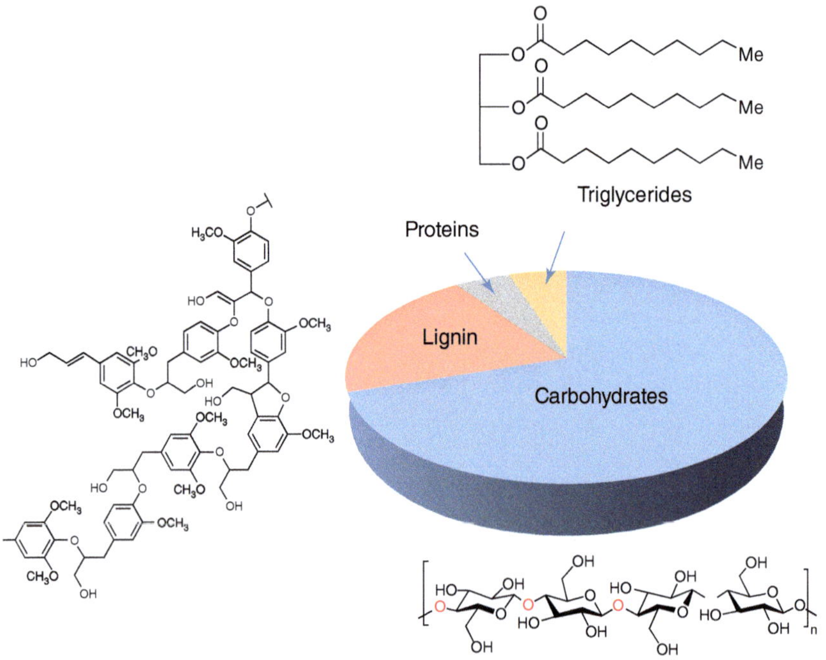

Figure 14.1 Composition of biomass.

of aromatic allylic alcohols that is resistant to hydrolysis and cannot be utilized by fermentation [9].

Biorefinery is the sustainable reforming of biomass into a variety of energy and marketable products. Cost-effective processes adapted to reform biomass molecules with abundant functional groups are required to make the quality and price of biomass-derived chemicals competitive with traditional commodities obtained from fossil fuels [10]. Various methods have been explored for the utilization of biomass, including microbially mediated reforming methods and thermochemical conversion approaches. The microbially mediated transformations, including anaerobic and aerobic digestion, are widely studied to convert biomass into valuable products [11–13]. The disadvantages of microbial methods are that they are time-consuming and difficult to control, because a series of metabolic reactions such as hydrolysis, acidogenesis, and methanogenesis are involved in the microbial processes [14]. Thermochemical methods can be used to convert biomass into biofuels or chemical feedstocks and biochar. Pyrolysis, gasification, and hydrothermal liquefaction are the main thermochemical technologies. They are usually conducted under harsh conditions, such as high pressure/temperature, thus requiring intensive thermal energy consumption [15]. For example, pyrolysis is often conducted at temperature range from 573 to 1073 K [16]. Hydrogenation of biomass is typically operated at high pressures ranging from 2.0 to 5.5 MPa [17, 18]. Besides, biomass conversion under these conditions often results in nonselective bond breaking and low-functionalized products [19].

Photochemistry continues to be an important component of modern chemistry in the twenty-first century [20]. An increasing interest are being paid to this field during the last three decades, especially titanium oxide chemistry [21]. Increasing amount of knowledge is obtained on mechanisms of photocatalysis, new technologies for storage and conversion of solar energy, and environmental remediation of liquid and gaseous ecosystems [22–24]. Recently, a new research avenue related to photocatalytic conversion of biomass has emerged as a promising alternative to conventional catalyzed processes [25]. Solar-energy-driven catalytic transformation, generally conducted under ambient conditions, appears to be a green strategy for the selective depolymerization of biomass [7, 26]. In addition, photogenerated electrons and holes can be used as reductants and oxidants simultaneously, enabling the accomplishment of one-step redox-neutral reactions [27]. Photoelectrochemical (PEC) reforming of biomass is an attractive research with many advantages. It is a technologically simple process that can be carried out under mild conditions and with high conversion [28]. Photoreforming of biomass mainly produces fuels including hydrogen and electricity and a few kinds of useful chemicals. In PEC fuel cell, biomass can be consumed directly, and solar energy can be converted into electricity or biofuel.

This chapter aims to provide an overview of recent work about the photocatalytic conversion of native lignocellulosic biomass, lignin, carbohydrates, and glycerol to corresponding chemicals of high value and hydrogen. Electricity, hydrogen, and some value-added chemical production from PEC reforming of biomass are also discussed.

14.2 Fundamentals of Photocatalytic and Photoelectrochemical Processes

14.2.1 Photocatalytic Process

An electron (e^-) will be excited from the valence band (VB) to the conduction band (CB), when a semiconductor photocatalyst is illuminated by light with energy exceeding its bandgap energy (E_{BG}), leaving a vacancy called *hole* (h^+) on the VB. The electrons on the CB can promote fuel-forming hydrogen (H_2) evolution reaction [29]. Biomass and its derivatives act as hole scavengers that can be used to consume photo-induced holes, thus suppressing e^-–h^+ recombination and increasing the efficiency of water-splitting processes [30]. In this process, CO_2 is produced when lignocellulosic biomass is completely oxidized [31]. Value-added products could be produced when biomass-derived intermediate compounds are selectively partly oxidized [32]. On the one hand, the holes (redox potential +2.53 V vs. standard hydrogen electrode [SHE]) generated in photocatalysis have a high potential to oxidize organic species directly. On the other hand, the positive holes can also react with water to form $^{\bullet}OH$ radical ($^{\bullet}OH$, redox potential +2.81 V vs. SHE) to oxidize the target compound [33]. Other reactive oxygen species (ROS) may also be formed in a photocatalytic process, such as $O_2^{\bullet-}$ (redox potential +0.89 V vs. SHE), O_3^-, and $^1O_2$ [25]. ROS with different oxidation potential may lead to different products in the photocatalytic process, due to different redox capability. For example, the oxidation potential of $O_2^{\bullet-}$ radical is relatively weaker than that of $^{\bullet}OH$. Thus, the oxidation reactions by $O_2^{\bullet-}$ radical could be incomplete. Therefore, the controllable generation of ROS is crucial in heterogeneous photocatalysis to design and predict pathways for selective organic photocatalytic oxidations.

14.2.2 Photoelectrochemical Process

PEC process is a new advanced technology, which could combine both the advantages of photocatalytic and electrocatalytic activities. PEC reactions are being studied to produce fuels and chemicals from biomass and its derivatives but still in the initial stage [34]. A typical PEC cell is comprised by a photoactive semiconductor electrode, a counterelectrode (e.g. Pt or p-type semiconductor) and electrolyte. The anode and cathode chambers are generally separated by an ion transfer membrane [35, 36]. Semiconductor electrodes can absorb light to generate e^-–h^+ pairs, and electrons could be excited to a higher energy level. These separated electrons and holes can be transported to the semiconductor/electrolyte interface, and redox reactions will be carried out [37]. An n-type semiconductor will serve as photoanode, while Pt or a p-type semiconductor will serve as a photocathode [38]. This n-type semiconductor electrode is usually named "photoanode," on which oxidation reactions take place and consume photogenerated holes. The potential of the cathode depends on the presence or absence of O_2, when the pH of electrolyte is zero. In the absence of O_2, the cathode behaves as a H_2 evolution electrode, the potential of which is conventionally taken equal to zero. In the presence of O_2, the cathode behaves as an O_2 electrode [39]. The potential of cathode is related to the following two reductive reactions [40]:

$$O_2 + 4H^+ + 4e^- \rightarrow 2H_2O\ (+1.23\ V) \tag{14.1}$$

$$O_2 + 2H^+ + 2e^- \rightarrow H_2O_2\ (+0.68\ V) \tag{14.2}$$

In a typical PEC cell, the performance is largely determined by the properties of the photoanode. For example, a photoanode with conduction band minimum (CBM) and valence band maximum (VBM) that could straddle the redox potentials of water can drive the overall PEC water splitting, while others could not. Additionally, fast charge transport and good absorption properties of the photoanode could significantly enhance the efficiency of water splitting. However, semiconductors that meet this requirement while also having a bandgap that enables a higher solar-to-hydrogen (STH) conversion efficiency than 10% are rare. PEC cells contain an anode and a cathode both of which are semiconductors would be more efficient, because both the photoanode and photocathode will contribute to the photon absorption and photovoltage generation [37]. Besides, an anode reaction that has more favorable kinetics and allows selective generation of value-added chemicals would benefit to increase the overall efficiency and utility of PEC cells [41].

14.3 Photocatalytic Reforming of Biomass

14.3.1 Photocatalytic Reforming of Lignin

The photoconversion processes of native and processed lignin to various valuable chemicals are concluded, as shown in Table 14.1. Titanium dioxide (TiO_2)-based photocatalysts were the most widely used for photocatalytic lignin reforming. Metal nanoparticles (NPs) deposition on the surface of TiO_2 was an effective way to optimize its photocatalytic performance. Bi and Pt co-modified TiO_2 was prepared and applied for photooxidation of lignin under solar light irradiation to produce guaiacol, vanillic acid, vanillin, and 4-phenyl-1-buten-4-ol. Bi and Pt NPs provided active sites on TiO_2 surface, which are essential for production of $O_2^{\bullet -}$ rather than $^{\bullet}OH$. Therefore, Bi/Pt–TiO_2 enhances selectivity toward desired products from lignin, as $O_2^{\bullet -}$ is milder than $^{\bullet}OH$ [42]. TiO_2/lignin-based carbon composite photocatalysts were studied for photocatalytic conversion of lignin [43]. The catalyst derived from TiO_2/lignin with ratio of 1 : 0.5 exhibited the best photocatalytic performance with the efficiency of lignin conversion up to 40.28% under UVA illumination, and commodity chemicals such as vanillin were produced.

Silver-based photocatalysts were also used for photocatalytic reforming of lignin. Silver-based photocatalysts exhibited good performance under visible light irradiation. A novel Ag_3PO_4/SnO_2/porcine bone composite showed good performance for conversion of lignosulfonate to small molecules (Table 14.1) [44].

Metal sulfide semiconductors are a class of photocatalysts with good visible light responses [50] and high photocatalytic activities, as well as good performance for selectively cleaving β-O-4 bond in lignin [51, 52]. Wang and coworkers reported the efficient cleavage of lignol β-O-4 bonds over $Zn_4In_2S_7$ upon visible light illumination [27]. Mild reaction conditions enable the preservation of functional groups such as phenol and methoxy groups in the substrate. Approximately 18.4 wt% of aromatic monomers were obtained for the photocatalytic conversion of dioxasolv birch lignin. The major soluble products detected are ketones, including syringyl- (S–) and guaiacyl- (G–) ketones and alcohols (sinapyl and coniferyl alcohols, belonging to S and G units, respectively) (Figure 14.2a). The S/G ratio of products is 3.2, which is close to that of the original lignin (2.8). Mechanistic studies revealed that cleavage of the β-O-4 bond over $Zn_4In_2S_7$ is through a one-step C_α radical intermediate pathway. C_α radical intermediate will be formed via the oxidation of C_α—H bond by photogenerated holes in the VB, which could then undergo reductive cleavage of the C_β—O bond by accepting photogenerated electrons from the CB (Figure 14.2b).

Photocatalysis has also been combined with biocatalytic, mechanical, chemical, and electrochemical approaches to enhance the conversion efficiency of lignin. For instance, photocatalytic processes (TiO_2/UV) and biocatalytic processes (laccase from *Trametes versicolor* served as the biocatalyst) were combined to degrade high-molecular-weight lignin. This dual-step process with successive photocatalysis and laccase treatment demonstrated outstanding performance [45]. A large amount of succinic and malonic acids were obtained in the TiO_2/UV system with the presence of H_2O_2. These by-products have potential to be used as precursors to

Table 14.1 Photocatalytic conversion of processed lignin in literatures.

Substrate	Photocatalyst	Light source	Media	Illumination time (h)	Conversion (%)	Main products (yield, %)	References
Lignosulfonate (100 mg l^{-1})	Bi1%/Pt1%–TiO_2 (1 g l^{-1})	Xe lamp 300 W	Water (100 ml)	1	84.5	Guaiacol (22.7%)	[42]
Kraft lignin (500 mg l^{-1})	TiO_2/carbon (1 g l^{-1})	Mercury lamp 400 W	Water/acetonitrile (20 : 80 V/V)	5	40.3	Vanillin (1.09%)	[43]
Lignosulfonate (50 mgl^{-1})	Ag_3PO_4/SnO_2/ porcine bone (50 mg)	Xe lamp 300 W	Water (100 ml)	7	NA	Alkyl acids	[44]
Lignin (1 g l^{-1})	TiO_2 (3 g l^{-1}) Laccase H_2O_2 (5.55 g l^{-1})	UV irradiation 24 W	Sodium acetate buffer solution (pH 5.0)	24	100	Organic acids Fatty acids Carbohydrates	[45]
1 : 1 w/w lignin-TiO_2 ball-milled mixtures (0.4 g)	0.4 g of 1 : 1 w/w lignin-TiO_2 ball-milled mixtures	125 W High pressure mercury light	Water (200 ml)	6	NA	Ethyl benzene, vanillin, acetovanillone, acetosyringone, syringaldehyde	[46]
Organosolv lignin black liquor (50 ml)	TiO_2 (0.1 g)	300 W UV lamp	60% (v/v) Ethanol solution, H_2SO_4 added (pH 2)	0.5	Oil (20%)	Syringaldehyde Syringol Pirocathecol Vanillin Sinapylaldehyde	[47]
Kraft lignin	$TiO_2/Ti/Ta_2O_5–IrO_2$ as working electrode and photocatalyst	Blue Wave TM 50 AS UV spot lamp	0.5 M NaOH	2	92	Vanillin and vanillic acid	[48]
Dioxasolv birch lignin (20 mg)	$Zn_4In_2S_7$ (10 mg)	Xe lamp (λ = 400–780 nm)	CH_3OH (5 ml)	24	NA	S-Ketone, G-ketone, S-alcohol, G-alcohol (18.4 wt%)	[27]

NA: not available.
Source: Liu et al. [49]. © 2019, The Royal Society of Chemistry.

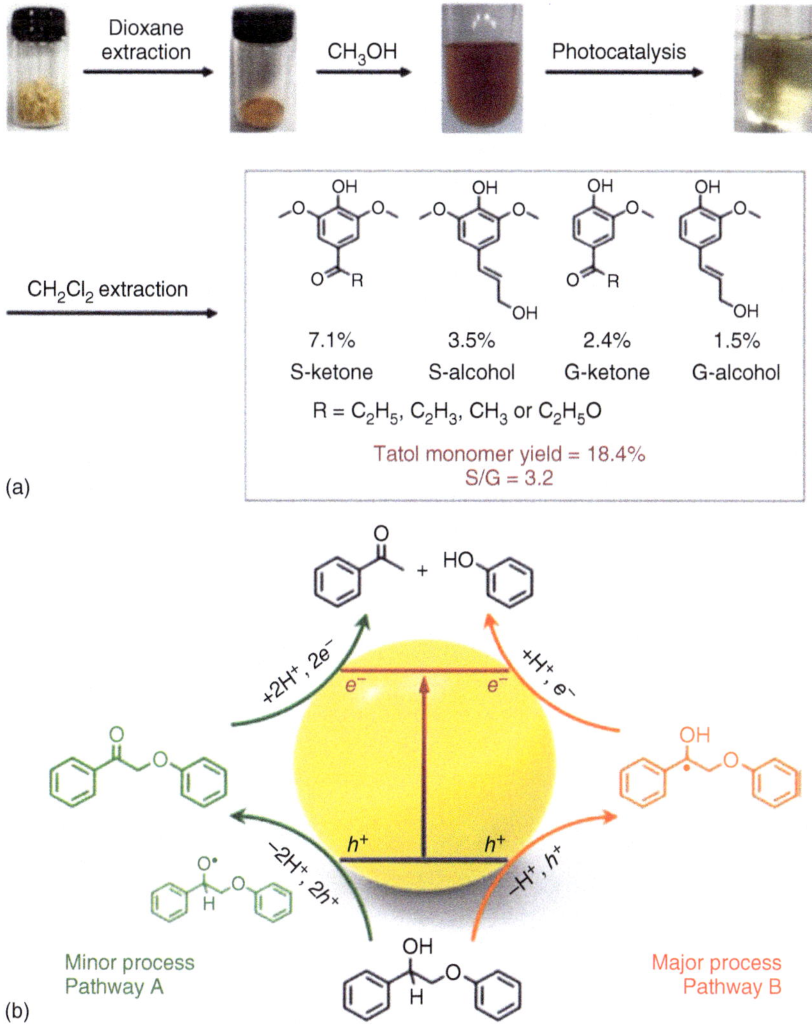

Figure 14.2 (a) Reaction process images, quantification, and distribution of products obtained from the photocatalytic depolymerization of birch lignin. (b) Proposed mechanism for β-O-4 bond cleavage in the photocatalytic conversion of lignin over the $Zn_4In_2S_7$ catalyst. Source: Reprinted with permission from Lin et al. [27].

produce a variety of useful chemicals. Li et al. demonstrated the feasibility of using photocatalytic degraded lignin for methane production. Lignin was degraded over Ag–AgCl/ZnO under natural solar light irradiation [53]. The lignin solution obtained from 120 minutes photocatalytic treatment produced more methane and biogas than the control, yielding 184 ml methane and 325 ml biogas for per gram of removed total organic carbon. $^{\bullet}$OH radical is the dominant oxidant for lignin degradation over Ag–AgCl/ZnO.

Lignin and TiO_2 mixtures were firstly treated by ball milling, followed by UV irradiation using TiO_2 as photocatalyst. This two-step process enabled the production of

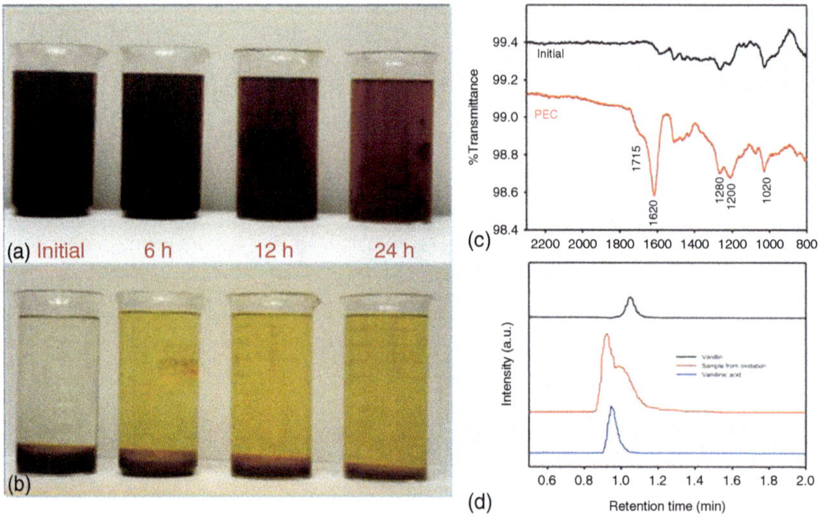

Figure 14.3 Photographs taken before (a) and after (b) precipitation of lignin during photoelectrocatalytic oxidation of 500 ppm lignin at different times, (c) Fourier transform infrared (FTIR) spectra of lignin and modified lignin, and (d) intermediates analyzed using HPLC. Source: Reprinted with permission from Tian et al. [48].

valuable phenolics and aromatic hydrocarbons from lignin. Wet milling using acetone and water resulted in a high yield of phenolic chemicals after three to four hours of UV exposure [46]. Pulping methods used include organosolv pulping and [Bmim][$MeSO_4$] pulping. UV irradiation time also has significant effects on photocatalytic cracking of lignin, and one hour of exposure time to UV light irradiation results in the highest yields of aromatic products [47]. The principal products obtained are syringaldehyde, pyrocatechol, and raspberry ketone.

Photochemical and electrochemical methods were also integrated to oxidize lignin. A Ta_2O_5–IrO_2 thin film was used as the electrocatalyst, and TiO_2 nanotube arrays were used as the photocatalyst [48]. When a photocatalyst is combined with an electrocatalyst, the applied potential not only drives the electrochemical oxidation on the electrocatalyst but also effectively suppresses the recombination of photogenerated electrons and holes on the photocatalyst face. Therefore, the combination of both methods showed a better lignin oxidation performance than either method alone, due to the strong synergistic effect. As depicted in Figure 14.3a, the lignin oxidation can be observed as lignin solution became lighter with increasing oxidation time. The lignin solution was acidified with HCl to pH 3.0 to precipitate lignin and determine its remaining amount. The remaining lignin was gradually consumed with the increase of oxidation time (Figure 14.3b). High performance liquid chromatography (HPLC) results showed that two primary intermediates were formed including vanillin and vanillic acid. Figure 14.3c showed the the FTIR spectra of lignin before and after photochemical-eletrochemical oxdiation. After oxidation, the carbonyl functionality was formed (Figure 14.3d).

Lignin received less attention as a substrate for photocatalytic hydrogen production [29]. Transition metal sulfides have been widely studied as photocatalysts for the photoconversion of lignin to H_2. One-dimensional (1D) NiS/CdS nanocomposite was fabricated and exhibited enhanced activity for photocatalytic H_2 generation, 5041 times than that of pristine CdS, with the coexistence of lignin and lactic acid as sacrificial agents for holes. NiS/CdS nanocomposite has enhanced performance than CdS, because NiS is beneficial to prolong average charge carrier lifetime by 97 times than that of CdS nanowires (NWs), potentially leading to more efficient charge separation and transfer [54]. A simultaneous lignin degradation and water splitting were reported using C-, N-, and S- co-doped ZnO/ZnS nanostructure materials as photocatalysts. C-, N-, and S-co-doped mixed phase ZnO/ZnS (sample S1) was prepared by calcination of a bis-thiourea zinc acetate (BTZA) complex at 500 °C, and calcination at 600 °C conferred a N- and S-doped single-phase ZnO (sample S3) [55]. Superior activity for both H_2 production (580 and 558 mmol h^{-1}) was observed via water splitting and the complete degradation of lignin under visible light illumination. This is due to the novel characteristics of sample S1, including its narrow bandgap of 2.83 eV and honeycomb-like nanostructure. In contrast, sample S3 exhibited higher activity for H_2 production (643 and 602 mmol h^{-1}) but less lignin degradation.

14.3.2 Photocatalytic Reforming of Carbohydrates

Carbohydrates, mainly originated from cellulose and hemicellulose, are the largest component of biomass [5]. Table 14.2 summarizes recent studies on photocatalytic reforming of cellulose to valuable chemicals and fuels. Kawai and Sakata first reported the photocatalytic conversion of carbohydrates such as saccharose, starch, and cellulose into H_2 fuel in 1980 [64]. The irradiation of carbohydrates, not only sugar or starch but also cellulose, leads to the efficient production of H_2 gas, in the presence of water and a $RuO_2/TiO_2/Pt$ photocatalyst powder. The photocatalyst was prepared by mixing powders of RuO_2, TiO_2, and Pt (weight ratio = 10 : 100 : 5) and grounding in an agate mortar.

For photocatalytic H_2 production on metal (e.g. Pt)-loaded TiO_2, cellulose hydrolysis to glucose is the necessary step [65]. However, cellulose hydrolysis hardly occurred in pure water even under light irradiation, which preferred to occur in acidic solution. A simple and efficient one-pot process (Figure 14.4) was firstly reported combining photocatalysis and acid hydrolysis to convert cellulose to H_2 and valuable chemicals [56]. The photocatalytic H_2 production was promoted by in situ generation of electron donors from cellulose hydrolysis, as the generated carbohydrates are very effective electron donors due to their low oxidation potentials.

Solid acid photocatalyst can be obtained by modification photocatalyst with acid groups. TiO_2 modified with nickel sulfide (Ni_xS_y) and chemisorbed sulfate (P25-S-Ni) was prepared in one step and used for the photoreforming of cellulose to H_2 in neutral water. P25-S-Ni showed a significantly elevated H_2 yield, which is almost 76-fold higher than P25 [59]. The authors inferred that SO_4^{2-} ions promoted cellulose hydrolysis and consequent accessibility of biomass to catalysts, and Ni_xS_y acted as a cocatalyst for photocatalytic H_2 production.

The deposition of plasmonic NPs such as Ag, Pt, and Au upon TiO_2 surface was efficient to extend light harvesting of photocatalyst to the visible light spectrum,

Table 14.2 Photocatalytic conversion of cellulose in literatures.

Substrate	Catalyst	Media	Light source	Temperature	Atmosphere	Time (h)	Product (selectivity, %)	References
Cellulose (2.5 mg)	TiO_2(Pt) (25 mg)	H_2SO_4 (0.6 M, 50 ml)	Iron-doped halide lamp 250 W	403 K	Ar	10	H_2 (123 μmol)	[56]
Cellulose (50 mg)	Ir/HY_3 (50 mg)	Water (450 mg) EMIMCl (4.5 g)	Xe lamp 300 W	363 K	NA	13	Cellobiose (10.9), glucose (40.4), HMF (24.0)	[57]
Cellulose (50 mg)	Au-HYT (50 mg)	Water (500 mg) EMIMCl (4.5 g)	Visible light 0.5 W cm^{-2}	413 K	NA	16	Glucose (48.1), HMF (10.6)	[58]
Cellulose (1 g)	P25-S-Ni (20 mg)	Water (100 ml) (ultranication for 5 min)	Xe lamp 500 W	353 K	N_2	3	H_2 (181.2 μmol)	[59]
Pt-loaded CdS/RC-4.5% (0.05 g)	Pt-loaded CdS/RC-4.5% (0.05 g)	Na_2S (0.25 M)–Na_2SO_3 (0.35 M) (mixed 100 ml)	Xe-illuminator 250 W	NA	Vaccum	5	H_2 (6.721 mmol g^{-1})	[60]
Cellulose@TiO_2(Pt) (15 mg)	Cellulose@TiO_2(Pt) (15 mg)	Water (50 ml)	Iron–halide lamp 250 W	293–313 K	Ar	42	H_2 (195.2 μmol)	[61]
Cellulose (200 mg)	Pt/TiO_2 (2.0 g l^{-1})	Water (30 ml)	UV-A irradiation (4 × 15 W)	NA	N_2	4	H_2 (54 μmol)	[62]
Cellulose (140 mg)	Pt/TiO_2 (2.0 g l^{-1})	Water (21 ml)	Natural sunlight (475 W m^{-2})	302–305 K	N_2	4	H_2 (54 μmol)	[62]
Cellulose (2 g)	TiO_2/NiO_x@C_g (20 mg)	Water (100 ml)	Xe lamp 500 W	298 K	N_2	5	H_2 (5.4 μmol h^{-1})	[63]
Cellulose (2 g)	TiO_2/NiO_x@C_g (20 mg)	Water (100 ml)	Xe lamp 500 W	333 K	N_2	5	H_2 (43.3 μmol h^{-1})	[63]
Cellulose (2 g)	TiO_2/NiO_x@C_g (20 mg)	Water (100 ml)	Xe lamp 500 W	353 K	N_2	5	H_2 (82.9 μmol h^{-1})	[63]

NA: not available.
Source: Liu et al. [49]. © 2019, The Royal Society of Chemistry.

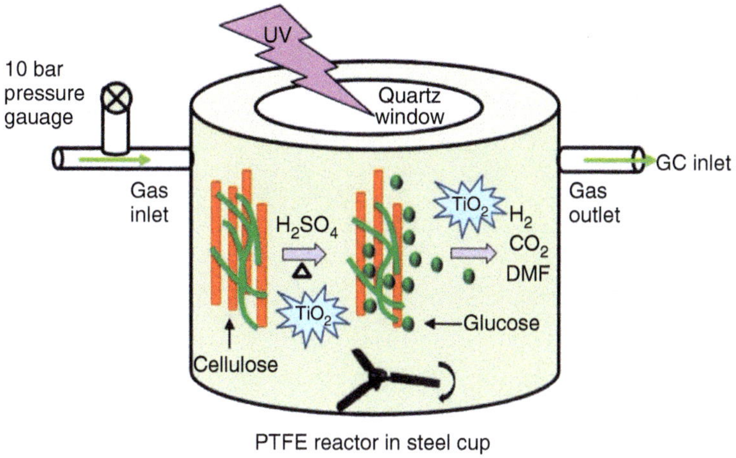

Figure 14.4 Illustration of experimental reactor used for photoreforming cellulose to hydrogen via combined photocatalysis and acid hydrolysis. Source: Zou et al. [56].

thanks to the localized surface plasmon resonance (LSPR). Ni-, Pd-, Pt-, or Au-impregnated TiO_2 were used for the photocatalytic reforming of cellulose to produce H_2. The H_2 production rate was approximately 173, 113, 77, and 134 μmol h^{-1} g^{-1} over Pd/TiO_2, Au/TiO_2, Ni/TiO_2, and Pt/TiO_2 with metal loading of around 0.2% [65].

Immobilizing cellulose particles onto photocatalysts would substantially enhance the photocatalytic efficiency. Ligand-to-metal-charge-transfer (LMCT) complex may be formed after immobilization, which allowed electrons to be injected directly on the CB of photocatalyst thus making them available to react. For example, Ke et al. prepared the CdS/regenerated cellulose (CdS/RC) nanocomposite films, and it exhibited superior performance compared with CdS NPs, including enhanced visible light photoactivity for H_2 production, prolonged photostability, and convenient regeneration [60]. Cellulose-immobilized TiO_2(Pt, 0.5 wt%) photocatalyst was prepared based on a facile "mixing and drying" deposition process [61]. The hybrid/cellulose system exhibited enhanced photocatalytic performance than bare TiO_2(Pt) for cellulose reforming to sugars and carbon dioxide, with simultaneous hydrogen production under UV or solar light irradiation. HPLC analysis of the solution during reaction indicated that cellobiose, glucose, and formic acid are the main soluble products.

Zhang et al. demonstrated the feasibility of direct hydrogen production from photocatalytic conversion of biomass [63]. Graphitic carbon layer grafted TiO_2/NiO_x NPs were prepared and used for H_2 production from direct photocatalytic conversion of cellulose. When TiO_2 is illuminated, NiO_x will be reduced by photo-excited electrons from the CB of TiO_2, resulting in reduced Ni. A Ni–H hydride is obtained when proton adsorbed on the Ni surface is reduced by photo-induced electron. Subsequently, two Ni–H hydrides afford one molecule of hydrogen.

Besides H_2 fuel, photocatalytic conversion of cellulose can provide a list of high value-added chemicals, such as some of the sugar-derived platform molecules

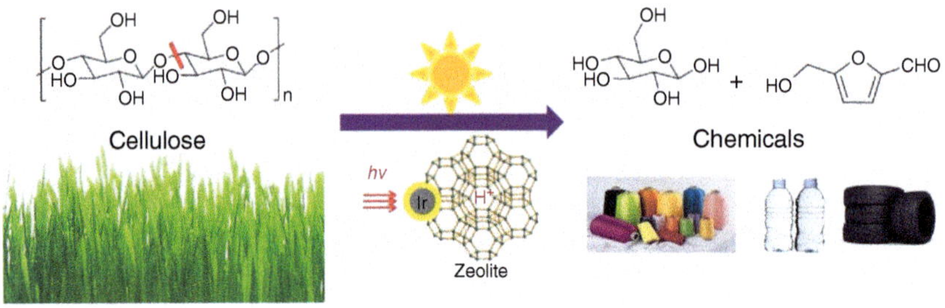

Figure 14.5 Photothermally promoted cleavage of β-1,4-glycosidic bonds of cellulose on Ir/HY catalyst. Source: Reprinted with permission from Zhang et al. [57].

(e.g. cellobiose, glucose, 5-hydroxymethylfurfural [HMF]). Homogeneous acid-assisted hydrolysis of cellulose has some drawbacks such as reactor corrosion and waste disposal, which could significantly limit its practical applications. Therefore, solid acid catalysts are widely studied due to their easy separation properties. Photocatalysts have also been combined with solid acid catalysts, such as metal oxides, metal phosphate, polymer-based acid resins, sulfonated carbonaceous acids, heteropoly acids, and H-form zeolites (HY) to achieve a synergistic effect. These solid acid catalysts possess a high affinity to polysaccharide structure, suitable for cellulose hydrolysis [66]. Plasmonic metals such as Au and Ir were deposited on HY to enhance cellulose hydrolysis under visible light irradiation, due to the synergistic effect of the photothermal plasmonic behavior of plasmonic NPs and acid catalysis of HY zeolite [57, 58]. Photocatalytic hydrolysis of non-pretreated crystalline cellulose was conducted over Ir/HY, and the main products obtained are cellobiose, glucose, and HMF (Figure 14.5) [57]. A new plasmonic nanostructure of TiO_2 nanofibers (NFs) decorated with Au-NPs (Au-HYT) supported on H-form Y-zeolites (HY) was synthesized and used for the photocatalytic hydrolysis of cellulose [58]. The yield of glucose and HMF was more than 60% at 130 °C for 24 hours.

Photocatalytic cellulose conversion to valuable chemicals can be integrated with H_2 evolution from water splitting, which could simultaneously boost H_2 production, as shown in Figure 14.6. Cellulose and its hydrolysis products act as scavengers for the VB holes and consume ROS and O_2 from water cleavage, which helps to suppress charge carrier recombination and/or O_2–H_2 back reaction. And at the same time, photo-induced electrons are readily available for H_2 evolution. Valuable intermediates were generated from polysaccharide depolymerization, among which HMF was identified [62]. Moreover, this procedure was successfully extended to cellulosic biomass, i.e. rice husk and alfalfa stems.

Other carbohydrates, mainly the hydrolysis products of cellulose, glucose, and HMF, are important intermediates in cellulose conversion process. And there are many researches about the photocatalytic conversion of glucose into smaller organic molecules, including formic acid, xylose, arabinose, and HMF to its corresponding aldehydes and acids [49].

One interesting research about the photocatalytic conversion of derivatives of HMF has been reported to obtain diesel fuel precursors (DFPs). 2,5-dimethylfuran (2,5-DMF) and 2-methylfuran (2-MF) can be obtained from HMF and furfural

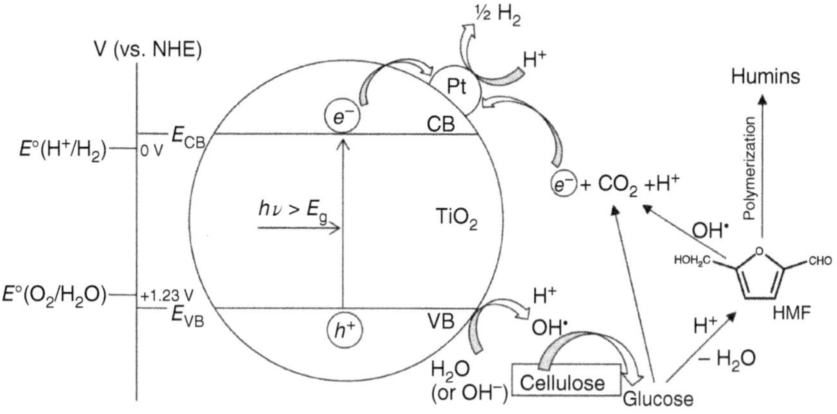

Figure 14.6 Schematic representation of the photocatalytic system for the H_2 evolution by water splitting over irradiated Pt/TiO_2 in the presence of cellulose as the sacrificial agent. Source: Speltini et al. [62], licensed under CC BY 3.0.

reforming. H_2 and DFPs could be simultaneously produced from 2,5-DMF and 2-MF over Ru-doped $ZnIn_2S_4$, under visible light illumination [67]. Photogenerated holes firstly activate the furfuryl C—H bond of 2,5-DMF/2-MF, delivering protons and furfuryl radicals, which then form the desired DFPs through C–C coupling. Meanwhile, the delivered protons are reduced by photo-induced electrons to produce H_2. Up to 1.04 g $g_{cat}{}^{-1}$ h^{-1} of DFPs are obtained with selectivity higher than 96%, together with 6.0 mmol g $g_{cat}{}^{-1}$ h^{-1} of H_2. Desired diesel fuels are produced after subsequent hydrodeoxygenation (HDO) reactions of DFPs, comprising ~C_{10}–C_{18} straight- and branched-chain alkanes. Moreover, feedstocks derived from hexosan and pentosan can also be converted into H_2 and DFPs over the Ru-$ZnIn_2S_4$ photocatalyst. Ru dopants, substituted in the position of indium ions in the $ZnIn_2S_4$ matrix, could improve light harvesting and facilitate charge separation, thus accelerating C–H activation for the coproduction of H_2 and DFPs.

14.3.3 Photocatalytic Reforming of Native Lignocellulose

Solar-driven photocatalytic reforming of lignocellulose to H_2 at ambient temperature offers a sustainable route toward its valorization to renewable fuels. In this process, lignocellulose acts as a hole scavenger, which could enable continuous supply of electrons for hydrogen production. It is particularly appealing to convert lignocellulose for H_2 generation without pretreatment.

Wakerley et al. reported a process of light-driven valorization of cellulose, hemicellulose, and lignin to produce H_2 over semiconducting cadmium sulfide quantum dots (QDs) in alkaline aqueous solution [31]. Basic conditions facilitate in situ formation of $Cd(OH)_2$/CdO (henceforth CdO_x) on the CdS surface, which significantly improves their photocatalytic performance because the obtained CdS/CdO_x QDs are more resistant to photocorrosion. In addition, the alkaline conditions promote biomass dissolution, making them more available to photocatalysis. More

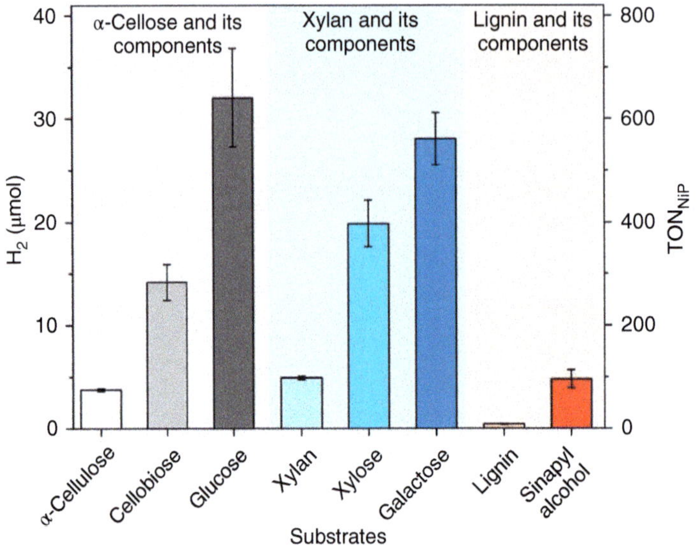

Figure 14.7 Photocatalytic H_2 production using activated $^{NCN}CN_x$ (5 mg) and Ni bis(diphosphine) (**NiP**) (50 nmol) with purified lignocellulose components (100 mg) in potassium phosphate (KPi) solution (0.1 M, pH 4.5, 3 ml) under AM 1.5 G irradiation for 24 hours at 25 °C. Source: Kasap et al. [68].

importantly, raw lignocellulose can be converted to H_2 at room temperature over CdS/CdO_x QDs, under visible light illumination.

Activated cyanamide-functionalized carbon nitride ($^{NCN}CN_x$) was prepared by ultrasonication. $^{NCN}CN_x$ showed respectful activities for the visible light-assisted reforming of purified (Figure 14.7) and native lignocellulose samples (e.g. sawdust and bagasse) to produce H_2, with the assistance of proton reduction cocatalysts. This room-temperature photoreforming process operated under benign conditions (pH ≈ 2–15), in the absence of toxic materials [68]. The conversion yields could reach 22% determined with different α-cellulose loadings, more than twice the yield of that over CdS/CdO_x QDs (9.7%) [31].

As mentioned before, metal sulfide semiconductors are a class of photocatalysts with good visible light responses [50] and high photocatalytic activities, as well as high potential for selectively cleaving the β-O-4 bond in lignin [51, 52]. A few studies were also reported for the valorization of native lignin in biomass to value-added aromatics over metal sulfide semiconductors. Wu et al. reported lignin-first approach to make the best of entire lignocellulose, which is driven by solar energy. CdS QDs with organic surfactants were of colloidal character, which are capable of penetrating better to solid biomass substrate. CdS QDs could catalyze the cleavage of β-O-4 bonds in lignin model compounds and native lignin efficiently, with cellulose/hemicellulose survived [69]. An electron–hole-coupled (EHCO) mechanism was proposed for the selective photocatalytic cleavage of β-O-4 bond via a C_α radical intermediate. The bond dissociation energy (BDE) of the β-O-4 bond

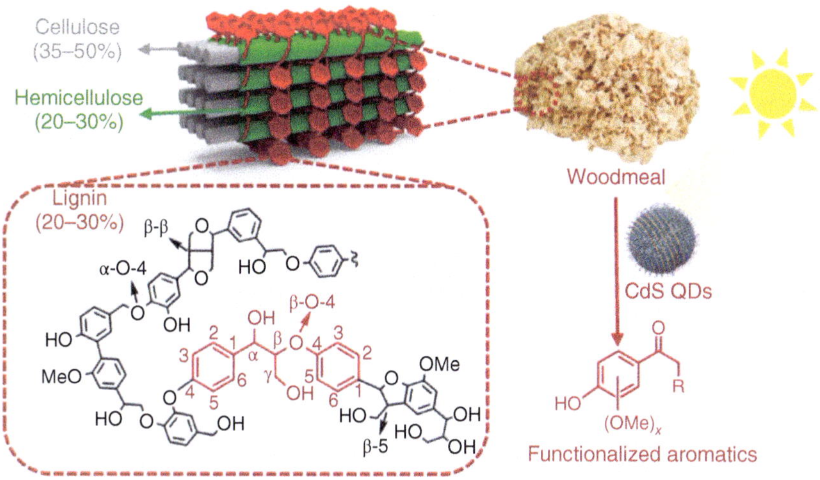

Figure 14.8 Schematic illustration of the lignocellulose structure and photocatalytic valorization of native lignin. Source: Reprinted with permission from Wu et al. [70].

decreased from 55 to 7.8 kal mol^{-1} in the C_α radical intermediate, which is the key reason for the high activity and selectivity of this system.

Ligand shell plays significant roles in photocatalytic conversion of lignin over CdS QDs (Figure 14.8) [70]. The formation of a QD colloidal solution by tuning the hydrophilic/hydrophobic property of organic ligands is essential to facilitate the penetration of QDs in lignin solution, leading to high conversion efficiency. Moreover, the ligands participate in the electron-transfer process, which is a crucial step of photocatalysis. The electron decay kinetic study of the process reveals a ligand-mediated electron-tunneling pathway. It was further found that CdS QDs with 3-mercaptopropionic acid ligand (CdS-C3 QDs) could convert native lignin and various technical lignins into functionalized aromatics, with the yield of aromatics increased with the content of β-O-4 linkages. The yield of aromatic monomers is 27 wt% for birch wood meal. In addition to aromatic monomer production, this lignin-first approach also showed good performance for the preservation of polysaccharides in birch wood meal.

14.3.4 Photocatalytic Reforming of Triglycerides and Glycerol

The formation of biodiesel from triglycerides is an exemplary real-world application for biomass conversion [7]. Glycerol is the main product in the biodiesel production process [71]. Additionally, glycerol as an important intermediate is widely used in organic synthesis and in the pharmaceutical industry. As a result, amounts of glycerol were found in wastewater from this industry. If glycerol could be used as a low-cost feedstock for the production of valuable commodities, biodiesel production would become more profitable and sustainable. For example, glycerol can be converted to dihydroxyacetone (DHA). It is worth to note that the price of DHA and purified glycerol are c. 150 and 0.6 USD kg^{-1}, respectively. Lots of attention has been paid to the conversion of glycerol to DHA due to this four order of magnitude difference in price.

Nguyen et al. reported a dual catalytic system for synthesis of alkenes from triglycerides and unrefined biomass [72]. This approach for dehydrodecarboxylation of carboxylic acids is enabled by a crossover of the photo-induced acridine-catalyzed O–H hydrogen atom transfer (HAT) and cobaloxime-catalyzed C–H-HAT processes. This approach can be applied for carboxylic acid conversion to a variety of alkenes. Moreover, the reaction can be embedded in a scalable triple-catalytic cooperative chemoenzymatic lipase–acridine–cobaloxime process, which could be used for production of long-chain terminal alkenes from the direct conversion of plant oils and biomass.

Only few studies about the selective oxidation of glycerol over semiconductor photocatalysts are available and only emerging during the past decade [73–81]. Maurino et al. firstly reported the photocatalytic oxidation of glycerol under aerated and ambient conditions [78]. P25 and Merck powders were used as photocatalysts, and the influence of surface modification by fluoride anions was further investigated. The main products from glycerol photooxidation were DHA and glyceraldehyde.

Recently, Augualiaro et al. found that TiO_2 can convert glycerol into DHA, with selectivity of 7.2% [73]. Guo et al. reported a rationally designed Au_xCu–CuS@TiO_2 heterostructures for the photocatalytic oxidation of glycerol. An integrated conversion rate of 72% for glycerol and selectivity of 66% for DHA were obtained on the sample with equimolar Au and Cu [82]. Moreover, it is worth to note that this efficient photocatalytic process was conducted in a neutral solution without external heating.

Plasmonic photocatalysts were applied to drive the photochemical reforming of glycerol to value-added products [83–85]. The LSPR properties of noble-metal NPs facilitate visible light absorption and an energy transfer to the semiconductor, which enabled activity for plasmonic photocatalysts. It has been reported that the catalytic performance of Au/TiO_2 catalyst for glycerol oxidation can be enhanced under neutral conditions by visible light illumination of the reaction solution [86]. Valuable products, such as DHA, glyceric acid and glycolic acid were obtained in the process. Visible light illumination could significantly enhance glycerol conversion efficiency. High selectivity up to 63% toward DHA was achieved over TiO_2 with 7.5 wt% Au loading. Photo-induced electrons would transfer from Au NPs to the CB of TiO_2. The remaining positive holes in Au NPs participated in glycerol oxidation. The studies were extended using monometallic Au and bimetallic AuCu catalysts deposited on mesoporous SiO_2, which supports as photocatalysts [87]. Au NP-loaded monodispersed mesoporous silica (MS) spheres demonstrated superior performance, compared with Au-deposited ordered mesoporous silica (SBA-15, KIT-6, and MCM-41). Moreover, bimetallic AuCu NP-loaded MS could further increase the performance by a factor of 2.5. And the selectivities for DHA could reach 80–90%.

It is important to explore other active semiconductors for photoreforming of glycerol. Xu and coworkers firstly reported flower-like Bi_2WO_6 (Figure 14.9a) as a

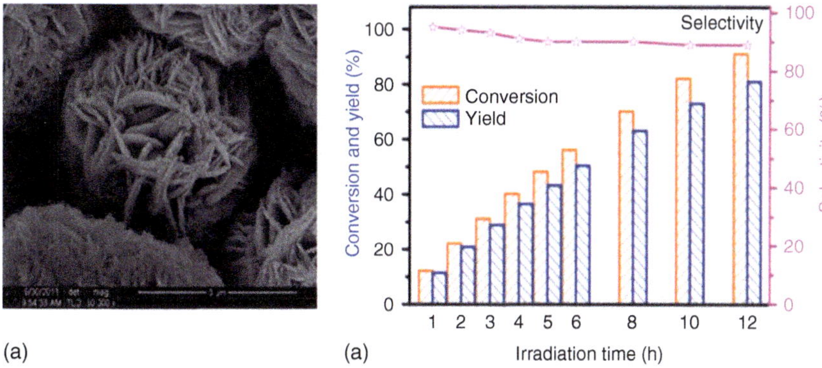

(a) (a)

Figure 14.9 (a) Scanning electron microscope (SEM) image of sample Bi_2WO_6; (b) Time-online photocatalytic performance toward selective oxidation of glycerol to DHA over Bi_2WO_6. Source: (a) Reprinted with permission from Zhang et al. [74]; (b) Zhang et al. [74].

highly active photocatalyst toward selectively aerobic oxidation of glycerol to DHA [74]. DHA was obtained with high yields (80–87%) after illumination for five hours over Bi_2WO_6 photocatalysts prepared by hydrothermal treatment of different times. Scale-up experiments were also conducted by enlarging the amount of glycerol from 0.1 to 0.5 mmol. As shown in Figure 14.9b, high selectivity for target product DHA could still be maintained at a high conversion rate over Bi_2WO_6. A possible mechanism for glycerol oxidation was proposed. The adsorbed glycerol on the surface of Bi_2WO_6 is firstly oxidized by photogenerated holes to form corresponding intermediate, which could further react with oxygen or $O_2^{\cdot-}$, producing target product DHA. And a number of factors lead to high selectivity for DHA, including the mild oxidation power of Bi_2WO_6, absence of nonselective $^{\cdot}OH$ radicals, regioselective oxidation of the second hydroxyl group in glycerol over Bi_2WO_6, weaker adsorption capacity of DHA than glycerol, and the stability of DHA over Bi_2WO_6. In a follow-up study, silica-entrapped Bi_2WO_6 photocatalyst showed significantly enhanced photocatalytic activity. A threefold increase for glycerol oxidation was obtained over silica-entrapped Bi_2WO_6 (10 wt% catalyst loading), compared with bare Bi_2WO_6 [88]. The transparent, stabilizing encapsulated silica matrix with high surface area enables light penetration, promotes photo-induced electron–hole separation, and enhances adsorption capacity toward glycerol. Moreover, tunable control of both selectivity and activity for photocatalytic oxidation of glycerol was available, due to the synergistic effects between the silica matrix and photoactive entrapped Bi_2WO_6.

Molybdena-modified bismuth tungstates (MoO_x/Bi_2WO_6) were prepared and applied for the photocatalytic oxidation of glycerol [89]. MoO_x/Bi_2WO_6 materials were found to have superior properties for the selective oxidation of glycerol. Contrary to the results reported by Xu and coworkers [74], unmodified Bi_2WO_6 exhibited no photocatalytic activity for glycerol oxidation. Pt NP deposition on MoO_x/Bi_2WO_6 was observed to be beneficial for the production of DHA, whereas the photocatalytic performance showed a general decrease with increasing MoO_x

loading, implying that MoO_x modification is detrimental for the photocatalytic performance of Pt/Bi_2WO_6. And Pt/Bi_2WO_6 showed the highest activity for DHA production. More researches are needed to explore mechanistic details of this behavior.

$Na_4W_{10}O_{32}$ was used as photocatalyst for the selective oxidation of glycerol under UV–Vis illumination [79]. $^{\cdot}OH$ radicals could be formed over dissolved $Na_4W_{10}O_{32}$ upon photo irradiation, which could effectively oxidize glycerol but with low selectivity to value-added products. The photocatalytic properties of $Na_4W_{10}O_{32}$ changed significantly after it was entrapped inside a silica matrix, producing glyceraldehyde (selectivity of 60%) and DHA (selectivity of 6%). EPR spectroscopy indicated that heterogenization did not influence the ability of $Na_4W_{10}O_{32}$ to oxidize water to hydroxyl radicals. The differences in selectivity critically depend on textural features because they would control surface interactions with substrates and intermediates, which have significant impacts on the photocatalytic properties of $Na_4W_{10}O_{32}$.

The other potential application is photocatalytic production of H_2 with glycerol as a sacrificial agent [90–93]. Kondarides et al. investigated the photocatalytic reforming of aqueous solution of glycerol employing Pt/TiO_2 as the photocatalyst [94]. Complete conversion of glycerol to H_2 and CO_2 was observed. Lalitha et al. studied the continuous production of H_2 from glycerol over Cu_2O/TiO_2 under visible light illumination [95]. A maximum H_2 production of 20 060 $\mu mol\,h^{-1}$ was obtained from photocatalytic reforming of 5% glycerol aqueous solution.

14.4 Photoelectrochemical Reforming of Biomass

The main products of PEC reforming of biomass are hydrogen, electricity, and useful chemicals [35]. In this chapter, we mainly focus on the PEC production of electricity, hydrogen, and useful chemicals from biomass and its derivatives.

14.4.1 Photoelectrochemical Conversion of Biomass to Produce Electricity

Photo fuel cells (PFCs) are comprised by a photoanode for biomass oxidation and a cathode for oxygen reduction, which offer an appealing PEC route to directly convert the two most abundant renewable resources into electricity [96]. A lot of attention has been paid toward the development of active photoanodes for efficient oxidation of biomass. Semiconductor materials such as $BiVO_4$, WO_3, α-Fe_2O_3, and TiO_2 were employed as the photoanodes to drive biomass-to-electricity conversion in the PFC device [36].

The phosphate–Fe_2O_3 photoanode decorated with nickel-phosphate (NiPi/Pi–Fe_2O_3) was prepared by decorating phosphate-modified hematite (Pi–Fe_2O_3) with nickel-phosphate complexes (NiPi) [97]. The photocurrent of NiPi/Pi–Fe_2O_3 was c. two times than that of Pi–Fe_2O_3 photoanode at 1.5 V (vs. reversible hydrogen electrode [RHE]). The maximum power generated from NiPi/Pi–Fe_2O_3 photoanode

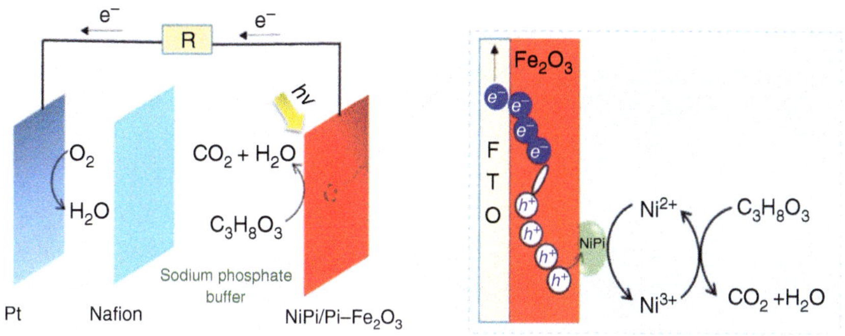

Figure 14.10 A simple solar-induced hybrid direct glycerol fuel cell consists of a Pt cathode, a Na-Pi buffer electrolyte with glycerol as fuel and NiPi/Pi-Fe_2O_3 as a photoanode, and the possible photo-generated charging process of glycerol process of glycerol oxidation over NiPi/Pi-Fe_2O_3. Source: Chong et al. [97].

was c. 2.4 times than that of Pi–Fe_2O_3. A possible mechanism for solar-driven electricity production from PEC glycerol fuel cell with NiPi/Pi–Fe_2O_3 photoanode was proposed (Figure 14.10). Photo-induced holes generated from α-Fe_2O_3 nanorods would migrate into NiPi complexes, where Ni^{2+} was oxidized to produce Ni^{3+} by h^+. Subsequently, glycerol was oxidized by Ni^{3+}, and Ni^{2+} was regenerated at the same time. Photogenerated electrons (e^-) migrated to counter electrode (Pt) for the oxygen reduction reaction, thus producing electricity in an external circuit.

PEC cells with TiO_2 photoanodes have been reported to improve photoreforming efficiency of glycerol [98]. Oxygen vacancies play a critical role in determining the surface and electronic properties of TiO_2 [99]. The redox behavior of glycerol over TiO_2 electrodes was reported [100]. It was found that the defectivity of TiO_2 not only significantly influences the photoactivity of the samples but also strongly affects the adsorption behavior of glycerol on TiO_2. The adsorbed glycerol was reduced by electrons injected in the system with sufficient V_o present. When the concentration of V_o available is limited, glycerol seems to be reduced by H incorporated in the lattice. TiO_2 was also composited with CdS to enhance the visible light activity for PEC glycerol reforming [98]. The efficiency of this PEC cell increased 10-fold in the presence of glycerol compared with the cell when there were only water and electrolytes.

Mohapatra et al. reported that organic additives could reduce e^-–h^+ recombination of the integrated TiO_2 nanotube (NTAs)/Ti photoanode, thus significantly improving its photocurrent density [101]. The open circuit potential (OCP) of the TiO_2 NTAs photoanode moved to −1.26 V (vs. Ag/AgCl), and a photocurrent density of 2.55 mA cm^{-2} was achieved at 0.2 V (vs. Ag/AgCl) in 1 M KOH solution with the presence of glycerol.

A membrane-free PFC was built consisting of dual photoelectrodes, including tungsten (W)-doped $BiVO_4$ photoanode and polyterthiophene (pTTh) photocathode for electricity production from biomass (Figure 14.11) [36]. Water-soluble glucose was converted using this PFC with the help of dissolved O_2 in the air, delivering a maximum power density of 82 μW cm^{-2}. As shown in Figure 14.11c, the CB of

Figure 14.11 SEM image of (a) W:$BiVO_4$ and (b) pTTh films. (c) The energy level diagram of the dual-photoelectrode PFC. Source: (a, b) Modified with permission from Zhang et al. [36]; (c) Zhang et al. [36].

W:$BiVO_4$ is more negative than the VB of pTTh; therefore an interior bias could be formed to spontaneously drive the chemical-to-electricity conversion under illumination. In addition, the PFC shows excellent photostability during a long-term discharging test of up to 20 hours.

A solar-induced hybrid fuel cell adopted a polyoxometalate (POM) photocatalyst was constructed to convert biomass into electricity. Natural polymeric biomasses, like starch, cellulose, lignin, even switchgrass, and wood powders, were used to power this process. This hybrid cell was resistant to be contaminated by most organic and inorganic compounds present in the fuels. A power density of 0.72 mW cm^{-2} was reached adopting cellulose as fuels, which is almost 100 times higher than that obtained from cellulose-based microbial fuel cells (MFCs) [102]. When this hybrid fuel cell was working, biomass was oxidized by PMo_{12} under solar irradiation, with Mo^{6+} reduced to Mo^{5+} simultaneously. Then Mo^{5+} could be oxidized to Mo^{6+} again by oxygen through an external circuit to produce electricity. In this process, the reduced POM (Mo^{5+}) released one electron to the carbon anode, and discharged a proton to the solution simultaneously, leading to the color change from deep blue to light yellow gradually. The highest power density of 45 mW cm^{-2}

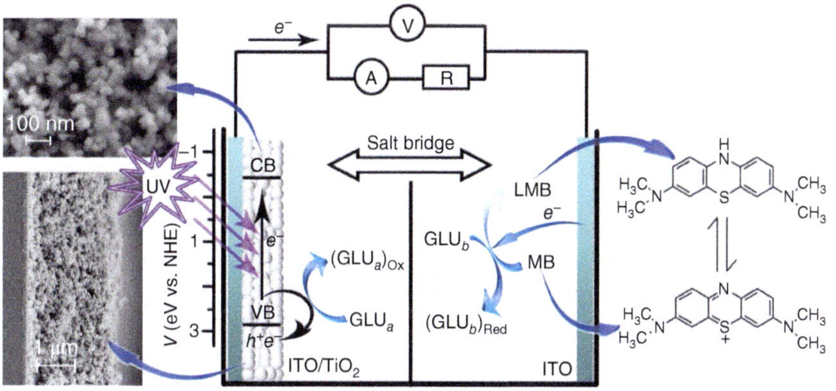

Figure 14.12 Schematic configuration and electron transfer processes of PEC glucose/glucose fuel cell employing TiO_2/ITO photoanode and ITO cathode upon incorporation of UV light and MB. Insets show SEM images of TiO_2/ITO and structures of MB and LMB. Source: Zhao et al. [104].

was obtained when sugars were adopted as fuels [103]. POM-I was oxidized to POM-II on the anode and regenerated on the cathode, with electricity generated by fulfilling this circuit. When irradiated by light or heated, POM-I could oxidize sugar and be stored in the reduced state, with Mo^{6+} reduced to Mo^{5+}.

A PEC fuel cell was launched by paired photoelectrocatalytic reactions of glucose, which is conducted by capturing the photogenerated electrons and holes from UV-excited TiO_2 NPs, with the assistance of methylene blue (MB) in neutral media [104]. MB in this mixed system could be rapidly reduced to colorless leucomethylene blue (LMB) by the photo-induced electrons, upon UV irradiation. Interestingly, MB could be recovered after UV light is turned off, which was used to represent reversible colorimetric reactions controlled by UV light. Moreover, glucose acted as the reductant and oxidant in anode and cathode chambers, respectively, while TiO_2/indium tin oxide (ITO) photoanode and ITO cathode were employed to assemble a UV-excited PEC fuel cell (Figure 14.12). The paired photoelectrocatalytic reactions were fulfilled by glucose oxidation and reduction reactions in anode and cathode chambers, respectively, with the indication of MB in neutral media. The cell current showed a linear response to glucose concentration ranging from 0.5 to 2.0 mol l^{-1}. When 2.0 mol l^{-1} glucose was employed as fuel and redox agents and 10.0 mol l^{-1} MB as the redox indicator, the assembled PEC fuel cell presented open-circuit photovoltage of 0.828 V, short-circuit photocurrent density of 29.2 μA cm^{-2}, and maximum power density of 7.2 μW cm^{-2} at 0.345 V under UV irradiation of 55 μW cm^{-2}.

Photoreforming of biomass was combined with MFCs to produce electricity. Firstly lignin was pretreated by photocatalysis [105]. Pretreating of lignosulfonate was performed over TiO_2 under UV illumination. The pretreated lignosulfonate was then used as fuels for MFCs. Maximum current and power densities of 6556 ± 360 mA m^{-3} and 1881 ± 103 mW m^{-3} were generated, when MFC was operated at 37 ± 1 °C.

14.4.2 Photoelectrochemical Conversion of Biomass to Produce Hydrogen

PEC reforming of biomass and water to hydrogen is an emerging promising approach, because it could be conducted under ambient condition with the help of light. Moreover, PEC reforming of water and biomass is more efficient than water splitting, because the calculated changes of Gibbs free energy for water and biomass conversion are obviously smaller than that of water splitting process. Biomass and its derivatives such as methanol, ethanol, glycerol, and sugars have been used for sustainable production of hydrogen [27]. Biomass and its derivatives are effective to consume photo-induced holes, favoring electron–hole separation and facilitating hydrogen evolution.

TiO_2-based photocatalysts are extensively studied as anode materials for PEC reforming of biomass. Ag/TiO_2 anatase films were used for photoelectrochemically hydrogen production from water/methanol decomposition. The overall hydrogen production rate over Ag/TiO_2 anatase films is $147.9 \pm 35.5\ \mu mol\ h^{-1}\ g^{-1}$. For anatase TiO_2 films, hydrogen production rate is $4.65 \pm 0.39\ \mu mol\ h^{-1}\ g^{-1}$ and decreases dramatically to $0.46 \pm 0.66\ \mu mol\ h^{-1}\ g^{-1}$ for amorphous TiO_2 films [106]. Nanocrystalline titania was used as photoanode for ethanol decomposition and hydrogen production. After adding ethanol, the flowing current and the rate of hydrogen evolution increased for this PEC cell [107]. Graphene-modified TiO_2 (G-TiO_2) was employed as an excellent photoanode for glycerol PEC reforming to produce hydrogen. Glycerol acts as a hole scavenger, suppressing the electron–hole recombination and enhance hydrogen production efficiency. Graphene incorporation could enhance the electron-transferring ability of G-TiO_2, therefore enhancing the photo current and hydrogen production efficiency of the PEC cell [108].

Thin film WO_3 was applied as photoelectrodes for PEC cell and exhibited excellent oxidation activity toward a variety of biomass derivatives [28]. WO_3-based PEC cell was feedstock flexible, with equal photocurrents achieved in the presence of methanol and ethanol. Renewable synthesis gas (H_2 + CO) was produced from this WO_3-based PEC cell.

14.4.3 Photoelectrochemical Conversion of Biomass to Produce Chemicals

A high-performance PEC based on $BiVO_4$ was developed with industrial by-product, i.e. glycerol as fuel. Glycerol of appropriate concentration can help to suppress surface charge recombination in $BiVO_4$ photoanode, thus enhancing PEC performance [109]. In addition, a remarkable cathodic shift of ~300 mV was observed after adding glycerol, and the current density was enhanced by four times, compared with the water oxidation. The incident photon-to-current efficiency (IPCE) was enhanced to 55%, three times higher than the system without glycerol. Moreover, not only hydrogen fuel can be produced from the PEC water splitting system, but also value-added DHA and formic acid (with the selectivity of 15% and 85%, respectively) are obtained from glycerol oxidation. This strategy could both boost the hydrogen production efficiency and make the biodiesel production more profitable and sustainable.

A PEC system with nanoporous $BiVO_4$ photoanode was constructed for selective oxidation of glycerol to DHA [110]. 56 mmol g_{cat}^{-1} h^{-1} of DHA can be produced in this system with selectivity of 51%, under AM 1.5 front illumination (100 mW cm^{-2}) in an acidic medium (pH = 2) without adscititious oxidant, at a potential of 1.2 V vs. RHE. The reason was that lower pH facilitated glycerol adsorption on $BiVO_4$, which could accelerate charge transfer and enhance photoelectrochemically conversion efficiency of glycerol to its derivatives. In addition, acidic environment also inhibited formic acid production, which helped to reduce further consumption of DHA and increase DHA selectivity.

Selective oxidation of biomass alcohols to the corresponding aldehydes is one of the most important chemical reactions in the fine chemical, pharmaceutical, and agrochemical industries [111]. Proper approaches to reform biomass alcohols are desired, which can be conducted under relatively mild conditions.

A PEC catalysis system with Au/CeO_2–TiO_2 NTs photocathodes is employed for selective oxidation of biomass alcohols to aldehydes, which are conducted in an O_2 atmosphere under mild conditions. Ninety-eight percent of benzyl alcohols is converted, with the selectivity of >99% for benzaldehyde at the bias potential of −0.8 V after reaction for eight hours under visible light illumination. As shown in Figure 14.13, the possible mechanism for PEC aerobic oxidation of benzyl alcohols is proposed. Benzyl alcohol adsorbed on the photocathode firstly interacted with photo-induced holes and benzyl alcohol cation radicals were produced. O_2 molecule reacted with photo-induced electrons to produce $O_2^{\bullet-}$. Then benzaldehyde was obtained from the reaction of $O_2^{\bullet-}$ radicals with benzyl alcohol cation radicals [112].

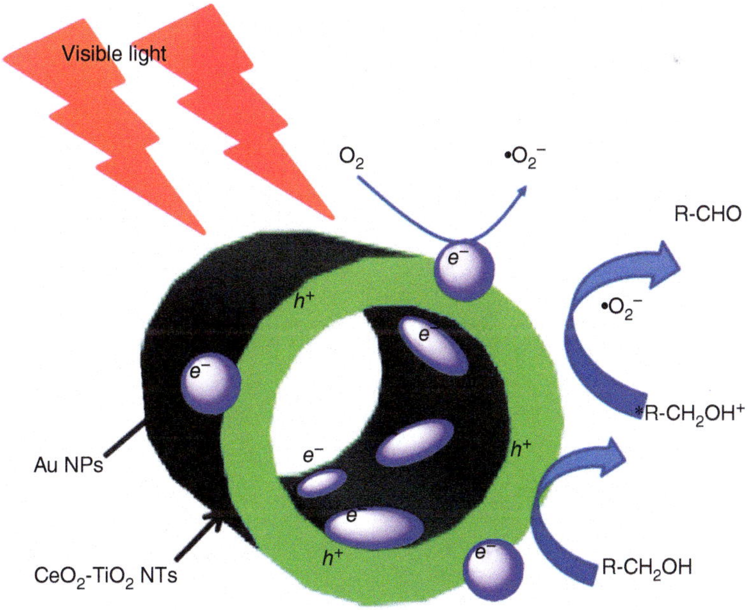

Figure 14.13 A plausible reaction mechanism of PEC selective aerobic oxidation of benzyl alcohols. Source: Zhang et al. [112].

A Z-scheme solar-driven dual photoelectrode PEC cell is constructed consisting of an Au/TiO_2 nanotube photonic crystals (Au/TiO_2 NTPC) photoanode and C/Cu_2O/Cu photocathode. This PEC cell was used for simultaneous H_2 production and selective oxidation of benzyl alcohol to benzaldehyde [113]. Up to 84.68% of benzyl alcohol is converted, with the selectivity of more than 99% for benzaldehyde. While in the Au/TiO_2 NTPC-Pt system, the conversion of benzyl alcohol is only 41.8%. The simultaneous H_2 production at the cathode is indeed promoted by the benzyl alcohol oxidation at the photoanode and reaches 143.83 μmol cm^{-2}, whereas only 10.42 μmol cm^{-2} of H_2 is produced without alcohol. Meanwhile, the amount of H_2 produced in the Au/TiO_2 NTPC-C/Cu_2O/Cu system is three times higher than that of the Au/TiO_2 NTPC-Pt system, which indicates that there is a synergistic effect resulting from the construction of Z-scheme PEC cell.

The PEC reforming of methanol was conducted for selective production of formate using Fe_2O_3 decorated with nanoporous nickel oxyhydroxide-borate (NiBi) film (Fe_2O_3/NiBi) as photoanode [114]. 86.5 μmol formate was produced on the Fe_2O_3/NiBi photoanode, and 226 μmol hydrogen was produced on Pt cathode, by using charge of 43.04 C.

HMF is an important intermediate in biomass conversion, and 2,5-Furandicarboxylic acid (FDCA) is a key monomer in polymer production industry. 2,5-bis-(hydroxymethyl)furan (BHMF) is a significant precursor for industrial production. Compared with electrochemical process, PEC process could significantly decrease the external voltage necessary for HMF reforming by using solar energy.

A highly efficient oxidation of HMF into FDCA is performed over the anode of a PEC with 2,2,6,6-Tetramethylpiperidine-1-oxyl (TEMPO) as a mediator [115]. In this PEC, electron–hole pairs are generated and separated on a n-type $BiVO_4$. The photo-induced electrons are transferred to the Pt cathode for H_2 production, whereas HMF is oxidized by the photogenerated holes on the photoanode.

A PEC was constructed consisting of a n-type $BiVO_4$ photoanode and a catalytic Ag cathode (Figure 14.14) [116]. Electron–hole pairs were generated and separated in $BiVO_4$ upon illumination. The photo-induced electrons were transferred to the Ag cathode for HMF reduction to BHMF. And photo-induced holes in the VB of $BiVO_4$ were consumed for water oxidation.

14.5 Conclusion Remarks and Perspectives

Producing bioenergy and value-added chemicals from biomass is a sustainable solution to meet the demands of highly populated areas and address biomass waste issues in urban and rural areas. This chapter reviews recent advances related to the photocatalytic and PEC reforming of biomass. The advantages of these two processes for biomass reforming are that they are generally conducted at ambient reaction conditions. This would enable the facile control of cleavage of dominant bonds in biomass to obtain most of value-added products. For photocatalytic conversion of biomass, the accomplishments are firstly summarized about depolymerization of robust native lignocellulose, processed lignin, carbohydrates,

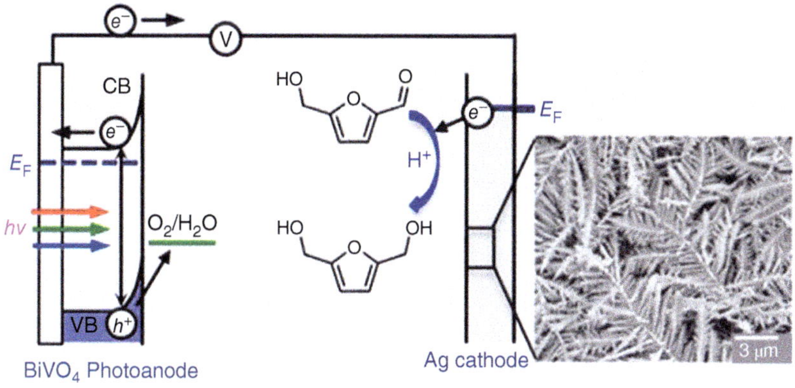

Figure 14.14 A photoelectrochemical cell for HMF reduction. Inset is the SEM image showing the surface morphologies of Ag_{gd}. Source: (a) Roylance et al. [116]; (b) Reprinted with permission from Roylance et al. [116].

and photocatalytic conversion of glycerol to useful chemicals and fuels. Enough attention has been paid to photocatalysts used, selective conversion of biomass to valuable chemicals, and photocatalytic hydrogen production processes. Afterward, the state-of-the-art accomplishments are reviewed with respect to PEC reforming of biomass to electricity, hydrogen and biomass-derived molecules including glycerol, alcohols, and 5-hydroxymethylfurfural to corresponding valuable chemicals. Reported photoelectrodes used and different types of PEC cells are discussed for energy and chemical production.

Regarding the works done, more efforts should be devoted to the following problems about photocatalytic and PEC reforming of biomass. Firstly, researches should pay more attention about the direct conversion of native biomass without pretreatment. Secondly, active photocatalysts are exploring to enhance biomass conversion selectivities for valuable chemicals to alleviate the pressure for isolating products.

Acknowledgments

This work was supported by an Australian Research Council (ARC) Future Fellowship (FT160100195), the FEIT Blue Sky Research Scheme 2019, and the UTS Early Career Research Development Grants.

References

1 Nel, W.P. and Cooper, C.J. (2009). *Energy Policy* 37: 166.
2 Alonso, D.M., Bond, J.Q., and Dumesic, J.A. (2010). *Green Chem.* 12: 1493.
3 Li, C., Zheng, M., Wang, A., and Zhang, T. (2012). *Energy Environ. Sci.* 5: 6383.

4 Hilgert, J., Meine, N., Rinaldi, R., and Schüth, F. (2013). *Energy Environ. Sci.* 6: 92.

5 Sheldon, R.A. (2014). *Green Chem.* 16: 950.

6 Corma, A., Iborra, S., and Velty, A. (2007). *Chem. Rev.* 107: 2411.

7 Granone, L.I., Sieland, F., Zheng, N. et al. (2018). *Green Chem.* 20: 1169.

8 Xu, C., Nasrollahzadeh, M., Selva, M. et al. (2019). *Chem. Soc. Rev.* 48: 4791.

9 Taarning, E., Osmundsen, C.M., Yang, X. et al. (2011). *Energy Environ. Sci.* 4: 793.

10 Gallezot, P. (2012). *Chem. Soc. Rev.* 41: 1538.

11 Yu, H. and Huang, G.H. (2009). *Bioresour. Technol.* 100: 2005.

12 Lesteur, M., Bellon-Maurel, V., Gonzalez, C. et al. (2010). *Process Biochem.* 45: 431.

13 Gunaseelan, V.N. (1997). *Biomass Bioenergy* 13: 83.

14 Khalid, A., Arshad, M., Anjum, M. et al. (2011). *Waste Manage. (Oxford)* 31: 1737.

15 Matson, T.D., Barta, K., Iretskii, A.V., and Ford, P.C. (2011). *J. Am. Chem. Soc.* 133: 14090.

16 Liu, X.-Q., Ding, H.-S., Wang, Y.-Y. et al. (2016). *Environ. Sci. Technol.* 50: 2602.

17 Wang, Y.-Y., Ling, L.-L., and Jiang, H. (2016). *Green Chem.* 18: 4032.

18 Espro, C., Gumina, B., Szumelda, T. et al. (2018). *Catalysts* 8: 313.

19 Ma, R., Hao, W., Ma, X. et al. (2014). *Angew. Chem.* 53: 7310.

20 Kamat, P.V. (1993). *Chem. Rev.* 93: 267.

21 Ni, M., Leung, M.K., Leung, D.Y., and Sumathy, K. (2007). *Renewable Sustainable Energy Rev.* 11: 401.

22 Konstantinou, I.K. and Albanis, T.A. (2004). *Appl. Catal., B* 49: 1.

23 Linic, S., Christopher, P., and Ingram, D.B. (2011). *Nat. Mater.* 10: 911.

24 Hao, Q., Wang, R., Lu, H. et al. (2017). *Appl. Catal., B* 219: 63.

25 Colmenares, J.C. and Luque, R. (2014). *Chem. Soc. Rev.* 43: 765.

26 Schultz, D.M. and Yoon, T.P. (2014). *Science* 343: 1239176.

27 Lin, J., Wu, X., Xie, S. et al. (2019). *ChemSusChem* 12: 5023.

28 Esposito, D.V., Forest, R.V., Chang, Y. et al. (2012). *Energy Environ. Sci.* 5: 9091.

29 Kuehnel, M.F. and Reisner, E. (2018). *Angew. Chem.* 57: 3290.

30 Colmenares, J.C. (2015). *Green Photo-Active Nanomaterials*, 168. RSC publishing.

31 Wakerley, D.W., Kuehnel, M.F., Orchard, K.L. et al. (2017). *Nat. Energy* 2: 17021.

32 Han, G., Jin, Y.H., Burgess, R.A. et al. (2017). *J. Am. Chem. Soc.* 139: 15584.

33 Li, S.-H., Liu, S., Colmenares, J.C., and Xu, Y.-J. (2016). *Green Chem.* 18: 594.

34 Ibrahim, N., Kamarudin, S.K., and Minggu, L.J. (2014). *J. Power Sources* 259: 33.

35 Lu, X., Xie, S., Yang, H. et al. (2014). *Chem. Soc. Rev.* 43: 7581.

36 Zhang, B., Fan, W., Yao, T. et al. (2017). *ChemSusChem* 10: 99.

37 Walter, M.G., Warren, E.L., McKone, J.R. et al. (2010). *Chem. Rev.* 110: 6446.

38 Lee, D.K., Lee, D., Lumley, M.A., and Choi, K.-S. (2019). *Chem. Soc. Rev.* 48: 2126.

39 Lianos, P. (2011). *J. Hazard. Mater.* 185: 575.

40 Ueno, H., Nemoto, J., Ohnuki, K. et al. (2009). *J. Appl. Electrochem.* 39: 1897.

41 Georgieva, J., Valova, E., Armyanov, S. et al. (2012). *J. Hazard. Mater.* 211: 30.

42 Gong, J., Imbault, A., and Farnood, R. (2017). *Appl. Catal., B* 204: 296.

43 Srisasiwimon, N., Chuangchote, S., Laosiripojana, N., and Sagawa, T. (2018). *ACS Sustainable Chem. Eng.* 6: 13968.

44 Chen, K., Cao, M., Ding, C., and Zheng, X. (2018). *RSC Adv.* 8: 26782.

45 Kamwilaisak, K. and Wright, P.C. (2012). *Energy Fuels* 26: 2400.

46 Nair, V., Dhar, P., and Vinu, R. (2016). *RSC Adv.* 6: 18204.

47 Prado, R., Erdocia, X., and Labidi, J. (2013). *Chemosphere* 91: 1355.

48 Tian, M., Wen, J., MacDonald, D. et al. (2010). *Electrochem. Commun.* 12: 527.

49 Liu, X., Duan, X., Wei, W. et al. (2019). *Green Chem.* 21: 4266.

50 Zhang, K. and Guo, L. (2013). *Catal. Sci. Technol.* 3: 1672.

51 Luo, N., Wang, M., Li, H. et al. (2017). *ACS Catal.* 7: 4571.

52 Luo, N., Wang, M., Li, H. et al. (2016). *ACS Catal.* 6: 7716.

53 Li, H., Lei, Z., Liu, C. et al. (2015). *Bioresour. Technol.* 175: 494.

54 Li, C., Wang, H., Naghadeh, S.B. et al. (2018). *Appl. Catal., B* 227: 229.

55 Kadam, S.R., Mate, V.R., Panmand, R.P. et al. (2014). *RSC Adv.* 4: 60626.

56 Zou, J., Zhang, G., and Xu, X. (2018). *Appl. Catal., A* 563: 73.

57 Zhang, B., Li, J., Guo, L. et al. (2018). *Appl. Catal., B* 237: 660.

58 Wang, L., Zhang, Z., Zhang, L. et al. (2015). *RSC Adv.* 5: 85242.

59 Hao, H., Zhang, L., Wang, W., and Zeng, S. (2018). *ChemSusChem* 11: 2810.

60 Ke, D., Liu, S., Dai, K. et al. (2009). *J. Phys. Chem. C* 113: 16021.

61 Zhang, G., Ni, C., Huang, X. et al. (2016). *Chem. Commun.* 52: 1673.

62 Speltini, A., Sturini, M., Dondi, D. et al. (2014). *Photochem. Photobiol. Sci.* 13: 1410.

63 Zhang, L., Wang, W., Zeng, S. et al. (2018). *Green Chem.* 20: 3008.

64 Qin, W., Meinhardt, K.A., Moffett, J.W. et al. (2017). *Environ. Microbiol. Rep.* 9: 250.

65 Caravaca, A., Jones, W., Hardacre, C., and Bowker, M. (2016). *Proc. R. Soc. A: Mathe., Phys. Eng. Sci.* 472: 20160054.

66 Geboers, J.A., Van de Vyver, S., Ooms, R. et al. (2011). *Catal. Sci. Technol.* 1: 714.

67 Luo, N., Montini, T., Zhang, J. et al. (2019). *Nat. Energy* 4: 575.

68 Kasap, H., Achilleos, D.S., Huang, A., and Reisner, E. (2018). *J. Am. Chem. Soc.* 140: 11604.

69 Wu, X., Fan, X., Xie, S. et al. (2018). *Nat. Catal.* 1: 772.

70 Wu, X., Xie, S., Liu, C. et al. (2019). *ACS Catal.* 9: 8443.

71 Monteiro, M.R., Kugelmeier, C.L., Pinheiro, R.S. et al. (2018). *Renewable Sustainable Energy Rev.* 88: 109.

72 Nguyen, V.T., Nguyen, V.D., Haug, G.C. et al. (2019). *ACS Catal.* 9: 9485.

73 Augugliaro, V., El Nazer, H.A.H., Loddo, V. et al. (2010). *Catal. Today* 151: 21.

74 Zhang, Y., Zhang, N., Tang, Z.-R., and Xu, Y.-J. (2013). *Chem. Sci.* 4: 1820.

75 Sakurai, H., Kiuchi, M., Heck, C., and Jin, T. (2016). *Chem. Commun.* 52: 13612.

76 Jedsukontorn, T., Meeyoo, V., Saito, N., and Hunsom, M. (2015). *Chem. Eng. J.* 281: 252.

77 Jedsukontorn, T., Meeyoo, V., Saito, N., and Hunsom, M. (2016). *Chin. J. Catal.* 37: 1975.

78 Maurino, V., Bedini, A., Minella, M. et al. (2008). *J. Adv. Oxid. Technol.* 11: 184.

79 Molinari, A., Maldotti, A., Bratovcic, A., and Magnacca, G. (2013). *Catal. Today* 206: 46.

80 Hermes, N.A., Corsetti, A., and Lansarin, M.A. (2014). *Chem. Lett.* 43: 143.

81 Zhou, B., Song, J., Zhou, H. et al. (2015). *RSC Adv.* 5: 36347.

82 Guo, L., Sun, Q., Marcus, K. et al. (2018). *J. Mater. Chem. A* 6: 22005.

83 Wang, C. and Astruc, D. (2014). *Chem. Soc. Rev.* 43: 7188.

84 Linic, S., Aslam, U., Boerigter, C., and Morabito, M. (2015). *Nat. Mater.* 14: 567.

85 Naldoni, A., Shalaev, V.M., and Brongersma, M.L. (2017). *Science* 356: 908.

86 Dodekatos, G. and Tüysüz, H. (2016). *Catal. Sci. Technol.* 6: 7307.

87 Schünemann, S., Dodekatos, G., and Tüysüz, H. (2015). *Chem. Mater.* 27: 7743.

88 Zhang, Y., Ciriminna, R., Palmisano, G. et al. (2014). *RSC Adv.* 4: 18341.

89 Dittmer, A., Menze, J., Mei, B. et al. (2016). *J. Phys. Chem. C* 120: 18191.

90 Daskalaki, V.M., Panagiotopoulou, P., and Kondarides, D.I. (2011). *Chem. Eng. J.* 170: 433.

91 Liu, R., Yoshida, H., Fujita, S.-i., and Arai, M. (2014). *Appl. Catal., B* 144: 41.

92 Jiang, X., Fu, X., Zhang, L. et al. (2015). *J. Mater. Chem. A* 3: 2271.

93 Cargnello, M., Gasparotto, A., Gombac, V. et al. (2011). *Eur. J. Inorg. Chem.* 2011: 4309.

94 Kondarides, D.I., Daskalaki, V.M., Patsoura, A., and Verykios, X.E. (2008). *Catal. Lett.* 122: 26.

95 Lalitha, K., Sadanandam, G., Kumari, V.D. et al. (2010). *J. Phys. Chem. C* 114: 22181.

96 Zhang, B., Shi, J., Ding, C. et al. (2015). *ChemSusChem* 8: 4049.

97 Chong, R., Wang, B., Li, D. et al. (2017). *Sol. Energy Mater. Sol. Cells* 160: 287.

98 Antoniadou, M. and Lianos, P. (2009). *J. Photochem. Photobiol., A* 204: 69.

99 Salari, M., Konstantinov, K., and Liu, H.K. (2011). *J. Mater. Chem.* 21: 5128.

100 Palmas, S., Da Pozzo, A., Mascia, M. et al. (2012). *J. Solid State Electrochem.* 16: 2493.

101 Mohapatra, S., Raja, K., Mahajan, V., and Misra, M. (2008). *J. Phys. Chem. C* 112: 11007.

102 Liu, W., Mu, W., Liu, M. et al. (2014). *Nat. Commun.* 5: 3208.

103 Liu, W., Gong, Y., Wu, W. et al. (2018). *ChemSusChem* 11: 2229.

104 Zhao, Q., Li, Z., Deng, Q. et al. (2016). *Electrochim. Acta* 210: 38.

105 Shewa, W.A., Lalman, J.A., Chaganti, S.R., and Heath, D.D. (2016). *Energy* 111: 774.

106 Alenzi, N., Liao, W.-S., Cremer, P.S. et al. (2010). *Int. J. Hydrogen Energy* 35: 11768.

107 Antoniadou, M., Bouras, P., Strataki, N., and Lianos, P. (2008). *Int. J. Hydrogen Energy* 33: 5045.

108 Ibadurrohman, M. and Hellgardt, K. (2014). *Int. J. Hydrogen Energy* 39: 18204.

109 Huang, L.-W., Vo, T.-G., and Chiang, C.-Y. (2019). *Electrochim. Acta* 322: 134725.

110 Liu, D., Liu, J.-C., Cai, W. et al. (2019). *Nat. Commun.* 10: 1779.

111 Enache, D.I., Edwards, J.K., Landon, P. et al. (2006). *Science* 311: 362.

112 Zhang, Y., Zhao, G., Zhang, Y., and Huang, X. (2014). *Green Chem.* 16: 3860.

113 Wu, Z., Wang, J., Zhou, Z., and Zhao, G. (2017). *J. Mater. Chem. A* 5: 12407.

114 Lin, C.-Y., Chueh, Y.-C., and Wu, C.-H. (2017). *Chem. Commun.* 53: 7345.

115 Cha, H.G. and Choi, K.S. (2015). *Nat. Chem.* 7: 328.

116 Roylance, J.J., Kim, T.W., and Choi, K.-S. (2016). *ACS Catal.* 6: 1840.

15

Reactors, Fundamentals, and Engineering Aspects for Photocatalytic and Photoelectrochemical Systems

Boon-Junn Ng, Xin Ying Kong, Yi-Hao Chew, Yee Wen Teh and Siang-Piao Chai

Monash University, Chemical Engineering Discipline, School of Engineering, Jalan Lagoon Selatan, Bandar Sunway, Selangor 47500, Malaysia

15.1 Fundamental Mechanisms of Photocatalytic and PEC Processes

Over the years, global energy consumption has reached an alarming rate, and the energy replenishment from the current energy pool is incapable to fulfill such needs. In other words, energy is the most pressing issue faced by humanity, and a diffusion of energy to more promising renewable carriers is anticipated in the near future. Among the renewable energy sources, solar energy displays the most promising avenue to solve the energy deficit attributed to the massive energy potential (173 000 TW) [1]. The highly abundant energy from the Sun is almost equivalent to 9600 times of the total recoverable energy from all the energy reserves in the Earth. This prompts efforts toward the development of solar energy harvesting, particularly in (i) electricity from solar photovoltaic (PV), (ii) heat from concentrating solar thermal power (CSP), and (iii) chemical solar fuels (hydrogen, H_2, and hydrocarbons). In this regard, solar fuels confer the upper hand in comparison with solar PV and CSP as it can be easily storable in a carrier form. Thus, solar-to-chemical conversion bestows a new horizon to harness energy from the Sun and is projected to play a major contribution to the future energy pool. With respect to this issue, artificial photosynthesis processes (photocatalytic and photoelectrochemical, PEC systems) show intriguing routes to harvest solar fuels. In this section, the fundamental rationales and mechanisms of photocatalytic and PEC systems will be discussed.

15.1.1 Rationales of Photocatalytic Systems

Among the solar fuels, H_2 and hydrocarbons are noteworthy as potential candidates to emerge as front runners in the future energy mix attributed to their high energy density [2]. The groundbreaking work from Fujishima and Honda in introducing

Solar-to-Chemical Conversion: Photocatalytic and Photoelectrochemcial Processes, First Edition.
Edited by Hongqi Sun.

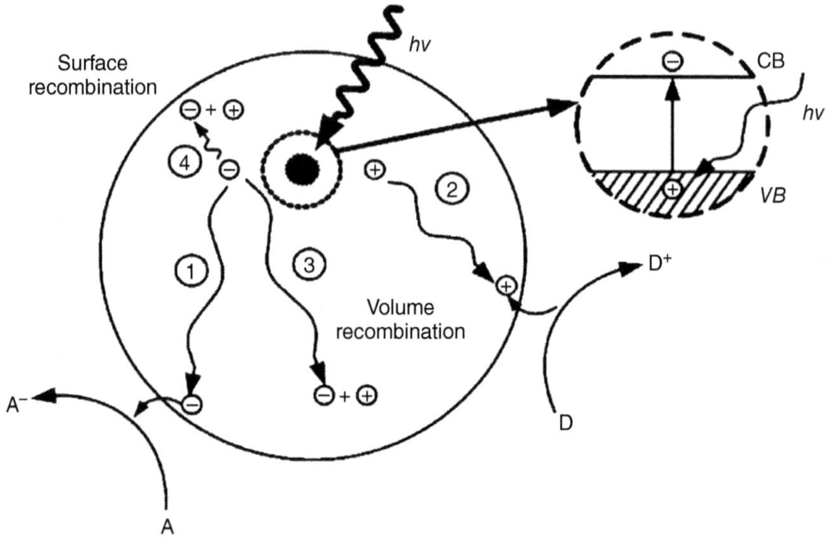

Figure 15.1 Schematic illustration of electron–hole pair generation in a semiconductor upon light irradiation. Possible pathways are labeled in 1– 4, where 1, reduction; 2, oxidation; 3, volume recombination; and 4, surface recombination; A, electron acceptor, and B, electron donor. Source: Linsebigler et al. [4].

titanium dioxide (TiO_2) photoelectrodes to split water into H_2 gas has opened up research interest in solar fuel production using artificial photosynthesis mean [3]. Generally, the major photocatalysis research focuses on photocatalytic water splitting for H_2 production and photoreduction of CO_2 into hydrocarbon fuels. In this context, photocatalysts provide the required photogenerated electrons and holes for the respective redox reaction. Unlike the continuum electronic states of metal, photocatalysts that are commonly made from semiconductors possess an energy void region known as band gap. This bandgap is extended from top of filled valance band (VB) to the bottom of vacant conduction band (CB) and is readily to be excited by light irradiation with energy greater or equal to the bandgap. As a result, the bandgap excitation process will generate electron–hole pairs with lifetime in nanosecond range for them to be migrated to the surface of photocatalysts. However, electron–hole pair recombination is hindering the efficiency of photocatalytic activity due to the flash recombination time, which is in the order of 10^{-9} s. As shown in Figure 15.1, recombination of charge carriers can happen in two possible pathways: (i) surface recombination and (ii) volume recombination. Hence, further modification of photocatalysts is imperative to suppress the undesired charge recombination.

15.1.1.1 Photocatalytic Water Splitting

Photocatalytic water splitting process is a light-driven heterogeneous reaction that converts water into H_2 and O_2 via the surface redox reaction, i.e. H_2-evolving reaction (HER) and O_2-evolving reaction (OER). According to overall water splitting stoichiometry in Eq. (15.1), Gibbs free energy of 237.2 kJ mol^{-1} or equivalent of an overpotential of 1.23 eV per electron is required to split water into H_2 and O_2 [5]. Besides, two electrons and four holes are necessary for the HER and OER process, as shown in Eqs. (15.2) and (15.3).

$$\text{Overall water splitting}: 2H_2O \rightarrow 2H_2 + O_2 \quad (\Delta G = 237.2\,\text{kJ mol}^{-1}) \tag{15.1}$$

$$\text{HER process}: 2H^+ + 2e^- \rightarrow H_2 \tag{15.2}$$

$$\text{OER process}: 2H_2O + 4\,h^+ \rightarrow O_2 + 4H^+ \tag{15.3}$$

As aforementioned, thermodynamic law places a constraint for overall water splitting in which an overpotential of 1.23 eV per electron is required for the redox reaction. This indicates that the potential for CB of photocatalysts should be more negative or analogously higher than the redox potential of H_2 production level, H^+/H_2 (−0.41 V vs. NHE at pH 7). On the other hand, the potential for VB of photocatalysts must be more positive or analogously lower than the redox potential of water oxidation, O_2/H_2O (+0.82 V vs. NHE at pH 7). In this context, an efficient photocatalytic water splitting system should render the following three properties: (i) small bandgap, (ii) slow charge recombination, and (iii) large overpotential to govern strong redox ability. However, it is clearly known that small bandgap and large overpotential are mutually exclusive. According to Figure 15.2a, one-step photoexcitation or single-component photocatalytic system cannot simultaneously exhibit the properties of small bandgap and large overpotential. Besides, single-component photocatalysts are often suffering from inevitable rapid charge recombination, which deteriorates the photocatalytic performance.

To overcome the limitations from one-step photoexcitation system, extensive researches are devoted to developing two-step photoexcitation system or known as Z-scheme. Inspired from the natural photosynthesis system, the coupling of two photocatalysts or two photosystems (PS) can improve the charge isolation efficiency and enhance the photocatalytic activity. Different from the conventional type-II heterojunction, Z-scheme system utilizes an electron mediator to interface two photosystems, i.e. PS I as H_2-evolving photocatalyst (HEP) and PS II as O_2-evolving photocatalyst (OEP). Hence, Z-scheme photocatalytic system confers cascade electron transfer profile in which photogenerated electrons are accumulated in PS I for HER while photogenerated holes are isolated in PS II for OER. Even though heterojunction-typed photocatalytic system can suppress charge recombination similar to Z-scheme system, the photogenerated electrons and holes are allocated in a narrower overpotential (Figure 15.2b). Thus, the presence of electron mediator is essential to govern vectorial electron transfer profile in Z-scheme system for strong redox potential (Figure 15.2c). Hitherto, Z-scheme system can be classified into three generations: (i) PS-A/D-PS system with redox mediator (first generation), (ii) PS-C-PS system with solid-state mediator (second generation), and (iii) PS–PS system without electron mediator or known as direct Z-scheme (third generation). In whole, the presence of Z-scheme system evokes new interest in developing highly efficient photocatalytic water splitting device and paves a future way for solar energy harvesting.

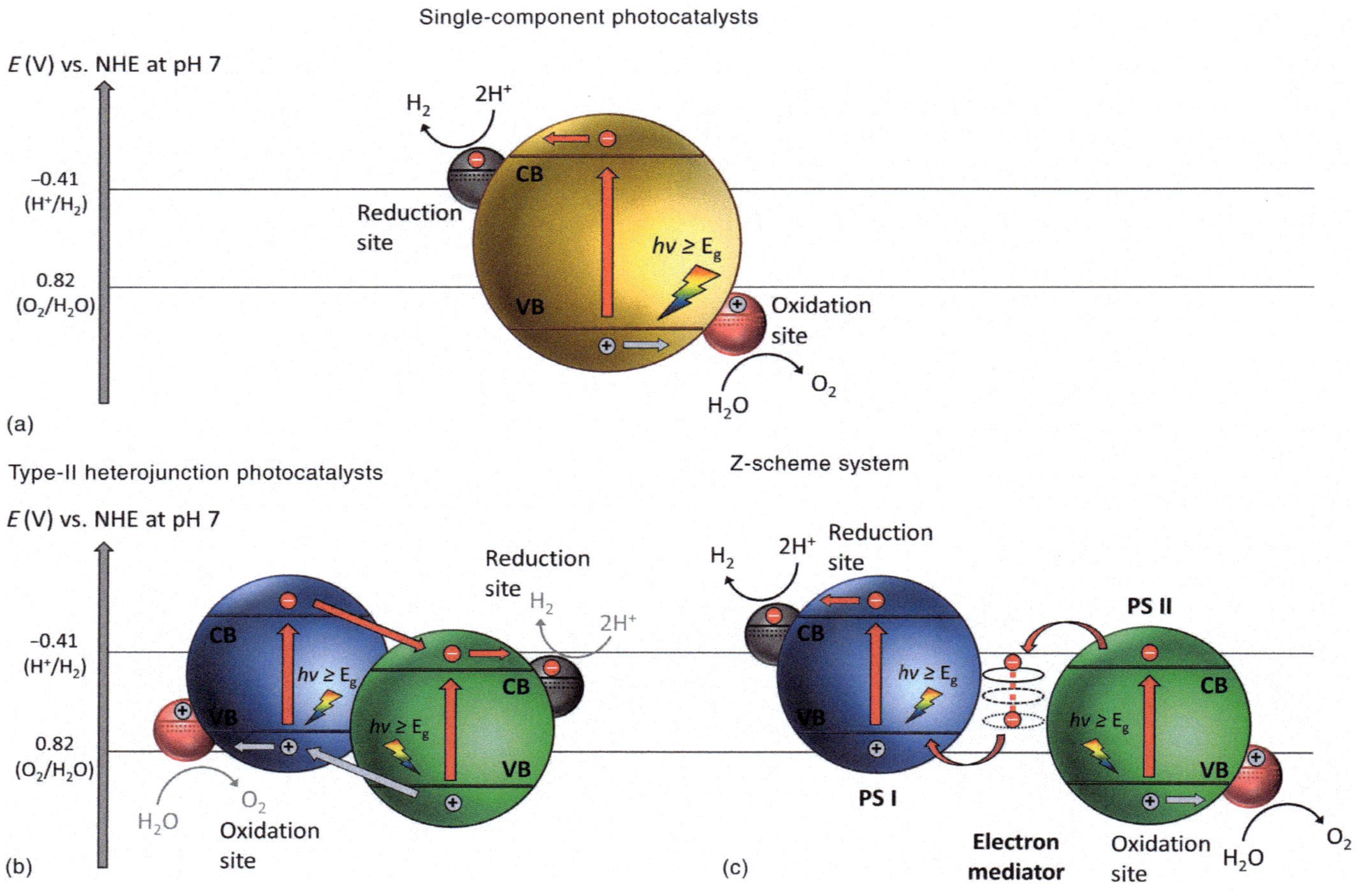

Figure 15.2 Schematic illustration of different photocatalytic water splitting systems. (a) One-step photoexcitation or single-component photocatalytic system. (b) Photocatalytic system with type-II heterojunction. (c) Two-step photoexcitation or Z-scheme photocatalytic system. Source: Chen et al. [6].

15.1.1.2 Photocatalytic CO_2 reduction

On the other hand, photocatalytic CO_2 reduction into value-added fuels is deemed to be a promising approach to concurrently alleviate the anthropogenic climate change and imminent energy crisis as (i) this process can be carried out at mild conditions, ambient temperature and atmospheric pressure; (ii) solar light, a form of clean, inexhaustible, and easy accessible energy, is the only input required to drive the process; and (iii) the in situ conversion of unwanted CO_2 into short-chain hydrocarbon fuels can maintain the carbon-neutral cycle without forming any secondary pollution [7]. Since photocatalytic reduction of CO_2 allows the undesirable CO_2 molecules to be integrated back into the utilization cycles in a form of carbon-neutral fuel, the process is regarded as "artificial photosynthesis" or "reverse combustion."

The development of this solar-to-fuel technology can be traced back to 1979, where Inoue et al. first reported on photocatalytic reduction of CO_2 into value-added chemicals including methanol (CH_3OH), formaldehyde (HCHO), formic acid (HCOOH), and trace amount of methane (CH_4) by using TiO_2, zinc oxide (ZnO), gallium phosphide (GaP), cadmium sulfide (CdS), tungsten trioxide (WO_3), and silicon carbide (SiC) powder as photocatalysts [8]. Since then, incessant research efforts have been devoted to developing novel photocatalysts. That being said, the state-of-the-art advances in this field are still far from commercialization and several breakthroughs must be made before this technology is economically sound. To improve the overall efficiency of this solar-to-fuel conversion, it is important to first understand the fundamental principles of photocatalytic CO_2 reduction.

As known, CO_2 is one of the most thermodynamically and chemically stable compounds of carbon, and hence, a large amount of input energy is needed to break the strong C=O bond for the formation of C—H bond, which can be utilized for hydrocarbon fuel production [9]. For the photocatalytic CO_2 conversion reaction to take place, multiple electrons and the corresponding number of protons are required. Owing to the fact that the chemical state of C atom in CO_2 compound is considerably high (C^{4+}), the CO_2 reduction process can only proceed with the support of reducing agents such as H_2O, H_2, SO_3^{2-}, S^{2-}, amines, and etc. [10]. Among all these reducing agents, H_2O is considered as the most suitable one attributed to its merits of abundance, nontoxicity and inexpensiveness.

Since photocatalytic conversion of CO_2 into short-chain hydrocarbon fuels such as CH_4 and CH_3OH are energetically uphill reactions that involve highly positive changes in Gibbs free energy, i.e. $CO_2 + 2H_2O \rightarrow CH_3OH + 3/2O_2$ ($\Delta G° = 702.2\,kJ\,mol^{-1}$) and $CO_2 + 2H_2O \rightarrow CH_4 + 2O_2$ ($\Delta G° = 818.3\,kJ\,mol^{-1}$), the input energy supplied by the incident light is used to overcome the reaction barriers. In a typical photocatalytic process, semiconductors, light irradiation, and reactant molecules are the necessities for photocatalytic reactions to take place. As aforementioned, semiconductors possess an energy bandgap – an energy interval void between typically existing electronic states [11]. Generally, the holes with oxidation ability migrate to the adsorbed H_2O molecule for H_2O oxidation into free H^+ (protons) and O_2, whereas the electrons with reducibility serve to reduce adsorbed CO_2 into carbon-neutral fuels such as CO, CH_4, and CH_3OH.

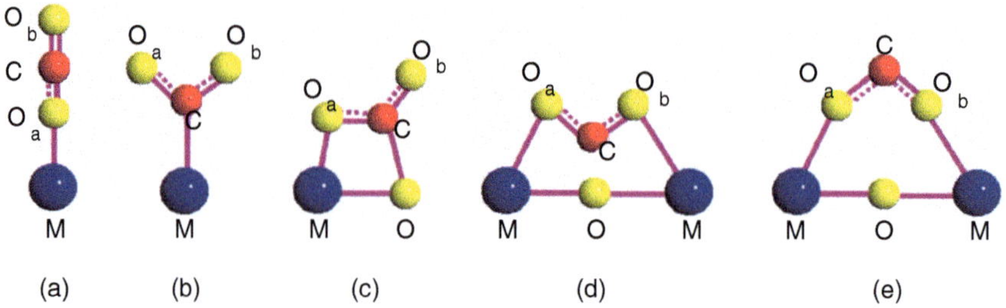

Figure 15.3 Possible configurations of CO_2 absorption on the surface of photocatalysts. Source: Liu et al. [12].

Other than suppressing the recombination of charge carrier, CO_2 adsorption and activation are also pivotal factors in deciding the photocatalytic activity. There are several types of CO_2 adsorption models, where each type gives different adsorption energy of the system. In most of the time, CO_2 molecule combines with the semiconductor photocatalysts via the O_a atom of CO_2 molecule through linear adsorption (Figure 15.3a) or via the C atom to produce a monodentate carbonate species (Figure 15.3b) [12]. In some cases, the CO_2 molecule interacts with the surface via both C and O_a atom, which leads to the bidentate carbonate species formation (Figure 15.3c). If the semiconductor surface possesses metal–oxygen–metal bond, bridged carbonate geometry will be generated where C atom of CO_2 pointing downward (Figure 15.3d) or upward (Figure 15.3e).

For the activation of thermodynamically stable CO_2 molecule, it involves a multistep process. In an ideal case, one of the photoinduced electrons from the excited photocatalyst transfer to the lowest unoccupied molecular orbital (LUMO) of CO_2, forming $\cdot CO_2^-$ species [13]. Accompanying the acceptance of an electron, the adsorption geometry of CO_2 changes from linear to bent structure, which resulted from the repulsion of unpaired electron located at the carbon atom with the two electron lone pairs sited on the oxygen atoms. Upon the bond angle of O—C—O to decrease from 180°, the energy of the LUMO of CO_2 is also reduced. This implies that the bent CO_2 molecule adsorbed on the photocatalyst surface is far less stable than the unadsorbed linear CO_2 molecule; thereby the reactivity for CO_2 reduction is greatly enhanced. However, as revealed by the scanning tunneling microscopy experiments, the transfer of an electron to the unadsorbed, free CO_2 is thermodynamically undesirable, owing to the extremely high LUMO position of CO_2, which requires a highly negative redox potential for the reaction of $CO_2 + e^- \rightarrow \cdot CO_2^-$ ($E^\circ_{redox} = -1.90\,V$ vs. NHE) to proceed [14]. Hence, a multicomponent and multistep process that involves numerous electrons and the corresponding number of protons is required to involve in the photocatalytic CO_2 reduction reaction [15]. Of note, the products formed are varied by the number of charge carriers involved in the reactions. For example, the formation of CO needs only two electrons and protons, whereas the production of CH_4 requires eight electrons and protons to drive the reaction. Basically, the reactivity and selectivity of the product yields are decided by several factors, including composition of photocatalysts, reduction potential of the photocatalysts, and the reaction conditions [16].

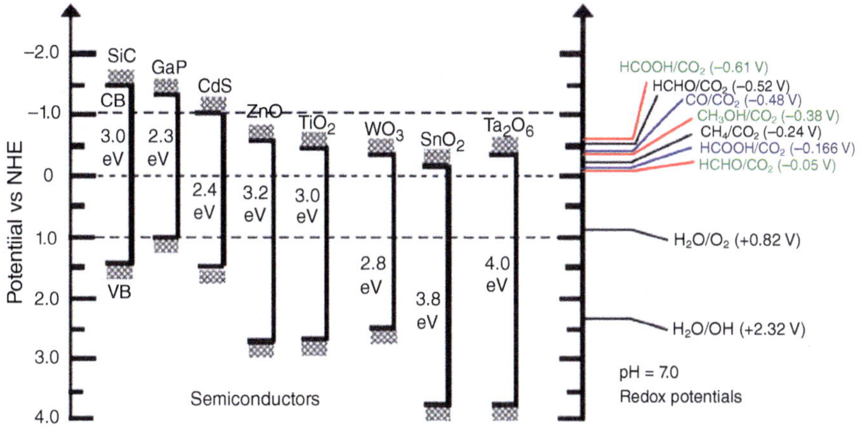

Figure 15.4 Bandgap energies and the band edge positions of some commonly reported photocatalysts as well as the products derived from photocatalytic CO_2 reduction with reference to normal hydrogen electrode (NHE). Source: Tu et al. [10].

Although different reaction conditions may yield different formation mechanisms, these two mechanisms of CO_2 into CH_4 are generally accepted: (i) $CO_2 \rightarrow CO \rightarrow C\cdot \rightarrow CH_2 \rightarrow CH_4$ and (ii) $CO_2 \rightarrow HCOOH \rightarrow HCHO \rightarrow CH_3OH \rightarrow CH_4$ [17]. As aforementioned, the photocatalytic transformation of CO_2 into CH_4 is a redox reaction, which involves (i) oxidation of H_2O for the formation of protons and O_2 and (ii) subsequent reduction of CO_2 into short-chain hydrocarbon using the generated protons from the oxidation reaction. Therefore, the employed photocatalysts must possess opportune energy band positions to drive the desired redox reactions. Functionally, the VB of the photocatalysts must be lower (or more positive) than the potential required to split water (0.82 V vs. NHE), whereas the CB of photocatalysts is required to be higher (or more negative) than the potential needed to form the desired product, which is −0.24 V vs. NHE for the case of CH_4. The reduction potentials required for other hydrocarbon fuels formation and the band edge positions of some commonly reported photocatalysts are shown in Figure 15.4.

In general, there are three main steps involved in photocatalysis: (i) photo-excited electrons transfer from CB to VB when photons with energy equal or larger than the bandgap of the photocatalyst are absorbed, (ii) separation of electrons and holes from each other and migration to the surface of semiconductor, and (iii) reduction and oxidation simultaneously occur at the active sites on the surface of photocatalysts. To achieve high photocatalytic activities, all the abovementioned crucial steps should be fully optimized. To this end, different types of photocatalysts have been developed to pave the ways for achieving high CO_2 conversion efficiency.

15.1.2 Rationales of PEC Systems

In view of PEC, the system is commonly fabricated in the configuration of photoelectrode–electrode manner. A typical PEC cell consists of a semiconducting photoelectrode, a metallic electrode and an electrolyte. Different from photovoltaic–electrolysis (PV-EC), PEC photoelectrode can contribute to both light absorption and redox reaction. The semiconductor is the main component of the PEC cell as it absorbs photons for the generation of electron–hole pairs [18]. When the energy absorbed is greater than the bandgap of the semiconductor, electrons are excited to the CB, leaving holes in the VB. When the semiconductor is connected with a metal, a built-in electric field is induced as the Fermi levels of the two materials equilibrate. The charge carriers are transferred from one phase to another across the semiconductor–electrolyte interface until they reach equilibrium, which results in band bending within the semiconductor. The space charge generated is crucial for solar energy conversion as it contributes to effective separation of the photo-induced charges. Without an efficacious system for charge separation, the inherent energy of the photo-induced charge carriers would be released through recombination. Moreover, an external bias is required to overcome the energy losses from the transfer of electrons and holes through the space charge region [19]. The pivotal advantage of the PEC theory is the spontaneous development of a heterojunction at the semiconductor–electrolyte interface [20].

Numerous configurations are established for a PEC cell as depicted in Figure 15.5. The two fundamental designs are delineated: (i) n-type photoanode, H_2 evolution catalyst (HEC) cathode (Figure 15.5a), and (ii) p-type photocathode, O_2 evolution catalyst (OEC) anode (Figure 15.5b). Generally, in a PEC cell, the photogenerated electrons are transported to the metallic electrode, where they reduce water to form H_2 gas. On the other hand, the photo-induced holes are accumulated at the semiconductor–electrolyte interface and are used to oxidize water to form O_2 gas. The type of semiconductor used for the working electrode plays a significant role in determining the efficiency of the PEC cell. Properties such as ideal bandgap, suitable band edge positions, and aqueous stability are desired for effective water splitting reactions.

More precisely, for an n-type semiconductor, electrons are collected at the photoanode upon light irradiation and are thereafter transported to the cathode where they are used to reduce H^+ into H_2. At the anode, the holes are used to oxidize water into O_2 and H^+. Hence, anodic photocurrents are produced when n-type semiconductors are utilized. When the work function of the metal used is higher than that of the n-type semiconductor, electrons are transferred from the semiconductor to the metal until the Fermi levels equilibrate. Band bending occurs at the metal-semiconductor interface, which is also known as the Schottky contact. However, when the work function of the semiconductor is larger than that of the metal used, an Ohmic contact is generated. Contrarily, for a p-type semiconductor, the electrons are used to reduce H^+ into H_2 at the photocathode, while oxidation of water occurs at the anode, thus generating a cathodic photocurrent.

Generally, p-type semiconductors are more stable as compared with n-type semiconductors as they have a greater resistance toward reduction reactions than oxidation reactions. On the other hand, n-type semiconductors have a lower tendency for photocorrosion as compared with p-type semiconductors as the charge transfer kinetics through the semiconductor–electrolyte interface are relatively quicker than

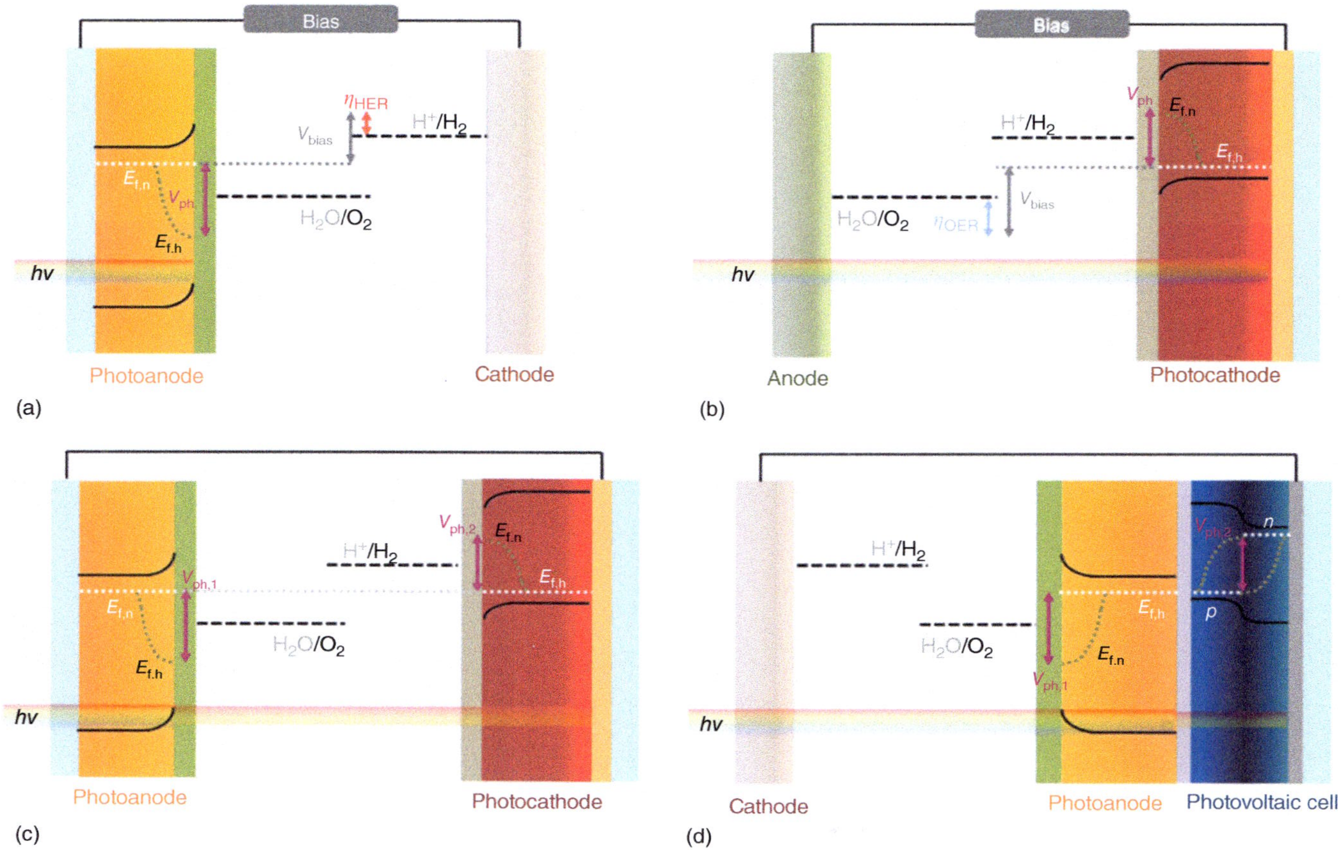

Figure 15.5 (a) Photoanode–HEC cathode, (b) photocathode–OEC anode, (c) photoanode–photocathode tandem cell, and (d) photoanode–photovoltaic (PEC-PV) tandem cell. Source: Kim et al. [1].

the oxidation reactions. However, the photostability of the semiconductors can be further enhanced by various methods such as surface passivation, intrinsic/extrinsic doping, and nanostructuring. Even so, an external applied bias is generally needed to offset the potential losses. An unassisted PEC water splitting system can be witnessed in a photocathode–photoanode dual light absorber tandem cell and a PV-PEC tandem cell, as shown in Figure 15.5c,d. Moreover, the redox potential differs according to the type of electrolyte used. The selection of electrolyte is highly governed by the material used for the photoelectrode. Equations (15.4) and (15.5) show the reduction and oxidations reactions and their redox potential referenced to NHE, respectively, when an alkaline electrolyte is used. Conversely, for an acidic electrolyte, the reactions are as shown in Equations (15.6) and (15.7) [21]. Whereas for PEC CO_2 reduction into hydrocarbons, the setup and configuration are similar to PEC water splitting system except the introduction of CO_2-saturated aqueous solution as electrolyte: [22]

$$4H_2O + 4e^- \rightleftarrows 2H_2 + 4OH^- - 0.828\ \text{V vs.NHE} \quad \text{Reduction} \tag{15.4}$$

$$4OH^- + 4\,h^+ \rightleftarrows 2H_2O + O_2 - 0.401\ \text{V vs.NHE} \quad \text{Oxidation} \tag{15.5}$$

$$4H^+ + 4e^- \rightleftarrows 2H_2 + 0.000\ \text{V vs.NHE} \quad \text{Reduction} \tag{15.6}$$

$$2H_2O + 4\,h^+ \rightleftarrows 4H^+ + O_2 - 1.229\ \text{V vs.NHE} \quad \text{Oxidation} \tag{15.7}$$

15.2 Reactor Design and Configuration

As a developing sustainable fuel generation technique that has yet to produce results sufficiently significant for industrial scale production, the reaction of photocatalytic water splitting and CO_2 reduction have hitherto been mostly conducted in lab scale. Nevertheless, small steps have been taken to slowly advance the application to larger scale. Throughout these progresses of development, various types of reactors have been utilized to accommodate the photocatalytic reactions. This section will summarize the designs of these reactors.

15.2.1 Reactors for Photocatalytic Systems

15.2.1.1 Reactors for Photocatalytic Water Splitting

Conventionally, lab-scale photocatalytic water splitting is often conducted in a powder suspension system where photocatalysts are dispersed in a reaction medium. Magnetic stirring is usually used to provide constant mixing throughout the reaction period and also to prevent sedimentation of reaction photocatalysts. In some setup, cooling water is circulated around the reactor to minimize the increase in temperature of the reaction medium, thus preventing the excessive evaporation of reaction medium. Moreover, depending on the method adopted to analyze the evolved gas, the reactors inlet and outlet design can be varied as well. For a batch process, the reactor is often vacuumed prior to the reaction to remove any residual dissolved air

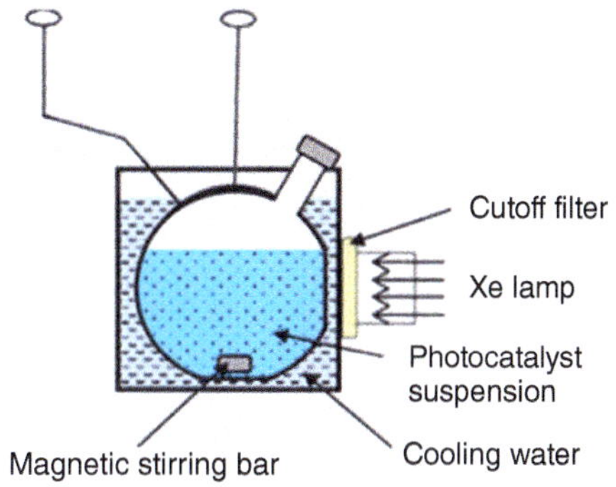

Figure 15.6 Schematic illustration of side-irradiated water splitting reactor. Source: Takata et al. [23].

in the solution. The evolved gas from the water splitting reaction is then collected for a desired period of time and further analyzed in an offline gas analyzing equipment, typically gas chromatography (GC). On the other hand, a continuous process features the utilization of online GC system where the evolved gas is directed straight into the analyzing column of GC. In this aspect, the reactor is usually purged thoroughly with an inert gas prior reaction. Upon initiation of the reaction, the inert gas stream continuously carries the evolved gas into the GC for analysis.

With such suspended particle reactor setup in mind, the most common configuration of the illumination source is either in side-irradiated or top-irradiated manner. Majority of researchers accommodate their reactions in side-irradiated reactors, which are relatively easier to set up and manage. Figure 15.6 shows the schematic graph of a typical side-irradiated reactor cooled by circulated water [23]. Although such a system carries the advantage of simple and low cost setup, one of the disadvantages is that the light transmission toward the reaction medium with photocatalyst suspension may be interfered by the refractive nature of the circulated water in the cooling water jacket.

Owing to this concern, top-irradiated reaction systems are preferred instead of side-irradiated systems [24]. The setup of this system is very similar to that of side-irradiated system, but the window of illumination is at the top of the reactor instead. Due to the top-irradiated configuration, the light source can directly illuminate onto the photocatalyst particles and not being affected by any form of cooling systems surrounding the reactor. A typical top-irradiated photocatalytic reactor is shown in Figure 15.7. The reaction vessel is equipped with a quartz lid to allow full wavelength range of light to be passed through and thus illuminating onto the reaction medium.

Another less common type of suspended particle photocatalytic reactor is the inner irradiated reactor. This type of reactor requires a specially designed cell where the intended light source can be inserted in the middle of the reactor [26]. An illustration of a typical inner irradiated reactor is shown in Figure 15.8a. Such configuration provides a comprehensive and uniform light illumination to the entire reaction medium, unlike the aforementioned configurations. Valadés-Pelayo et al. have also proposed the scaling up of an inner irradiated reactor by using the Monte Carlo radiative model [28]. The reactor set up (Figure 15.8b) includes

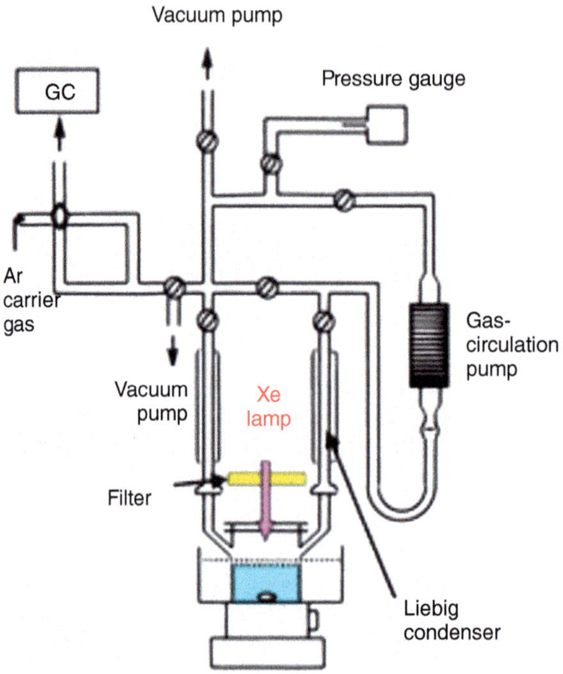

Figure 15.7 Schematic illustration of top-irradiated water splitting reactor. Source: Kudo et al. [25].

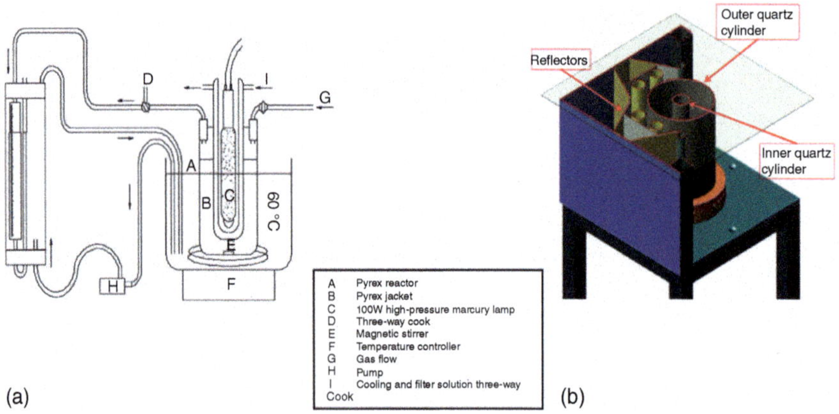

Figure 15.8 (a) Schematic illustrations of inner-irradiated reactor in lab scale. Source: Wei et al. [27]. (b) Schematic illustrations of inner-irradiated reactor with reflectors. Source: Valadés-Pelayo et al. [28].

four reflectors placed around the annular section to maximize the utilization of light from the source placed within the inner quartz cylinder. Despite a more superior performance on delivering uniform incident light distribution throughout the reactor, the inner irradiated reactor is generally uncommon and less used by researchers due to considerations on the design complication, which usually leads to a higher cost of fabrication.

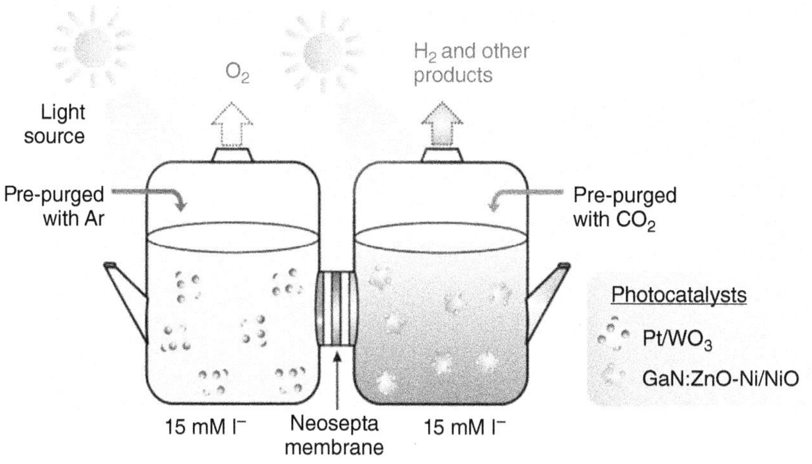

Figure 15.9 Schematic illustration of a twin photo-reactor system. Source: Sasaki et al. [29].

The major concern on photocatalytic overall water splitting reaction is the separation of H_2 and O_2 gases. The on-site separation of these resultant gases is preferred over post separation for storage purposes. Hence, the twin photo-reactor system has been used specifically to accomplish this objective [29]. A schematic diagram of the twin photo-reactor system is presented in Figure 15.9. The twin reactor system requires the separation of HEP and OEP in a different compartment of the reactor. An aqueous phase electron mediator is then used as a medium for charge transfer across these two compartments, and an anion-exchanged membrane is required to prevent blending of both photocatalysts and the evolved gases while allowing movement of anions across the compartments. This configuration effectively allows simultaneous generation of H_2 and O_2 gases, yet on-site isolating them from mixing.

Thus far, all reactors mentioned are suitable to accommodate particle suspension type of reaction. In recent years, Domen's group started extending the field of research toward immobilizing photocatalysts into the form of particulate sheets as an alternative to conventional powder suspension [30]. Initially, the studies were of smaller scale, where the immobilized photocatalyst sheets are placed at the bottom of a typical top-irradiated reactor for the administration of water splitting reaction. When the research work was moving toward scaling up of the photocatalyst sheet system, water splitting panels incited by the design of classic solar cells were proposed. The detailed scheme and photo of the water splitting panels reported by Goto et al. are manifested in Figure 15.10 [30]. The $1 \times 1\ m^2$ panel was made up of nine photocatalyst sheets of size $33 \times 33\ cm^2$, and a hydrophilized acrylic plate was adopted to serve as the illumination window. A 4 mm thin layer of water functioning as reaction medium was then introduced into the panel before the reaction. The water splitting panel was then subjected under natural sunlight illumination, and solar-to-hydrogen (STH) of 0.4% was achieved. This strategy of mimicking a solar panel is beneficial for large-scale photocatalytic water splitting

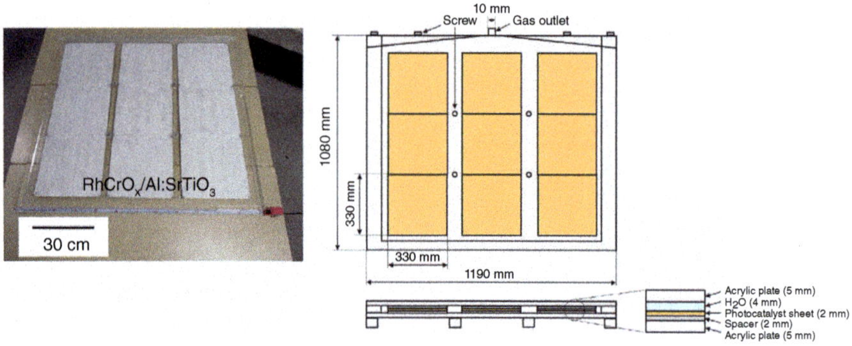

Figure 15.10 Schematic illustration of photocatalytic water splitting panel. Source: (a) Reproduced with permission Goto et al. [30]. Copyright 2018, Elsevier; (b) Goto et al. [30].

from many aspects. First of all, the immobilized photocatalysts setup allows the usage of minimum volume of reaction medium, unlike suspended particles system where the photocatalysts need to be stirred continuously in a larger volume. In the panel system reported by Goto et al., only 4 mm thick of water is required. This not only lowers the hydraulic pressure that the gases needed to overcome to evolve and being collected; the overall setup is also much lighter in weight, which simplifies the process of setting up and scaling up the panel-like reactor system. Not only that, the photocatalyst sheets can be easily recycled too as compared with the suspended particles reactors. Nevertheless, separation of the evolved H_2 and O_2 gases in this panel-like reactor is crucial in the future when to realize augmentation of overall water splitting system.

15.2.1.2 Reactors for Photocatalytic CO_2 Reduction

In this part, the reactors designed for photocatalytic CO_2 reduction will be discussed. Since CO_2 is a thermodynamically stable compound owing to its strong C=O bond (bond enthalpy of +805 kJ mol^{-1}), the conversion of CO_2 into carbon-neutral fuels requires substantial amount of energy input to break the bond [31]. Typically, the clean and exhaustible solar energy serves as the input energy to drive the photocatalytic CO_2 reduction reaction, where CO_2 is the reactant and semiconductor photocatalysts are required to provide the surface active sites. Other than the abovementioned necessities, reducing agent such as H_2O, NaOH, NH_3, or C_3H_7OH is also needed for the reactions to take place [32]. Among all, H_2O is the most commonly used reducing agent for photocatalytic CO_2 reduction owing to its naturally abundant, inexpensive, and readily available. Therefore, the main focus of this section will be on the reactor design with H_2O as the reducing agent and CO_2 as the feedstock in the presence of photocatalysts and light illumination to drive the photoreactions.

Photo-reactor plays a pivotal role in determining the overall efficiency of photocatalytic reactions. Basically, a photo-reactor is a vessel that allows the intimate contact of reactants and photocatalysts for the formation of reaction products under light irradiation [33]. For an ideal reactor, it is essential to have uniform

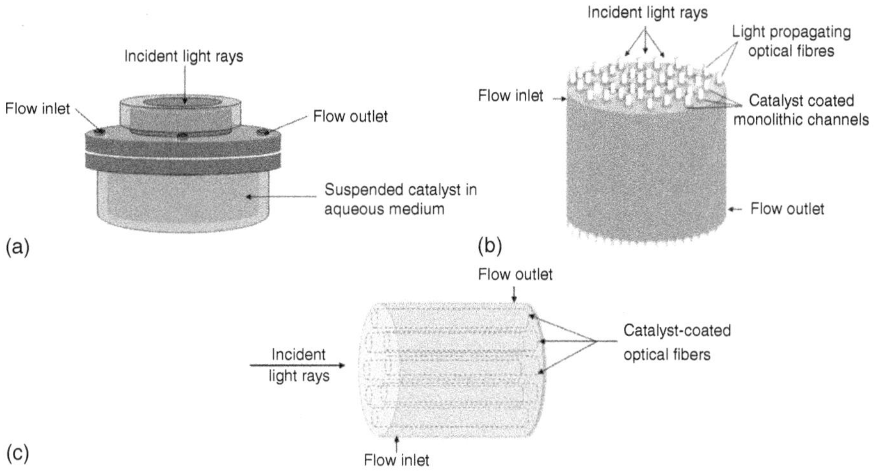

Figure 15.11 Schematic illustration of (a) slurry reactor, (b) internally illuminated reactor, and (c) optical fiber reactor. Source: Ola et al. [35], licensed under CC BY 4.0.

light distribution in order to maximize the utilization of the incident light so as to initiate the surface reactions over the photocatalysts. There are two major types of photo-reactors for CO_2 reduction, which are slurry reactor (i.e. fluidized bed reactor) and fixed bed reactor. Before designing a photo-reactor for CO_2 reduction, there are several key parameters to consider, including reaction phases (i.e. gas–solid, liquid–solid, or gas–liquid solid) and operation modes (i.e. continuous, batch, or semibatch) [34]. Generally, photocatalysts are placed in the photo-reactors in two ways, which are either in suspension or immobilized form. For suspension, the photocatalysts are homogeneously dispersed in an aqueous medium prior to the photoreaction (Figure 15.11a). On the other hand, the more conventional immobilization technique is through coating the photocatalysts on supports such as glass, monolithic, or optical fiber, as shown in Figure 15.11b,c.

For photocatalytic CO_2 reduction, the most commonly used reactor is the slurry reactor that involves three phases, in which (i) CO_2 in gas phase, (ii) H_2O in liquid phase, and (iii) photocatalyst in solid phase. Prior to the reaction, the photo-reactor is evacuated to remove the background gas, whereas the photocatalysts are suspended in an aqueous medium through magnetic stirring or continuous gas bubbling to remain the particles in suspension state. Since the particles are in mobilization form, slurry reactor is often referred as fluidized bed reactor [36]. Generally, slurry reactors can operate in both batch and continuous mode. For the case of batch operation, the CO_2 is channeled into the reactor with the presence of photocatalysts and followed by sealing the reactor for reactions to take place under light irradiation. On the contrary, a continuous operation mode allows incessant CO_2 gas bubbling through water where the gaseous product is continuously collected and analyzed by a GC, whereas the liquid product is collected by a syringe at certain time interval for analysis. Since the photocatalyst particles are dispersed or suspended in liquid medium, the rate of reaction is determined by several key factors, which are types and photoresponse of the photocatalysts, surface light intensity, and absorption properties of the reactants and nonreactants in the solution [37]. There are several advantages possessed by a slurry reactor, including (i) high photocatalysts loading, (ii) ease of reactor construction, and (iii) minimal mass transfer constraint due to agitated movement of solid particles. However, the major drawbacks of a slurry reactor include (i) hard separation of photocatalyst from the reaction mixture, (ii) vigorous agitation leading to the attrition and erosion by abrasion of the photocatalyst particles, (iii) restricted processing capacity resulting from the mass transport limitation, and (iv) low intensity of light achieving the photocatalyst surface due to hard penetration of light through bulk liquid phase [38].

Opportunely, the major drawback of difficult separation of photocatalysts exhibited by the slurry reactor could be addressed by a fixed bed reactor. As the name suggests, the photocatalysts in a fixed bed reactor are immobilized onto fixed supports such as glass, optical fiber, monolith, beads, etc. At the meantime, a fixed bed reactor allows continuous process operation, and light photons can reach the exposed photocatalyst surface directly without any hindrance in a fixed bed reactor. In this system, photocatalysts are coated on a support matrix, which then placed in the reactor directly under light illumination. Hence, light distribution on the photocatalysts is a deciding factor as it is greatly affected by the orientation of light source and the spatial distance between photocatalysts and light source [39]. With coating the photocatalysts in a thin film on the support, the surface area of light-illuminated photocatalysts will be greatly enlarged, and the light penetration problem that exists in fluidized bed reactor could be avoided.

Figure 15.12 depicts a schematic diagram of fixed bed reactor in gas–solid phase with a continuous flow. CO_2 inlet is bubbled through H_2O, and the CO_2/H_2O gas mixture is channeled to the reactor for intimate contact between the reactant gas and photocatalysts immobilized on glass support. Similar to fluidized bed reactor, there are also several factors that may affect the efficiency of CO_2 reduction in a fixed bed reactor, which are the type and particle size of photocatalysts, thickness of photocatalyst film, gas flowrate, incident light orientation, and intensity. Therefore, all these parameters are important to operate a fixed bed reactor.

15.2.2 Reactors for PEC Systems

Different from a photocatalytic system, PEC process is constructed in a two-electrode cell configuration, either in (i) single light absorber (photoelectrode–metal electrode) or (ii) dual light absorber (photocathode–photoanode tandem cell). A typical PEC reactor is equipped with a membrane that separates HER and OER in two compartments, and thus the backward reaction can be perfectly prevented [1]. In general, PEC reactors are designed in two fashions: (i) wired PEC system and (ii) wireless PEC system. As shown in Figure 15.13, PEC system with single light absorber is constructed in a two-compartment reactor cell with photoelectrode (photocathode/photoanode) and metal electrode (cathode/anode) connected by an external wiring. The distance between the cathode and anode should be optimized to reduce the potential losses and in the same time to render uniform current

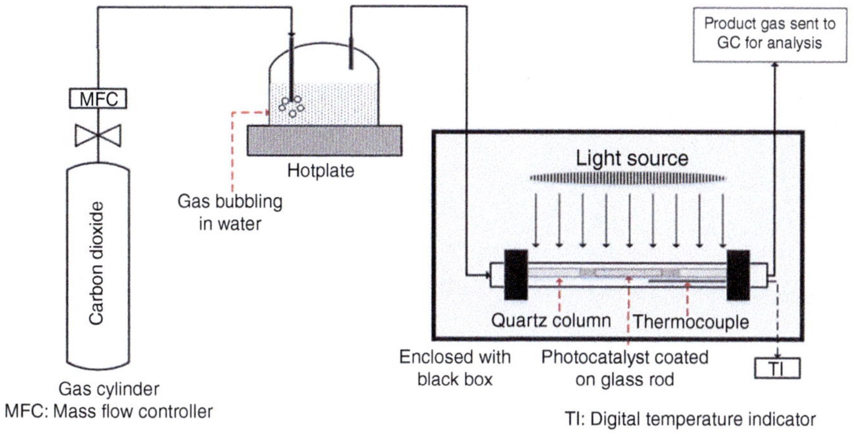

Figure 15.12 Schematic diagram of photocatalytic CO_2 reduction in a continuous gas flow fixed bed reactor. Source: Kong et al. [7].

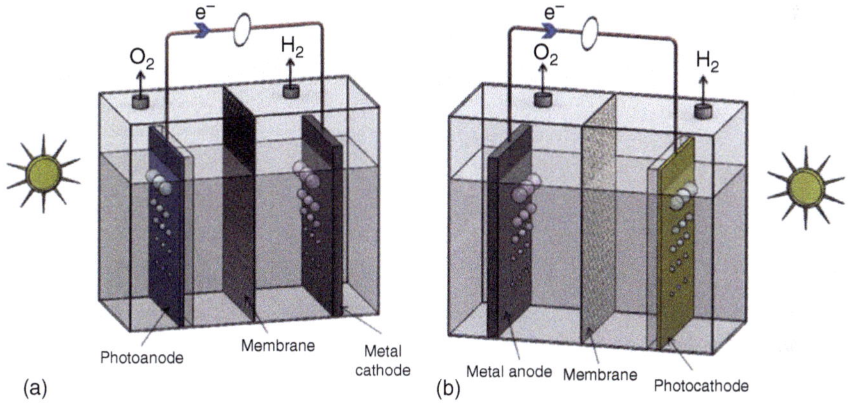

Figure 15.13 Schematic diagram of PEC water splitting system in the configuration of (a) n-type photoanode-HEC and (b) p-type photocathode-OEC. Source: Ahmed et al. [40].

distribution [40]. Attributed to the presence of a membrane, H_2 and O_2 can be separated and collected at the outlets of the reactor. This reactor design is also applicable to wired PEC tandem cell, but no applied bias is needed to compensate the overpotential losses.

Ascribed to the groundbreaking idea of assembling photoanode and photocathode in the same miniature plane, a wireless PEC system that is known as monolithic device can be fabricated. The details of monolithic device will be discussed henceforth in Section 15.4, while this section will focus on the reactor designed for wireless PEC system. As delineated in Figure 15.14a, the integrated wireless PEC system that is composed of two light absorbers are mounted on the same acrylic frame [41]. The device is set up in a reactor containing membrane for the gas separation. Similar to the reactor setup in wired PEC system, H_2 and O_2 gas can be collected separately at

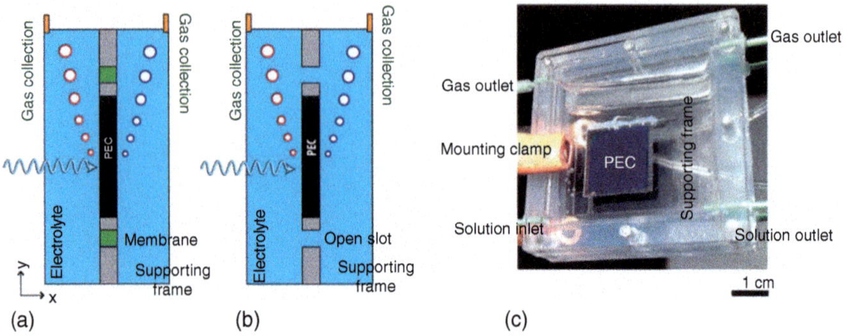

Figure 15.14 Schematic diagram of PEC water splitting devices: (a) with membrane and (b) without membrane. (c) Photograph of an integrated membrane-free device with internal wiring. Source: Reproduced with permission Jin et al. [41]. Copyright 2014, Royal Society of Chemistry.

the different outlets. Besides, incessant efforts are also devoted to create PEC reactor without any membrane. In Figure 15.14b, the clever geometry of the PEC reactor enables the separated evolution of H_2 and O_2 gases. Hence, the design presents a new concept on the design of PEC reactor to achieve isolated H_2 and O_2 evolution (Figure 15.14c). On the other hand, the design of reactor for PEC CO_2 reduction into hydrocarbons is very similar to those in water splitting system. According to Figure 15.15, photoanode (Pt-modified TiO_2 nanotubes) is connected to photocathode (Pt-modified RGO) via an external wiring [42]. The two photoelectrodes are separated by a Nafion membrane. Consequently, the light-driven reaction converts H_2O into O_2 on the photoanode and CO_2 into hydrocarbons (C_2H_5OH and CH_3COOH) over the photocathode.

15.3 Engineering Aspects of Photocatalytic and PEC Processes

With the ever-increasing demand from global energy consumption and the incompetency of current energy pool to supply such needs, it is timely to consider on scaling up of photocatalytic and PEC systems in solar fuel harvesting. This section will focus on the engineering aspects of photocatalytic and PEC processes, particularly on the key demonstration of scaling attempts devoted to photocatalytic water splitting using photocatalyst sheets and monolithic devices for H_2 production.

15.3.1 Photocatalyst Sheets: Scaling-up of Photocatalytic Water Splitting

For many years, photocatalytic water splitting has been conducted in a powder suspension form, which greatly limits the practicability of large-scale production. The first demonstration of large-scale solar H_2 production using photocatalyst

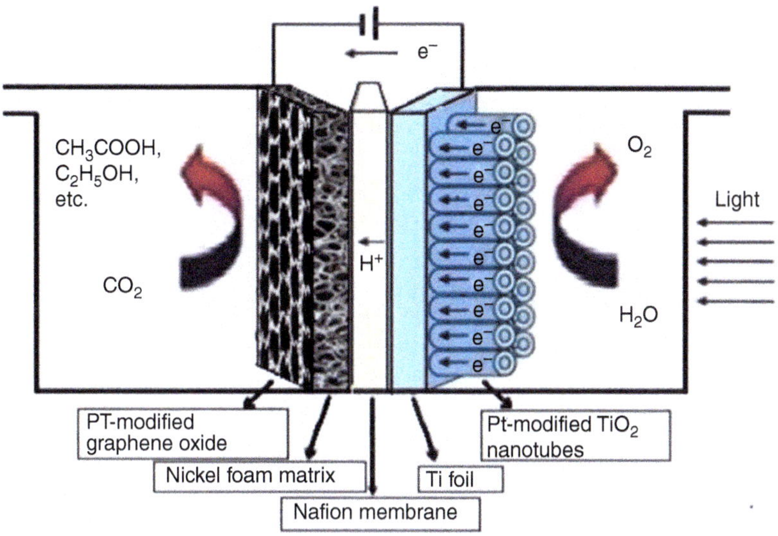

Figure 15.15 Schematic diagram of PEC CO_2 reduction system. Source: Cheng et al. [42].

sheets by Schröder et al. utilized the immobilization of Pt/g-C_3N_4 onto stainless steel plates [43]. In this context, Nafion was employed as polymeric binder for the drop casting of Pt/g-C_3N_4 and in the same time acting as second conducting layer to reduce the mass transfer resistance between the photocatalysts. As shown in Figure 15.16, the large-scale reactor was set up with nine photocatalyst sheets with immobilized Pt/g-C_3N_4, with a total irradiation active area of about 0.756 m^2. Besides, the reactor was equipped with plexiglass window, which is UV transparent ($\lambda > 300$ nm). In addition, the window was balanced by cross and longitudinal bracings ascribed to the high hydrostatic pressure inside the reactor. The photocatalytic H_2 production was carried out using circulated 10 vol% TEOA solution by a peristaltic pump, and the reactor was subjected to natural sunlight irradiation. The energy of the irradiated light was measured to be 83.8 kWh using a light sensor. This system produces a total of 18.21 of H_2 gas under long-term measurement (30 days) with STH efficiency recorded at 0.12%. Hence, the demonstration of large-scale photocatalytic reactor with immobilized photocatalyst sheets opens a new research interest in scaling-up of solar fuel production by photocatalysis. However, due to the limitation of single-component g-C_3N_4 to drive pure water splitting, a sacrificial reagent such as TEOA is needed for the half reaction. Very recently, Goto et al. demonstrated the concept of water splitting panel using immobilized $RhCrO_x$/Al:$SrTiO_3$ photocatalyst sheets for H_2 evolution [30]. In this study, the influence of water pressure was studied to ensure uninterrupted evolution of H_2 gas. The presence of midgap impurity state in Al:$SrTiO_3$ can improve the light absorption properties and prolong the lifetime of photogenerated charge carriers. Consequently, the water splitting panel by Goto et al. rendered H_2 evolution from pure water with measured STH of 0.4% under natural sunlight irradiation.

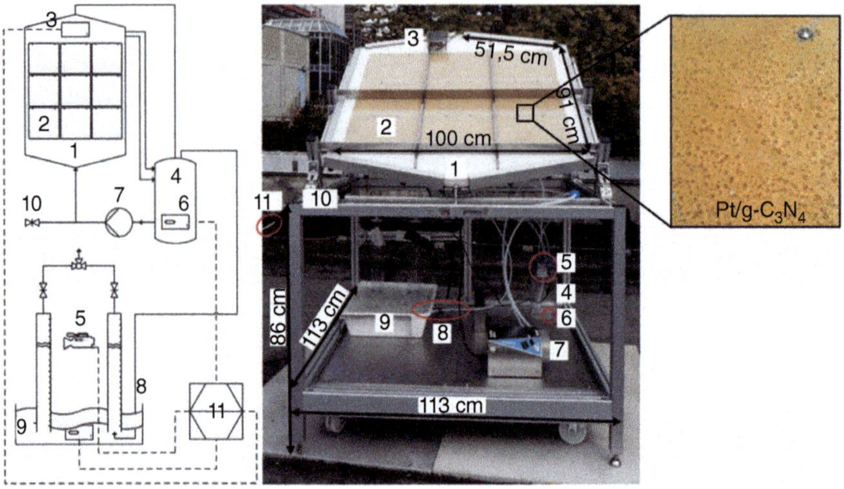

Figure 15.16 Large-scale reactor for photocatalytic H_2 production containing immobilized Pt/*g*-C_3N_4 working under natural sunlight. Source: Reproduced with permission Schröder et al. [43]. Copyright 2015, Wiley.

Attributed to the cascade electron profile that mimics natural photosynthesis of green plant, Z-scheme photocatalytic system can simultaneously confer the properties that are lacking in the single-component photocatalysts that include (i) small bandgap, (ii) large overpotential, and (iii) slow charge recombination. In 2016, Wang et al. introduced the concept of photocatalyst sheets in particulate Z-scheme system by developing a dual-layer-structured film [30]. As delineated in Figure 15.17a, the two-layered photocatalyst sheets were prepared via particle transfer method: (i) top layer consisting of photocatalysts ($SrTiO_3$:La,Rh as PS I and $BiVO_4$:Mo as PS II) and (ii) bottom layer of electron mediator (Au). The ingenious arrangement of PS I and II on the top layer can allow vectorial electron transfer to be accomplished via the underlying electron mediator layer. The electron transfer profile of $SrTiO_3$:La,Rh/Au/$BiVO_4$:Mo is shown in Figure 15.17b. As compared with powder suspension system, particulate Z-scheme photocatalyst sheets encompass the advantages of potential scalability and suppression of the effect of H^+/OH^- concentration potential and pH gradient. However, the thickness of the photocatalysts layer should be adjusted and optimized ascribed to the increase in electrical resistance when the number of grain boundaries is increased. Besides, the photocatalysts layer

Figure 15.17 (a) Schematic illustration of synthesis protocol of $SrTiO_3$:La,Rh/Au/$BiVO_4$:Mo Z-scheme photocatalyst sheets using particle transfer method. (b) Electron transfer mechanism of $SrTiO_3$:La,Rh/Au/$BiVO_4$:Mo Z-scheme system. (c) Photograph of the ink used for screen printing and the corresponding printed photocatalyst sheets. Source: (a, c) Reproduced with permission Wang et al. [30]. Copyright 2016, Nature Publishing Group; (b) Wang et al. [30].

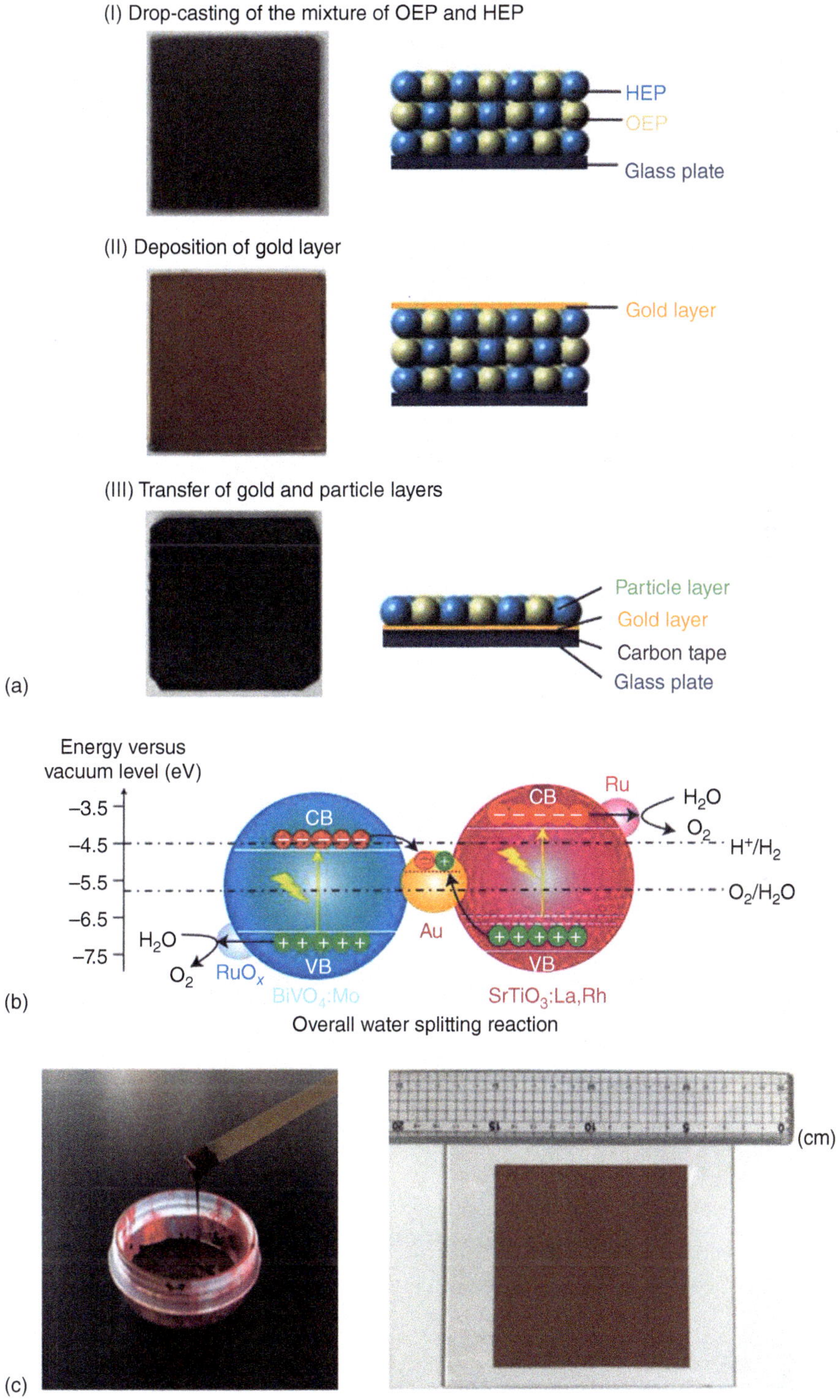
(I) Drop-casting of the mixture of OEP and HEP
HEP
OEP
Glass plate
(II) Deposition of gold layer
Gold layer
(III) Transfer of gold and particle layers
Particle layer
Gold layer
Carbon tape
Glass plate
(a)
Energy versus vacuum level (eV)
−3.5
−4.5
−5.5
−6.5
−7.5
CB
VB
Au
Ru
CB
VB
H2O
O2
H+/H2
O2/H2O
H2O
O2
RuOx
BiVO4:Mo
SrTiO3:La,Rh
Overall water splitting reaction
(b)
(cm)
(c)

should be thick enough to maximize the courage of photoactive area. In the work by Wang et al., $SrTiO_3$:La,Rh/Au/$BiVO_4$:Mo Z-scheme photocatalyst sheets were further subjected to post calcination to enhance the contact interface between photocatalysts layer and the conducting film. On top of that, core–shell Ru/Cr_2O_3 co-catalysts were incorporated into the Z-scheme sheets to promote forward reaction with minimal backward reaction. This is attributed to the nature of Cr_2O_3 as a protective shell that prevent the produced O_2 to permeate through the layer, which effectively suppressed competitive oxygen reduction reaction (ORR). In addition, the fabrication of particulate Z-scheme photocatalyst sheets can be further extended to the employment of screen printing using colloidal suspension of PS I and II, as shown in Figure 15.17c. Very recently, carbonaceous materials such as graphene and carbon nanotubes (CNTs) have been a relative new addition to the family of electron mediator to govern Z-scheme system [2]. Wang et al. firstly reported the implementation of carbon conducting layer as electron mediator in $SrTiO_3$:La,Rh/C/$BiVO_4$:Mo photocatalyst sheets system [30]. In contrary to metallic conductor that imposes reverse reaction, the employment of carbon layer provides a more promising avenue to operate Z-scheme photocatalytic water splitting system ascribed to its abundancy and inert for backward process. Besides, the work function of carbon layer (5.2 eV) is relatively high and comparable with metallic conductor. Thus, Z-scheme photocatalyst sheets with carbonaceous materials as electron mediator are feasible for large-scale water splitting application. As a result, the developed system demonstrated a STH of 1.0% under ambient pressure, which is the highest among all the reported Z-scheme systems that are operated under the similar condition.

Apart from photocatalyst sheets with conductor as electron mediator, a fully integrated nanosystem of Si/TiO_2 film was developed by Liu et al. for solar overall water splitting [44]. As delineated in Figure 15.18a,b, it can be clearly observed that the miniature photocatalyst sheets of Si/TiO_2 resembles a tree-like configuration in

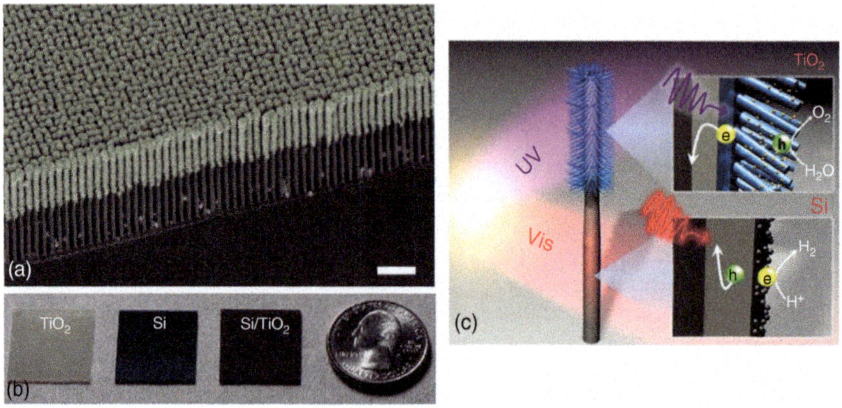

Figure 15.18 (a) False-colored SEM image of Si/TiO_2 nanotree arrays photocatalyst sheets. (b) Photographs of Si/TiO_2 photocatalyst sheets. (c) Charge transfer mechanism in Si/TiO_2 photocatalyst sheets. Source: Reproduced with permission Liu et al. [44]. Copyright 2013, American Chemical Society.

nanoscale. Different from Z-scheme system with external electron mediator, coupling of PS I of higher Fermi level with another PS II can confer a unique electronic structure of direct Z-scheme system. This is attributed to the band bending between PS I and II that generates an internal electric field that features Ohmic contact properties. In other words, the space charged region between the interfaces of PS I and II can act as an electron transporting channel with cascade profile. In the work by Liu et al., the Si/TiO_2 photocatalyst sheets were further incorporated with Pt as HER co-catalysts and IrO_2 as OER co-catalysts to promote overall water splitting (Figure 15.18c). In overall, this finding extends the applicability of photocatalyst sheets to a new horizon and provides the potential scalability of photocatalytic water splitting in the near future.

15.3.2 Monolithic Devices: Wireless Approach of PEC Reaction

Among all the solar-to-chemical conversion methods, PEC process is regarded as one of the most widely studied technologies especially in solar H_2 harvesting. As aforementioned, PEC system is generally constructed in the photoelectrode–electrode mean: either in n-type photoanode–HEC cathode or p-type photocathode–PEC anode. However, an external bias is usually required in PEC system to compensate for the overvoltages and other losses. Even though the groundbreaking idea of tandem cell with two light absorbers system can realize unassisted PEC water splitting, the complexity of wired system is still hindering the potential for large-scale application.

With respect to this issue, monolithic water splitting devices are introduced by compact assembly of light absorbers, thus creating a wireless system. Monolithic systems (also known as artificial leaf) are generally classified into three different configurations: (i) photoanode–photocathode (PEC tandem cell), (ii) PV-EC, and (iii) PV-PEC tandem cell [1]. One of the typical examples of monolithic system with PV-PEC tandem cell configuration can be witnessed from the work by Nocera et al., in which (3jn) Si solar cell (3jn-a-Si) was assembled with Co-Bi (OEC) and NiMoZn (HEC) [45]. As shown in Figure 15.19a, it can be observed that the electron flow in wireless cell is driven by Ohmic contact. The unique design of PV-PEC monolithic tandem cell can omit the needs of external connection and reduce the complexity of the water splitting system. As a result, the Co-Bi/3jn a-Si/NiMoZn system was able to demonstrate a STH of 2.5% after an irradiation time of two hours. Besides, Okamoto et al. reported a modulated III–V triple-junction solar cell for wireless monolithic system [46]. In the study, the path length of light was investigated, and it was found out that the distance of light travel should be short enough to activate the bottom layer of the solar cell. The configuration of the monolithic device is shown in Figure 15.19b, in which the triple-junction solar cell was connected to Pt as HEC and IrO_2 as OEC. Consequently, the $Pt/AgAu/InGaP/GaAs/Ge/IrO_2$ monolithic system demonstrated a STH of 11.2% under neutral pH condition.

In 2017, Shoji et al. reported a $SrTiO_3$-based artificial leaf with Cu_xO as HER co-catalysts and CoPi as OER co-catalysts for the application of selective CO_2 reduction [47]. The developed monolithic device can function as a free-standing system

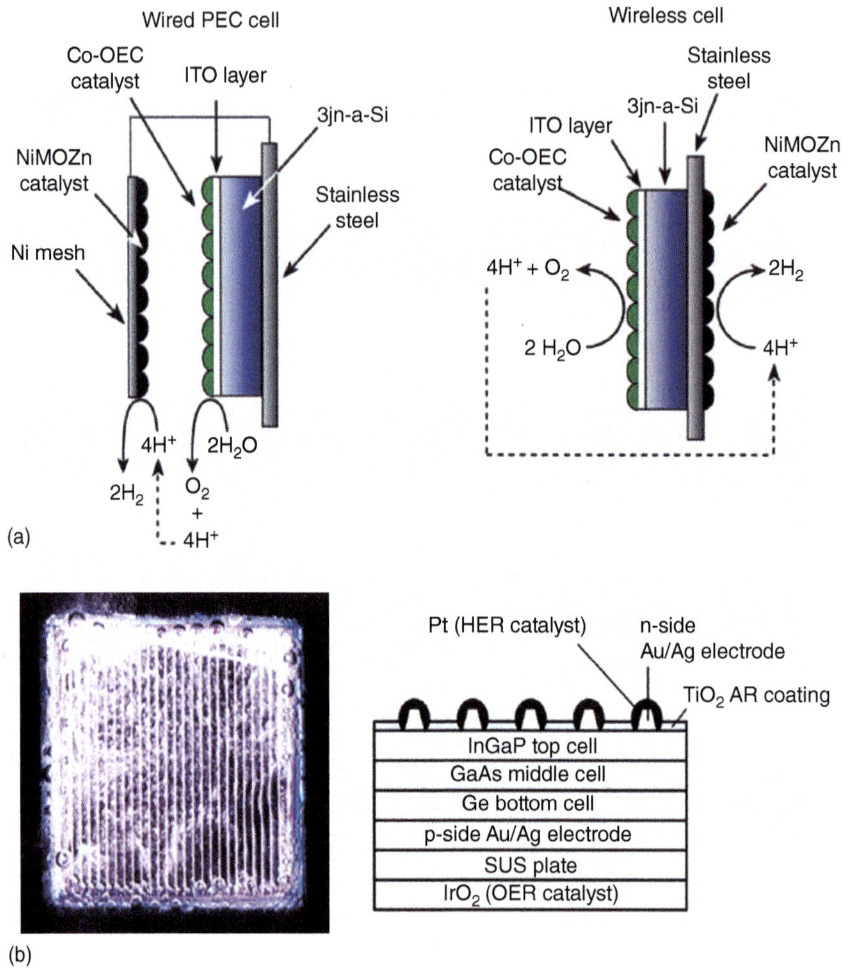

Figure 15.19 (a) Wired PEC cell and monolithic wireless cell of Co-Bi/3jn a-Si/NiMoZn system. Source: Reece et al. [45]. Copyright 2011, American Association for the Advancement of Science. (b) Modulated III–V triple-junction solar cell (Pt/AgAu/InGaP/GaAs/Ge/IrO_x) and the corresponding photograph of H_2 and O_2 evolution. Source: Reproduced with permission Okamoto et al. [46]. Copyright 2017, American Chemical Society.

without the needs of any applied bias (Figure 15.20a). The UV-driven photocatalytic activity of $SrTiO_3$ artificial leaf was tested by bubbling CO_2 gas into aqueous solution containing the film, and the product was collected from the headspace of the reactor. In the photocatalytic experiment, the product was detected to be CO, H_2 and O_2. Since the chemical stoichiometry of CO_2 reduction is a two-electron transfer pathway while the O_2 generation involves four electrons transfer, the ratio of CO and H_2 to O_2 should be in 2 : 1. Besides, the selectivity of CO production was measured to be c. 80%, which is the highest selectivity of CO_2 reduction using a monolithic system. On top of that, a TiO_2-based artificial leaf produced from 3D printing method was documented by Chen et al. for photo-reduction of CO_2 into

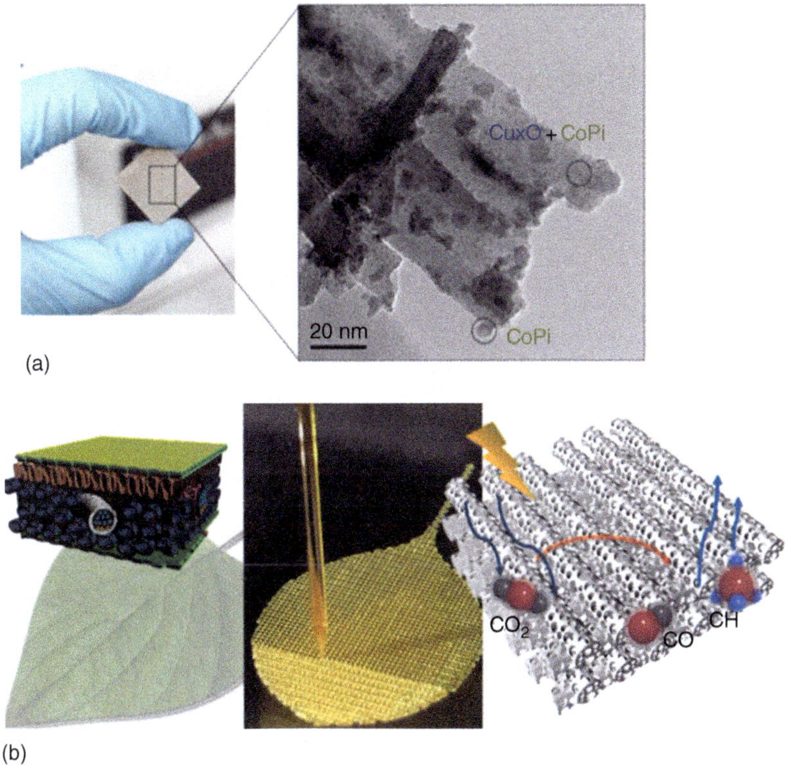

Figure 15.20 (a) $SrTiO_3$-based artificial leaf with Cu_xO as HER co-catalysts and CoPi as OER co-catalysts. Source: Reproduced with permission Chen et al. [48]. Copyright 2017, American Chemical Society. (b) TiO_2-based artificial photosynthetic systems (APSs) produced from 3D printing. Source: Reproduced with permission Chen et al. [48]. Copyright 2018, American Chemical Society.

CO and CH_4 [48]. As shown in Figure 15.20b, the artificial leaf was fabricated in 3D architectures with magnitude from nanometer to centimeter in scalable configuration. TiO_2-based ink was employed as the host material for the production of 3D artificial leaf with tunable porosity and high surface area. The presence of macropores renders efficient mass transfer and rapid diffusion of product gas. In whole, the strategy of 3D printing presents a viable method for potential scale-up application of monolithic device.

15.4 Conclusions and Outlook

Solar fuel production has been widely studied and investigated over the past 50 years mainly via PV-EC, PEC, and photocatalysis. These technologies have achieved significant progressive improvements over the years. However, the cost for renewable solar fuels is outweighed by their conversion efficiency. Since the world decides on the next giant step for solar energy augmentation, revolutionary research studies on the development of efficient materials, reactor design, and scaling-up attempts are required.

Among solar fuel harvesting technologies, photocatalysis and PEC show enormous potentials and compelling strategies to produce clean fuel. This chapter provides an insight on the key factors in mass production of solar fuels, which includes (i) fundamental kinetics and mechanisms, (ii) the design and setup of reactors, and (iii) engineering aspects of photocatalytic and PEC systems with potential scalability. First and foremost, an in-depth understanding on the fundamental rationales and mechanisms of solar water splitting to produce H_2 and photocatalytic CO_2 reduction into hydrocarbons is critical and serves as the steppingstone for the development of these technologies toward commercialization. An efficient photocatalytic system requires photocatalysts to fulfill three important criteria: (i) small bandgap for broad light absorption, (ii) efficient charge separation, and (iii) strong redox ability associated to large overpotential. In this regard, development of new system such as Z-scheme is a breakthrough research to improve photocatalysis to a new horizon. In view of PEC, single light absorber system requires external applied bias to compensate for the losses of overpotential. The design of two light absorber system or known as tandem cell can operate without the needs of applied bias. For reactor design, photocatalytic reactor normally focuses on the planar panel configuration to allow more exposure of light into the suspension mixture. Very recently, the fabrication of water splitting panel with photocatalyst sheets opens up a new avenue to conduct water splitting experiment. Different from photocatalytic reactor, the PEC reactor bestows the advantage of separated gas evolution, i.e. H_2 and O_2 evolution at different chamber. This is attributed to the configuration of anode and cathode in two different compartments separated by a membrane. Lastly, recent advancement in photocatalysis and PEC processes has prompted the groundbreaking ideas of photocatalyst sheets and monolithic system, which shows potentials in large-scale applications. In conclusion, revolutionary research efforts on the development of photocatalytic materials, processes, and the design of reactors are to be made in order to realize augmentation of photocatalysis and PEC systems.

Acknowledgments

This work was funded by the Ministry of Education (MOE) Malaysia under Fundamental Research Grant Scheme (FRGS) – Malaysia Research Star Award (MRSA) (Ref. no.: FRGS-MRGS/1/2018/TK02/MUSM/01/1).

List of Abbreviations

PV	Photovoltaic
CSP	Concentrating thermal solar power
H_2	Hydrogen
PEC	Photoelectrochemical

CB	Conduction band
VB	Valence band
HER	H_2 evolution reaction
OER	O_2 evolution reaction
PS	Photosystem
HEP	H_2 evolution photocatalyst
OEP	O_2 evolution photocatalyst
LUMO	Lowest unoccupied molecular orbital
PV-EC	Photovoltaic-electrolysis
HEC	H_2 evolution catalyst
OEC	O_2 evolution catalyst
STH	Solar-to-hydrogen conversion efficiency
GC	Gas chromatography

References

1 Kim, J.H., Hansora, D., Sharma, P. et al. (2019). *Chem. Soc. Rev.* 48: 1908–1971.

2 (a) Ng, B.-J., Putri, L.K., Tan, L.-L. et al. (2017). *Chem. Eng. J.* 316: 41–49. (b) Kong, X.Y., Tan, W.L., Ng, B.-J. et al. (2017). *Nano Res.* 10: 1720–1731.

3 Wang, Q., An, N., Chen, W. et al. (2012). *Int. J. Hydrogen Energy* 37: 12886–12892.

4 Linsebigler, A.L., Lu, G.Q., and Yates, J.T. Jr., (1995). *Chem. Rev.* 95: 735–758.

5 Maeda, K. and Domen, K. (2010). *J. Phys. Chem. Lett.* 1: 2655–2661.

6 Chen, S., Takata, T., and Domen, K. (2017). *Nat. Rev. Mater.* 2: 17050.

7 (a) Chen, D., Zhang, X., and Lee, A.F. (2015). *J. Mater. Chem. A* 3: 14487–14516. (b) Li, K., Peng, B., and Peng, T. (2016). *ACS Catal.* 6: 7485–7527. (c) Kong, X.Y., Lee, W.Q., Mohamed, A.R., and Chai, S.-P. (2019). *Chem. Eng. J.* 372: 1183–1193. (d) Kong, X.Y., Ng, B.-J., Tan, K.H. et al. (2018). *Catal. Today* 314: 20–27.

8 Inoue, T., Fujishima, A., Konishi, S., and Honda, K. (1979). *Nature* 277: 637–638.

9 Nocera, D.G. (2012). *Acc. Chem. Res.* 45: 767–776.

10 Tu, W., Zhou, Y., and Zou, Z. (2014). *Adv. Mater.* 26: 4607–4626.

11 Hoffmann, M.R., Martin, S.T., Choi, W., and Bahnemann, D.W. (1995). *Chem. Rev.* 95: 69–96.

12 Liu, L., Fan, W., Zhao, X. et al. (2012). *Langmuir* 28: 10415–10424.

13 Kong, X.Y., Lee, W.P.C., Ong, W.-J. et al. (2016). *ChemCatChem* 8: 3074–3081.

14 Tan, S., Zhao, Y., Zhao, J. et al. (2011). *Phys. Rev. B: Condens. Matter* 84.

15 Gagliardi, C.J., Westlake, B.C., Kent, C.A. et al. (2010). *Coord. Chem. Rev.* 254: 2459–2471.

16 Tan, L.-L., Ong, W.-J., Chai, S.-P., and Mohamed, A.R. (2017). *Chem. Eng. J.* 308: 248–255.

17 Liu, G., Hoivik, N., Wang, K., and Jakobsen, H. (2012). *Sol. Energy Mater. Sol. Cells* 105: 53–68.

18 Butler, M. and Ginley, D. (1980). *J. Mater. Sci.* 15: 1–19.

19 Jiang, C., Moniz, S.J., Wang, A. et al. (2017). *Chem. Soc. Rev.* 46: 4645–4660.

20 Mayer, M.T. (2017). *Curr. Opin. Electrochem.* 2: 104–110.

21 Van de Krol, R. and Grätzel, M. (2012). *Photoelectrochemical Hydrogen Production*, vol. 90. Springer.

22 Ikeda, S., Fujikawa, S., Harada, T. et al. (2019). *ACS Appl. Energy Mater.* 2: 6911–6918.

23 Takata, T., Pan, C., and Domen, K. (2015). *Sci. Technol. Adv. Mater.* 16: 033506.

24 (a) An, C., Feng, J., Liu, J. et al. (2017). *Inorg. Chem. Front.* 4: 1042–1047. (b) Putri, L.K., Ng, B.-J., Tan, K.H. et al. (2018). *Catal. Today* 315: 93–102. (c) Qiu, B., Zhu, Q., Du, M. et al. (2017). *Angew. Chem. Int. Ed.* 56: 2684–2688. (d) Sun, X., Mi, Y., Jiao, F., and Xu, X. (2018). *ACS Catal.* 8: 3209–3221.

25 Kudo, A. and Miseki, Y. (2009). *Chem. Soc. Rev.* 38: 253–278.

26 (a) Huang, Y., Wu, J., Wei, Y. et al. (2008). *J. Alloys Compd.* 456: 364–367. (b) Rayalu, S.S., Jose, D., Joshi, M.V. et al. (2013). *Appl. Catal., B* 142–143: 684–693.

27 Wei, Y., Li, J., Huang, Y. et al. (2009). *Sol. Energy Mater. Sol. Cells* 93: 1176–1181.

28 Valadés-Pelayo, P.J., Guayaquil Sosa, F., Serrano, B., and de Lasa, H. (2015). *Chem. Eng. Sci.* 126: 42–54.

29 Sasaki, Y., Kato, H., and Kudo, A. (2013). *J. Am. Chem. Soc.* 135: 5441–5449.

30 (a) Goto, Y., Hisatomi, T., Wang, Q. et al. (2018). *Joule* 2: 509–520. (b) Hisatomi, T., Yamamoto, T., Wang, Q. et al. (2018). *Catal. Sci. Technol.* 8: 3918–3925. (c) Pan, Z., Hisatomi, T., Wang, Q. et al. (2016). *ACS Catal.* 6: 7188–7196. (d) Sun, S., Hisatomi, T., Wang, Q. et al. (2018). *ACS Catal.* 8: 1690–1696. (e) Wang, Q., Hisatomi, T., Jia, Q. et al. (2016). *Nat. Mater.* 15: 611–615. f) Wang, Q., Hisatomi, T., Suzuki, Y. et al. (2017). *J. Am. Chem. Soc.* 139: 1675–1683.

31 Jiang, Z., Xiao, T., Kuznetsov, V.L., and Edwards, P.P. (2010). *Philos. Trans. A Math. Phys. Eng. Sci.* 368: 3343–3364.

32 (a) Liu, B.-J., Torimoto, T., and Yoneyama, H. (1998). *J. Photochem. Photobiol., A* 113: 93–97. (b) Wu, H.-Y., Bai, H., and Wu, J.C.S. (2014). *Ind. Eng. Chem. Res.* 53: 11221–11227.

33 Bideau, M., Claudel, B., Faure, L., and Kazouan, H. (1995). *J. Photochem. Photobiol., A* 91: 137–144.

34 Schiavello, M., Augugliaro, V., and Palmisano, L. (1991). *J. Catal.* 127: 332–341.

35 Ola, O. and Maroto-Valer, M.M. (2015). *J. Photochem. Photobiol., C* 24: 16–42.

36 Khan, A.A. and Tahir, M. (2019). J. CO_2 Util. 29: 205–239.

37 Ray, A.K. and Beenackers, A.A.C.M. (1998). *Catal. Today* 40: 73–83.

38 (a) Du, P., Carneiro, J.T., Moulijn, J.A., and Mul, G. (2008). *Appl. Catal., A* 334: 119–128. (b) Nguyen, T.-V. and Wu, J.C.S. (2008). *Appl. Catal., A* 335: 112–120.

39 Howe, R.F. (1998). *Dev. Chem. Eng. Miner. Process.* 6: 55–84.

40 Ahmed, M. and Dincer, I. (2019). *Int. J. Hydrogen Energy* 44: 2474–2507.

41 Jin, J., Walczak, K., Singh, M.R. et al. (2014). *Energy Environ. Sci.* 7: 3371–3380.

42 Cheng, J., Zhang, M., Wu, G. et al. (2014). *Environ. Sci. Technol.* 48: 7076–7084.

43 Schröder, M., Kailasam, K., Borgmeyer, J. et al. (2015). *Energy Technol.* 3: 1014–1017.

44 Liu, C., Tang, J., Chen, H.M. et al. (2013). *Nano Lett.* 13: 2989–2992.

45 Reece, S.Y., Hamel, J.A., Sung, K. et al. (2011). *Science* 334: 645–648.

46 Okamoto, S., Deguchi, M., and Yotsuhashi, S. (2017). *J. Phys. Chem. C* 121: 1393–1398.

47 Shoji, S., Yamaguchi, A., Sakai, E., and Miyauchi, M. (2017). *ACS Appl. Mater. Interfaces* 9: 20613–20619.

48 Chen, L., Tang, X., Xie, P. et al. (2018). *Chem. Mater.* 30: 799–806.

16

Prospects of Solar Fuels

Hongqi Sun

Edith Cowan University, School of Engineering, 270 Joondalup Drive, Joondalup, WA 6027, Australia

Scientists have made fantastic progresses in the production and utilization of solar fuels. Thanks to the interdisciplinary research, many "dream reactions" have been successfully demonstrated in the laboratory. It can be fairly confident to say that tremendous opportunities exist for human race having a more sustainable future. The conversion of solar to chemicals indeed has twofold means: transforming solar energy to chemicals and converting solar fuel-based chemicals to electricity. To expand the scope, applying light, photovoltaics, and/or solar heat to conventional chemical processes also possesses a great promise for solar energy conversion and utilization.

Inspired by photoelectrochemical water splitting for hydrogen generation, theoretical fundamentals and bench-scale tests can fully show the potentials of converting water, carbon dioxide, nitrogen, and oxygen, via solar energy utilization, to a wide range of chemicals, such as hydrogen, oxygen, carbon monoxide, formaldehyde, methanol, formic acid, ethanol (and more C2), ammonia, hydrazine, and hydrogen peroxide. Reforming natural gases and biomass can produce value-added or more environmentally benign chemicals/fuels. Conventional C1 chemistry, referring to chemical reactions for carbon monoxide, carbon dioxide, methane, and methanol etc., is also a great platform for solar to chemicals. Not too surprisingly, light has been introduced to Fischer–Tropsch synthesis, water–gas shift, methane dry or steam reforming, carbon dioxide hydrogenation, and methanol reforming, etc.

Everything looks promising and not just theoretically. In 2017, the first 200 L of synthetic fuels were produced by solar energy conversion through Fischer–Tropsch using carbon dioxide in air under the SOLETAIR project, as conducted in a spinoff of Karlsruhe Institute of Technology (KIT). Very recently, a solar fuel synthesis pilot plant, at a thousand-tonne scale, commenced operation in Lanzhou, China. In the process, carbon dioxide, water, and solar energy are converted into transportable liquid fuels such as menthol. The processes include (i) solar photovoltaics to generate electricity, (ii) electrolyzer to split water to produce hydrogen, and (iii) carbon dioxide hydrogenation to produce methanol.

Solar-to-Chemical Conversion: Photocatalytic and Photoelectrochemcial Processes, First Edition.
Edited by Hongqi Sun.

A question that may have already been there and will continue to exist is: are solar fuels feasible? Despite the research advances in literature and pilot plants, the answer might not be that straightforward.

Feasibility, which might be one level lower than sustainability, should at least meet four requirements: theory, technology, economics, and environment. This book presents comprehensive fundamentals in solar-to-chemical processes; thus no doubts exist questioning the theories. However, it is far before all theories are mature. As for the technology, the situation is even worse. The identification, experimentation, investigation, discussion, validation, conclusion, and dissemination of the problems in theories and technologies might cover the whole processes. This exactly represents the extensive research activities in the area, as illustrated in the book. Besides the role as the foundation, or the basis for being cautiously optimistic, the research advances, like shown in this book, can identify the challenges as well.

The dimension of challenges is not smaller than that of the progresses we have achieved. Heterogeneous reactions using suspended powders for water splitting or carbon dioxide reduction might be the simplest systems for solar fuel production. The variables for optimization of this kind of systems can be the catalyst materials (type, shape, size, crystalline, textile, and surface), the typical semiconductor-based photocatalysis (light absorption, activation, electron-hole transportation and recombination, and surface reactions), reactor (type, size, and irradiation), reaction (stirring, temperature, solution chemistry, light scattering, liquid–solid or liquid–gas or gas–solid or three-phase interfaces), and collection (analysis, separation, and purification). Some other parameters, such as co-catalysts, sacrifice reagents, and reaction atmosphere, can also be included. Once the research on some specific parameters cannot be implemented by experiments, due to either insufficient instrumentation or skills, computational studies are then required. Density functional theory calculations have been usually used for materials chemistry and catalysis, to validate or guide experimental studies. What makes things more complicated is that the published results and proposed mechanisms are sometimes controversial. Not exaggeratingly, any singular parameter mentioned above (and the sub-parameter in the brackets) can fulfill the content of a whole book! This is for the simplest solar fuel production system.

Introducing the available photovoltaic technology to enhance the efficiency of photocatalysis is a promising approach, as converting solar energy to electricity is less complicated than the solar-driven chemical processes. In this book, we discussed the photoelectrochemical reactions for hydrogen evolution, oxygen evolution, overall water splitting, carbon dioxide reduction, nitrogen fixation, hydrogen peroxide formation, methane reforming, and biomass reforming. Any research topic in suspended powder systems can be a research topic in photoelectrochemistry. The emerging parameters are (i) the electrode fabrication (catalysts, substrate, cocatalysts, and surface, etc.), (ii) configuration (single or dual photoelectrode, bias, and solutions, etc.), and (iii) engineering with conventional photovoltaics.

As a typical C1 chemical reaction, this book introduces the photocatalytic and photoelectrochemical methane reforming. The ideas can be easily extended to other similar reactions, for example, Fischer–Tropsch synthesis, water–gas-shift, and carbon dioxide hydrogenation. The most significant change to these conventional reactions is the need of light, which demands the extra care to the reactor design, in addition to the heat supply. It is still not clear yet how the light will change the reaction mechanism as well as the conversion and selectivity.

A very important prospect in this filed is photothermal catalysis, which is, due to the limit of progress, not included in this book. This area is of the capacity similar to, or not less than photoelectrochemical catalysis. The nature of using electrochemical or thermochemical process to enhance photocatalysis is the extended light utilization, which cannot be achieved by the sole semiconductor-based photocatalysis. The heat sustaining conventional thermal catalysis for upgrading fossil fuels or fine chemical synthesis can be feasibly obtained by infrared capture. Meantime photocatalysis at accelerated temperatures can easily archive higher solar conversion efficiencies.

Research endeavors to validate the theoretical and technological feasibility of solar to chemicals are continuing. The economic analysis, i.e. the investment costs and the returns, of solar fuels is also important. Some primary studies indicate that exciting opportunities exist in producing renewable energies at competitive costs in the medium term. Without such a driving force and more convincing evidences for the profits, the future of solar fuels will become very tough. Industry will be attracted if there are real profits at present, or at least in very near future. We cannot blame there is no investment to fundamental research, pilot tests and large scale of solar fuels. Capitals are driven by genuine profits. Engineering aspects should be ready before it can be presented to industries. One of the main issues in the research is that the solar fuel production is being downgraded to a test method for the development of new materials. Without the active involvements from engineers, the commercial solar fuel production can never be ready for industries.

Don't forget the initial incentive of the exploration of the renewable energies after this long way: the environment. It can never take granted that the production and use of renewable energies will certainly protect the environment. The installation of solar panels, hydroturbos, and wind turbos has already induced many issues to ecosystem. The manufacture, use, and disposal of materials for electrochemical applications have led to new contaminations. Comprehensive studies should be conducted to evaluate the ratio of environmental benefits to environmental pollutions from the production and use of solar fuels.

There are no doubts that government can play a vital role in exploring the economic and environmental feasibility of the solar fuels. Climate change, a term that might be known by most kids in primary school, is not being recognized by some governments, even in developed countries. This might be because of some unknown considerations, but, as a very popular saying, you can never wake up a person who is pretending to be asleep. Without such a recognition of the true crisis facing human being's future, the foundation of solar fuels is broken. Therefore, for scientists the research is not enough, and more advocation efforts should be done.

To be optimistic, but with precautions!

Index

Solar-to-Chemical Conversion: Photocatalytic and Photoelectrochemcial Processes, First Edition.
Edited by Hongqi Sun.

d

e

t